Advanced Asphalt Materials and Paving Technologies

Special Issue Editors

Zhanping You
Qingli Dai
Feipeng Xiao

MDPI • Basel • Beijing • Wuhan • Barcelona • Belgrade

Special Issue Editors
Zhanping You
Michigan Technological University
USA

Qingli Dai
Michigan Technological University
USA

Feipeng Xiao
Tongji University
China

Editorial Office
MDPI
St. Alban-Anlage 66
Basel, Switzerland

This edition is a reprint of the Special Issue published online in the open access journal *Applied Sciences* (ISSN 2076-3417) from 2017–2018 (available at: http://www.mdpi.com/journal/applsci/special_issues/paving_technologies).

For citation purposes, cite each article independently as indicated on the article page online and as indicated below:

Lastname, F.M.; Lastname, F.M. Article title. *Journal Name* **Year**, *Article number*, page range.

First Editon 2018

ISBN 978-3-03842-889-3 (Pbk)
ISBN 978-3-03842-890-9 (PDF)

Table of Contents

About the Special Issue Editors

Zhanping You, Professor, received his Ph.D. in civil engineering from the University of Illinois at Urbana-Champaign. He has been a Professor of Civil and Environmental Engineering at Michigan Technological University since 2005. Professor You has over 20 years of practical and research experience in pavement engineering and materials. In his capacity as a Professor and a professional engineer, he has completed research projects on a wide range of subjects including pavement design, asphalt materials and the micromechanics of pavement materials. He has also served as a principal investigator and director of the Center of Excellence for Transportation Materials, which is a partnership between the Michigan Department of Transportation and Michigan Technological University in the USA. His contribution to pavement and materials research has been featured in newspapers, magazines, and other media. Professor You has published over 260 papers. In 2004 and 2005, he was a recipient of the U.S. Department of Transportation's Dwight David Eisenhower Transportation Faculty Fellowship.

Qingli Dai, Professor, has made significant achievements in the area of material design, characterization, testing and analysis for sustainable civil infrastructure applications, especially regarding self-healing abilities, damage mechanism diagnosis and multi-physical interactions in asphalt mixtures and Portland cement concrete. Her research involves multidisciplinary areas, such as micromechanics, the finite element method, fracture mechanics, molecular dynamic simulation, sensing for damage mechanisms, self-healing materials, active-material actuator design and aerodynamic and aeroelastic simulation. She has taught classes such as matrix structural analysis, finite element analysis and the advanced mechanics of materials. Dr. Dai has been the principle investigator on research projects funded by the National Science Foundation and the State Department of Transportation. She has authored and co-authored more than 150 peer-reviewed papers in prestigious journals and conferences.

Feipeng Xiao is a Professor and Department Chair of the Department of Road and Airport Engineering at Tongji University, China. As a registered civil engineer in Maryland, USA, he has almost 20 years of work experience in pavement engineering, structural engineering, etc. Dr. Xiao has been appointed as an Editor of Construction and Building Materials, and as Associate Editor and Editorial Board Member of four other international journals. He has been widely recognized in the field of sustainable pavement materials research, with over 150 peer-reviewed articles in international journals and conference proceedings and has also worked as the principle investigator on research projects sponsored by many organizations and government agencies. He is a member of many international professional societies by nomination or election.

Editorial

Advanced Paving Materials and Technologies

Zhanping You [1,*], Qingli Dai [1] and Feipeng Xiao [2]

1 Department of Civil and Environmental Engineering, Michigan Technological University, Houghton 49931, MI, USA; qingdai@mtu.edu
2 Key Laboratory of Road and Traffic Engineering of the Ministry of Education, Tongji University, Shanghai 200000, China; fpxiao@tongji.edu.cn
* Correspondence: zyou@mtu.edu

Received: 30 March 2018; Accepted: 2 April 2018; Published: 9 April 2018

There has been tremendous amount of research advances in the area of pavement materials and paving technologies in the past decade. These include the use of warm mix asphalt technologies, rubber asphalt, bioasphalt, nanomaterial applications, new construction technologies, new concrete materials, and the application of pavement mechanistic-empirical design. With all these developments, a collection of peer-reviewed articles with a theme of advanced asphalt materials and paving technologies is necessary for the industry, researchers, government agencies, and other stakeholders. This collection promotes new, low-cost technologies with high durability, environmental friendliness, and effective resource usage in the area of advanced asphalt materials and paving technologies. These papers include

1. Towards an Alternate Evaluation of Moisture-Induced Damage of Bituminous Materials [1]
2. Property Analysis of Exfoliated Graphite Nanoplatelets Modified Asphalt Model Using a Molecular Dynamics (MD) Method [2]
3. Adhesion Evaluation of Asphalt-Aggregate Interface Using a Surface Free Energy Method [3]
4. Laboratory and On-Site Tests for Rapid Runway Repair [4]
5. Tire–Pavement Friction Characteristics with Elastic Properties of Asphalt Pavements [5]
6. Technologies and Principles of Hot Recycling and Investigation of Preheated Reclaimed Asphalt Pavement Batching Process in an Asphalt Mixing Plant [6]
7. Evaluation of Adhesion and Hysteresis Friction of a Rubber–Pavement System [7]
8. Research on the Performance of a Dense Graded Ultra-Thin Wearing Course Mixture [8]
9. Improving Asphalt Mixture Performance by Partially Replacing Bitumen with Waste Motor Oil and Elastomer Modifiers [9]
10. Mechanical Resilience of Modified Bitumen at Different Cooling Rates: A Rheological and Atomic Force Microscopy Investigation [10]
11. Using a Molecular Dynamics Simulation to Investigate Asphalt Nano-Cracking under External Loading Conditions [11]
12. An Evaluation of Mechanical Properties of Recycled Material for Utilization in Asphalt Mixtures [12]
13. A Study of Surfactant Additives for the Manufacture of Warm Mix Asphalt: From Laboratory Design to Asphalt Plant Manufacture [13]
14. Laboratory Evaluation of Rejuvenating Agent on Reclaimed SBS Modified Asphalt Pavement [14]
15. Three-Dimensional Digital Sieving of Asphalt Mixture Based on X-ray Computed Tomography [15]
16. Permeability and Stiffness Assessment of Paved and Unpaved Roads with Geocomposite Drainage Layers [16]
17. An Evaluation of Aging Resistance of Graphene-Oxide-Modified Asphalt [17]

18. Application of a Finite Layer Method in Pavement Structural Analysis [18]
19. A New Life for Cross-Linked Plastic Waste as Aggregates and Binder Modifier for Asphalt Mixtures [19]
20. Study of the Diffusion of Rejuvenators and Its Effect on Aged Bitumen Binder [20]
21. Simulation of Permanent Deformation in High-Modulus Asphalt Pavement with Sloped and Horizontally Curved Alignment [21]
22. Fatigue Life Prediction of High Modulus Asphalt Concrete Based on the Local Stress–Strain Method [22]
23. Low Temperature Performance Characteristics of Reclaimed Asphalt Pavement (RAP) Mortars with Virgin and Aged Soft Binders [23]
24. The Effect of Fibers on the Mixture Design of Stone Matrix Asphalt [24]]
25. Ultrasonic Techniques for Air Void Size Distribution and Property Evaluation in Both Early-Age and Hardened Concrete Samples [25]
26. Thermal and Fatigue Evaluation of Asphalt Mixtures Containing RAP Treated with a Bio-Agent [26]
27. Numerical Study on the Asphalt Concrete Structure for Blast and Impact Load Using the Karagozian and Case Concrete Model [27]
28. Steady-State Creep of Asphalt Concrete [28]

These 28 papers have been peer reviewed under the journal's rigorous review criteria. The collection includes invited papers from experts in international communities, and articles have been selected from the 2017 World Transport Convention (WTC) in Beijing held in June 2017.

Conflicts of Interest: The authors declare no conflict of interest.

References

1. Diab, A.; You, Z.; Yang, X.; Hasan, M.R.M. Towards an Alternate Evaluation of Moisture-Induced Damage of Bituminous Materials. *Appl. Sci.* **2017**, *7*, 1049. [CrossRef]
2. Yao, H.; Dai, Q.; You, Z.; Bick, A.; Wang, M.; Guo, S. Property Analysis of Exfoliated Graphite Nanoplatelets Modified Asphalt Model Using Molecular Dynamics (MD) Method. *Appl. Sci.* **2017**, *7*, 43. [CrossRef]
3. Ji, J.; Yao, H.; Liu, L.; Suo, Z.; Zhai, P.; Yang, X.; You, Z. Adhesion Evaluation of Asphalt-Aggregate Interface Using Surface Free Energy Method. *Appl. Sci.* **2017**, *7*, 156. [CrossRef]
4. Leonelli, F.; Di Mascio, P.; Germinario, A.; Picarella, F.; Moretti, L.; Cassata, M.; De Rubeis, A. Laboratory and On-Site Tests for Rapid Runway Repair. *Appl. Sci.* **2017**, *7*, 1192. [CrossRef]
5. Yu, M.; Wu, G.; Kong, L.; Tang, Y. Tire-Pavement Friction Characteristics with Elastic Properties of Asphalt Pavements. *Appl. Sci.* **2017**, *7*, 1123. [CrossRef]
6. Sivilevičius, H.; Bražiūnas, J.; Prentkovskis, O. Technologies and Principles of Hot Recycling and Investigation of Preheated Reclaimed Asphalt Pavement Batching Process in an Asphalt Mixing Plant. *Appl. Sci.* **2017**, *7*, 1104. [CrossRef]
7. Al-Assi, M.; Kassem, E. Evaluation of Adhesion and Hysteresis Friction of Rubber–Pavement System. *Appl. Sci.* **2017**, *7*, 1029. [CrossRef]
8. Geng, L.; Ma, T.; Zhang, J.; Huang, X.; Hu, P. Research on Performance of a Dense Graded Ultra-Thin Wearing Course Mixture. *Appl. Sci.* **2017**, *7*, 800. [CrossRef]
9. Fernandes, S.; Peralta, J.; Oliveira, J.R.; Williams, R.C.; Silva, H.M. Improving Asphalt Mixture Performance by Partially Replacing Bitumen with Waste Motor Oil and Elastomer Modifiers. *Appl. Sci.* **2017**, *7*, 794. [CrossRef]
10. Rossi, C.O.; Ashimova, S.; Calandra, P.; Santo, M.P.D.; Angelico, R. Mechanical Resilience of Modified Bitumen at Different Cooling Rates: A Rheological and Atomic Force Microscopy Investigation. *Appl. Sci.* **2017**, *7*, 779. [CrossRef]
11. Hou, Y.; Wang, L.; Wang, D.; Qu, X.; Wu, J. Using a Molecular Dynamics Simulation to Investigate Asphalt Nano-Cracking under External Loading Conditions. *Appl. Sci.* **2017**, *7*, 770. [CrossRef]

12. Tahmoorian, F.; Samali, B.; Tam, V.W.; Yeaman, J. Evaluation of Mechanical Properties of Recycled Material for Utilization in Asphalt Mixtures. *Appl. Sci.* **2017**, *7*, 763. [CrossRef]
13. Sol-Sánchez, M.; Moreno-Navarro, F.; Rubio-Gámez, M.C. Study of Surfactant Additives for the Manufacture of Warm Mix Asphalt: From Laboratory Design to Asphalt Plant Manufacture. *Appl. Sci.* **2017**, *7*, 745. [CrossRef]
14. Wang, J.; Zeng, W.; Qin, Y.; Huang, S.; Xu, J. Laboratory Evaluation of Rejuvenating Agent on Reclaimed SBS Modified Asphalt Pavement. *Appl. Sci.* **2017**, *7*, 743. [CrossRef]
15. Hu, C.; Ma, J.; Kutay, M.E. Three Dimensional Digital Sieving of Asphalt Mixture Based on X-ray Computed Tomography. *Appl. Sci.* **2017**, *7*, 734. [CrossRef]
16. Li, C.; Ashlock, J.; White, D.; Vennapusa, P. Permeability and Stiffness Assessment of Paved and Unpaved Roads with Geocomposite Drainage Layers. *Appl. Sci.* **2017**, *7*, 718. [CrossRef]
17. Wu, S.; Zhao, Z.; Li, Y.; Pang, L.; Amirkhanian, S.; Riara, M. Evaluation of Aging Resistance of Graphene Oxide Modified Asphalt. *Appl. Sci.* **2017**, *7*, 702. [CrossRef]
18. Liu, P.; Xing, Q.; Dong, Y.; Wang, D.; Oeser, M.; Yuan, S. Application of Finite Layer Method in Pavement Structural Analysis. *Appl. Sci.* **2017**, *7*, 611. [CrossRef]
19. Costa, L.; Peralta, J.; Oliveira, J.R.; Silva, H.M. A New Life for Cross-Linked Plastic Waste as Aggregates and Binder Modifier for Asphalt Mixtures. *Appl. Sci.* **2017**, *7*, 603. [CrossRef]
20. Xiao, Y.; Li, C.; Wan, M.; Zhou, X.; Wang, Y.; Wu, S. Study of the diffusion of rejuvenators and its effect on aged bitumen binder. *Appl. Sci.* **2017**, *7*, 397. [CrossRef]
21. Zheng, M.; Han, L.; Wang, C.; Xu, Z.; Li, H.; Ma, Q. Simulation of Permanent Deformation in High-Modulus Asphalt Pavement with Sloped and Horizontally Curved Alignment. *Appl. Sci.* **2017**, *7*, 331. [CrossRef]
22. Zheng, M.; Li, P.; Yang, J.; Li, H.; Qiu, Y.; Zhang, Z. Fatigue Life Prediction of High Modulus Asphalt Concrete Based on the Local Stress-Strain Method. *Appl. Sci.* **2017**, *7*, 305. [CrossRef]
23. Xiao, F.; Li, R.; Zhang, H.; Amirkhanian, S. Low Temperature Performance Characteristics of Reclaimed Asphalt Pavement (RAP) Mortars with Virgin and Aged Soft Binders. *Appl. Sci.* **2017**, *7*, 304. [CrossRef]
24. Sheng, Y.; Li, H.; Guo, P.; Zhao, G.; Chen, H.; Xiong, R. Effect of Fibers on Mixture Design of Stone Matrix Asphalt. *Appl. Sci.* **2017**, *7*, 297. [CrossRef]
25. Guo, S.; Dai, Q.; Sun, X.; Sun, Y.; Liu, Z. Ultrasonic Techniques for Air Void Size Distribution and Property Evaluation in Both Early-Age and Hardened Concrete Samples. *Appl. Sci.* **2017**, *7*, 290. [CrossRef]
26. Kowalski, K.J.; Król, J.B.; Bańkowski, W.; Radziszewski, P.; Sarnowski, M. Thermal and fatigue evaluation of asphalt mixtures containing RAP treated with a bio-agent. *Appl. Sci.* **2017**, *7*, 216. [CrossRef]
27. Wu, J.; Li, L.; Du, X.; Liu, X. Numerical study on the asphalt concrete structure for blast and impact load using the Karagozian and case concrete model. *Appl. Sci.* **2017**, *7*, 202. [CrossRef]
28. Iskakbayev, A.; Teltayev, B.; Oliviero Rossi, C. Steady-state creep of asphalt concrete. *Appl. Sci.* **2017**, *7*, 142. [CrossRef]

Perspective

Towards an Alternate Evaluation of Moisture-Induced Damage of Bituminous Materials

Aboelkasim Diab [1], Zhanping You [2,*], Xu Yang [3] and Mohd Rosli Mohd Hasan [4]

1 Department of Civil Engineering, Aswan University, Aswan 81542, Egypt; adiab@aswu.edu.eg or daali@mtu.edu
2 Department of Civil and Environmental Engineering, Michigan Technological University, Houghton, MI 49931, USA
3 Department of Civil Engineering, Monash University, Clayton, VIC 3800, Australia; xu.yang@monash.edu
4 School of Civil Engineering, Universiti Sains Malaysia, Engineering Campus, Nibong Tebal Penang 14300, Malaysia; cerosli@usm.my
* Correspondence: zyou@mtu.edu; Tel.: +1-906-487-1059

Received: 12 September 2017; Accepted: 10 October 2017; Published: 13 October 2017

Abstract: Moisture-induced damage is widely known to cause multiple distresses that affect the durability of constructed pavements and eventually lead to the costly maintenance of pavement structures. The reliability and practicality of the assessment protocol to evaluate moisture susceptibility of flexible pavements presents a dilemma within the asphalt community that arises from the complexity and interrelation of moisture mechanisms in the asphalt–aggregate system. Researchers worldwide are continuously trying to develop suitable evaluation methods to simulate the combined destructive field-induced effects of moisture in the laboratory to help practitioners identify and alleviate this complex problem. The main objective of this article is to provide insights and highlight the challenges and opportunities of this important topic in order to extend and share knowledge towards finding a realistic assessment protocol of moisture damage in the laboratory. Two scenarios are proposed in this article: (1) a damage rate concept that accounts for the change of mechanical property (e.g., indirect tensile strength) with respect to the conditioning time, and (2) the establishment of a database using a surface free energy concept to help stakeholders select appropriate asphalt–aggregate combinations without the need to run additional moisture susceptibility tests.

Keywords: asphalt pavement; moisture damage; damage rate; surface free energy

---

1. Outlook

The distresses related to moisture-induced damage still remain one of the most common but complex issues of bituminous pavements [1,2]. Mixtures that are not properly designed or evaluated against ever increasing traffic loading alongside the expected exposure to moisture in the field could potentially affect the pavement durability that can lead to progressive degradation in the form of several distresses, such as raveling, rutting, or cracking [3–5]. In service life, the bituminous pavement is subjected to different environmental conditions combined with traffic loading, which both affect the durability and life-cycle cost of the constructed structure. Displacement, detachment, spontaneous emulsification, pore-pressure-induced damage, hydraulic scour, pH instability, and climatic conditions are considered contributing mechanisms associated with moisture damage in the field [6,7]. So far, the moisture damage of asphalt pavements is not well understood, as it is a complex phenomenon that is affected by the physicochemical properties of the constituents forming the asphalt mixture. In the design stage, the asphalt materials are usually characterized for the moisture effect at the macro scale using compacted asphalt samples. Although the evaluation of compacted asphalt mixtures is commonly used in the pavement engineering community, there are several problems still unsolved, and different perspectives should be considered towards this issue. During the mixture design, the asphalt mixture must pass the tensile strength ratio (TSR) test to ensure the moisture susceptibility of the designed mixture. As a supporting step, in turn, this perspective article is intended to highlight the current issue of moisture damage evaluation and provide an overview of the opportunities to be considered for future researches.

2. Challenges

Various, predominantly empirical, mechanical test methods to evaluate the moisture susceptibility of asphalt mixtures have been developed and are discussed elsewhere [8]. The AASHTO T 283 test method is frequently used by highway agencies worldwide for the sake of evaluation of moisture damage of compacted asphalt concrete mixtures. Basically, in the T 283 test, the TSR, which is calculated as the ratio of the indirect tensile strength (ITS) of conditioned specimens to that for dry specimens, is used to determine the resistance of the asphalt mixture to moisture damage. A minimum TSR of 80% is usually used as a threshold to ensure that the mixture is moisture-resistant, while the mixture is considered moisture-prone if the TSR is any lower. However, several concerns of this ratio have been discussed [9,10]:

- Does this value guarantee a satisfactory resistance to moisture damage in the long-term? The current practice of the TSR concept is based on short-term aged samples, but the moisture susceptibility is a concern throughout the service life of asphalt pavement.
- What is the proper conditioning method to mimic the moisture damage in real life? Freeze–thaw is used in AASHTO T 283, and the moisture induced stress tester (MIST) was introduced in recent years. Can these methods, or some other method, better mimic real moisture damage?
- As the unconditioned or dry compacted specimens are different from the conditioned or wet specimens, does specimen-to-specimen variability affect the results? Even for the same mixture, the prepared compacted samples would have different distributions of aggregate particles and/or air voids, which affects the measured property. Therefore, the test still lacks repeatability, and its accuracy is questionable.
- Does the saturation level while conditioning affect the repeatability of the results?
- Based on previous studies on the AASHTO T 283 test, a number of successful cases that pass in the laboratory would fail in the field.

3. Opportunities

As for suggestions, two scenarios for the evaluation of moisture damage are discussed in this article, the future studies may evaluate.

3.1. First Scenario

The moisture-induced damage of asphalt mixtures can be considered as a time-dependent damage due to the combined effects of moisture and traffic loading. The mechanical property $MP(t)$ at time t is affected by the initial mechanical property $MP(0)$ (i.e., dry specimen) and the induced damage $D(t)$ during conditioning. The progressive damage due to moisture and/or traffic loading can be represented as

$$MP(t) = MP(0) - D(t). \tag{1}$$

Based on Equation (1), Figure 1 depicts a schematic of the mechanical property (such as ITS) with respect to the time of conditioning. As examples, the figure compares possible trends of two different mixes.

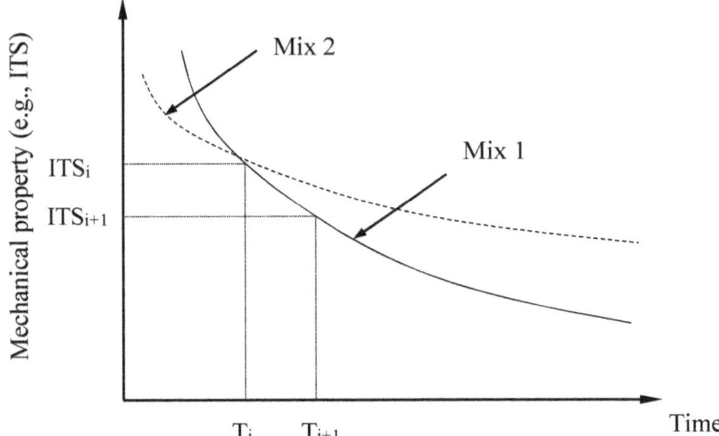

Figure 1. Mechanical property as a function of conditioning time due to the effect of environmental conditions and traffic loading.

This method should involve the conditioning and evaluation steps. In the conditioning step, realistic or field-like conditions should be adopted to induce reasonable effects of moisture and traffic. In a recently published work [9], the authors suggested a conditioning procedure by exposing compacted asphalt specimens to a temperature as high as the highest performance grading (PG) of the used asphalt binder, while the traffic loading was selected based on realistic effects from the literature to simulate the pressure of vehicle wheel on the wet surface. These effects were integrated into the proposed conditioning system in the laboratory. In regard to the evaluation, rather than using a single criterion (i.e., TSR), a model was developed to identify the time-dependent damage of the mixture using the damage rate D_r concept:

$$D_r = \frac{ITS_i - ITS_{i+1}}{T_{i+1} - T_i}. \quad (2)$$

In this model, one can quantify the moisture damage over time and determine how much the mixture degrades with respect to time of conditioning. Using the concept of damage rate, the moisture effects can be evaluated at different times, which is more promising as the mixture might have different behaviors over time. Although its usefulness to discriminate between insusceptible and moisture susceptible mixtures in a rational manner, the concept necessitates the development of a threshold to help quantitatively perform the judgment.

3.2. Second Scenario

The second suggested scenario is beneficial to pavement contractors and can save more time. Under the effects of moisture and traffic, the susceptibility to moisture damage of asphalt–aggregate systems is related to cohesive failure within the matrix and/or adhesive failure at the interface between the asphalt binder and aggregate (Figure 2).

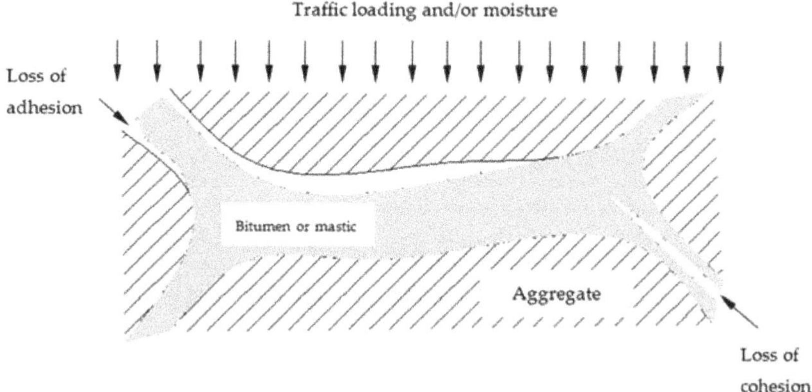

Figure 2. Possible failures due to moisture damage and traffic loading.

The moisture damage of asphalt mixtures can be studied through fundamental approaches to understand the mechanisms causing the susceptibility to moisture damage. Based on the theory of thermodynamics, the surface free energy (SFE) can be used to evaluate the adhesive bond between asphalt binder and aggregate and the cohesive bond between binder molecules. The SFE concept can be applied to select compatible asphalt–aggregate combination(s) based on a satisfactory resistance to moisture damage without the need to performing additional tests (e.g., TSR). It is recommended that highway agencies establish a database of the compatible combinations of the available aggregate and asphalt binders and provide an appropriate strategy to minimize the moisture susceptibility of the asphalt pavement to a satisfactory level by using appropriate adhesion promoters (e.g., hydrated lime or liquid anti-stripping agents) for the incompatible combinations. The existing successful pavements with minimal or no stripping can be used as a reference to establish this database. The minimum cohesive and adhesive bonding between asphalt binder and aggregate can be used to provide a criterion to distinguish between compatible and incompatible combinations based on this concept; hereby, a threshold can be developed based on successfully performing pavements to evaluate different combinations of aggregates and asphalt binders. The importance of this strategy is related to the fact that the practitioners can access the developed database for further implementation and help save cost and time because in this case there is no need to run additional tests to assess the moisture susceptibility of the mixtures for different combinations.

Author Contributions: Aboelkasim Diab wrote the manuscript, while Zhanping You, Xu Yang, and Mohd Rosli Mohd Hasan took care of the revision.

Conflicts of Interest: The authors declare no conflict of interest.

References

1. Tarefder, R.A.; Zaman, A. Characterization of Asphalt Materials for Moisture Damage Using Atomic Force Microscopy and Nanoindentation. In *Nanotechnology in Civil Infrastructure: A Paradigm Shift*; Gopalakrishnan, K., Birgisson, B., Taylor, P., Attoh-Okine, N.O., Eds.; Springer: Berlin/Heidelberg, Germany, 2011; pp. 237–256.
2. Gubler, R.; Partl, M.N.; Canestrari, F.; Grilli, A. Influence of water and temperature on mechanical properties of selected asphalt pavements. *Mater. Struct.* **2005**, *38*, 523–532. [CrossRef]
3. Alam, M.M.; Vemuri, N.; Tandon, V.; Nazarian, S.; Picornell, M. *A Test Method for Identifying Moisture Susceptible Asphalt Concrete Mixes. Research Project 0–1455: Evaluation of Environmental Conditioning System (ECS) for Predicting Moisture Damage Susceptibility of HMAC*; The University of Texas at El Paso: El Paso, TX, USA, 1998.

4. Cho, D.-W.; Kim, K. The mechanisms of moisture damage in asphalt pavement by applying chemistry aspects. *KSCE J. Civ. Eng.* **2010**, *14*, 333–341. [CrossRef]
5. Mohd Hasan, M.R.; You, Z.; Porter, D.; Goh, S.W. Laboratory moisture susceptibility evaluation of WMA under possible field conditions. *Constr. Build. Mater.* **2015**, *101*, 57–64. [CrossRef]
6. Kiggundu, B.M.; Roberts, F.L. *Stripping in HMA Mixtures: State-of-the-Art and Critical Review of Test Methods*; National Center for Asphalt Technology: Auburn, AL, USA, 1988.
7. Hamzah, M.; Hasan, M.; van de Ven, M.; Voskuilen, J. Development of Dynamic Asphalt Stripping Machine for Better Prediction of Moisture Damage on Porous Asphalt in the Field. In *7th RILEM International Conference on Cracking in Pavements*; Springer: New York, NY, USA; pp. 71–81. [CrossRef]
8. Solaimanian, M.; Fedor, D.; Bonaquist, R.; Soltani, A.; Tandon, V. Simple performance test for moisture damage prediction in asphalt concrete. *Assoc. Asph. Paving Technol. AAPT* **2006**, *75*, 345–380.
9. Diab, A.; Sangiorgi, C.; Enieb, M.; You, Z. A conditioning method to evaluate moisture influence on the durability of asphalt mixture materials. *Can. J. Civ. Eng.* **2016**, *43*, 943–948. [CrossRef]
10. Epps, J.; Sebaaly, P.E.; Penaranda, J.; Maher, M.R.; McCann, M.B.; Hand, A.J. *Compatibility of a Test for Moisture-Induced Damage with Superpave Volumetric Design*; National Academy Press: Washington, DC, USA, 2000.

Article

Property Analysis of Exfoliated Graphite Nanoplatelets Modified Asphalt Model Using Molecular Dynamics (MD) Method

Hui Yao [1], Qingli Dai [1], Zhanping You [1,*], Andreas Bick [2], Min Wang [3] and Shuaicheng Guo [1]

1 Department of Civil and Environmental Engineering, Michigan Technological University, Houghton, MI 49931, USA; huiyao@mtu.edu (H.Y.); qingdai@mtu.edu (Q.D.); sguo3@mtu.edu (S.G.)

2 Scienomics SARL, 16 rue de l'Arcade, Paris 75008, France; andreas.bick@scienomics.com

3 Department of Mathematical Sciences, Michigan Technological University, Houghton, MI 49931, USA; minwang@mtu.edu

* Correspondence: zyou@mtu.edu; Tel.: +1-906-487-1059

Academic Editor: Feipeng Xiao

Received: 18 November 2016; Accepted: 18 December 2016; Published: 3 January 2017

Abstract: This Molecular Dynamics (MD) simulation paper presents a physical property comparison study between exfoliated graphite nanoplatelets (xGNP) modified and control asphalt models, including density, glass transition temperature, viscosity and thermal conductivity. The three-component control asphalt model consists of asphaltenes, aromatics, and saturates based on previous references. The xGNP asphalt model was built by incorporating an xGNP and control asphalt model and controlling mass ratios to represent the laboratory prepared samples. The Amber Cornell Extension Force Field (ACEFF) was used with assigned molecular electro-static potential (ESP) charge from NWChem analysis. After optimization and ensemble relaxation, the properties of the control and xGNP modified asphalt models were computed and analyzed using the MD method. The MD simulated results have a similar trend as the test results. The property analysis showed that: (1) the density of the xGNP modified model is higher than that of the control model; (2) the glass transition temperature of the xGNP modified model is closer to the laboratory data of the Strategic Highway Research Program (SHRP) asphalt binders than that of the control model; (3) the viscosities of the xGNP modified model at different temperatures are higher than those of the control model, and it coincides with the trend in the laboratory data; (4) the thermal conductivities of the xGNP modified asphalt model are higher than those of the control asphalt model at different temperatures, and it is consistent with the trend in the laboratory data.

Keywords: molecular dynamics (MD); exfoliated graphite nanoplatelets; asphalt; glass transition temperature; viscosity; thermal conductivity

---

1. Introduction

1.1. Asphalt Material

Asphalt is a byproduct of petroleum refinement and is also widely applied to many fields such as transportation, recreation, building construction, etc. Around 90% of asphalt consists of carbon and hydrogen. Based on the Corbett method, asphalt can be separated into four components: asphaltenes, saturates, napthene aromatics, and polar aromatics. Asphalt is composed of asphaltenes, paraffins, first acidiffins, second acidiffins, and nitrogen bases using the Rostler method [1]. These different types of molecules in asphalt interact with each other and affect the chemo-physical properties of asphalt [1].

Due to their special properties, nanomaterials have been introduced and used in different fields to enhance composite materials [2]. Some of their special properties include the dominance of interfacial phenomena, size and quantum effects, etc. [3]. Nanomaterials are used for electronics, agriculture, construction, food and medical technologies. Also, different types of nanoclay have been widely used in the modification of asphalt. The test results show that the layered structure of nanoclay improved the high-temperature performance of asphalt and the resistance to rutting and fatigue cracking [4,5]. The nanosilica material was also used and added to the asphalt matrix to improve performance. The micro images of nanosilica modified asphalt were observed, and the test results indicate that the resistance to permanent deformation in the modified asphalt improved [6]. The literature shows that graphite was used for the modification of asphalt, and the addition of graphite improved the electrical property of asphalt [7,8]. In this study, the multi-layer graphite sheets were applied to modify the asphalt in consideration of the high thermal stability, self-lubrication, and high electrical conductivity of multi-layer graphite sheets [9–12]. This is also the motivation to use the material for modification of the asphalt model.

1.2. Molecular Dynamics (MD) Method

Molecular Dynamics (MD), originating in theoretical physics, was applied widely in materials science [13,14]. MD is a kind of computer program to simulate the movements of atoms in materials, and the atoms and molecules interact for a designated time based on the Newton's law of motion. The trajectories of atoms and molecules are monitored, and the energy of the system is computed. In physics, MD was used to examine physical properties [13,14]. The evolution of dynamics in a single molecule is used to determine the macro properties of the system. The "statistical mechanics by number" and "Laplace's vision of Newtonian mechanics" were also used to describe the molecular dynamics. The simulation size and total duration were selected so that the calculation can be finished within a reasonable amount of time [15]. The Large-scale Atomic/Molecular Massively Parallel Simulator (LAMMPS) [16] and the Monte Carlo for Complex Chemical Systems (MCCCS) program [17] were commonly used for MD simulations. The computation algorithm of the MD simulation is shown in Figure 1. In addition, compared to other methods (such as Finite Element Method (FEM) and Discrete Element Method (DEM)), the MD method helps address specific problems or principles of atoms or materials, and also, a specific property can be studied by altering specific contributions. Moreover, the material behaviors or response can be analyzed under extreme conditions on a nanoscale.

Recently, three components of mixtures (asphaltenes, aromatics and saturates) were chosen to represent the asphalt using MD simulation. In this reference model, 1,7-dimethylnaphthalene ($C_{12}H_{12}$) and docosane (n-$C_{22}H_{46}$) were represented as naphthene aromatics and saturates, respectively [18]. Two kinds of asphaltene structures were used, and the density of each component was calculated using the MD method. From the simulation results, there are still many differences between the test and the simulation data in the calculations of the glass transition temperature and bulk modulus [18]. A new asphalt model with four components (asphaltenes, polar aromatics, naphthene aromatics and saturates) was created, and the density and thermal expansion coefficient of the asphalt model were calculated. The MD simulation results agreed with the laboratory data [19]. In addition, polystyrene was added to this asphalt model for polymer modification analysis. The radial distribution functions g(r) of components of the asphalt model were computed [19]. The MD simulation was also recently applied to study the self-diffusivity properties of asphalt binders. The effect of healing on the fatigue performance of binders was studied, and the MD model of an asphalt binder was created to predict the healing effect. The self-diffusion effect caused the binder molecules to flow across the crack interface. The correlation between the length and branching of molecules and self-diffusion of asphalt molecules was investigated [20]. The relationship between the asphalt and aggregate was established and the asphalt-quartz structure model of the interface was used in the system. The Consistent-Valence Force Field (CVFF_aug) was considered in this simulation to characterize the inter-atom interactions [21]. Based on a literature review, few researchers used MD to simulate modified asphalt and studied the physical properties of modified asphalt models in engineering disciplines. Graphite has a high thermal stability, high electrical conductivity, good self-lubricating and dry lubricating properties. The graphite was used to modify asphalt binder, and thermal conductivity and anti-aging properties improved after the addition of graphite in the asphalt binder [22,23]. Due to these improvements, in this study, the common multi-layer graphite model was adopted, and the components of the control asphalt model was composed of asphaltenes, saturates and aromatics. The Amber Cornell Extension Force Field (ACEFF) was developed and used to simulate the asphalt modified with exfoliated multi-layered graphite nanoplatelets (xGNP).

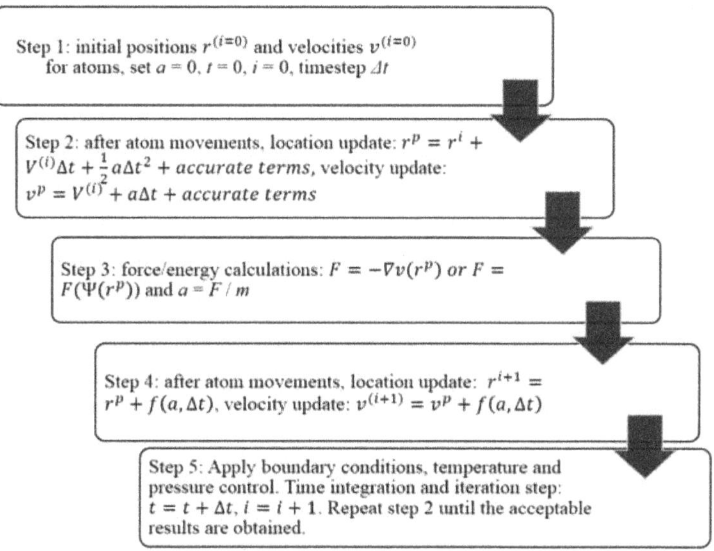

Figure 1. Computation algorithm of Molecular Dynamics (MD) simulation.

1.3. Objectives and Sections

The objectives of this study are to use the MD method to simulate and compare the properties of the xGNP modified and control asphalt model. The three-component control asphalt model consists of asphaltenes, aromatics, and saturates based on previous work [24]. The MD simulation and optimization methods were described in Section 2, as well as the force field. The common multi-layer graphite model was incorporated in the xGNP modified asphalt model to represent the xGNP modifier. The Amber Cornell Extension Force Field and Electrostatic Potential (ESP) charges were assigned to the components of the control and modified asphalt models, as described in Section 3. Different properties of the asphalt models were computed including the density, glass transition temperature (Tg), viscosity and thermal conductivity. The MD simulation data of the control and modified models and their laboratory results were compared in Sections 4–6.

2. Force Field and Optimization Methods

2.1. Classical MD Simulation Methods

Different ensembles can be used in the MD method, including the Microcanonical ensemble (NVE ensemble), Canonical ensemble (NVT ensemble), Isothermal-isobaric ensemble (NPT ensemble), Isoenthalpic-isobaric ensemble (NPH ensemble), and Generalized ensembles [25]. For instance, in the NVE ensemble, the number of moles (N), volume (V) and energy (E) in the isolated system are not be changed. The system experiences the adiabatic process and no heat exchange would occur.

2.2. Force Field

A force field, presented as parameters of mathematical functions in molecular mechanics, was used to describe the energies of the atoms. Force field parameters and functions are obtained from experimental tests and quantum mechanical calculations. Many force fields were developed and introduced by researchers, such as the Chemistry at HARvard Molecular Mechanics (CHARMM) Force Field [26], Assisted Model Building and Energy Refinement (AMBER) Force Field [27], Condensed-phase Optimized Molecular Potential for Atomistic Simulation Studies (COMPASS) Force Field [28], Optimized Potential for Liquid Simulation (OPLS) Force Field [22] and DREIDING Force Field [29]. In this study, the Amber Cornell Extension Force Field (ACEFF) was used to define the movement in the molecular system based on the Amber Cornell Force Field [30], and the experimental parameters in this force field were referenced from the General Amber Force Field (GAFF) [31]. The formula is shown in Equation (1).

$$\begin{gathered} E_{total} = \sum_{bonds} K_r\left(r - r_{eq}\right)^2 + \sum_{angles} K_\theta\left(\theta - \theta_{eq}\right)^2 + \\ \sum_{dihedrals} \frac{V_n}{2}[1 + \cos(n\varphi - \gamma)] + \sum_{i<j}\left[\frac{A_{ij}}{R_{ij}^{12}} - \frac{B_{ij}}{R_{ij}^{6}} + \frac{q_i q_j}{\epsilon R_{ij}}\right] \end{gathered} \tag{1}$$

where r_{eq} and θ_{eq} are the equilibrium structural data from an X-ray test; K_r is the force coefficient determined by linear interpolation between the single and double bond values; K_θ is the force coefficient from vibrational analysis of a simple sp2 atom; n is the multiplicity for dihedrals; γ is the phase angle for the torsional angle parameters; A, B and q are the non-bonded potentials between atom pairs; R_{ij} is the distance between the atoms; and ϵ is the well depth for van der Waals energy.

2.3. Optimization Methods

When the molecular systems are built, energy optimization and data smoothing are required to optimize the molecular systems and output results, respectively. These procedures help the researchers understand more about the systems. The following methods were used in this study.

(1) Conjugate Gradient Method

The conjugate gradient method is a kind of iterative algorithm to solve the partial differential equations. The solutions for unconstrained optimization problems like energy minimizations were also developed by Hestenes and Stiefel [32]. The formula is shown in Equation (2). The iterative method was essential and required for energy optimization (lowest energy) of large system, and an initial guess at the solution was the start of the iteration approach. The iterative method does not provide the exact solution, but can improve the approximation after a certain number of iterations. This function was also restricted by computational resources.

$$f(x) = \frac{1}{2}x^T Ax - x^T b,\ x \in R^n \text{ (iterative method)} \tag{2}$$

where A is symmetric, positive and real; b is a known coefficient; and vectors n and T are non-zero.

(2) Particle-Particle-Particle-Mesh Method (PPPM)

The Particle-Particle-Particle-Mesh Method (PPPM or P^3M) is used to compute long-rang electrostatic force, which can be divided into two parts: short- and long-range interparticle forces [33]. Short-range interactions are computed from the particle-particle (PP) calculation, and the long-range interactions are processed by the particle-mesh method [34]. The formula is shown in Equation (3). During the energy or force calculation in the system, the particles normally are forced to occupy a low spatial resolution, and it may cause errors in the results. The P^3M method is designed to calculate potential through a direct sum for particles. The P^3M method was adopted to consider the speed and accuracy of simulations in this study.

$$F_{ij} = F_{ij}^{sr} + F_{ij}^{m} \quad (3)$$

where F_{ij} is the interparticle forces in the system; F_{ij}^{sr} is the rapidly varying short-range component; F_{ij}^{m} is the slowly varying component.

3. Model Generation

Based on previous work of the authors [24] and reference [18], three components were used in the control asphalt including the asphaltenes, aromatics, and saturates at a ratio of 5:27:41. The ratio of asphaltenes, aromatics and saturates was cited form the reference [18], and this ratio (asphaltenes, aromatics and saturates at 5:27:41) in the asphalt model was based on the asphaltene mass fraction (20.7 wt%) and alkane/aromatic carbon ratio (59.6 wt%, 19.7 wt%; wt% is weight percentage) [35]. The 1,7-Dimethylnaphthalene [36] (Figure 2a) and docosane [37] (Figure 2b) were used to represent the aromatics and saturates, respectively. 1,7-Dimethylnaphthalene ($C_{12}H_{12}$) was reported by Zhang [18] and Groenzin [36] based on the ratio of alkane and aromatic in the asphalt. The docosane ($C_{22}H_{46}$) was reported by Zhang [18] and Kowalwski [37] based on the properties (melting and boiling points) of docosane and saturates. The asphaltene structure (Figure 2c) is from the references [18,38], $C_{64}H_{52}S_2$. In addition, the ESP was calculated and assigned to the components by NWChem. The control asphalt model was built through the compression of three components with NPT ensemble, and running time is around 1 ns. The start density of the control asphalt system is around 0.1 g/cm^3, and the target density is 1.0 g/cm^3, which is similar to that of the real asphalt. The low start density is good for relocations of atoms or components of systems during the energy optimization. Based on the previous property calculation of the asphalt model with the same components [18,24], the properties of the asphalt model are similar to those of the real asphalt tested in the laboratory. In this study, the ACEFF and ESP were assigned in the MD systems, and it is expected that the improvement of properties will be observed during the calculations. Therefore, the control asphalt model with ACEFF was generated to represent the control asphalt PG 58-28.

In the laboratory tests, the modifier, xGNP graphene nanoplatelets, used in this study is produced by XG Sciences Inc., and its micro-image (Figure 2d) was obtained by the field emission scanning electron microscope (FE-SEM). The distance between the graphene layers is around 3.35 Å, and the mole mass content of xGNP nanoplatelets in the modified asphalt is 2% by the weight of the control asphalt. During the preparation of xGNP modified asphalt in the laboratory, 2% xGNP multi-layer graphite particles were slowly added in the asphalt matrix at the temperature of 145 °C. The modified asphalt was sheared in the high shear machine for two hours to ensure that particles were well dispersed. Similarly, in the simulation test, the common multi-layer graphite model (Figure 2e) was used to represent the xGNP nanoplatelets, and 2% xGNP nanoplatelets by the weight of control model were randomly added to the control model. Mass mentioned in this study is based on 1 mole of the simulation box. The xGNP model with four layers was placed in the control asphalt model, and NPT ensemble was employed to compress the modified system. The xGNP modified asphalt model was generated and is shown in Figure 2f. The composition of the modified asphalt system is shown in Table 1. Different optimization methods mentioned above, the conjugated gradient method and the PPPM method, were adopted during the energy optimizations. The optimized system with the lowest energy was stable for calculations.

Table 1. The composition of xGNP modified asphalt system.

Modified Asphalt Model	Mass (g/mol)	Sum Formula	Number of Atoms per Molecule	Number of Bonds per Molecule	Number of Molecules	Total Mass (g)	Mass Fraction (%)
Asphaltene	885.23	$C_{64}H_{52}S_2$	118	132	20	17,704.59	20.25
1,7-Dimethylnaphthalene	156.22	$C_{12}H_{12}$	24	25	108	16,872.16	19.30
Docosane	310.6	$C_{22}H_{46}$	68	67	164	50,938.50	58.30
xGNP	472.53	$C_{38}H_{16}$	54	65	4	1890.13	2.15
Modified asphalt model	-	-	-	-	296	87,405.39	-

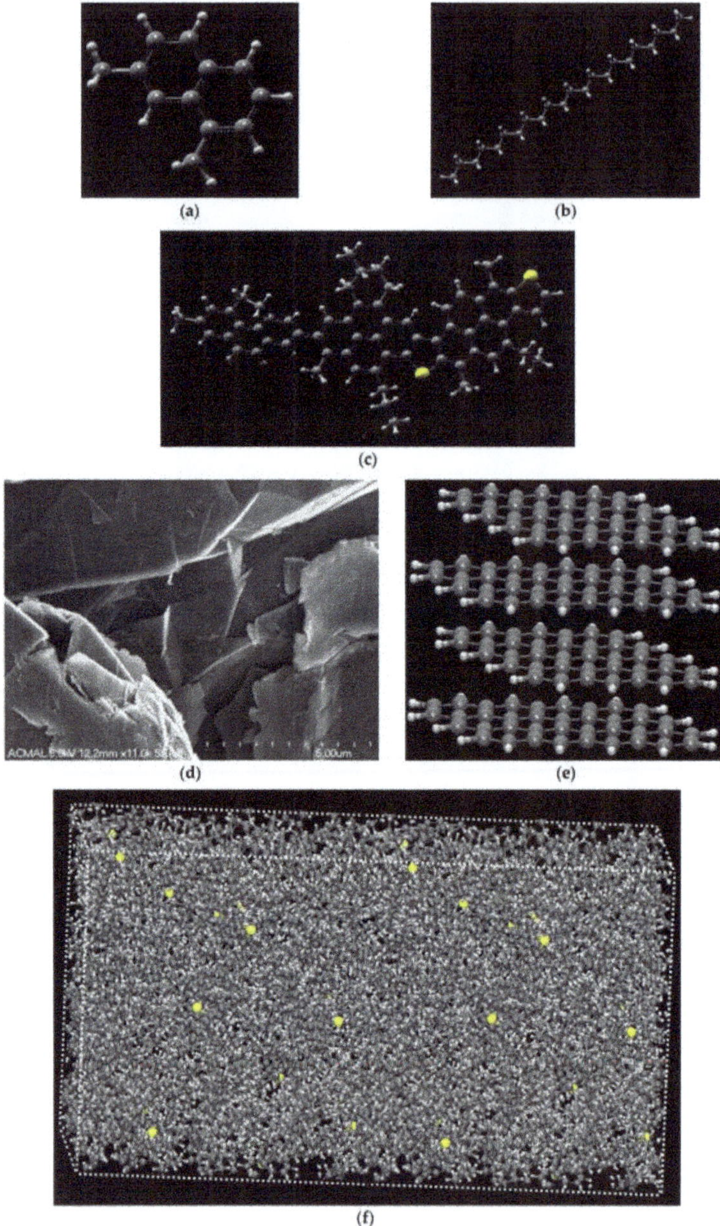

Figure 2. SEM image of multi-layer xGNP particles, structure of multi-layer xGNP model, components of control asphalt model and xGNP modified asphalt model. (a) The structure of 1,7-Dimethaylnaphtalene; (b) the structure of docosane: white color for hydrogen atom; grey color for carbon atom; (c) The structure of asphaltene: white color for hydrogen atom; grey color for carbon atom and yellow color for sulfur atom; (d) FE-SEM image of multi-layer xGNP nanoplatelets; (e) Multi-layer graphite model: white color for hydrogen atom; grey color for carbon atom; (f) Molecular structures of xGNP modified asphalt model: white color for hydrogen atom; grey color for carbon atom and yellow color for sulfur atom.

4. Physical Properties of the Control and xGNP Modified Asphalt Models

4.1. Density

When the control asphalt and modified asphalt models were generated using molecular dynamics, the densities of these models were computed at the conditions of room temperature and standard atmosphere pressure to evaluate the molecular model. The LAMMPS and optimization methods mentioned above were used to conduct the experimental MD simulations. The NPT ensemble simulations were employed to compress or relax the unit cell. The temperatures, pressures, and densities of the control and xGNP modified asphalt models are shown in Figure 3.

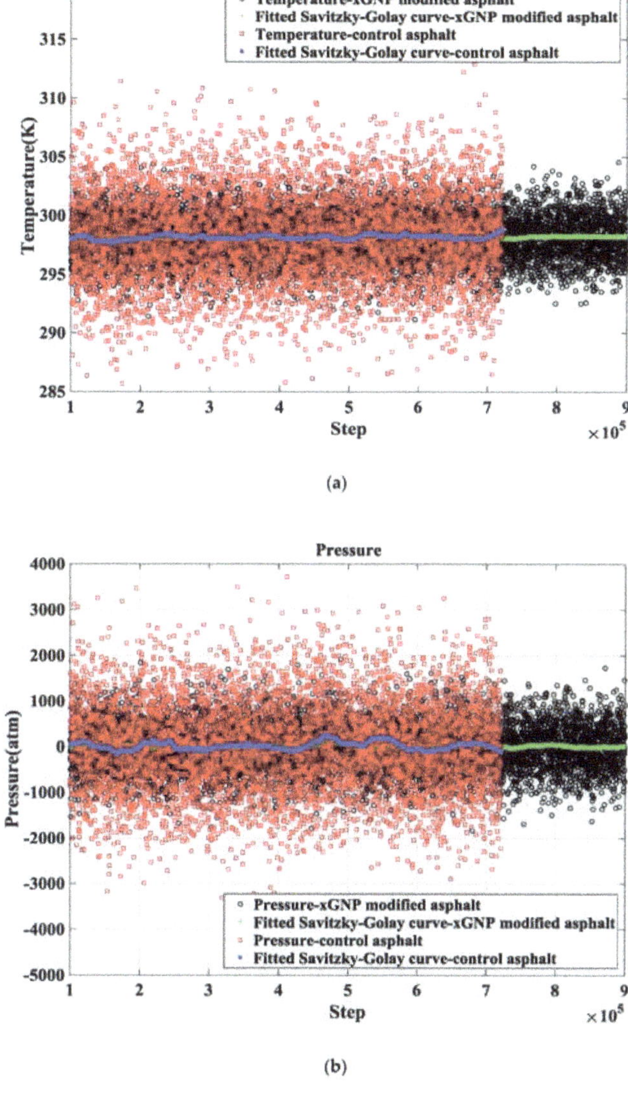

Figure 3. *Cont.*

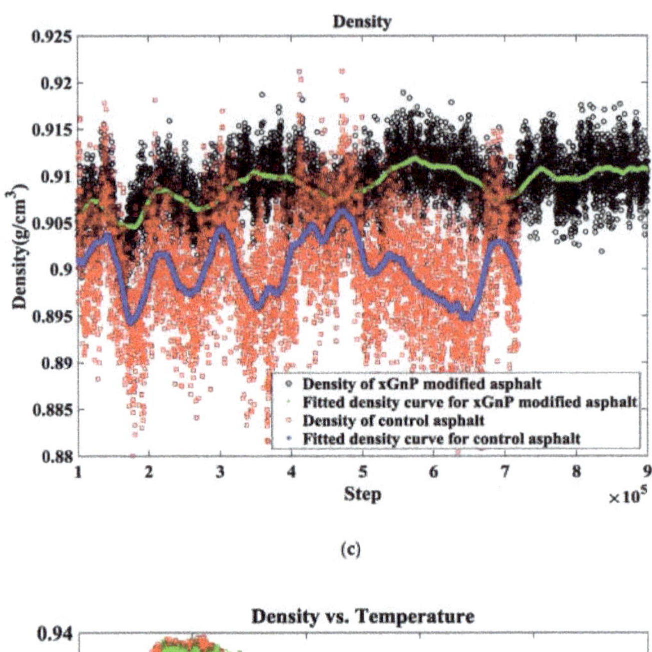

(c)

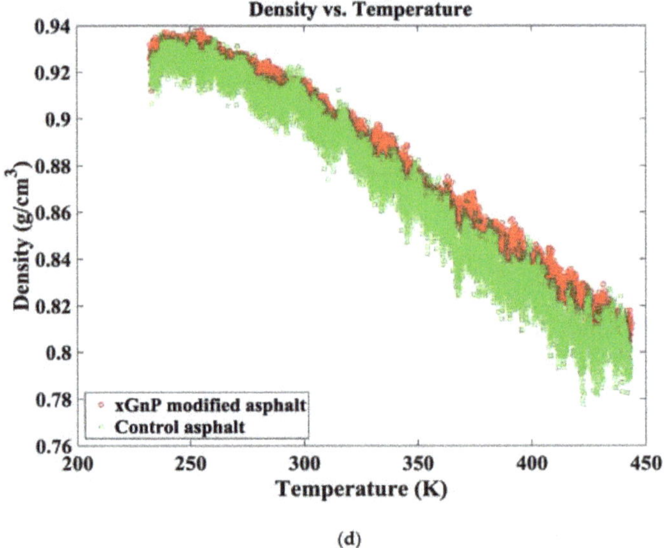

(d)

Figure 3. Density curves of the control and xGNP modified asphalt models. (**a**) Temperatures in the xGNP modified asphalt system through these steps; (**b**) Pressures in the xGNP modified asphalt; (**c**) Densities of the control and xGNP modified asphalt systems system through these steps; (**d**) Densities of the control and xGNP modified asphalt systems at different temperatures.

Figure 3a,b show the temperatures and pressures in the control and modified asphalt systems under the NPT ensemble. The systems were run more than 1 ns to be stable and optimized. Some of the results with the simulation steps are shown in Figure 3. The data were fitted by a Savitzky-Golay filter [39] (a kind of generalized moving average) with a span (a parameter to control the average) of 10%. The temperatures of the control and xGNP modified asphalt are close to 298 K during these steps, and the pressures fluctuate around 1 atmosphere (atm). Meanwhile, it is interesting to note that the data fluctuation of the control model is obviously greater than that of the modified asphalt model due to the difference in the molecular number. The fluctuated data range varies with a $1/\sqrt{N}$ variation based on the baseline of the moving average, where N is the number of molecules in the MD system [40]. Figure 3c shows densities of the control and xGNP modified asphalt systems, and the data was fitted by a Savitzky-Golay filter [39] with a span of 10%. The densities of the xGNP modified asphalt system are larger than those of the control asphalt model. It is reasonable that the addition of xGNP particles in the control system increases the density of the modified system. It is apparent that the density data amplitude of the control system is larger than that of the xGNP modified asphalt model, as well as the stability, due to their being more molecules. In addition, the densities of the control and xGNP modified asphalt systems are also close to the laboratory (0.95 g/cm^3–1.05 g/cm^3) [18,40,41] and reference data [18]. Therefore, the modified asphalt model can be deemed reasonable by obtaining a comparable mass density. Figure 3d displays the density curve of the control and xGNP modified asphalt with different temperatures, which range from 233.15 K to 443.15 K. The densities of the asphalt models decrease with the increase in temperature, and the density of xGNP modified system is slightly greater than that of the control model under different temperatures. Meanwhile, the density trends of the control and xGNP modified models are similar to that of the reference model [18]. In addition, the data fluctuation amplitude of xGNP modified model is smaller than that of the control system.

4.2. Glass Transition Temperature

The glass transition temperatures (Tg) of materials are influenced by their components, and the addition of a new component in the material results in a difference in the glass transition temperature. The formula [42] of the glass transition temperatures of composite materials is shown in Equation (4). The asphalt transfers from the viscoelastic state to a brittle one. The internal stress increases and thermal energy are insufficient below the glass transition temperature in the materials. Therefore, the glass transition temperature is an important parameter or property for amorphous materials, and it should be low for a good low-temperature performance [42]. The glass transition temperature is the temperature where two asymptotes intersect on the specific volume-temperature curve. In the laboratory, the glass transition temperature of materials can be tested by differential scanning calorimetry (DSC). In this MD simulation study, the control and xGNP modified asphalt systems were relaxed under the NPT ensemble with a temperature range of 233.15 K–443.15 K. The specific volumes of these systems were calculated and the glass transition temperatures of the models were determined. The MD simulation results of the control and xGNP modified asphalt models are shown in Figure 4.

$$\frac{1}{T_g} = \sum \frac{w_i}{T_{gi}} \tag{4}$$

where T_g and T_{gi} are the glass transition temperature of the composite material and its component, respectively; and w_i is the mass fraction of the component.

Figure 4a or Figure 4b shows the specific volumes of the control system or xGNP modified asphalt model, respectively. Figure 4c shows the specific volumes of both control and xGNP modified asphalt models. The specific volumes increase with the increasing temperatures of the models. The glass transition temperature of the control asphalt model is around 300 K [24]. As shown in the Figure, it is deduced that the Tg of the xGNP modified asphalt system is around 250 K. Based on the laboratory data in the reference [43], the Tg of asphalt ranges from 223 K to 303 K. The Tg of the modified asphalt model is within the laboratory data range, and it is also better than the reference data [18] (298 K–358 K) of the asphalt model. In order to get better thermal relaxation in the asphalt, a low Tg of asphalt is expected. Figure 4d shows the comparisons of glass transition temperatures of the references and MD simulations. The glass transition temperatures of the Strategic Highway Research Program (SHRP) asphalt binders are all around 250 K [42], including SHRP asphalt AAA-1, AAB-1, AAC-1, AAD-1, AAF-1, AAG-1, AAK-1 and AAM-1. The Tg of xGNP modified asphalt model is close to the glass transition temperatures of SHRP asphalt binders. Therefore, the Tg of xGNP modified asphalt is reasonable and better than the results of the control [24] and reference asphalt models [18].

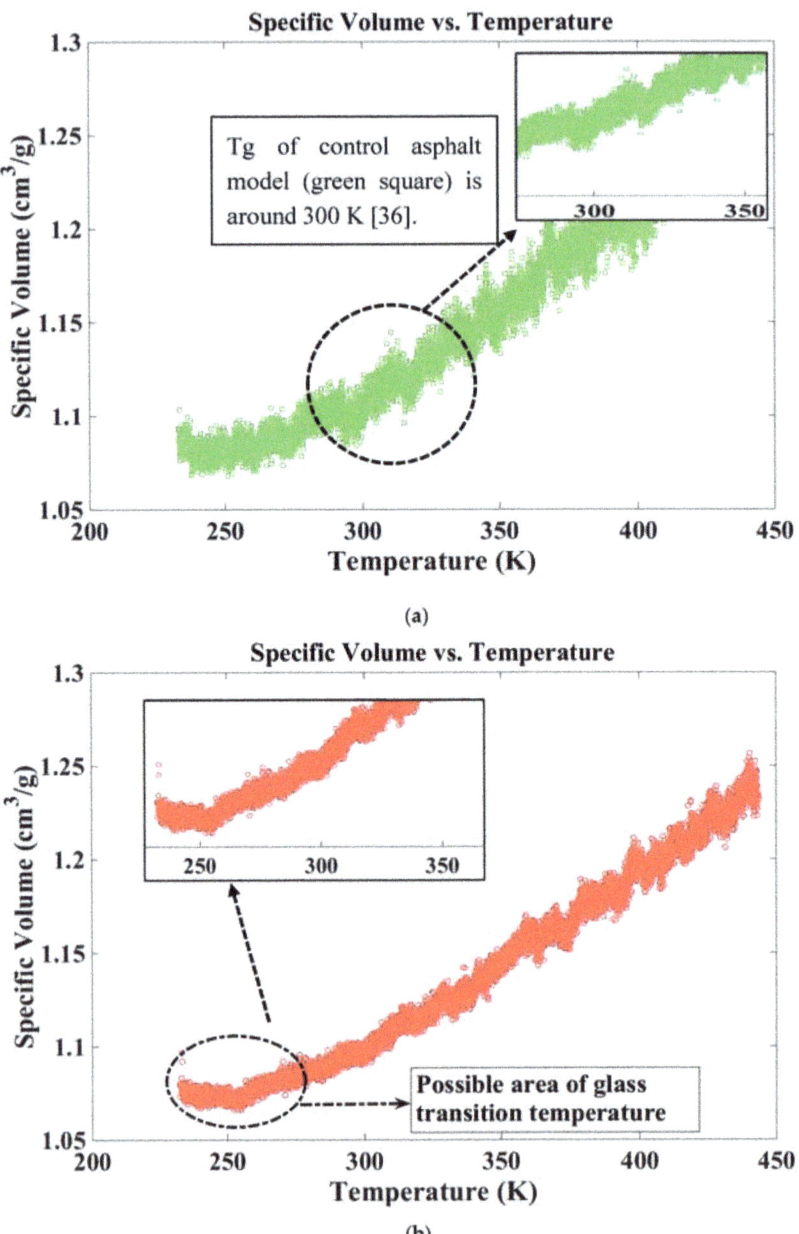

Figure 4. *Cont.*

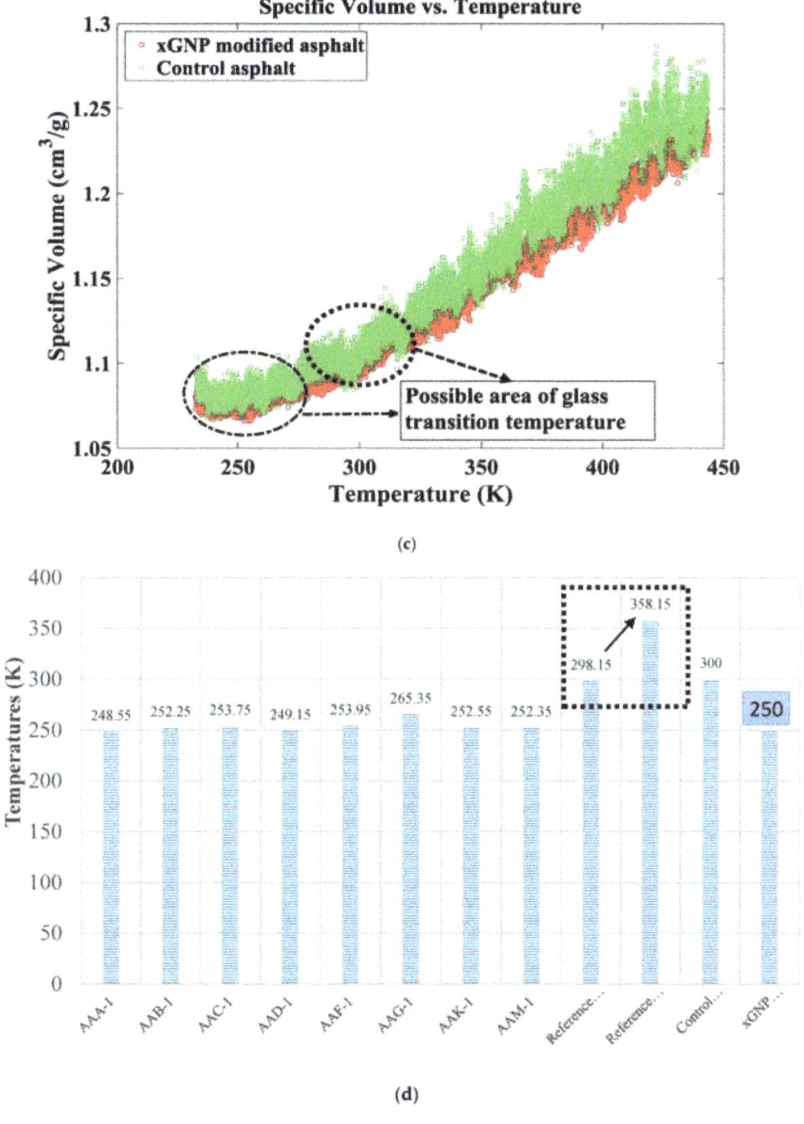

(d)

Figure 4. Specific volumes and temperatures of the control and xGNP modified asphalt systems. (**a**) specific volumes of the control asphalt system: amplified section to show the Tg temperature with asymptotes intersect for control asphalt model; (**b**) specific volumes of the xGNP modified asphalt system: amplified section to show the Tg temperature with asymptotes intersect for modified asphalt system; (**c**) specific volumes of the xGNP modified and control asphalt system; (**d**) Glass transition temperatures of different asphalt types and models: the data of the glass transition temperature of different binders is from the reference [18,24,42]: 248.55 K for AAA-1 binder, 252.25 K for AAB-1, 253.75 K for AAC-1, 249.15 K for AAD-1, 253.95 K for AAF-1, 265.35 K for AAG-1, 252.55 K for AAK-1, 252.35 K for AAM-1, which were tested by Usman (1997) [42]; 300 K for control asphalt model which is from the MD simulation by Yao at al. 2016 [24]; and temperature range from 298.15 K to 358.15 K for the reference model from MD simulation by Zhang and Greenfield [18]; Tg temperature of xGNP modified asphalt model is 250 K.

5. Rheological Properties of the Control and xGNP Modified Asphalt Models

5.1. Viscosity Measurement Method and Results

Dynamic shear viscosity is an important parameter of fluids to measure the resistance to gradual deformation induced by shear stress. If the shear speed caused by the external force is appropriate, the fluid particles move parallel to the particles sheared. The speed varies linearly from the sheared layers to different layers. The resistance between these layers is caused by friction. The formula to calculate viscosity is shown in Equation (5).

$$\eta = \frac{Fy}{Au} \tag{5}$$

where F is the applied force; A is the area of the plate; η is the dynamic shear viscosity; and u/y is the shear gradient.

During the construction of asphalt pavement, the viscosity determines the mixing and compaction temperatures, which relates to the pump ability, mix ability and workability of asphalt. Based on the American Society for Testing and Materials (ASTM) D4402, the Brookfield DV-II plus viscometer (Figure 5a) was selected to test the viscosity of asphalt in the laboratory. The viscosity test results are shown in Figure 5b.

(a)

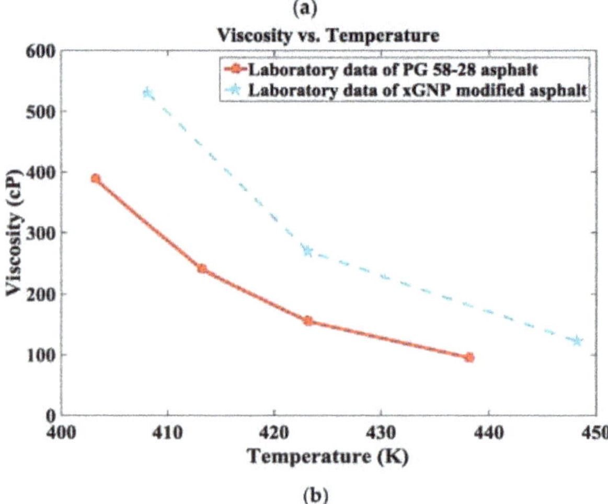

(b)

Figure 5. Laboratory results of viscosity of the control and xGNP modified asphalt binders. (**a**) Brookfield DV-II plus viscometer for testing viscosity of asphalt; (**b**) viscosities of the control (PG 58-28) and xGNP modified asphalt binders.

From the test data, the viscosities of the xGNP modified asphalt binder (light blue line, Figure 5b) are higher than those of the control asphalt binder (red line). It indicates that multi-layer xGNP particles increase the viscosity of the modified asphalt binder. In addition, the viscosity decreases with the increase in temperatures of the asphalt binder. Exponential trends were also observed in the test data. The MD simulation for viscosity of the asphalt binder model was discussed in the following Section 5.2.

5.2. MD Viscosity Aimulation Methods and Results

In the MD experimental simulation, there are four common methods to calculate the dynamic shear viscosity in the MD systems [16]: (1) a non-equilibrium MD (NEMD): the unit cell is sheared by "fix deform" and the temperature is controlled; (2) a NEMD: the viscosity is computed through the velocity and pressure in the systems; (3) a reverse non-equilibrium MD (rNEMD): the momentum flux is transferred between different layers in the unit cell through the Muller-Plathe algorithm; (4) equilibrium MD (EMD): the Green-Kubo (GK) formula is used to compute the viscosity, and continuous momentum flows are applied in the unit cell.

In this MD study, the Muller-Plathe method was used to calculate the viscosities of the control and xGNP modified asphalt systems. The unit cell of asphalt models was split into 20 layers. The viscosity calculation is shown in Equation (6). During the calculation of viscosity in the MD simulation, unit conversion is needed. The viscosity unit in the MD simulation is gram/mol/angstrom/femtosecond, but the unit in the laboratory test is kilogram/meter/second. Avogadro's constant is used to convert from microscopic to macroscopic states. In addition, the momentums transferred in the control and xGNP modified systems at the temperature of 443 K are shown in Figure 6a. The viscosities of the control and xGNP modified asphalt systems at a temperature of 423 K are shown in Figure 6b,c. The MD and laboratory viscosity results of the control and xGNP modified asphalt systems at different temperatures are shown in Section 5.4.

$$j_z(p_x) = -\eta \frac{\partial v_x}{\partial_z} \text{ and } j(p_x) = \frac{P_x}{2tA} \quad (6)$$

where η is the dynamic shear viscosity; $\frac{\partial v_x}{\partial_z}$ is the shear rate; $j(p_x)$ is the input momentum flux; P_x is the input momentum; t is the simulation time; $A = L_x L_y$, L_x is the length of unit cell in the x direction; and L_y is the length of unit cell in the y direction.

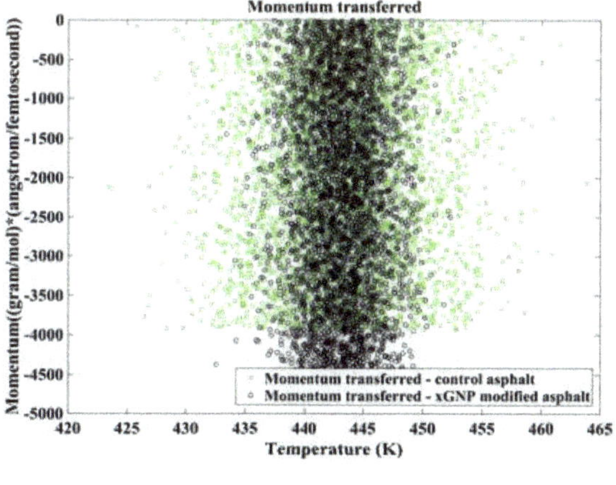

(a)

Figure 6. *Cont.*

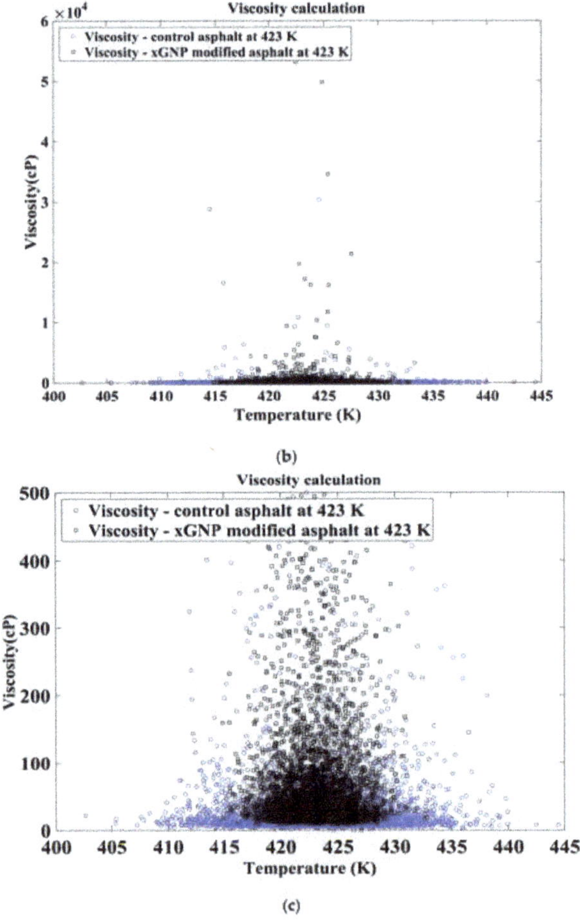

Figure 6. Viscosity test and MD calculation for the control and xGNP modified asphalt systems. (**a**) Momentum transferred in the control and xGNP modified asphalt binder systems at the temperature of 443 K; (**b**) Viscosities of the control and xGNP modified asphalt binder systems at the temperature of 423 K; (**c**) Viscosities (from 0 to 500 cp) of the control and xGNP modified asphalt binder systems at the temperature of 423 K.

Figure 6a displays the momentum transferred in the control and xGNP modified asphalt binder systems at a temperature of 443 K, as well as some of the results from different temperatures for repeatability. The momentum transferred in the xGNP modified asphalt binder model is more than that of the control model. The temperatures in the control asphalt binder model fluctuate more than those of the xGNP modified asphalt binder model due to there being fewer molecules. The temperatures are also around 443 K, and do not have large variations. It indicates that more molecules in the MD system lead to less data variation and a stable structure of materials. It coincides with the conclusion that the vibrating range of MD data is within a $1/\sqrt{N}$ variation [40] (N is the number of molecules in the system). Figure 6b,c demonstrates the viscosities of the control and xGNP modified asphalt binder systems at a temperature of 423 K, and it is also part of data analysis under different temperatures. The temperature fluctuation of the control model is larger than that of xGNP modified asphalt binder model due to fewer molecules compared to xGNP modified asphalt binder system. The temperatures are also centered at 423 K with a 20 K variation. It coincides with the temperature setting of simulations. Furthermore, the statistical analysis of viscosities of the control and xGNP modified asphalt models was performed to analyze the distributions of viscosity data in MD simulation.

5.3. Statistical Analysis for Viscosities of the Control and xGNP Modified Asphalt Models

The variation of data in the viscosity calculation is observed in the last section. The statistical analysis was used to better understand the data distribution, and it is also good for describing and reproducing the data. It is well known that many experimental or observational data arising in engineering is shaped by a lognormal distribution due to its various appealing properties. If a random variable x follows a lognormal distribution, the random variable $Y = \log(X)$ is distributed as a normal. The probability density function (PDF) of a lognormal distribution with parameters μ and σ is given by

$$f(x) = \frac{1}{x\sigma\sqrt{2\pi}} e^{-\frac{(\log x-\mu)^2}{2\sigma^2}} \tag{7}$$

where $x > 0$, $-\infty < \mu < \infty$, and $\sigma > 0$. In this study, we consider the lognormal distribution, because it provides heavier tails compared with the normal one and is thus more flexible to experimental data when studying robustness to outliers. Due to large variations of data from 7.375894 to 53,239.325870, we consider the more appropriate fits based on data ranging from 0 to 300, and from 0 to 500. The parameter estimates of μ and σ with their standard errors in parenthesis are presented in Table 2. The histograms with the fitted lognormal distributions are depicted in the top two figures (Figure 7). It can be seen from the two figures that with different choices of truncations, the lognormal distribution provides more flexible fits to xGNP data. A similar conclusion can also been drawn for the fitness of the lognormal distribution to control data (Figure 7). Consequently, we may conclude that the data departure from the lognormality is acceptable or slight, the lognormal distribution is a more robust and flexible model, allowing a better fit as shown above.

Table 2. Parameter estimates of μ and σ with their standard errors.

Type	μ	σ
xGNP (<500)	3.81542596 (0.01393144)	0.85901818 (0.00985102)
Control (<500)	3.01951678 (0.01529361)	0.95801986 (0.01081422)

Note: xGNP: viscosity data of the xGNP modified asphalt model; Control: viscosity data of the control modified asphalt model.

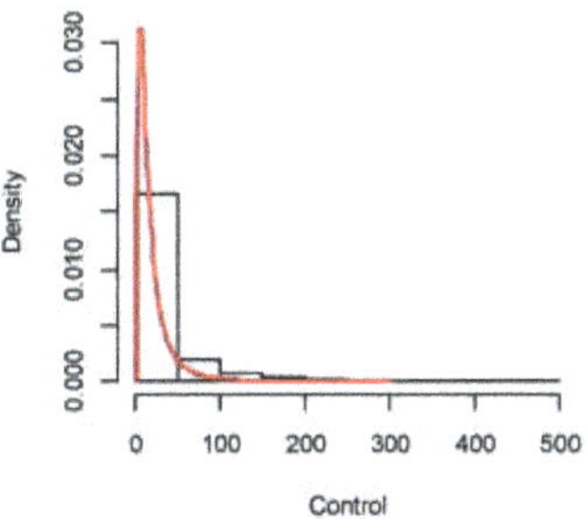

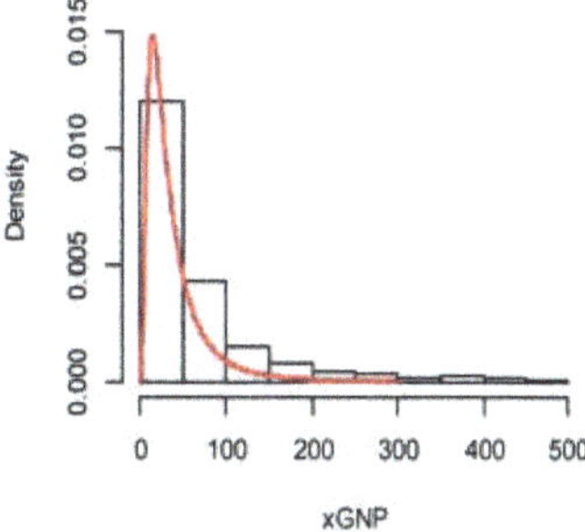

Figure 7. The histogram with the fitted lognormal distributions (the right plot for xGNP data, and the left plot from Control data; Bars represent the MD simulation data, and the red lines represent fitted lognormal curves; the data number in x-axis represents the viscosities of the asphalt binder model (viscosity unit: cp), and the distribution density was shown in y-axis; "control" in this figure means "control asphalt binder model" and "xGNP" means "xGNP modified asphalt binder model).

5.4. Comparison of Viscosity Predictions of the Control and xGNP Modified Asphalt Models

Based on the data analysis from the MD simulation and laboratory data, the viscosity data between the control and xGNP modified asphalt binder models was compared and analyzed, as well as the experimental results. The exponential regressions were used to fit the viscosity data of the control and xGNP modified asphalt binder models. The comparisons between the control and xGNP modified asphalt binder models were conducted, as shown in Figure 8, including the reference data.

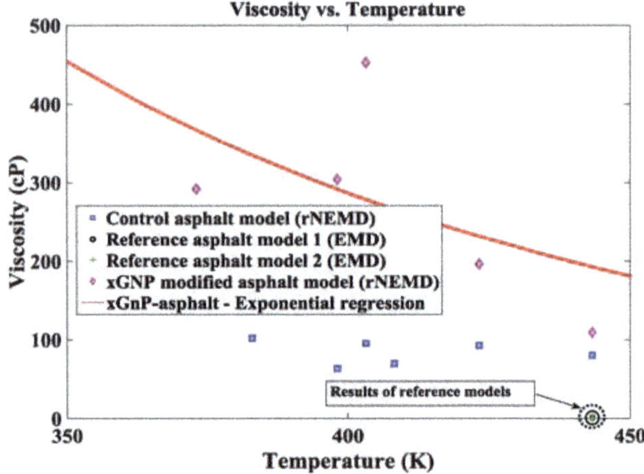

Figure 8. MD viscosity results of the control and xGNP modified asphalt binder systems.

Figure 8 shows the viscosities of the control and xGNP modified asphalt binder models. The viscosities of MD simulations were averaged from the calculations under each separate temperature. The MD simulation viscosities of the xGNP modified asphalt binder model (red line in Figure 8) are also higher than those of the control asphalt binder model. It is similar to the trend in the laboratory data. The exponential trends are observed to be fitted for the MD simulation results due to the same trend in the laboratory data [44]. In this figure, it is obvious that the viscosities of the control and xGNP modified asphalt binder models are higher than those of the reference models [19] (0.65 cp and 1.35 cp at 443.15 K) using Green-Kubo and Einstein (Ein) EMD methods. The relatively flat line is observed in the viscosity data of the control asphalt binder model, and the viscosity results (92.47 cp and 80.13 cp) at 423.15 K and 443.15 K are close to the laboratory data (155.0 cp at 423.15 K and 95.0 cp at 438.15 K in Figure 5b) at two different temperatures for the control asphalt binder model, respectively. It is noticed that there is some improvement between viscosities of the control asphalt binder model (Figure 8) and the asphalt model in the reference [24]. Furthermore, for the xGNP modified asphalt model, the viscosities (452.42 cp, 195.98 cp and 108.96 cp) at 403.15 K, 423.15 K, and 443.15 K, respectively, are very close to the laboratory data (530.0 cp, 270.0 cp and 122.5 cp). The viscosity simulation results of the xGNP modified asphalt binder system are better than those of the control asphalt binder model, and the trend in viscosity of the xGNP modified asphalt binder model is similar to that of the laboratory data. With regard to this improvement, it is caused by the increase in the molecular number in the xGNP modified asphalt model compared to the control asphalt model. It is also an improvement to use the Muller-Plathe method to calculate the viscosity of the asphalt binder model, as well as the optimization methods used in the simulations. It is expected that more molecules in the MD asphalt system improve the accuracy of data prediction. Therefore, the viscosity calculation of MD simulations in the xGNP modified asphalt system with the Amber Cornell Extension Force Field provides a better prediction using the Muller-Plathe method compared to the results of the references [19] and the control asphalt model.

6. Thermal Property of the Control and xGNP Modified Asphalt Models

6.1. Thermal Conductivity Measurement Methods and Results

The thermal conductivity is a kind of measure of materials to transmit the heat energy in a diffusive manner based on Fourier's law. The materials with a high thermal conductivity are applied to the heat sink, and the materials with a low thermal conductivity are manufactured for thermal insulation. In the laboratory, the xGNP modified asphalt was mixed with ultrasonic stirring during the process of high shear so that the xGNP particles can be homogenously dispersed in the asphalt matrix. The thermal properties' analyzer (KD2 Pro) was employed to measure the thermal conductivity of asphalt based on the transient line heat source method [45]. The asphalt was placed in the glass tube as shown in Figure 9a,b, and the single needle TR-1, with a 2.4 mm diameter and 60 mm length, was used to test the thermal conductivity. During the heating and cooling processes, the temperature-time relationship is monitored by the sensor located in the needle. The thermal conductivity is calculated with the parameters from the fitted curve for temperature-time. The formula for thermal conductivity is shown in Equation (8). The thermal conductivities of asphalt under different temperatures in the laboratory are shown in Figure 9c.

$$\begin{gathered} T = m_0 + m_2 t + m_3 lnt \text{ (Heating process)} \\ T = m_1 + m_2 t + m_3 ln\frac{t}{(t-t_h)} \text{ (Cooling process) and } k = \frac{q}{4\pi m_3} \end{gathered} \tag{8}$$

where m_0 and m_1 are the ambient temperatures in the heating and cooling processes, respectively; m_2 is the rate of drift of the background temperature; m_3 is the slope of a line relating temperature rise to the logarithm of temperature; q is the heat input and k is the thermal conductivity.

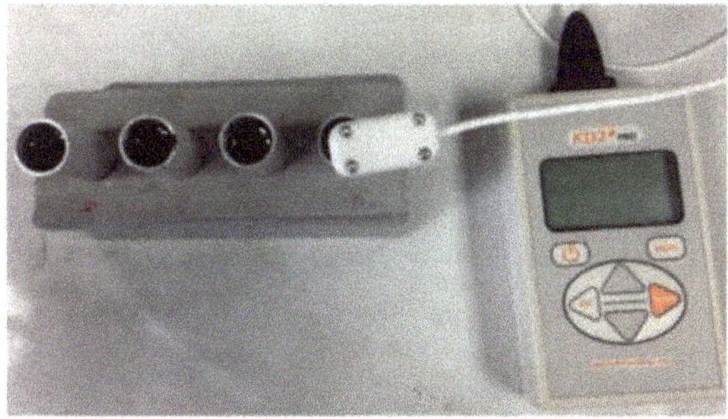

(a)

(b)

Figure 9. *Cont.*

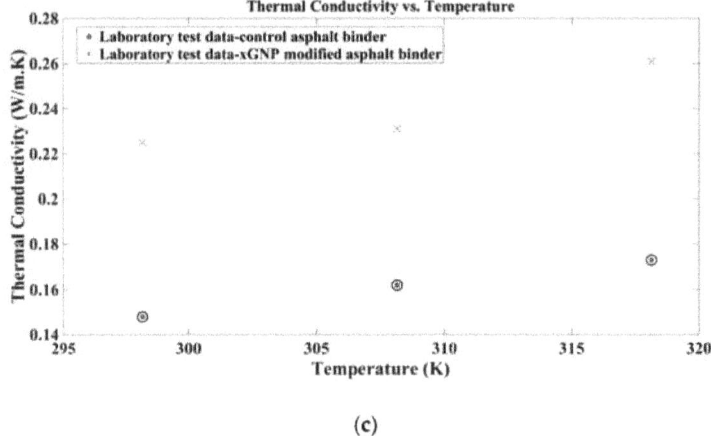

(c)

Figure 9. Thermal conductivity test apparatus and results. (**a**) KD2 Pro thermal conductivity tester in the laboratory; (**b**) the chamber for temperature control in the thermal conductivity test; (**c**) Thermal conductivity results of the control and xGNP modified asphalt binders.

Figure 9a,b display the KD2 Pro thermal conductivity apparatus and the temperature control chamber used in the laboratory, respectively. Figure 9c demonstrates the thermal conductivity results tested in the laboratory. The thermal conductivity of the control asphalt binder at room temperature is 0.148 W/m·K, and it is close to the reference data of asphalt [22] (0.170 W/m·K). The thermal conductivity of the xGNP modified asphalt binder at room temperature is 0.226 W/m·K, and it is near to the reference data [22] from 0.396 W/m·K to 0.934 W/m·K with different amounts of graphite (different types from the multi-layer graphite used in this study). Through the test results, the addition of xGNP particle increases the thermal conductivity of the asphalt binder. It is likely that high thermal conductivity of xGNP particles (around 3000 W/m·K) enhances the thermal transfer in the asphalt binder matrix based on the thermal conductivity data of xGNP particles from xgsciences.com [46]. It can also be expected that the light absorption is improved after the addition of xGNP particles in the asphalt binder.

6.2. MD Simulation Methods and Results

In the MD simulations, there are four methods to compute the thermal conductivity for MD systems: (1) NEMD: energy is applied to the hot region, and an equal amount of energy is subtracted from the cold area in the simulation cell. The heat flux is monitored between different temperature layers; (2) NEMD: Energy is added or subtracted in two regions, and the temperature difference of the intermediate region is monitored; (3) rNEMD: The kinetic energies of two atoms in different layers are swapped, and the temperature gradient is monitored; (4) EMD: the heat flux can be computed from the fluctuations of per-atom potential and kinetic energies, as well as the stress tensor. It is common in NEMD (non-equilibrium MD) for calculating the thermal conductivity of systems to impose the temperature gradient and the responded heat flux is measured. However, a reverse non-equilibrium MD (rNEMD) algorithm is used in the Muller-Plathe method [47]. The heat flux is applied in the system and the temperature gradient is measured. When the heat flux is imposed in the simulation cell, which is divided into N slabs (N is an even number, 20 was used in this study) with identical thickness. Energy transfer (Figure 10a) is produced from hot to cold slabs through the z-direction and it causes the temperature difference (Figure 10b) between these two slabs. Velocity exchange occurs in two particles, and the energy conservation is satisfied. The formulas for thermal conductivity and heat flux are shown in Equation (9). In addition, the Avogadro constant was used to complete the unit conversion due to different scales from microscopic to macroscopic states based on Equation (9). The mass/volume effect in the MD simulation was also considered in the unit conversion. The results of thermal conductivity of the xGNP modified and control asphalt binder models are presented in Figure 10c.

$$J = -\lambda \nabla T \text{ and } \lambda = -\frac{\sum_{transfers} \frac{m}{2}\left(v_{hot}^2 - v_{cold}^2\right)}{2tL_xL_y\langle \partial T/\partial z\rangle} \quad (9)$$

where ∇T is the temperature gradient (scalar) in the simulation cell; J is the energy transferred (scalar) through the surface of layers; λ is the thermal conductivity; t is the simulation time; v_{hot} is the velocity of the hot particle; v_{cold} is the velocity of the cold particle; m is the identical mass of particles; L_x is the length of the simulation box in the x-direction; and L_y is the length of the simulation box in the y-direction; and $\partial T/\partial z$ is the temperature gradient in the z-direction.

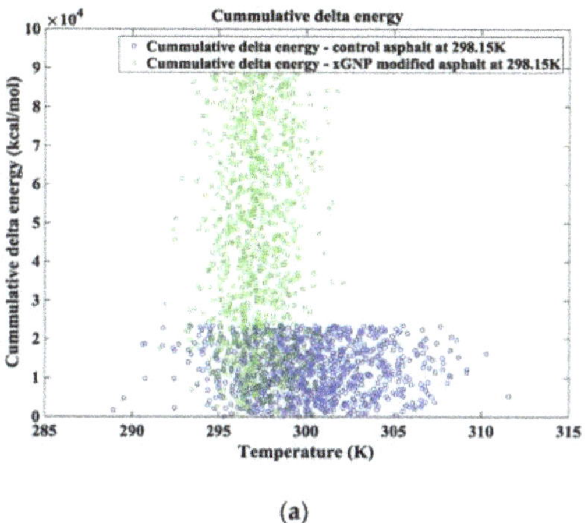

(a)

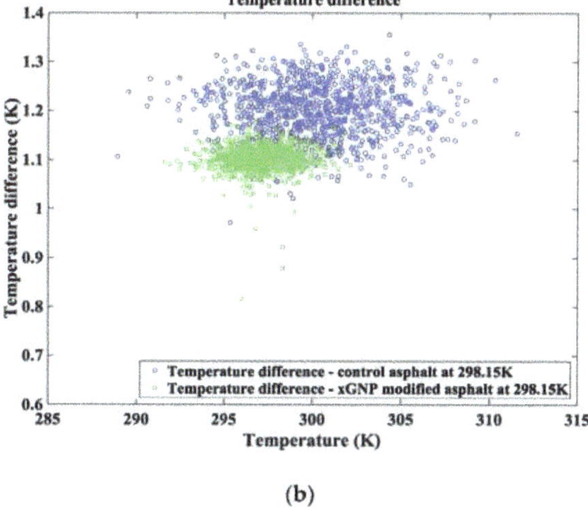

(b)

Figure 10. Thermal conductivity calculations in the MD simulations. (**a**) Cumulative delta energies of the control and xGNP modified asphalt binder systems at a temperature of 298.15 K; (**b**) Temperature difference of the control and xGNP modified asphalt binder systems at a temperature of 298.15 K.

Figure 10a shows the cumulative energy input in different molecular binder systems at a temperature of 298.15 K. The control asphalt binder system has a relatively low energy input in contrast to the energy of the xGNP modified asphalt binder system due to the small number of molecules and low volume in the control binder system. The energy of the xGNP modified asphalt binder system is four times greater than that of the control asphalt binder system, and it is the same as the mass ratio of the xGNP modified asphalt binder system to the control binder system. The temperature variation of the xGNP modified asphalt binder model is also lower than that of the control binder system due to the large number of molecules. Figure 8d shows the temperature difference in different asphalt binder systems after the input of the heat flux at a temperature of 298.15 K. The variation in the temperature difference of the xGNP modified binder model is less than that of the control binder model, as well as the temperature variation in the MD simulation. It is likely that more molecules in the system result in better stress and heating responses and produce a stable system. The cumulative energy and temperature difference of the control and xGNP modified asphalt binder models at different temperatures were calculated using MD simulations, and based on Equation (9), the thermal conductivity results of the asphalt binder models are shown in Figure 11 (next section).

6.3. Comparison of Thermal Conductivity of the Control and xGNP Modified Asphalt Binder Models

Figure 11 demonstrates the thermal conductivity results through MD simulations. The addition of xGNP model in the asphalt binder model increases the thermal conductivity of the modified asphalt binder model. It is consistent with the laboratory data. It is reasonable that the thermal conductivity of the control and xGNP modified asphalt models at room temperature is around 0.275 W/m·K and 1.146 W/m·K, respectively, compared to the reference data of graphite modified asphalt binders [22] from 0.396 W/m·K to 0.934 W/m·K. There is an insignificant difference between the laboratory data and MD simulation results. The xGNP particles in the control binder improves the thermal conductivity of the modified asphalt binder from the experimental data. The same trend of thermal conductivity is also observed in the data from the MD simulation after the addition of the multi-layer graphite xGNP model in the control asphalt binder model. The thermal conductivities of the control and xGNP modified asphalt binders increase with the increase in temperatures of the experimental tests, and the thermal conductivities of the control and xGNP modified asphalt binder models also increase by increasing the temperatures of the systems. However, there are minor differences between the experimental data and MD simulation results. There may be a few reasons for this: (1) in the preparation of the samples and laboratory testing, the xGNP particles in the modified asphalt were not perfectly dispersed in the tested area due to the mixing method and not due to operational errors, and this causes inhomogeneous heating of the modified asphalt during testing; (2) the test area for thermal conductivity is relatively small; (3) the multi-layer graphite xGNP model does not fully represent the xGNP particles in the asphalt binder matrix for the calculation of thermal conductivity, and there are some improvements needed for models of xGNP particles and asphalt binder. After the analysis of laboratory and MD data, it is confirmed that xGNP particles can improve the thermal conductivity of asphalt, and the multi-layer graphite xGNP model can also enhance the thermal conductivity of the control asphalt binder model. The trend in temperature versus thermal conductivity of the MD simulation results is the same as the trend in the experimental data.

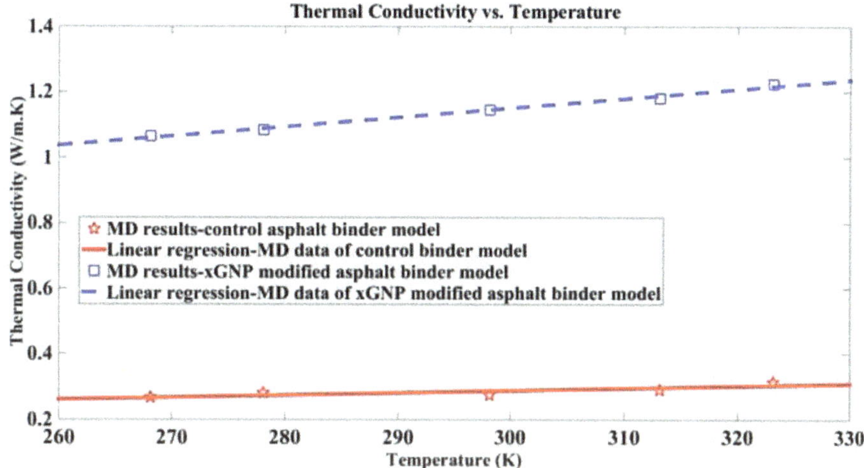

Figure 11. Thermal conductivity results of MD simulations.

7. Discussion and Conclusions

The MD model of multi-layer graphite xGNP nanoplatelets was created and used for the investigation of the effect of modification on the control asphalt binder model. The control asphalt binder model was composed of three components: asphaltenes, aromatics, and saturates at a certain ratio. The xGNP modified asphalt binder model was generated from the addition of the xGNP model in the control asphalt binder model. The conjugate gradient method and PPPM were used for energy optimization, and the Savitzky-Golay filter was used to smooth data. The Amber Cornell Extension Force Field and ESP charge were used in these asphalt models, and the physical properties of the MD binder models were calculated including density, the glass transition temperature, viscosity, and thermal conductivity. The following conclusions may be made.

(1) The densities of these asphalt binder models were computed, and the addition of the multi-layer xGNP model increased the density of the xGNP modified binder model compared to that of the control binder model. The molecular number in MD systems significantly affects the data variation for density calculation. The density of MD asphalt binder systems decreases with the increase in temperatures.

(2) The glass transition temperature of the xGNP modified asphalt binder model is around 250 K, and it is better than the results of the reference, 298 K–358 K [18]. This glass transition temperature is better than previous results (around 300 K [24]) for the control asphalt binder model, because it is the same as the glass transition temperature of SHRP asphalt binders, around 250 K, from laboratory results [42].

(3) The Muller-Plathe method was used to calculate the viscosity of the control and xGNP modified asphalt binder models. The 20 layers in the MD asphalt models were separated for this calculation. The addition of xGNP particles in the control asphalt binder matrix improves viscosities of the modified asphalt at different temperatures, and the same effect of multi-layer xGNP models in the control asphalt binder model was observed. Compared to the experimental viscosities of the xGNP modified asphalt binder, the viscosities of the MD simulation is close to the experimental results at the temperatures of 403 K, 423 K, and 443 K. The relationship between viscosities and temperatures in the data of the MD simulations is also the same as that of the laboratory results.

(4) The experimental data shows that the xGNP particles in the control asphalt binder increase the thermal conductivity of the modified binder at room temperature. During the calculation of thermal conductivity, the Muller-Plathe method was used in these MD simulations, and the multi-layer xGNP

model in the control binder model also improves the thermal conductivity of the control binder model at room temperature. The thermal conductivities of the control and xGNP modified asphalt binders increase with increasing temperatures, and the same trend is observed in the data of MD simulations.

Therefore, the multi-layer xGNP graphite particles in the asphalt binder can improve viscosity and thermal conductivity of the asphalt binder, and the xGNP model in the control asphalt binder model can also enhance the density, glass transition temperature, viscosity and thermal conductivity of the control binder model. It is obvious that the same trend of experimental data and MD results is observed during the testing and MD calculations of different properties of asphalt. It is likely that the xGNP particles can be utilized and generalized for pavement construction and heat sinks. The contributions of this paper include (1) the use of the xGNP graphite particles to enhance the performance of the asphalt binder; (2) the generation of the xGNP model for the modification of the asphalt model; (3) the application of the Muller-Plathe method to compute the thermal conductivity of the asphalt models; and (4) the use of the correlation analysis to reveal the linear relationship in MD simulation data. In addition, more properties of the xGNP modified asphalt binder and its models will be tested and calculated for future research.

Acknowledgments: The authors appreciate the financial support of the U.S. National Science Foundation (NSF) under grant 1300286. The computational studies were performed using Scienomics MAPS software suite (MAPS, Version 3.4, Scienomics, Paris, France, 2014) and computer cluster (Superior research center) at Michigan Technological University. The last author also acknowledges the fellowship support from China Scholarship Council under No. 201406370141. Any opinion, finding, and conclusion expressed in this paper are those of the authors and do not necessarily represent the view of any organization.

Author Contributions: Hui Yao, Qingli Dai and Zhanping You conceived of this project and revised this manuscript. Hui Yao wrote this manuscript. Andreas Bick provided the technical support for the generation of the models. Min Wang helped statistical information and support, and Shuaicheng Guo helped with the test of thermal conductivity.

Conflicts of Interest: The authors declare no conflict of interest.

References

1. Asphalt Institute. *Superpave Performance Graded Asphalt Binder Specification and Testing*; Superpave Series No 1 (SP-1); Asphalt Institute: Lexington, KY, USA, 2003.
2. Yao, H.; Li, L.; Xie, H.; Dan, H.-C.; Yang, X.-L. Microstructure and Performance Analysis of Nanomaterials Modified Asphalt. In Proceedings of the American Society of Civil Engineers—Geo Hunan International Conference, Hunan, China, 9–11 June 2011; pp. 220–228.
3. Ahmad, S.; Tripathy, D.B.; Mishra, A. Sustainable Nanomaterials. In *Encyclopedia of Inorganic and Bioinorganic Chemistry*; John Wiley & Sons: New York, NY, USA, 2016.
4. Yao, H.; You, Z.; Li, L.; Goh, S.W.; Lee, C.H.; Yap, Y.K.; Shi, X. Rheological properties and chemical analysis of nanoclay and carbon microfiber modified asphalt with Fourier transform infrared spectroscopy. *Const. Build. Mater.* **2013**, *38*, 327–337. [CrossRef]
5. Yao, H.; You, Z.; Li, L.; Shi, X.; Goh, S.W.; Mills-Beale, J.; Wingard, D. Performance of asphalt binder blended with non-modified and polymer-modified nanoclay. *Const. Build. Mater.* **2012**, *35*, 159–170. [CrossRef]
6. Yao, H.; You, Z.; Li, L.; Lee, C.H.; Wingard, D.; Yap, Y.K.; Shi, X.; Goh, S.W. Rheological Properties and Chemical Bonding of Asphalt Modified with Nanosilica. *J. Mater. Civ. Eng.* **2013**, *25*, 1619–1630. [CrossRef]
7. Liu, X.; Wu, S. Study on the graphite and carbon fiber modified asphalt concrete. *Const. Build. Mater.* **2011**, *25*, 1807–1811. [CrossRef]
8. Wu, S.P.; Mo, L.T.; Shui, Z.H. Piezoresistivity of Graphite Modified Asphalt-Based Composites. *Key Eng. Mater.* **2003**, *249*, 391–396. [CrossRef]
9. Bonaccorso, F.; Colombo, L.; Yu, G.; Stoller, M.; Tozzini, V.; Ferrari, A.C.; Ruoff, R.S.; Pellegrini, V. Graphene, related two-dimensional crystals, and hybrid systems for energy conversion and storage. *Science* **2015**, *347*. [CrossRef] [PubMed]
10. Girit, Ç.Ö.; Meyer, J.C.; Erni, R.; Rossell, M.D.; Kisielowski, C.; Yang, L.; Park, C.-H.; Crommie, M.F.; Cohen, M.L.; Louie, S.G.; et al. Graphene at the Edge: Stability and Dynamics. *Science* **2009**, *323*, 1705–1708. [CrossRef] [PubMed]

11. Geim, A.K. Graphene: Status and Prospects. *Science* **2009**, *324*, 1530–1534. [CrossRef] [PubMed]
12. Ghosh, S.; Calizo, I.; Teweldebrhan, D.; Pokatilov, E.P.; Nika, D.L.; Balandin, A.A.; Bao, W.; Miao, F.; Lau, C.N. Extremely high thermal conductivity of graphene: Prospects for thermal management applications in nanoelectronic circuits. *Appl. Phys. Lett.* **2008**, *92*, 151911. [CrossRef]
13. Rahman, A. Correlations in the Motion of Atoms in Liquid Argon. *Phys. Rev.* **1964**, *136*, A405–A411. [CrossRef]
14. Alder, B.J.; Wainwright, T.E. Studies in Molecular Dynamics. I. General Method. *J. Che. Phys.* **1959**, *31*, 459–466. [CrossRef]
15. Schlick, T. Pursuing Laplace's Vision on Modern Computers. In *Mathematical Approaches to Biomolecular Structure and Dynamics*; Mesirov, J.P., Schulten, K., Sumners, D.W., Eds.; Springer: New York, NY, USA, 1996; Volume 82, pp. 219–247.
16. Plimpton, S. Fast Parallel Algorithms for Short-Range Molecular Dynamics. *J. Comput. Phys.* **1995**, *117*, 1–19. [CrossRef]
17. Martin, M.G. MCCCS Towhee: A tool for Monte Carlo molecular simulation. *Mol. Simul.* **2013**, *39*, 1212–1222. [CrossRef]
18. Zhang, L.; Greenfield, M.L. Analyzing Properties of Model Asphalts Using Molecular Simulation. *Energy Fuels* **2007**, *21*, 1712–1716. [CrossRef]
19. Zhang, L.; Greenfield, M.L. Effects of Polymer Modification on Properties and Microstructure of Model Asphalt Systems. *Energy Fuels* **2008**, *22*, 3363–3375. [CrossRef]
20. Bhasin, A.; Bommavaram, R.; Greenfield, M.; Little, D. Use of Molecular Dynamics to Investigate Self-Healing Mechanisms in Asphalt Binders. *J. Mater. Civ. Eng.* **2011**, *23*, 485–492. [CrossRef]
21. Lu, Y.; Wang, L. Nanoscale modelling of mechanical properties of asphalt–aggregate interface under tensile loading. *Int. J. Pavement Eng.* **2010**, *11*, 393–401. [CrossRef]
22. Pan, P.; Wu, S.; Xiao, Y.; Wang, P.; Liu, X. Influence of graphite on the thermal characteristics and anti-ageing properties of asphalt binder. *Constr. Build. Mater.* **2014**, *68*, 220–226. [CrossRef]
23. Yao, H.; Dai, Q.; You, Z.; Ye, M.; Yap, Y.K. Rheological Properties, Low-Temperature Cracking Resistance, and Optical Performance of Exfoliated Graphite Nanoplatelets Modified Asphalt Binder. *Constr. Build. Mater.* **2016**, *13*, 988–996. [CrossRef]
24. Yao, H.; Dai, Q.; You, Z. Molecular dynamics simulation of physicochemical properties of the asphalt model. *Fuel* **2016**, *164*, 83–93. [CrossRef]
25. Bandyopadhyay, A. Molecular Modeling of EPON 862-DETDA Polymer. Ph.D. Thesis, Michigan Technological University, Houghton, MI, USA, 2012.
26. Li, Z.; Yu, H.; Zhuang, W.; Mukamel, S. Geometry and excitation energy fluctuations of NMA in aqueous solution with CHARMM, AMBER, OPLS, and GROMOS force fields: Implications for protein ultraviolet spectra simulation. *Chem. Phys. Lett.* **2008**, *452*, 78–83. [CrossRef] [PubMed]
27. Schweizer, S.; Bick, A.; Subramanian, L.; Krokidis, X. Influences on the stability of collagen triple-helix. *Fluid Phase Equilib.* **2014**, *362*, 113–117. [CrossRef]
28. Martin, M.G. Comparison of the AMBER, CHARMM, COMPASS, GROMOS, OPLS, TraPPE and UFF force fields for prediction of vapor–liquid coexistence curves and liquid densities. *Fluid Phase Equilib.* **2006**, *248*, 50–55. [CrossRef]
29. Mayo, S.L.; Olafson, B.D.; Goddard, W.A. DREIDING: A generic force field for molecular simulations. *J. Phys. Chem.* **1990**, *94*, 8897–8909. [CrossRef]
30. Cornell, W.D.; Cieplak, P.; Bayly, C.I.; Gould, I.R.; Merz, K.M.; Ferguson, D.M.; Spellmeyer, D.C.; Fox, T.; Caldwell, J.W.; Kollman, P.A. A Second Generation Force Field for the Simulation of Proteins, Nucleic Acids, and Organic Molecules. *J. Am. Chem. Soc.* **1995**, *117*, 5179–5197. [CrossRef]
31. Wang, J.; Wolf, R.M.; Caldwell, J.W.; Kollman, P.A.; Case, D.A. Development and Testing of a General Amber Force Field. *J. Comput. Chem.* **2004**, *25*, 1157–1174. [CrossRef] [PubMed]
32. Hazewinkel, M. *Encyclopaedia of Mathematics (set)*; Springer: Berlin, Germany, 1994.
33. Hockney, R.W.; Eastwood, J.W. *Computer Simulation Using Particles*; Taylor & Francis Group: New York, NY, USA, 1988.
34. Sadus, R.J. *Molecular Simulation of Fluids: Theory, Algorithms, and Object-Orientation*; Elsevier: Amsterdam, The Netherlands, 2002.

35. Storm, D.A.; Edwards, J.C.; DeCanio, S.J.; Sheu, E.Y. Molecular Representations of Ratawi and Alaska North Slope Asphaltenes Based on Liquid- and Solid-Sate Nmr. *Energy Fuels* **1994**, *8*, 561–566. [CrossRef]
36. Groenzin, H.; Mullins, O.C. Molecular Size and Structure of Asphaltenes from Various Sources. *Energy Fuels* **2000**, *14*, 677–684. [CrossRef]
37. Kowalewski, I.; Vandenbroucke, M.; Huc, A.Y.; Taylor, M.J.; Faulon, J.L. Preliminary Results on Molecular Modeling of Asphaltenes Using Structure Elucidation Programs in Conjunction with Molecular Simulation Programs. *Energy Fuels* **1996**, *10*, 97–107. [CrossRef]
38. Artok, L.; Su, Y.; Hirose, Y.; Hosokawa, M.; Murata, S.; Nomura, M. Structure and Reactivity of Petroleum-Derived Asphaltene†. *Energy Fuels* **1999**, *13*, 287–296. [CrossRef]
39. Savitzky, A.; Golay, M.J.E. Smoothing and Differentiation of Data by Simplified Least Squares Procedures. *Anal. Chem.* **1964**, *36*, 1627–1639. [CrossRef]
40. Greenfield, M.L.; Zhang, L. *Final Report-Developing Model Asphalt Systems Using Molecular Simulation*; URITC Project No. 000216; University of Rhode Island Transportation Center, Department of Chemical Engineering, University of Rhode Island: Kingston, RI, USA, 2009; pp. 1–109.
41. Li, D.D.; Greenfield, M.L. Chemical compositions of improved model asphalt systems for molecular simulations. *Fuel* **2014**, *115*, 347–356. [CrossRef]
42. Usmani, A. *Asphalt Science and Technology*; Taylor & Francis: Boca Raton, FL, USA, 1997.
43. Tabatabaee, H.A.; Velasquez, R.; Bahia, H.U. Predicting low temperature physical hardening in asphalt binders. *Constr. Build. Mater.* **2012**, *34*, 162–169. [CrossRef]
44. Chilingarian, G.V.; Yen, T.F. *Asphaltenes and Asphalts*, 1st ed.; Elsevier Science: Amsterdam, The Netherlands, 1994.
45. Vacquier, V. The measurement of thermal conductivity of solids with a transient linear heat source on the plane surface of a poorly conducting body. *Earth Planet. Sci. Lett.* **1985**, *74*, 275–279. [CrossRef]
46. *xGnP Graphene Nanoplatelets—Grade H*; MSDS No. CSSS-TCO-010-112126; XG Sciences: Lansing, MI, USA, October 2012. Available online: http://xgsciences.com/products/graphene-nanoplatelets/grade-h/ (accessed on 1 May 2014).
47. Müller-Plathe, F. A simple nonequilibrium molecular dynamics method for calculating the thermal conductivity. *J. Chem. Phys.* **1997**, *106*, 6082–6085. [CrossRef]

Article

Adhesion Evaluation of Asphalt-Aggregate Interface Using Surface Free Energy Method

Jie Ji [1], Hui Yao [2,3,*], Luhou Liu [1], Zhi Suo [1], Peng Zhai [1], Xu Yang [2] and Zhanping You [2]

1 School of Civil Engineering and Transportation, Beijing University of Civil Engineering and Architecture, Beijing 100044, China; jijie@bucea.edu.cn (J.J.); liuluhou@163.com (L.L.); suozhi@bucea.edu.cn (Z.S.); zhaipeng@bucea.edu.cn (P.Z.)
2 Department of Civil and Environmental Engineering, Michigan Technological University, 1400 Townsend Drive, Houghton, MI 49931, USA; xyang2@mtu.edu (X.Y.); zyou@mtu.edu (Z.Y.)
3 School of Traffic and Transportation Engineering, Changsha University of Science and Technology, Changsha 410114, China
* Correspondence: huiyao@mtu.edu; Tel.: +1-906-487-1059

Academic Editor: Jorge de Brito
Received: 30 November 2016; Accepted: 3 February 2017; Published: 9 February 2017

Abstract: The influence of organic additives (Sasobit and RH) and water on the adhesion of the asphalt-aggregate interface was studied according to the surface free energy theory. Two asphalt binders (SK-70 and SK-90), and two aggregate types (limestone and basalt) were used in this study. The sessile drop method was employed to test surface free energy components of asphalt, organic additives and aggregates. The adhesion models of the asphalt-aggregate interface in dry and wet conditions were established, and the adhesion work was calculated subsequently. The energy ratios were built to evaluate the effect of organic additives and water on the adhesiveness of the asphalt-aggregate interface. The results indicate that the addition of organic additives can enhance the adhesion of the asphalt-aggregate interface in dry conditions, because organic additives reduced the surface free energy of asphalt. However, the organic additives have hydrophobic characteristics and are sensitive to water. As a result, the adhesiveness of the asphalt-aggregate interface of the asphalt containing organic additives in wet conditions sharply decreased due to water damage to asphalt and organic additives. Furthermore, the compatibility of asphalt, aggregate with organic additive was noted and discussed.

Keywords: surface free energy; adhesion; asphalt-aggregate interface; organic additive

---

1. Introduction

The compaction temperatures of hot mix asphalt (HMA) are usually above 160 °C, which consumes a large amount of fuel energies and results in the emission of CO_2. Warm mix asphalt (WMA) technology has been generalized to the asphalt pavement industry for a few years. The warm mix technologies can reduce the asphalt production temperature by as much as 30 °C. There are two widely used warm mix technologies: adding organic additives and applying water foaming. In the water foaming, due to the lower compaction temperature of WMA, water cannot be completely evaporated out of aggregates. The remaining water can impact the adhesiveness of the asphalt-aggregate interface and lead to moisture damage. Therefore, many researchers start to study problems of water damage in WMA.

Currently, the theories for studying the adhesiveness of the asphalt-aggregate interface include the molecular orientation theory [1], chemical reaction theory [2], surface free energy theory [3,4] and molecular dynamics [5], etc. The surface free energy theory has been applied to research the adhesion of asphalt-aggregate interface. The theory of surface free energy could be used to evaluate water damage and fatigue cracking of HMA [6]. The surface free energies between the asphalt and aggregates were measured using the Wilhelmy plate and absorption methods, respectively, and calculated the adhesiveness of asphalt-aggregate interface with and without water, and it was feasible to use the surface free energy theory to analyze the water damage of HMA [7]. The surface free energy components of asphalt-aggregate interface were analyzed by Wilhelmy plate and adsorption methods, respectively, noting that the surface energy theory could be useful in analyzing water damage in HMA [8]. The surface free energies between the asphalt and aggregates were tested and the adhesion energy ratio was established to predict the adhesion of the asphalt-aggregate interface [9]. The evaluation of the surface energy and moisture susceptibility of various combinations of aggregates and asphalt binders were analyzed [10]. The adhesive properties could be used to estimate adhesiveness of the asphalt-aggregate interface [11]. The asphalt-aggregate interaction for moisture-induced damage mechanisms was studied using surface free energy and predicted moisture-induced damage in HMA [12]. In 2010, the surface free energies of asphalt and aggregate were tested, and the adhesion trends of asphalt-aggregate with and without water were analyzed and calculated [13]. The adhesion models of additive (SAK)-asphalt-aggregate were established and the adhesiveness was used to predict water damage of WMA [14]. The treatment of aggregate surface with hydrated lime narrows down the energy difference under dry and wet conditions, and it helps resist moisture damage [15]. Two different waxes and three kinds of aggregates were used to study the physico-chemical surface characteristics between the aggregates and asphalt. The Dynamic Contact Angle (DCA) and Dynamic Vapour Sorption Devices (DVSD) were used to measure and calculate the components of surface energy. The analysis results indicate that the waxes can adversely affect the adhesion between the aggregates and asphalt [16]. However, there are only a few studies to predict the adhesiveness of organic additive-asphalt-aggregate under the dry and wet conditions according to the surface free energy theory. In the paper, the adhesiveness of asphalt-aggregate and organic additive-asphalt-aggregate systems with and without water is studied based on the surface free energy theory.

2. Raw Materials and Methodology

The raw materials in this study include asphalt binder, aggregate, organic additives. Two types of asphalt binders were used: SK-70 and SK-90. The organic additives were Sasobit and RH, and the aggregates were limestone and basalt. The organic additive-modified asphalt was produced by adding 3% Sasobit and RH by weight of the base asphalt in the asphalt matrix. The sessile drop method was conducted to test the surface free energy components of base asphalts, organic additives, organic additive-modified asphalts and aggregates. After that, the adhesion models of asphalt-aggregate and organic additive-asphalt-aggregate in dry and wet conditions were established, and their adhesion works were calculated subsequently according to the surface free energy theory. Energy ratios were established to evaluate the adhesion of asphalt-aggregate and organic additive-asphalt-aggregate interfaces under dry and wet conditions. Finally, the influence of organic additives and water on asphalt-aggregate adhesion was evaluated based on the adhesion works and energy ratios obtained.

3. Properties of Raw Materials

3.1. Base Asphalt

According to the Standard Test Methods of Bitumen and Bituminous Mixtures for Highway Engineering of China (JTG E20-2011) [17], the properties of base asphalts, SK-70 and SK-90, were measured, and the results are shown in Figure 1. In addition, three replicates for each test was adopted in this paper.

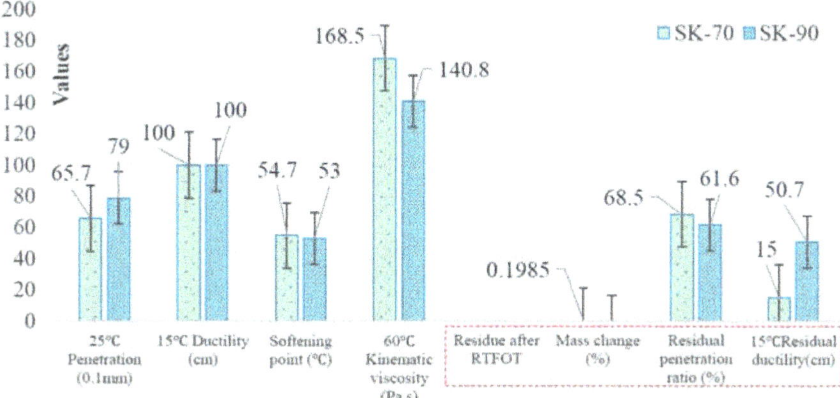

Figure 1. Properties of SK-70 and SK-90 asphalt (RTFOT: Rolling Thin Film Oven Test).

The Performance Grade (PG) classifications of SK-70 and SK-90 were evaluated based on the results of the Dynamic Shear Rheometer (DSR) and Bending Beam Rheometer (BBR) tests of the Strategic Highway Research Program (SHRP). The PG of SK-70 and SK-90 are determined to be PG64-24.

3.2. Organic Additives

Sasobit is a WMA product of Sasol Wax, located in South Africa (CAS number: 8002-74-2). "RH" is a kind of WMA additive, which is developed by the Research Institute of China Highway Ministry of Transport. The properties of Sasobit and RH were analyzed, and the test results are shown in Figure 2.

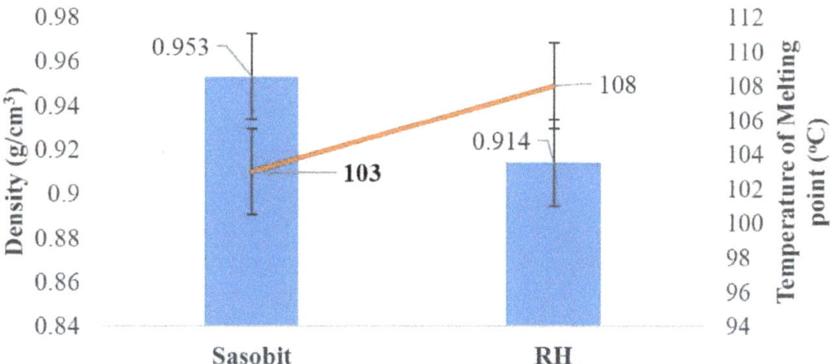

Figure 2. Properties of Organic Additives.

3.3. Organic Additive-Modified Asphalt

Four modified asphalts are processed by adding 3% Sasobit and RH (by mass of asphalt) into base asphalts, respectively. Sasobit and RH organic waxes can be dissolved easily into base asphalts at a temperature above 100 °C. In this paper, Sasobit and RH were blended into base asphalts at a temperature of 120 °C and stirred manually for 15 min. The properties of organic additive-modified asphalts were tested. The test results are shown in Figure 3.

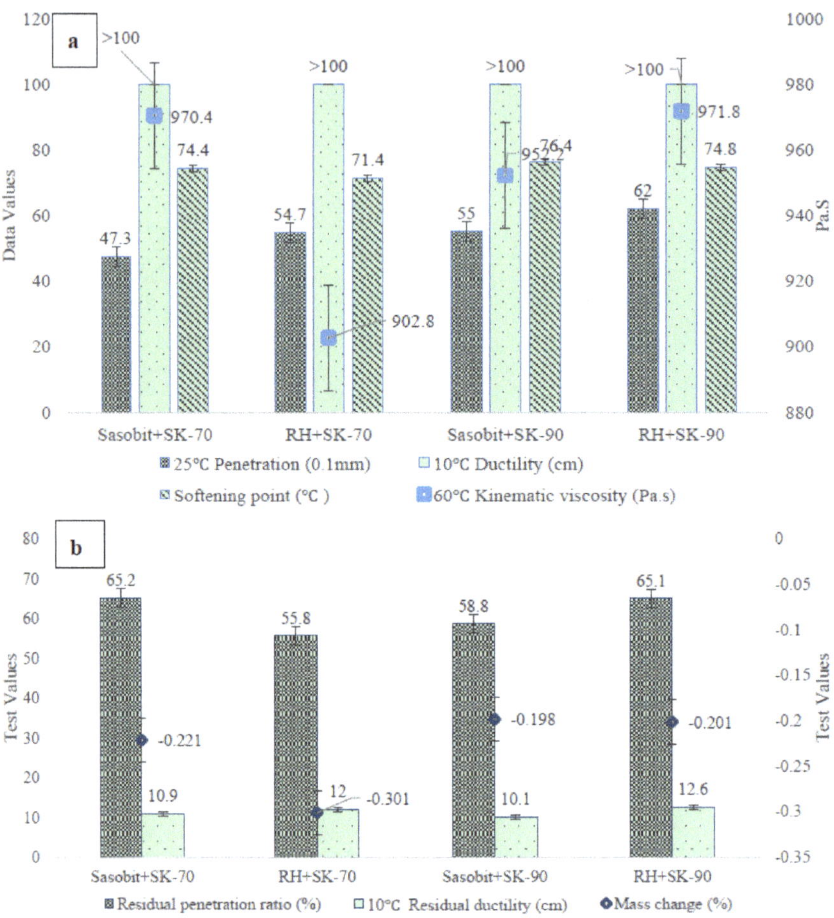

Figure 3. Properties of Organic Additive-Modified Asphalts, (**a**) Penetration, Ductility, Softening Point, Kinematic Viscosity of asphalt binders; (**b**) Residual Penetration Ratio, Residual ductility, and Mass Change of asphalt binders.

PG classifications of the four organic additive-modified asphalts are tested according to the DSR and BBR tests of SHRP, and presented in Table 1.

Table 1. PG Classifications of Organic Additive-Modified Asphalts.

Items	SK-70 and SK-90	Sasobit + SK-70	RH + SK-70	Sasobit + SK-90	RH + SK-90
PG	PG 64-24	PG64-24	PG64-18	PG64-24	PG64-24

3.4. Aggregates

According to (JTG E42-2005) the Standard Test Methods of Aggregate for Highway Engineering in China [18], the properties of limestone and basalt aggregates are tested. The specific gravities of limestone and basalt aggregates are 2.667 and 2.655, respectively.

3.5. Surface Free Energy of Raw Materials

There are many methods for testing the surface free energy of different materials, including the capillary method, rings method, drop weight method, Wilhelmy plate method, sessile drop method, atomic force microscopy and the nuclear magnetic resonance method. Little and Bhasin [19] measured the surface free energy components of asphalt pavement materials using the sessile drop method. Murat Koc et al. presented a sessile drop device for measuring the surface energy components of both asphalt binders and aggregates [20,21]. In this manuscript, a sessile drop device is employed to measure contact angles on the surface of raw materials. The surface energy components of raw materials are calculated using the measured contact angles.

The test liquids are distilled water, glycerin and formamide. The surface free energy components [22] of all test liquids are known in advance, as shown in Figure 4.

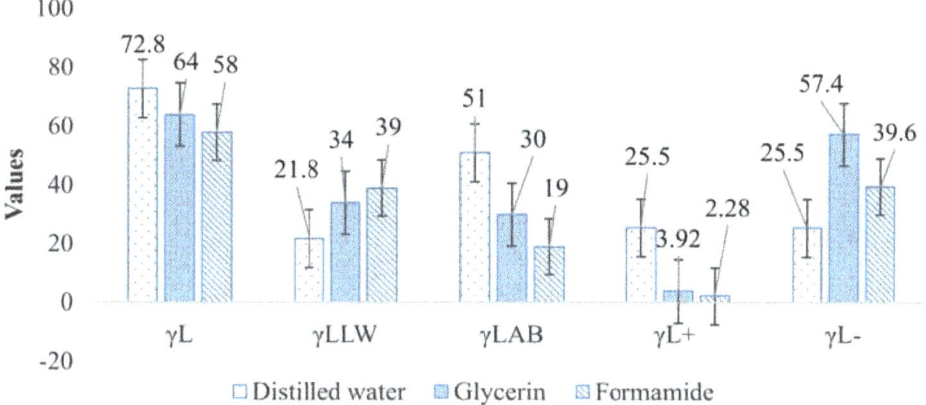

Figure 4. The Surface Free Energy Components of Test Liquids (mJ/m^2) (Note: γL—the surface free energy of test liquid; γL^{LW}—the nonpolar part of the surface free energy of test liquid; γL^{AB}—the polar part of the surface free energy of test liquid; γL^{+}—the acidic effect part of the surface free energy of test liquid; γL^{-}—the basic effect part of the surface free energy of test liquid.).

The sessile drop method was conducted to test contact angles between the surface of the raw materials and test liquids. Based on the simplified matrix formula of the Young-Dupre Equation (see Equation (1)) [23,24] the surface free energy components of raw materials are calculated, as shown in Table 2.

$$\begin{bmatrix} \sqrt{\gamma_{L1}^{LW}} & \sqrt{\gamma_{L1}^{+}} & \sqrt{\gamma_{L1}^{-}} \\ \sqrt{\gamma_{L2}^{LW}} & \sqrt{\gamma_{L2}^{+}} & \sqrt{\gamma_{L2}^{-}} \\ \sqrt{\gamma_{L3}^{LW}} & \sqrt{\gamma_{L3}^{+}} & \sqrt{\gamma_{L3}^{-}} \end{bmatrix} \begin{bmatrix} \sqrt{\gamma_{a}^{LW}} \\ \sqrt{\gamma_{a}^{-}} \\ \sqrt{\gamma_{a}^{+}} \end{bmatrix} = \begin{bmatrix} \frac{\gamma_{L1}(1+\cos\theta_1)}{2} \\ \frac{\gamma_{L2}(1+\cos\theta_2)}{2} \\ \frac{\gamma_{L3}(1+\cos\theta_3)}{2} \end{bmatrix} \tag{1}$$

where γ_{L1}, γ_{L2}, γ_{L3}—the surface free energy of distilled water, glycerin, and formamide, respectively; γ_{L1}^{LW}, γ_{L2}^{LW}, γ_{L3}^{LW}—the nonpolar part of surface free energy of distilled water, glycerin, and formamide, respectively; γ_{L1}^{+}, γ_{L2}^{+}, γ_{L3}^{+}—the acidic effect part of surface free energy of distilled water, glycerin, and formamide, respectively; γ_{L1}^{-}, γ_{L2}^{-}, γ_{L3}^{-}—the basic effect part of surface free energy of distilled water, glycerin, and formamide, respectively; θ_1, θ_2, θ_3—the contact angle between raw material and distilled water, glycerin, and formamide, respectively; the rest of the parameters are the same as above.

Table 2. Surface free energy components of raw materials (mJ/m^2).

Items	γ_a	$\gamma_a{}^{LW}$	$\gamma_a{}^{AB}$	$\gamma_a{}^{+}$	$\gamma_a{}^{-}$
SK-70	25.570	25.541	0.028	0.000	4.416
SK-90	21.512	21.082	0.430	0.016	2.915
Sasobit	40.210	38.370	1.840	0.129	6.543
RH	26.94	21.63	5.32	1.63	4.34
Sasobit + SK-70	17.353	14.709	2.643	2.023	0.863
Sasobit + SK-90	24.150	24.144	0.006	0.419	0.000
RH + SK-70	16.75	15.10	1.65	0.21	3.22
RH + SK-90	13.17	9.91	3.26	1.60	1.65
Limestone	48.351	46.427	1.924	0.108	8.603
Basalt	53.140	52.016	1.123	2.313	0.136

The method proposed by Fwoke is applied to further verify the validity of the test methods and results [25,26]. Fwoke pointed out that there is a good linear relationship between $\gamma_L \cos\theta$ and γ_L of test liquid. If the correlation coefficient is higher than 0.95, it indicates that the test method and results are effective. However, when the correlation coefficient is lower than 0.95, the test method and results are ineffective. In this paper, the correlation coefficients of $\gamma_L \cos\theta$ and γ_L of test liquid are all above 0.95, which indicates that using the sessile drop method to test the surface free energy components of raw materials is feasible.

4. Adhesion of Asphalt-Aggregate Interface

4.1. Adhesion Models of the Asphalt-Aggregate Interface

Adhesion models of the asphalt-aggregate interface in different conditions are established respectively based on surface free energy theory [27–29]. The adhesion model of asphalt-aggregate without water (dry condition) is established, and is as shown in Equation (2).

$$W_{as} = 2(\sqrt{\gamma_a^{LW}\gamma_s^{LW}} + \sqrt{\gamma_a^{+}\gamma_s^{-}} + \sqrt{\gamma_a^{-}\gamma_s^{+}}) \tag{2}$$

where W_{as} denotes the adhesion of the asphalt-aggregate interface, respectively. The adhesion model of the organic additive-asphalt-aggregate system without water (dry condition) is also established as follows:

$$W_{ase} = 2(\gamma_e^{LW} + 2\sqrt{\gamma_e^{+}\gamma_e^{-}}) - 2\sqrt{\gamma_a^{LW}\gamma_e^{LW}} + \sqrt{\gamma_a^{+}\gamma_e^{-}} + \sqrt{\gamma_a^{-}\gamma_e^{+}} - 2(\sqrt{\gamma_s^{LW}\gamma_e^{LW}} + \sqrt{\gamma_s^{+}\gamma_e^{-}} + \sqrt{\gamma_s^{-}\gamma_e^{+}}) \tag{3}$$

where W_{ase} denotes the adhesion of organic additive-asphalt-aggregate without water; the rest of the parameters are the same as above. The adhesion model of asphalt-aggregate with water (wet condition) can be written as follows:

$$\begin{aligned} W_{asw} = \; & -(2\sqrt{\gamma_a^{LW}\gamma_w^{LW}} + 2\sqrt{\gamma_s^{LW}\gamma_w^{LW}} + 2\sqrt{\gamma_w^{+}}(\sqrt{\gamma_a^{-}} + \sqrt{\gamma_s^{-}}) \\ & +2\sqrt{\gamma_w^{-}}(\sqrt{\gamma_a^{+}} + \sqrt{\gamma_s^{+}}) - 2\gamma_w^{LW} - 2\sqrt{\gamma_a^{LW}\gamma_s^{LW}} - 4\sqrt{\gamma_w^{+}\gamma_w^{-}} \\ & -2\sqrt{\gamma_a^{+}\gamma_s^{-}} - 2\sqrt{\gamma_a^{-}\gamma_s^{+}}) \end{aligned} \tag{4}$$

where W_{asw} denotes the adhesion energy of asphalt-aggregate with water; the rest of the parameters are the same as above. The adhesion model of organic additive-asphalt-aggregate with water (wet condition) is built and is expressed as:

$$\begin{aligned} W_{asew} = \; & -(4\sqrt{\gamma_a^{LW}\gamma_e^{LW}} + 4\sqrt{\gamma_s^{LW}\gamma_e^{LW}} - 4\sqrt{\gamma_e^{LW}\gamma_w^{LW}} - 2\sqrt{\gamma_a^{LW}\gamma_s^{LW}} \\ & -2\sqrt{\gamma_a^{LW}\gamma_w^{LW}} - 2\sqrt{\gamma_s^{LW}\gamma_w^{LW}} + 2\sqrt{\gamma_e^{+}}(\sqrt{\gamma_a^{-}} + \sqrt{\gamma_s^{-}}) \\ & +2\sqrt{\gamma_e^{-}}(\sqrt{\gamma_a^{+}} + \sqrt{\gamma_s^{+}}) - 2\sqrt{\gamma_w^{+}}(\sqrt{\gamma_a^{-}} + \sqrt{\gamma_s^{-}} + 2\sqrt{\gamma_e^{-}}) \\ & -2\sqrt{\gamma_w^{-}}(\sqrt{\gamma_a^{+}} + \sqrt{\gamma_s^{+}} + 2\sqrt{\gamma_e^{+}}) + 2\gamma_w^{LW} - 2\gamma_e^{LW} \\ & +4\sqrt{\gamma_w^{+}\gamma_w^{-}} - 4\sqrt{\gamma_e^{+}\gamma_e^{-}} - 2\sqrt{\gamma_a^{+}\gamma_s^{-}} - 2\sqrt{\gamma_a^{-}\gamma_s^{+}} + 2\sqrt{\gamma_a^{+}\gamma_e^{-}} \\ & +2\sqrt{\gamma_a^{-}\gamma_e^{+}} + 2\sqrt{\gamma_s^{+}\gamma_e^{-}} + 2\sqrt{\gamma_s^{+}\gamma_e^{-}}) \end{aligned} \tag{5}$$

where W_{asew} denotes the adhesion of organic additive-asphalt-aggregate with water; the rest of the parameters are the same as above.

4.2. Energy Ratios

Energy ratios (EP_1 and EP_2) are built using the adhesive properties of the asphalt-aggregate interface in different conditions. EP_1 is used to evaluate the adhesion of the asphalt-aggregate interface affected by water, and is calculated using Equation (6):

$$EP_1 = \frac{W_{as}}{W_{asw}} \; or \; \frac{W_{ase}}{W_{asew}} \tag{6}$$

When the EP_1 value is higher than 1, the adhesive property of the asphalt-aggregate interface with water is lower than that of the asphalt-aggregate interface without water. It indicates that water has a negative influence on the adhesion of the asphalt-aggregate interface. When the EP_1 value is equal to 1, it predicts that water has no interaction with the asphalt-aggregate interface. When the EP_1 value is lower than 1, the adhesion of the asphalt-aggregate interface with water is greater than that of the asphalt-aggregate interface without water. It predicts that water promotes adhesion in the asphalt-aggregate interface. EP_2 is used to characterize the adhesion of the asphalt-aggregate interface affected by organic additive, and is calculated using Equation (7):

$$EP_2 = \frac{W_{as}}{W_{ase}} \; or \; \frac{W_{asw}}{W_{asew}} \tag{7}$$

When the EP_2 value is greater than 1, the adhesion of the organic additive-asphalt-aggregate is lower than that of the asphalt-aggregate. It means that organic additive has a negative influence on the adhesion of the asphalt-aggregate interface. When the EP_2 value equals 1, the organic additive has no influence on the adhesion of the asphalt-aggregate. When the EP_2 value is less than 1, the adhesion of the organic additive-asphalt-aggregate with or without water is higher than that of asphalt-aggregate with or without water. It indicates that organic additives improve the adhesion of asphalt-aggregate.

4.3. Adhesion of Asphalt-Aggregate

The adhesion energies of asphalt-aggregate interfaces in different conditions are calculated, and the results are shown in Figure 5. In the dry condition, it was observed that the adhesion of the asphalt-aggregate interface of asphalt containing organic additives was higher than that of the base asphalt. This is because the surface free energy of organic additives modified asphalts is lower than that of base asphalt. As a result, the stability of organic additive-asphalt-aggregate is greater than that of asphalt-aggregate. Therefore, the surface free energy of asphalt can be reduced by adding organic additives so that the adhesion of asphalt-aggregate interface in a dry condition can be enhanced. In the wet condition, it was observed that the adhesion of the base asphalt and the asphalt containing organic additives decreased significantly. The reason for this is that water has a higher surface energy that prevents effective bonding between asphalt and aggregate. The asphalt-aggregate adhesive energy of the base asphalt, the Sasobit-modified asphalt, and the RH-modified asphalt, was reduced by 10.8%, 47.9% and 32.9% on average, respectively. This indicates that water has a great influence on the adhesion of asphalt-aggregate. In addition, it was found that water has a greater effect on the asphalt-aggregate adhesion of the asphalt containing organic additives compared to that of the base asphalt. This indicates that after the addition of organic additives, moisture damage can be more severe.

In fact, previous studies have shown that asphalt mixes containing organic additives have a greater moisture susceptibility than the conventional asphalt mix [30]. This is due to the hydrophobic characteristics of organic additives. As a result, the asphalts containing organic additives refuse to form effective bonding between asphalt and the aggregate surface. The comparison between the two organic additives showed that the asphalt containing Sasobit is more susceptible to moisture damage than the asphalt containing RH. The comparison between the two asphalt binders showed that the SK-70 exhibited greater adhesion than the SK-90. This may be because of the higher surface energy of SK-70, as shown in Table 2. In addition, it was found that the SK-70 is more resistant to water damage than the SK-90. The average reduction in adhesion of SK-70-aggregate and SK-90-aggregate interfaces were 33.4% and 27.7%, respectively. The comparison between the two types of aggregates showed that the adhesion of the limestone-asphalt interface was lower than that of the basalt-asphalt interface. This can also be attributed to the higher surface free energy of basalt. However, it was found that the limestone-asphalt interface was more resistant to water damage than the basalt-asphalt interface. The average reductions in adhesion of limestone-asphalt and basalt-asphalt interfaces were 27.0% and 34.1%, respectively.

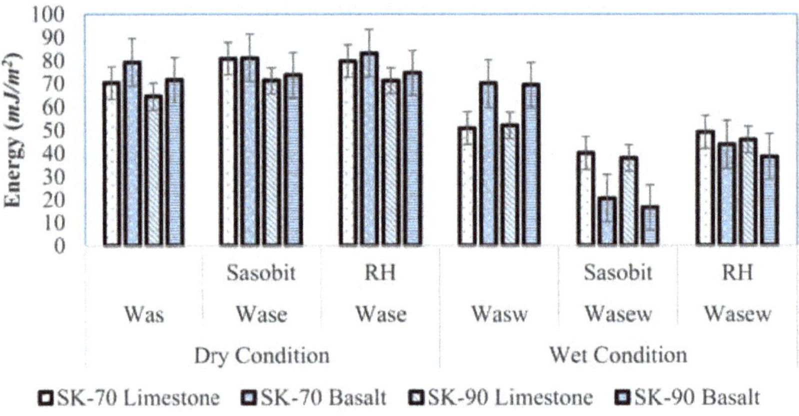

Figure 5. Adhesion of the Asphalt-Aggregate Interface in Different Conditions (mJ/m^2).

4.4. Adhesion of the Asphalt-Aggregate Interface

The results of energy ratios (EP_1 and EP_2) are presented in Figure 6. As mentioned above, the EP_1 value is an indication of how water impacts the adhesion of an asphalt-aggregate or organic additive-asphalt-aggregate interface. All the EP_1 values were greater than 1, indicating that the adhesive properties of asphalt-aggregate or organic additive-asphalt-aggregate reduce with residual water in the asphalt mixture, and thus the adhesion energy of asphalt-aggregate or organic additive-asphalt-aggregate in a wet condition declines. In addition, the adhesion energy of asphalt-aggregate or organic additive-asphalt-aggregate in a wet condition decreases more rapidly, when the EP_1 value becomes higher. It can be found that the EP_1 values of organic additive-asphalt-aggregate were always higher than that of asphalt-aggregate. It indicates that the adhesion of organic additive-asphalt-aggregate in the wet condition is far lower than that of asphalt-aggregate with water. This means that the organic additive has significantly negative influences on asphalt-aggregate interface adhesion. Because the two organic additives are also organic waxes, it can form an isolating layer to prevent aggregate from absorbing asphalt when the aggregate interface contains some residual water.

Therefore, the organic additive lowers the adhesiveness of asphalt-aggregate sharply in a wet condition. The EP_1 value of asphalt-basalt in a dry condition is less than that of asphalt-limestone, which indicates that the adhesiveness of asphalt-basalt in a dry condition is relatively better. However, the EP_1 value of asphalt-basalt in a wet condition is greater than that of asphalt-limestone, indicating that limestone has a positive influence on the adhesiveness of asphalt-aggregate in a wet condition; or limestone is less sensitive to water; or that limestone has a good compatibility with asphalt. The compatibility between asphalt and aggregate should therefore be studied and improved to reach a high adhesion energy of the asphalt-aggregate interface.

The EP_2 value indicates that organic additives impacted the adhesiveness of asphalt-aggregate. All the EP_2 values of asphalt-aggregate or organic additive-asphalt-aggregate in dry conditions were lower than 1. However, all the EP_2 values of asphalt-aggregate or organic additive-asphalt-aggregate in wet conditions were higher than 1. Once there is some residual water in the asphalt mixture, the organic additive has a negative influence on the adhesion of asphalt-aggregate. It can also be concluded that the organic additive is very sensitive to water and has hydrophobic characteristics. The adhesion of organic additive-asphalt-aggregate in wet conditions can be reduced due to the dual action of water and organic additive. For preventing water damage to WMA, the residual water in aggregate should be excluded. The EP_2 value of Sasobit-asphalt-aggregate is higher than that of RH-asphalt-aggregate, in either the dry or wet condition, which indicates that Sasobit is more sensitive to water and decreases the adhesiveness of asphalt-aggregate. For ensuring the high adhesion of the asphalt-aggregate interface, the compatibility of asphalt, aggregate and organic additive should be strictly observed.

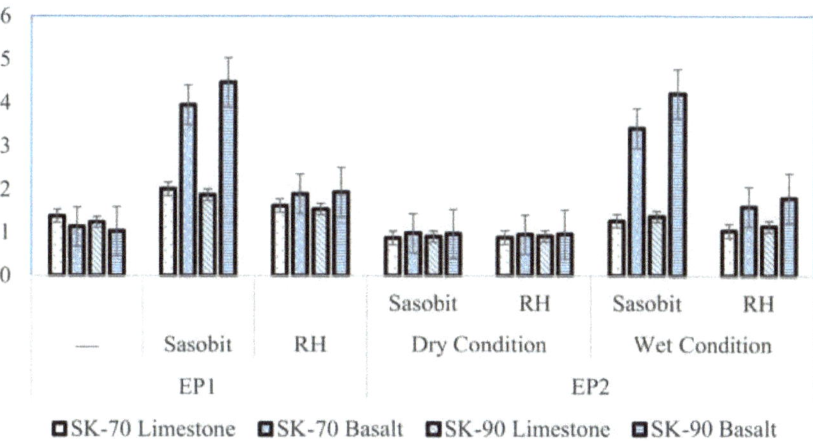

Figure 6. The EP_1 and EP_2 Values of Asphalt-Aggregate Interface in Different Systems.

5. Conclusions

The effect of organic additives on the strength of adhesion between the aggregate and asphalt was investigated. The components of the surface free energy of materials were tested and calculated. The surface energy test was used to evaluate the adhesive strength. Based on the test results and analysis, the conclusions can be obtained.

(1) Energy ratio values (EP_1 and EP_2) can be used to estimate the adhesiveness of asphalt-aggregate affected by water or organic additive. When EP_1 and EP_2 values increase, the adhesion of the asphalt-aggregate interface influenced by water or organic additive decreases.

(2) Organic additives improve the adhesiveness of asphalt-aggregate interface in dry conditions since the organic additives have hydrophobic characteristics and high surface free energy, although the adhesiveness of the asphalt-aggregate interface in a wet conditions decreases dramatically.
(3) The properties of asphalt and aggregate have some negative or positive impacts on the adhesion of asphalt-aggregate. If the asphalt and aggregate have a good compatibility, the adhesiveness of the asphalt-aggregate interface can be promoted.

Acknowledgments: This project was supported by the Importation and Development of the High-Caliber Talents Project of Beijing Municipal Institutions (Grant No. PXM2013-014210-000165). The authors would like to express their gratitude to Jinqi Gao (Beijing University of Civil Engineering and Architecture) and Aboelkasim Diab (Michigan Technological University) for their assistance in the laboratory work and revision.

Author Contributions: Jie Ji, Hui Yao and Zhi Suo conceived the experiments; Luhou Liu and Peng Zhai performed the experiments; Jie Ji and Hui Yao wrote the paper; Xu Yang and Zhanping You helped revise the paper.

Conflicts of Interest: The authors declare no conflict of interest.

References

1. Ensley, E.K. Multilayer adsorption with molecular orientation of asphalt on mineral aggregate and other substrates. *J. Appl. Chem. Biotechnol.* **1975**, *25*, 671–682. [CrossRef]
2. Bagampadde, U.; Isacsson, U.; Kiggundu, B.M. Classical and Contemporary Aspects of Stripping in Bituminous Mixes. *Road Mater. Pavement Des.* **2004**, *5*, 7–43. [CrossRef]
3. Al-Rawashdeh, A.; Sargand, S. Performance Assessment of a Warm Asphalt Binder in the Presence of Water by Using Surface Free Energy Concepts and Nanoscale Techniques. *J. Mater. Civ. Eng.* **2013**, *26*, 803–811. [CrossRef]
4. Khodaii, A.; Khalifeh, V.; Dehnad, M.H.; Hamedi, G.H. Evaluating the Effect of Zycosoil on Moisture Damage of Hot-Mix Asphalt Using the Surface Energy Method. *J. Mater. Civ. Eng.* **2013**, *26*, 259–266. [CrossRef]
5. Yao, H.; Dai, Q.; You, Z. Chemo-physical analysis and molecular dynamics (MD) simulation of moisture susceptibility of nano hydrated lime modified asphalt mixtures. *Constr. Build. Mater.* **2015**, *101*, 536–547. [CrossRef]
6. Elphingstone, G.M. *Adhesion and Cohesion in Asphalt-Aggregate System*; Texas A&M University: College Station, TX, USA, 1997.
7. Cheng, D. *Surface Free Energy of Asphalt-Aggregate System and Performance Analysis of Asphalt Concrete Based on Surface Free Energy*; Texas A&M University: College Station, TX, USA, 2002.
8. Zheng, X.; Wang, J.; Yang, Q. Study on Water Stability of Asphalt Mixture Based on Surface Free Energy. *J. Chin. Foreign Highw.* **2004**, *40*, 88–90.
9. Zollinger, C.J. Application of Surface Energy Measurements to Evaluate Moisture Susceptibility of Asphalt and Aggregates. Master's Thesis, Texas A&M University, College Station, TX, USA, 2005.
10. Jonathan, H.; Eyad, A.M. *System for the Evaluation of Moisture Damage Using Fundamental Materials Properties*; Report 0-4524-1; Texas Transportation Institute, Texas A&M University: College Station, TX, USA, 2006.
11. Xiao, Q.; Hao, P.; Xu, O. The Testing Method of Adhesion of Asphalt-Aggregate. *J. Chang'an Univ. (Sci. Ed.)* **2007**, *27*, 19–22.
12. Nazimuddin Mohammad, W. *Effect of Additives on Surface Free Energy Characteristics of Aggregates and Binders in Hot Mix Asphalt*; University of Oklahoma: Norman, OK, USA, 2007.
13. Liu, Y.; Han, S.; Li, B. Research on Adhesion between Asphalt and Aggregate Based on Surface Free Theory. *J. Build. Mater.* **2010**, *13*, 769–772.
14. Hui, D. *The Water Stability of Warm Asphalt Mixture Based on Surface Free Energy Theory*; Lanzhou Jiao Tong University: Lanzhou, China, 2012.
15. Neiad, F.M.; Hamedi, G.H.; Azarhoosh, A.R. Use of Surface Free Energy Method to Evaluate Effect of Hydrate Lime on Moisture Damage in Hot-Mix Asphalt. *J. Mater. Civ. Eng.* **2013**, *25*, 1119–1126.
16. Lamperti, R.; Grenfell, J.; Sangiorgi, C.; Lantieri, C.; Airey, G.D. Influence of Waxes on Adhesion Properties of Bituminous Binders. *Constr. Build. Mater.* **2015**, *76*, 404–412. [CrossRef]
17. *(JTG E20-2011) Standard Test Methods of Bitumen and Bituminous Mixtures for Highway Engineering*; Renmin Communication Press: Beijing, China, 2011.

18. *(JTG E42-2005) Standard Test Methods of Aggregate for Highway Engineering*; Renmin Communication Press: Beijing, China, 2005.
19. Little, D.N.; Bhasin, A. *Using Surface Energy Measurements to Select Materials for Asphalt Pavement*; NCHRP Web-Only Document 104; Texas Transportation Institute: College Station, TX, USA, 2006.
20. Murat, K. *Development of Testing Protocols for Direct Measurements of Contact Angles on Aggregate and Asphalt Binder Surfaces Using a Sessile Drop Device*; Oklahoma State University: Stillwater, OK, USA, 2013.
21. Murat, K.; Rifat, R. Assessment of a Sessile Drop Device and a New Testing Approach Measuring Contact Angles on Aggregates and Asphalt Binders. *J. Mater. Civ. Eng.* **2012**, *26*, 391–398.
22. Cheng, D.X.; Little, D.N.; Lytton, R.L.; Holste, J.C. Use of Surface Free Energy Properties of the Asphalt-Aggregate System to Predict Moisture Damage Potential (with Discussion). *Assoc. Asph. Paving Technol.* **2002**, *71*, 59–88.
23. Wasiuddin, N.M.; Fogle, C.M.; Zaman, M.M.; O'Rear, E.A. Effect of Anti-strip Additives on Surface Free Energy Characteristics of Asphalt Binders for Moisture-Induced Damage Potential. *J. Test. Eval.* **2007**, *35*, 123–130.
24. Bhasin, A.; Masad, E.; Little, D.; Lytton, R. Limits on Adhesive Bond Energy for Improved Resistance of Hot-Mix Asphalt to Moisture Damage. *J. Transp. Res. Board* **2007**, *170*, 3–31. [CrossRef]
25. Lyton, R.L.; Masad, E.A.; Zollinger, C.; Bulut, R.; Little, D.N. *Measurement of Surface Energy and Its Relationship to Moisture Damage*; Report No. FHWA/TX-05/0-4524-2; Texas Transportation Institute, Texas Department of Transportation, Research and Technology Implementation Office, Federal Highway Administration: Austin, TX, USA, 2005.
26. Fwoke, D.Y.; Neumann, A.W. Contact Angle Measurement and Contact Angle Interpretation. *Adv. Colloid Interface Sci.* **1999**, *81*, 167–249.
27. Bahramian, A. *Evaluating Surface Energy Components of Asphalt Binders Using Wilhelmy Plate and Sessile Drop Techniques*; Royal Institute of Technology (KTH): Stockholm, Sweden, 2012.
28. Wang, Y. *Application Research on Adhesion between Asphalt and Aggregate Based on Surface Free Energy Theory*; Chongqing Jiao Tong University: Chongqing, China, 2012.
29. Arno, H.; Dallas, N.L. *Adhesion in Bitumen Aggregate System and Quantification of the Effect of Water on the Adhesive Bond*; Report No. ICAR/505-1; International Center for Aggregates Research, Texas Transportation Institute, The Texas A&M University System, College Station: Austin, TX, USA, 2005; pp. 59–63.
30. Caro, S.; Beltran, D.P.; Alvarez, A.E.; Estakhri, C. Analysis of moisture damage susceptibility of warm mix asphalt (WMA) mixtures based on Dynamic Mechanical Analyzer (DMA) testing and a fracture mechanics model. *Constr. Build. Mater.* **2012**, *35*, 460–467. [CrossRef]

Article

Laboratory and On-Site Tests for Rapid Runway Repair

Federico Leonelli [1], Paola Di Mascio [1], Antonello Germinario [2], Francesco Picarella [3], Laura Moretti [1,*], Mauro Cassata [3] and Alberto De Rubeis [4]

1 Department of Civil, Constructional and Environmental Engineering, Sapienza University of Rome, Via Eudossiana 18, 00184 Rome, Italy; fede.leonelli@gmail.com (F.L.); paola.dimascio@uniroma1.it (P.D.M.)
2 Laboratorio Principale Prove e Sperimentazioni (ITAF Infrastructure Department), Viale di Marino snc, 00043 Ciampino, Italy; laboratorioprove@tiscalinet.it
3 2 Reparto Genio A.M. (ITAF Infrastructure Department), Viale di Marino snc, 00043 Ciampino, Italy; francesco.picarella@aeronautica.difesa.it (F.P.); mauro.cassata@aeronautica.difesa.it (M.C.)
4 Servizio Infrastrutture A.M. (Chief of ITAF Infrastructure Department), Viale dell'Università 4, 00185 Rome, Italy; alberto.derubeis@aeronautica.difesa.it
* Correspondence: laura.moretti@uniroma1.it; Tel.: +39-06-4458-5114

Received: 15 October 2017; Accepted: 16 November 2017; Published: 19 November 2017

Featured Application: The results of this study offer a broad vision for the rapid runway repair, giving a useful quantitative and objective tool for airport management body; moreover, they could be applied also for road pavements.

Abstract: The attention to rapid pavement repair has grown fast in recent decades: this topic is strategic for the airport management process for civil purposes and peacekeeping missions. This work presents the results of laboratory and on-site tests for rapid runway repair, in order to analyse and compare technical and mechanical performances of 12 different materials currently used in airport. The study focuses on site repairs, a technique adopted most frequently than repairs with modular elements. After describing mechanical and physical properties of the examined materials (2 bituminous emulsions, 5 cement mortars, 4 cold bituminous mixtures and 1 expanding resin), the study presents the results of carried out mechanical tests. The results demonstrate that the best performing material is a one-component fast setting and hardening cement mortar with graded aggregates. This material allows the runway reopening 6 h after the work. A cold bituminous mixture (bicomponent premixed cold asphalt with water as catalyst) and the ordinary cement concrete allow the reopening to traffic after 18 h, but both ensure a lower service life (1000 coverages) than the cement mortar (10,000 coverages). The obtained results include important information both laboratory level and field, and they could be used by airport management bodies and road agencies when scheduling and evaluating pavement repairs.

Keywords: rapid runway repair; cement mortar; cold bituminous mixture; expanding resin; aircraft classification number (ACN); pavement classification number (PCN)

---

1. Introduction

Transport infrastructures ensure transport mobility and accommodate infrastructures which provide the essential needs of the population, as food, energy, telecommunications, waterworks, health and safety networks, sewage systems [1]. Therefore, transport network is a lifeline [2], and its vulnerability exposes people to additional risks. Air transport needs for high priority because it is a strategic infrastructure. Often it is the only alternative to link remote territories, especially when it comes to emergency [1,3]. For example, airports are strategic when other transport infrastructures are not usable as consequence of a natural disaster (e.g., earthquake, flooding, storm) [4] or when tactical transport should support a peacekeeping mission [5]. For a rescue operation to succeed, a fully functioning system is of the essence [6–8]; therefore, repair and maintenance works should be fast and effective to ensure the opportune evenness during the operations [9,10].

For a long time, the Air Force studied in many countries the rapid runway repair (RRR) because its strategic importance; in 2016 the North Atlantic Treaty Organization (NATO) published the Standardization Agreement (STANAG) 2929 about this issue [11]. It provides for data and elements useful also for civil sector when rapid repair needs, and considers a standard pavement damage as a crater with a real diameter of 12 m and a maximum depth of 3 m. The term "real diameter" refers to not only the real crater caused by a warp, but also to the surrounding affected pavement. Under such conditions, two main categories of repairs could be applied: site repairs and repairs with prefabricated elements. Repairs with prefabricated elements refer to application of modular prefabricated elements on roller compacted granular materials. This solution ensures greater strength than the on-site ones, and prevent Foreign Objects Debris (FOD).

Three types of modular repairs are currently used:

- fiberglass mats: the pavement discontinuity is filled with controlled granulometry stones, rolled and levelled with the unpaired part of the pavement (Figure 1). A fiberglass mat, composed of two or more layers of fiberglass impregnated with polyurethane or polyester resins, is laid upon to prevent FOD risk [12]. Finally, the mat is anchored to the pavement with bolts and plugs. This method is simple and rapid: its longest procedure is fixing to the ground [13].
- precast concrete slabs: precast concrete slabs 1.5×1.5 m wide and 15 cm thick are laid down on a foundation levelled, which is 15 cm under the final pavement level (Figure 2) [14]. Slabs have a steel containment profile around their perimeter, double internal reinforcement, and two slots for lifting [15]. The system guarantees bearing capacity and durability, but has several operational difficulties: existing pavement should be cut to contain exactly a proper number of slabs; hot mastics, resins or hot bitumen should be applied to finish the joints [16].
- metal mats: prefabricated metal elements are suitable for both recess and covering execution (Figure 3). Usually, aluminium mats 4 cm thick are used because their high strength and low weight. Joints are simple and exact, outside elements are tied to avoid removal of elements and risk of FOD.

Figure 1. Anchoring the fiberglass mat.

Figure 2. Precast concrete slabs with a steel containment profile.

Figure 3. Metal mats.

Site repairs restore pavement continuity creating a structural package filling the crater or the discontinuity. Quickly hardening cement/resin-bounded mortars or bituminous emulsions are used. Three types of interventions are possible:

- recovery by percolation: the crater is filled with part of the material (10–70 mm diameter) spilled from the crater itself, then different materials of suitable granulometry complete the filling up to reach the ground level. Percolation of cementitious or bituminous binders finishes the upper thickness of the pavement. The granulometry of filling material varies with the used binder: it is 10–25 mm when it is bitumen, 25–70 mm when it is cement;
- recovery by surface filling: the crater is filled as in the previous case, but no more than 5–8 cm from the final level; a surface layer of cold bituminous asphalt composes the new upper layer and completes the repair;
- recovery by deep filling: the crater is filled as before, up to 20–30 cm from the final level; the restoration will be realized using ordinary concrete.

Site repairs are more frequent than modular repairs because they are more versatile: they allow repair under various conditions (e.g., extension of pavement to be repaired, volume to be filled . . .). Their technique could be used also in road sector, where site repairs are just about the only used and rapid ones are desirable [17–19]. Under such conditions, the study focused on evaluation and identification of the best technical solution for easy and rapid pavement repair. As consequence of this condition, the study examined 12 materials and mixtures having with reference the NATO standard STANAG 2929 [11], which defines the maximum time for airport reopening after a repair work. Laboratory and in situ [20,21] tests have been performed to find the best choice from a technical point of view; finally, the examined materials have been compared respect to their installation costs. The results from this work provide interesting information useful to design a RRR, both for airport and road pavements.

2. Materials and Methods

Four different categories of materials used for repairing airport pavements have been examined in the study: bituminous emulsions, quick-hardening cement mortars, ordinary cement concretes, cold bituminous conglomerates, and an expanding resin (Table 1).

Table 1. Materials examined in the study.

Category	Notation
Bituminous emulsions	E1 E2
Cement mortars	M1 M2 M3
Cement concretes	M4 M5
Cold bituminous mixes	B1 B2 B3 B4
Expanding resin	R

Two bituminous emulsions have been tested:

- E1 is an over-stabilized cationic emulsion composed of 60% styrene-butadiene-styrene (SBS)-modified bitumen. Table 2 lists technical characteristics of bitumen extracted from the emulsion.

Table 2. Technical characteristics of bitumen extracted from the emulsion E1.

Characteristic	Value	Unit of Measure
Penetration at 25 °C	55	dmm
Softening point	62	°C
Fraas breaking point	−16	°C

Its correct temperature use ranges from 5 and 80 °C; moreover, it contains structural regenerative additives, therefore it is suitable for cold state repairs.

- E2 is a bicomponent modified, workable cold bitumen: it is useful for pavement maintenance when temperature ranges between 10 and 30 °C. Its maturity time is not more than 45 min after mixing the two components.

Five quick-hardening cement mortars (see Table 1) have been tested:

- the first and second cement mortars are composed of a Portland cement respectively 32.5 (M1) and 42.5 (M2) compliant with the standard EN 197-1 [22]. Mortars have been mixed with a water/powder (w/p) ratio equal to 0.45;
- the third cement mortar (M3) is a one-component fast setting and hardening cement with silica fume mortar [23–25]. It is fibre reinforced, suitable for smoothing, filling, and repairing concrete surfaces. Its correct w/p ratio is 0.13. Its elastic modulus evaluated according to the standard EN 13412 [26] is 32.6 GPa.
- the fourth cement mortar (M4) is a one-component fast setting and hardening cement mortar with graded aggregates. It is suitable for smoothing, filling, and repairing concrete surfaces; for thickness over 5 cm it is suitable for casting with 6/10 aggregates without segregation. Its correct w/p ratio is 0.125. Its elastic modulus evaluated according to the standard EN 13412 [26] is 32.6 GPa.

- the fifth cement mortar (M5) is a thixotropic, non-shrink, fibre reinforced, fast setting and hardening hydraulic mortar with graded aggregates. It is suitable for filling and repairing concrete surfaces. Its correct w/p ratio is 0.22. Its elastic modulus evaluated according to the standard EN 13412 [26] is 22 GPa.

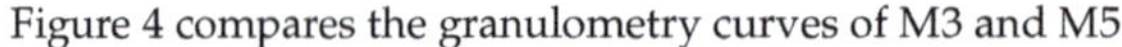
Figure 4 compares the granulometry curves of M3 and M5.

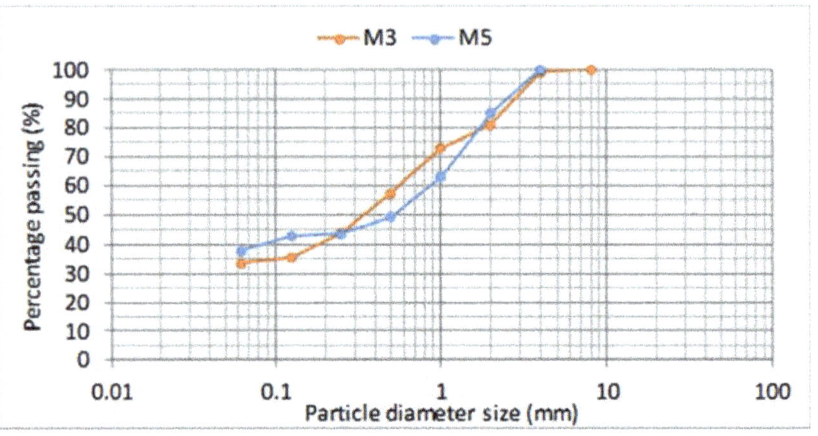

Figure 4. Granulometry curve of M3 and M5.

The cement mortars M3 and M5 have similar granulometric curve: both have over 33% percentage passing at 63 µm, and over 98% percentage passing at 4 mm. Nevertheless, M5 has more fine content than M3, while it has less fine sand than M3: these differences explain the different w/p ratios above listed.

Figure 5 compares the compressive strength of cement mortars M3, M4 and M5 whose time, temperature, and procedures for laying are compliant with those adopted for ordinary cement concrete. Compressive strength has been evaluated after maturity at 20 °C according to the standards: EN 12190 [27] for M3 and M4, and UNI EN 1015-11 for M5 [28].

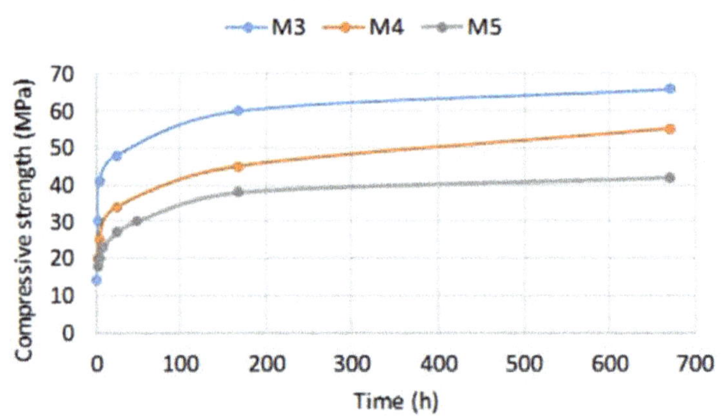

Figure 5. Compressive strength evolution of M3, M4 and M5.

Figure 6 compares the flexural strength of cement mortars M3, M4 and M5. Flexural strength has been evaluated after maturity at 20 °C according to the standards: EN 196-1 [29] and EN 13813 [30] for M3 and M4, and EN 1015-11 [28] for M5.

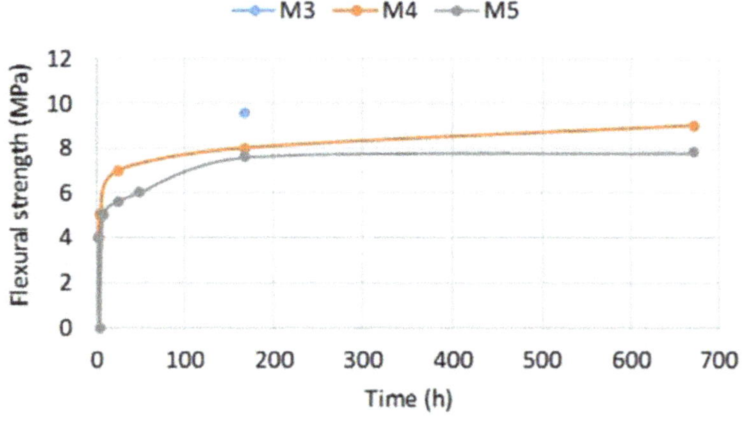

Figure 6. Flexural strength evolution of M3, M4 and M5.

Four cold bituminous mixtures (see Table 1) have been considered. Time, temperature, and procedures for their laying are compliant with those adopted for ordinary asphalt.

- B1 is a premixed cold bituminous mixture composed of fine aggregates and bitumen (6% by volume). After its application, the pavement can be immediately re-opened to traffic. The traffic itself settles the material, ensuring adhesion to the existing paving. This feature reduces time and costs for compaction, which cannot be overlooked using ordinary asphalt mixes [31]. Table 3 lists mechanical characteristics of B1 evaluated according to the standard ASTM D6927 (75 blows on each side) [32].
- B2 is a one-component premixed asphalt: it permits to repair 20–70 mm thick potholes with a single layer application. Table 4 lists technical characteristics of B2.
- B3 is a bicomponent premixed cold asphalt: it permits to repair bituminous and concrete pavements with up to 7 cm thick layers. Water is its catalyst for the hardening process.
- B4 is a premixed cold asphalt composed of bitumen (5.5% by weight of aggregates), vegetal oils, plasticizer additives and aggregates. It permits to repair bituminous and concrete pavements with layers not less than 2 cm and up to 6 cm thick.

Table 5 lists technical characteristics of B4.

Table 3. Mechanical characteristics of bituminous mixture B1.

Characteristic	Value	Unit of Measure
Marshall stability at 25 °C after 24 h	>3	kN
Marshall stiffness at 25 °C after 24 h	>1.5	kN/mm
Residual voids	<10	%
Indirect tensile strength of Marshall specimen at 25 °C after 24 h	>55	kPa

Table 4. Technical characteristics of bituminous mixture B2.

Characteristic	Value	Unit of Measure
Volumetric mass density	2.3	g/cm^3
Aggregate size	0–8	mm
Bitumen content	7.4–8.4	%
Voids content (after 75 blows Marshall)	7–9	%
Marshall stability after 24 h under water at 60 °C	≥4	kN
Marshall flow after 24 h under water at 60 °C	2–5	mm

Table 5. Technical characteristics of bituminous mixture B4.

Characteristic	Value	Unit of Measure
Volumetric mass density	>2.10	g/cm^3
Voids content (after 75 blows Marshall)	<10	%

Figure 7 compares the granulometry curves of B1, B2, B3, and B4.

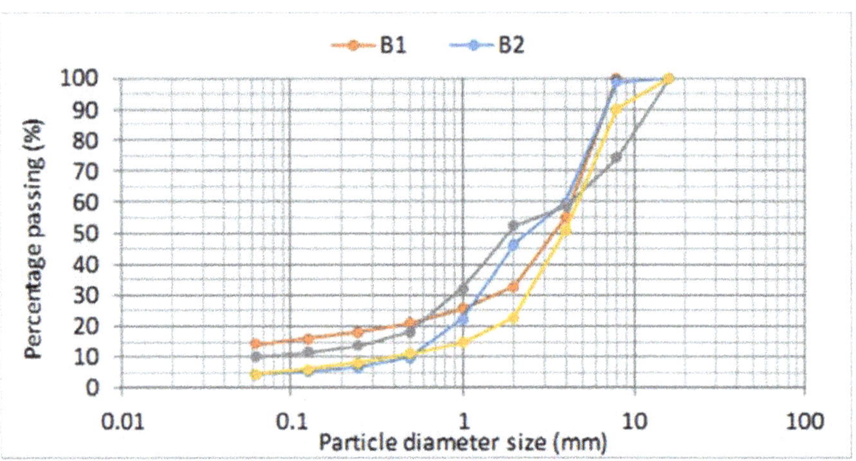

Figure 7. Granulometry curve of B1, B2, B3, and B4.

Expanding resins are used for transport infrastructures when the bearing capacity of soil does not satisfy the design requirements. Their expansion allows consolidation treatment by mean removal of the interstitial water and/or the filling of voids. The consolidation treatment involves executing injections through small metal cannulas placed on a regular mesh on the area to be treated.

In the study, a high-density bi-component polyurethane resin R (see Table 1) has been tested: its starting time is 40 ± 2 s and its expansion time is 85 ± 10 s. Table 6 lists its technical characteristics.

Table 6. Technical characteristics of the examined expanding resin.

Characteristic	Value	Unit of Measure
Compressive strength at 28 days	5.0	MPa
Shear strength at 28 days	5.0	MPa
Percentage closed cells	91.5 ± 1.5	%

In the first phase of the study, the presented 12 materials were tested in the Laboratory of the Italian Air Force (2nd Department of Genio located in Ciampino-Rome) to verify:

- ease of mixing and laying;
- percolability through a D40-70 grain size class (for emulsions and mortars);
- self-levelling properties;
- Marshall stability of cold conglomerates;
- cubic compressive resistance of cement mortars and expanding resin.

According to the need for fast reopening of the airport, mechanical resistance tests were conducted with the timing of 3 h, 6 h, 12 h, 24 h, and 48 h, 3days, 7days, and 28 days to evaluate the maturity of the products.

The texts performed were:

2.1. *Percolation Tests with Bituminous Emulsions and Cement Mortars*

The percolation test is a laboratory test designed by the 2nd Department of Genio of the Italian Air Force to verify the performance of a granular material bounded by a percolated binder. The procedure involved 7–12 mm granular materials with bituminous emulsions and 40–70 mm aggregates with cementitious mortars. In all cases, the tests aimed to find the binder composition (i.e., content of water for E1, content of cement for E2 and the mortars) which allows a percolation of about 20 cm and therefore it is defined "optimum"; otherwise the binder consistency is "fluid" (if the percolation thickness is more than 20 cm) or "plastic" (if the percolation thickness is lower than 20 cm).

2.2. *Structural Strength of Concrete Obtained from Cement Mortars and Standard 32.5/42.5 Cement Concrete*

Concrete made from cement mortars was obtained by percolating mortar inside 15 cm cubic moulds where the aggregates were previously located. Specimens were tested according to the standard EN 12390-2 [33].

In this phase specimens with ordinary concrete mixed with cement 32.5 and 42.5 and water/cement ratio equal to 0.500 were also tested. During the mixing process, the authors observed a rather aggressive gripping phenomenon in cement concrete 42.5, which could lead to difficulties during works.

2.3. *Marshall Stability of Cold Bituminous Mixtures*

Different series of Marshall specimens [34] have been made to evaluate the increments of resistance to time: for each time and cold bituminous mixture, 4 specimens have been tested.

2.4. *On-Site Tests*

On-site repairs were designed using the software FAARFIELD 1.41 (Federal Aviation Administration Rigid and Flexible Iterative Elastic Layered Design) (Federal Aviation Administration, Washington, DC, USA) [35]. It provides possible configurations of flexible and rigid pavement layers by simulating the number of coverages of the traffic mix. In this study, the layers thickness was calculated considering a reference aircraft, the C-130J (Lockheed Martin, Bethesda, MD, USA), typical for civil and military operations of the Italian Air Force. For this aircraft, the load distribution at the time of landing is 5% on the front gear and 95% on the rear one. 500 annual coverages during 20-year service life were considered: this volume traffic permit to design a permanent repair, as defined by the NATO criteria [36]. The flexible pavement model was used for bituminous materials, while the rigid pavement model was used for cementitious materials and the expanding resin [37,38].

The on-site tests involved the best performing materials found during the laboratory experimentation.

3. Results

3.1. Percolation Tests with Bituminous Emulsions and Cement Mortars

The results of percolation tests are listed in Table 7.

Figure 8 shows some percolation specimens obtained during the study.

Table 7. Results of percolation tests.

Product	Condition and Results			
E1	percentage of cement -	0% fluid	25% fluid	50% optimum
E2	as it is	optimum	-	-
M1	water percentage	40% plastic	45% optimum	50% fluid
M2	water percentage	40% plastic	45% optimum	50% fluid
M3	water percentage	12% plastic	13% optimum	14% fluid
M4	water percentage	12% plastic	12.5% optimum	13% fluid
M5	water percentage	16% plastic	20% plastic	24% plastic

Figure 8. Example of percolation specimens.

3.2. Structural Strength of Concrete Obtained from Cement Mortars and Standard 32.5/42.5 Cement Concrete

The results of compression tests on cementitious mixes are listed in Table 8.

Figure 9 highlights M3 and M4 have mechanical features that can be used as solutions for RRR, especially when percolation needs. M1 and M2 exhibited a high percentage of voids and, consequently, lower mechanical characteristics than the ordinary concretes mixed with 32.5 and 42.5 cements (both CEM I type). Ordinary concretes have mechanical characteristics that can be used as solutions for RRR, especially for deep filling.

Table 8. Compressive cubic strength of cementitious mixes.

Product	Compressive Cubic Strength (MPa)							
	Water/Cement Ratio	Time (h)						
		3	6	12	24	72	168	672
M1	0.450	0	0.5	1.8	3.5	13.4	18.6	27.2
M2	0.450	0	0.5	2.5	4.3	13.1	17.6	22.2
M3	0.130	32.9	37.5	41.0	44.8	52.6	57.4	65.8
M4	0.125	21.7	34.7	36.0	38.4	48.1	54.3	64.4
M5	-	-	-	-	-	-	-	-
R	-	4.59	4.60	4.72	4.74	4.77	5.45	5.42
Concrete with cement 32.5	0.500	0.13	1.8	11.5	19.2	30.2	33.5	48.5
Concrete with cement 42.5	0.500	0.31	2.3	12.6	20.4	32.9	38.4	54.0

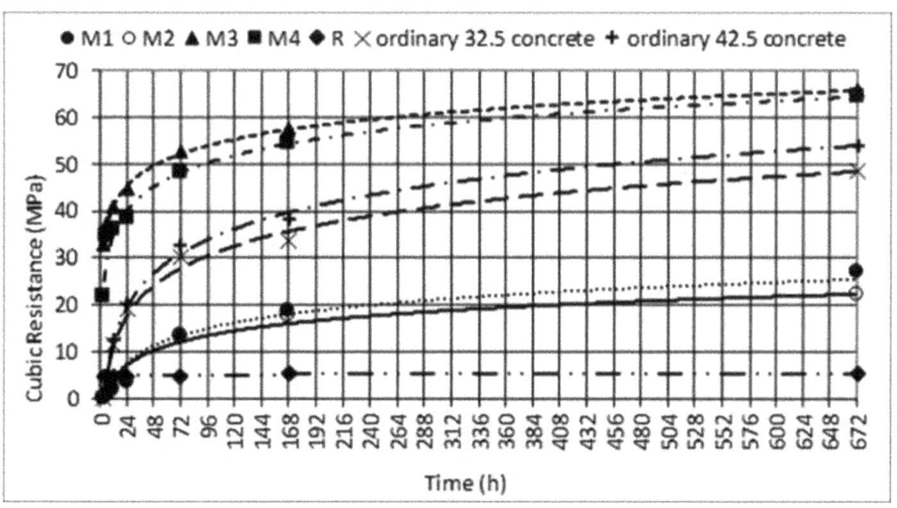

Figure 9. Structural strength depending on time.

3.3. *Marshall Stability of Cold Bituminous Mixtures*

Table 9 shows the Marshall stability of the examined cold bituminous mixtures at different times.

Table 9. Marshall stability.

Product	Marshall Stability (daN)					
	Time (h)					
	3	6	12	24	48	72
B1	222	238	241	240	266	270
B2	36	-	-	-	-	-
B3	599	850	1041	1232	1327	1213
B4	-	-	-	-	-	-

Figure 10 shows that B1 had insufficient mechanical characteristics to allow the rapid repair of the runways; B2 had only one specimen able to be tested; B4 did not have consistency: its specimen broke even before being inserted into the press. Only B3 showed mechanical characteristics appropriate for rapid repair.

At the end of this laboratory experimentation, the more reliable products which could be used for the runway rapid repair were: M4, B3, and M1. These materials and the expanding resin were used for on-site tests.

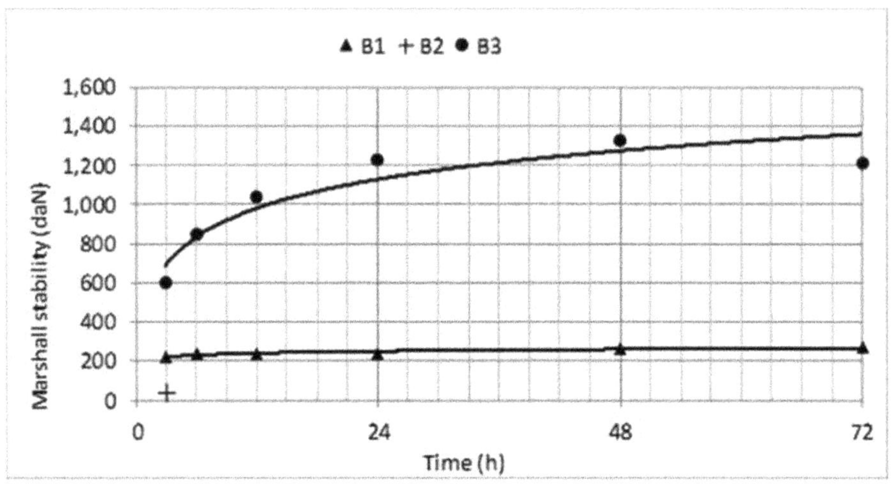

Figure 10. Marshall stability depending on time.

3.4. On-Site Tests

The on-site experimentation involved four test fields to be repaired, arranged in a 12-m diameter circle (Figure 11).

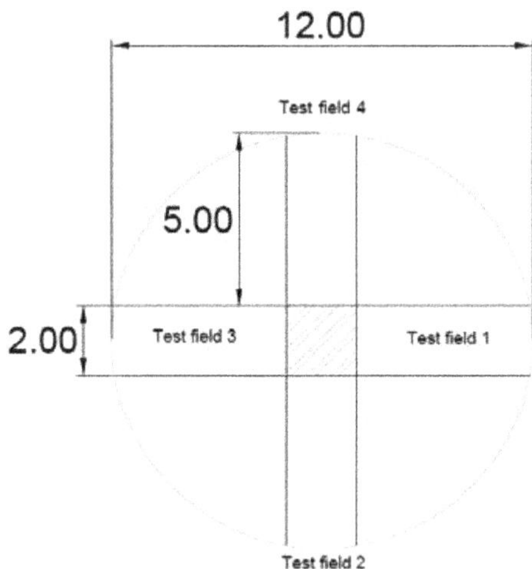

Figure 11. Diagrams of test fields (unit of measure: m).

The test fields were sized to avoid the disturbance effects due to the loads from the adjacent test fields. Their characteristics were:

- Test field 1: consisting of a D40-70 stone foundation and B3 wearing (flexible pavement);

- Test field 2: consisting of a D40-70 stone foundation and M1 wearing (rigid pavement);
- Test field 3: consisting of a D40-70 stone foundation up to 20 cm from the ground level, overlaid by a geotextile and a further layer of crushed stone with M4 used by percolation, (rigid pavement);
- Test field 4: consisting of an expanding resin injected into a stone foundation (40–70 mm) to increase its bearing characteristics and to create a support for concrete slabs.

During the execution of the expanding resin injection into the layer of stone, there was a strong expansion of the mixture, even up to 20 cm (Figure 12).

Figure 12. Expansion of expanded resin.

The on-site experimentation allowed the verification of the load bearing during time through Light Weight Deflectometer (LWD) and Heavy Weight Deflectometer (HWD) tests. LWD was used to evaluate the pavement Deformation Modulus M_d under the resin-treated area [39]. HWD technology allowed to back calculate the elastic modules of the repair layers by mean the software Elmod® 6.1.75 (Dynatest: Søborg, Denmark) [40]. According to the measured site deflection data, the software gave back an approximated deflection basin with decay curves according to the material under study.

Table 10 lists the results of LWD tests on three resin-treated points represented in Figure 13.

Figure 13. Resin-treated points of light weight deflectometer (LWD) tests.

Table 10. Deformation modulus of the resin-treated area.

Time (h)	M_d (MPa)							
	0	3	6	24	48	72	168	672
Point 1	7.37	2.87	4.19	4.5	2.9	3.41	4.1	2.71
Point 2	13.38	6.9	9.58	10.3	10.9	8.66	7.4	7.83
Point 3	14.99	10.13	11.42	7.6	9.8	10.73	15.3	10.11

The results listed in Table 10 highlight high variable values of M_d, as confirmed by the statistical results listed in Table 11. The set of values for conducting this analysis corresponds to the M_d values of the different tested points at different times. For every testing time, the mean, standard deviation, and coefficient of variation (CV, defined as the ratio of the standard deviation to the mean) were calculated.

Table 11. Statistical analysis of the M_d results of the resin-treated area.

Time (h)	0	3	6	24	48	72	168	672
Average (MPa)	11.91	6.63	8.40	7.47	7.87	7.60	8.93	6.88
Standard deviation (MPa)	4.02	3.64	3.76	2.90	4.34	3.77	5.76	3.79
CV	33.7%	54.8%	44.7%	38.9%	55.1%	49.6%	64.4%	55.1%

The high values of CV reveal that this technology is not reliable, as confirmed by the decreasing trend of M_d in Figure 14.

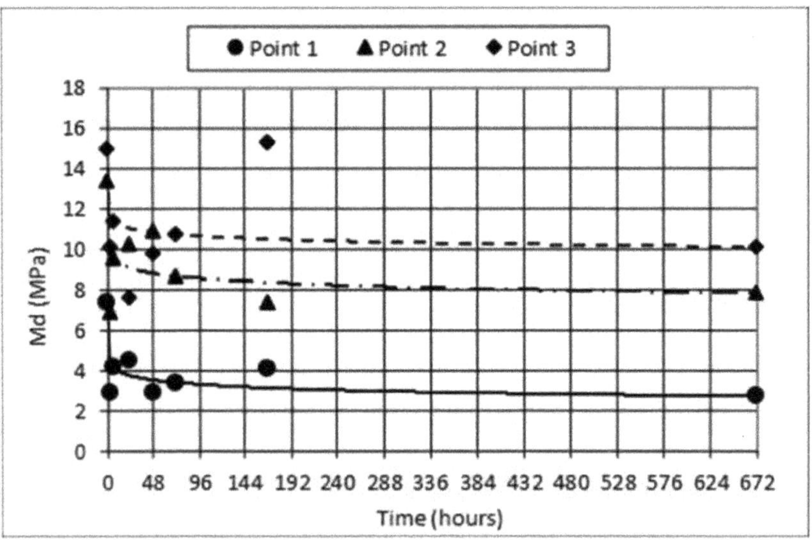

Figure 14. Deformation moduli after the resin injection.

The observed decreasing of M_d is caused by the viscous properties of the resin and its pronounced expansion which mobilizes the granular material. Indeed, this process alters the lithic skeleton because the resin replaces the stones instead of occupying the gaps between them. The mechanical performances shown in Table 10 and Figure 14 prevent the use of the expanding resin as material useful for RRR.

As done for LWD tests, the results of HWD tests and back calculation were performed for different times. During the analysis of the data some inconsistencies were found in the evaluation of the elastic moduli. For a more accurate evaluation of the moduli, filtering of data measured by the tenth geophone, which was always on the extreme edge of the test fields, was carried out (Figure 15).

Figure 15. The HWD machine over the M4-treated area.

The filtering process gave more precise and realistic results in terms of elastic modulus E of the pavement layers. In the first measurements (3 h), only the contribution of five geophones was filtered, according to the Dynatest guidelines [41].

A layered elastic model has been modelled to compute stresses, strains and deflections caused by surface load at any point in the pavement structure. According to the layered elastic theory [42], the model assumed that each pavement structural layer was homogeneous, isotropic, and linearly elastic. Table 12 lists the geometrical properties of the pavements modelled with the software Elmod® 6.1.75 [41].

Table 12. Geometrical properties of pavement models.

Layer	Test Field 1		Test Field 2		Test Field 3	
	Name	**Thickness (cm)**	**Name**	**Thickness (cm)**	**Name**	**Thickness (cm)**
Top layer	$E_{1,B3}$	10	$E_{1,M1}$	20	$E_{1,M4}$	20
Bottom layer	$E_{2,B3}$	90	$E_{2,M1}$	80	$E_{2,M4}$	80
Subgrade	E_{sub}	infinite	E_{sub}	infinite	E_{sub}	infinite

Top layers are composed of tested binders percolated into D40-70 granular bottom layers. The subgrade is the natural material underneath the test fields.

Tables 13–15 list the elastic moduli obtained from back analysis respectively for test field with B3, M1 and M4 materials.

Table 13. Elastic moduli of test field with B3.

Time (h)	3	6	24	72	168	672
$E_{1,B3}$ (MPa)	1193	1398	2797	2070	2607	3336
$E_{2,B3}$ (MPa)	450	450	450	450	450	450
E_{sub} (MPa)	350	340	323	322	301	350

Table 14. HWD test on test field with M1.

Time (h)	3	6	24	72	168	672
$E_{1,M1}$ (MPa)	-	-	14,783	21,724	28,193	38,609
$E_{2,M1}$ (MPa)	-	-	450	450	450	450
E_{sub} (MPa)	-	-	350	324	336	350

Table 15. Elastic moduli of test field with M4.

Time (h)	3	6	24	72	168	672
$E_{1,M4}$ (MPa)	15,146	25,614	27,000	27,000	27,150	28,461
$E_{2,M4}$ (MPa)	350	384	405	450	407	427
E_{sub} (MPa)	300	330	398	347	398	360

In the test fields 1 and 2, the values of the bottom layers elastic modulus ($E_{2,B3}$ and $E_{2,M1}$) were the same, equal to 450 MPa. This value of E_2 has been calculated before laying the top materials B3 and M1, and it has been assumed as constant during the subsequent analysis. This choice avoided anomalous back calculation results due to the small tests areas (10 m^2 each). Indeed, under ordinary conditions, the HWD tests are performed on areas larger than those prepared in this study, therefore the wave transmission in the study differs from usual. The most important differences involve the lower layers, whose elastic moduli are deduced from the responses of the geophones most distant from the loading plate. In this case, the response of the most distant geophones is also the most affected by side effects. On the other hand, this assumption has not been possible due to the technical characteristics of the third test field. M4 is a percolated material, thus the thicknesses and mechanical properties of layers are not defined and constant as those of test fields with materials B3 and M1.

Figure 16 shows the evolution of the elastic modulus of top layers paved with M1, M4 and B3.

Figure 16 highlights the mechanical performance of B3 are not comparable with those obtained using the ordinary cement concrete (M1) and the cementitious mortar M4. M4 has the most rapid rate of increase of elastic modulus: its E value is about constant after 1 day. On the contrary, for M1 the increase of the elastic modulus is slower, but its elastic modulus is comparable to that of M4 at 7 days and is growing until at least 28 days, when the difference is appreciable (38.6 GPa for M1 versus 28.4 GPa for M4).

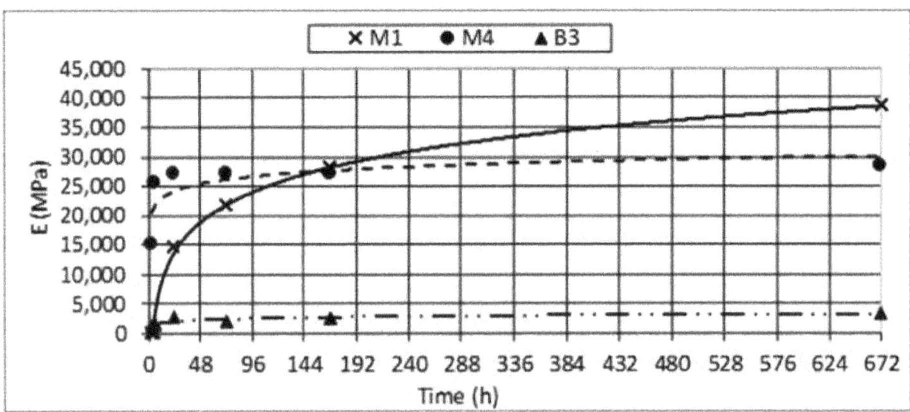

Figure 16. Elastic moduli of top layers.

4. Discussion

At the end of HWD tests, the ACN/PCN method was used to evaluate the load bearing capacity of the tested materials [43,44]. It is a system of rating airport pavements designated by the International Civil Aviation Organization [45] to compare the airport pavement strength (Pavement Classification Number, PCN) to the operation conditions of the traffic mix considering its gross weight and the subgrade bearing capacity (Aircraft Classification Number, ACN). The ACN value is twice the derived single-wheel load expressed in thousands of kilograms, with single-wheel tire pressure standardized at 1.25 MPa, which requires the same pavement thickness required by the examined aircraft. The calculation methods of ACN consider for flexible pavements 10,000 coverages and for rigid pavements 2.75 MPa concrete working stress [44]. Both values are representative of typical airport pavements.

Two criteria allow the evaluation of PCN: the "Using" aircraft or the "Technical" evaluation methods [44]. The first one is based on the experience, the second one on analytical procedures. In this study, the technical method proposed by the software Elmod® 6.1.75 has been used to calculate the PCN value. It considers the elastic modulus of layers (evaluated by mean back calculation), the configuration of the main gear of the design aircraft, and the number of coverages during the service life.

An airport pavement is verified when Equation (1) is satisfied:

$$PCN \geq ACN. \tag{1}$$

Data obtained in the HWD tests allowed the calculation of the PCN, using the software Dynatest Elmod® 6.1.75 [40]: different values were calculated for B3, M1, and M4 test fields at different maturity ages and for three traffic levels. 100, 1000 and 10,000 coverages during the service life were considered to simulate expedient, temporary, and permanent repairs according to the standard STANAG 7208 [36].

In the study, the subgrade under the test fields (Figure 11) was classified as "C" (low strength), therefore the ACN of the reference aircraft is 35.

Tables 16–18 list PCN values obtained for test field respectively with B3, M1 and M4 materials: the red cells indicate PCN<ACN, while green ones PCN ≥ ACN.

Table 16. Pavement Classification Number (PCN) values test field with B3.

Time (h)	3	6	24	72	168	672
Allowable Coverages	-	-	-	-	-	-
100	14	18	32	42	57	63
1000	10	11	26	35	42	53
10,000	3	7	14	22	33	44

Table 17. Pavement Classification Number (PCN) values test field with M1.

Time (h)	3	6	24	72	168	672
Allowable Coverages	-	-	-	-	-	-
100	-	-	39	43	50	50
1000	-	-	32	35	40	41
10,000	-	-	26	28	35	36

Table 18. Pavement Classification Number (PCN) values test field with M4.

Time (h)	3	6	24	72	168	672
Allowable Coverages	-	-	-	-	-	-
100	35	53	57	61	57	62
1000	29	44	47	50	47	51
10,000	24	37	39	42	39	43

The results reveal M4 has the most rapid PCN evolution: after 6 h maturity time, the verification ACN/PCN is satisfied. B3 and M1 ensure comparable results, but their reopening time (when Equation (1) is satisfied) is longer (7.5 days) than that of M4.

At the end of the study, an economic analysis has been carried out to estimate the unit costs for repairing a runway discontinuity [46,47]. Unit costs derive from the official price list used by the Italian Air Force (internal document) (Rome, Italy). Table 19 lists the results of the economic analysis.

Table 19. Unit cost of the examined rapid runway repair (RRR).

Material	Cost (€/m^2)
B3	180
M1	50
M4	75

According to the results listed in Table 19, M4 is the material that best suits needs for RRR: its technical and economic characteristics balance the conflicting objectives of budget and performance. Using M4, the airport could be reopened 3 h after the conclusion of the emergency repair (100 coverages during the service life), while 6 h after the RRR works it is possible to achieve a permanent repair, which allow 10,000 coverages during the service life.

5. Conclusions

The need for rapid repairing runway pavements has grown fast over the years, both for civil purposes and for peacekeeping missions. Therefore, there is the need for a technology that could be applied balancing conflicting objectives of resistance and rapidity. At this purpose, on-site repairs are more versatile than modular repairs currently used in the military sector because they can be adapted to different conditions. The specific sector of runway construction currently needs the comparison of technical performances offered by the innovative materials used for on-site repairs. In Italy, the Laboratory of the Italian Air Force (2nd Department of Genio) analysed technical and economic performances of several types of rapid runway repairs having the C-130J aircraft was as design aircraft. 12 materials currently used for pavement repair have been tested to evaluate their laboratory and on-site performances at different times.

This paper presents the first experimental results obtained on the following materials:

- 2 bituminous emulsions,
- 5 cement mortars,
- 4 cold asphalt mixes,
- 1 expanding resin.

The experimental study involved two phases:

1. laboratory tests on all materials except resin to focus on:
    - ease of mixing and application;
    - percolability (for emulsions, mortars and concretes);
    - Marshall stability (for cold asphalt mixes).
2. on-site experiments on the materials which performed the best for repairing. A cold asphalt mix, a cement mortar, an ordinary cement concrete and the expanding resin were used to repair four 10 m^2 widespread areas. All three tested mixes required time, temperature, and procedures compliant with those used to lay ordinary cement and bituminous runway pavements. On the contrary, the expanding resin required dedicated instruments to be injected. The bearing performances have been evaluated using the ACN/PCN system (for the three mixes) or LWD (for the expanding resin).

The on-site results highlighted that:

- it is not possible to repair runway pavement with expanding resin for two reasons: its strong expansion, even up to 20 cm over the desired level, and its bearing performances. Indeed, the LWD tests showed a decreasing value of deformation moduli over maturity time;
- the tested cement mortar M4 (one-component fast setting and hardening cement mortar with graded aggregates) ensures the fastest runway reopening in presence of a permanent repair of the damaged concrete pavement: 6 h after the work, the runway can be opened to traffic for 10,000 coverages of the design aircraft;
- surface filling (up to 6 cm) with the cold bituminous mixture B3 (bicomponent premixed cold asphalt with water as catalyst) and deep filling repair with the concrete M1 (ordinary concrete mixed with CEM I 32.5) have similar performances: both allow the reopening to traffic after 18 h maturity time, but for only 1000 coverages of the C-130J aircraft. Indeed, M1 and B3 require 7.5 days-age to support 10,000 coverages. B3 and M1 are both suitable for repairing flexible and rigid pavements.

Finally, it was also observed that the unit costs of the examined materials widely vary: B3 has a higher economic impact than that of M1 and M4, mainly due to the material since the costs of machines and works to have the rehabilitated pavement are comparable.

The results also provide a framework and a reference for any further study into specific cases using additional materials. All test methodologies can be applied to simulate with the software FAARFIELD (Version 1.41, Federal Aviation Administration, Washington, DC, USA) a fleet of almost any civil aircraft, therefore the procedure could be replicated for further analysis. As observed in this study, the contribution of each considered material to the overall repair work can be specialized to maximize the performances:

- percolated resin can be used in bottom layers as foundation and base,
- cold bituminous mixtures and cement mortars can be used for top layers.

Author Contributions: Alberto De Rubeis, Mauro Cassata, Paola Di Mascio and Laura Moretti conceived and designed the experiments; Federico Leonelli, Antonello Germinario and Francesco Picarellaperformed the experiments; all the authors analyzed the data and wrote the paper.

Conflicts of Interest: The authors declare no conflict of interest.

References

1. Organisation for Economic Co-Operation and Development. *Infrastructure to 2030. Volume 2 Mapping Policy for Electricity, Water and Transport*; Organisation for Economic Co-Operation and Development: Paris, France, 2007.
2. Cirianni, F.; Fonte, F.; Leonardi, G.; Scopelliti, F. Analysis of lifelines transportation vulnerability. *Procedia Soc. Behav. Sci.* **2012**, *53*, 29–38. [CrossRef]
3. Tong, S.Q.; Wang, N.; Song, N.Q. Emergency evacuation capability evaluation and optimization for an offshore airport: The case of Dalian Offshore Airport, Dalian, China. *Saf. Sci.* **2017**, *92*, 128–137. [CrossRef]
4. Shibayama, T. Japan's transport planning at national level, natural disasters, and their interplays. *Eur. Transp. Res. Rev.* **2017**, *9*, 44. [CrossRef]
5. Ripley, T. *Runway Repair Delays threaten Dutch F-16 deployment*; Jane's Defence Weekly: London, UK, 2006; pp. 353–354.
6. Loprencipe, G.; Zoccali, P. Comparison of methods for evaluating airport pavement roughness. *Int. J. Pavement Eng.* **2017**, 1–10. [CrossRef]
7. Loprencipe, G.; Cantisani, G. Evaluation methods for improving surface geometry of concrete floors: A case study. *Case Stud. Struct. Eng.* **2015**, 14–25. [CrossRef]
8. Bonin, G.; Cantisani, G.; Loprencipe, G.; Ranzo, A. Dynamic effects in concrete airport pavement joints. *Ind. Ital. Cem.* **2007**, *77*, 590–607.
9. Miccoli, S.; Finucci, F.; Murro, R. Assessing project quality: A multidimensional approach. *Adv. Mater. Res.* **2014**, *1030–1032*, 2519–2522. [CrossRef]

10. Miccoli, S.; Finucci, F.; Murro, R. Criteria and procedures for regional environmental regeneration: A European strategic project. *Appl. Mech. Mater.* **2014**, *675–677*, 401–405. [CrossRef]
11. NATO. *STANAG 2929, AATMP-03 (Allied Air Traffic Management Publication) "Airfield Damage Repair (ADR) Capability"*; North Atlantic Treaty Organization: Brussels, Belgium, 2016.
12. Dukes, P.E. Rapid runway repair: Fiberglass mats pass the test. *Mil. Eng.* **1988**, *80*, 453–455.
13. Dover, D.; Anderson, M.; Brown, R.W. Recent Advances in Matting Technology for Military Runways. In Proceedings of the 27th Annual International Air Transport Conference, Orlando, FL, USA, 30 June–3 July 2002.
14. Ashtiani, R.S.; Jackson, C.J.; Saeed, A.; Hammons, M.I. *Pre-Cast Concrete Panels for Rapid Repair of Airfield Rigid Pavements*; Air Force Research Laboratory Materials and Manufacturing Directorate Airbase Technologies Division: Washington, DC, USA, 2012.
15. Bull, J.W.; Woodfors, C.H. Design of precast concrete pavement units for rapid maintenance of runways. *Comput. Struct.* **1997**, *64*, 857–864. [CrossRef]
16. Sander, T.C.; Roesler, J.R. Case study: Runway 12L-30R keel suction rehabilitation, Lambert—St. Louis International Airport. In Proceedings of the 2006 Airfield and Highway Pavement Specialty Conference, Atlanta, GA, USA, 30 April–3 May 2006; pp. 872–884.
17. Wang, Y.; Kong, L.; Chen, Q.; Lau, B.; Wang, Y. Research and application of a black rapid repair concrete for municipal pavement rehabilitation around manholes. *Constr. Build. Mater.* **2017**, *150*, 204–213. [CrossRef]
18. Guan, Y.; Gao, Y.; Sun, R.; Won, M.C.; Ge, Z. Experimental study and field application of calcium sulfoaluminate cement for rapid repair of concrete pavements. *Front. Struct. Civ. Eng.* **2017**, *11*, 338. [CrossRef]
19. Shanahan, N.; Bien-Aime, A.; Buidens, D.; Meagher, T.; Sedaghat, A.; Riding, K.; Zayed, A. Combined effect of water reducer-retarder and variable chloride-based accelerator dosage on rapid repair concrete mixtures for jointed plain concrete pavement. *J. Mater. Civ. Eng.* **2016**, *28*, 04016036. [CrossRef]
20. Kavussi, A.; Abbasghorbani, M.; Moghadas Nejad, F.; Bamdad Ziksari, A. A new method to determine maintenance and repair activities at network-level pavement management using falling weight deflectometer. *J. Civ. Eng. Manag.* **2017**, *23*, 338–346. [CrossRef]
21. Chou, C.P.; Wang, S.Y.; Tsai, C.Y. Methodology of applying heavy weight deflectometer for calculation of runway pavement classification number. *Transp. Res. Rec.* **1990**, 57–64. [CrossRef]
22. EN (European Committee for Standardization). *EN 197-1: 2000. Cement Part 1: Composition, Specifications and Conformity Criteria for Common Cements*; European Committee for Standardization: Brussels, Belgium, 2000.
23. Sadowski, Ł.; Stefaniuk, D. Microstructural evolution within the interphase between hardening overlay and existing concrete substrates. *Appl. Sci.* **2017**, *7*, 123. [CrossRef]
24. Williams, M.; Ortega, J.M.; Sánchez, I.; Cabeza, M.; Climent, M.A. Non-destructive study of the microstructural effects of sodium and magnesium sulphate attack on mortars containing silica fume using impedance spectroscopy. *Appl. Sci.* **2017**, *7*, 648. [CrossRef]
25. Ortega, J.M.; Esteban, M.D.; Rodríguez, R.R.; Pastor, J.L.; Ibanco, F.J.; Sánchez, I.; Climent, M.A. Influence of silica fume addition in the long-term performance of sustainable cement grouts for micropiles exposed to a sulphate aggressive medium. *Materials* **2017**, *10*, 890. [CrossRef] [PubMed]
26. EN (European Committee for Standardization). *EN 13412:2006. Products and Systems for the Protection and Repair of Concrete Structures. Test Methods. Determination of Modulus of Elasticity in Compression*; European Committee for Standardization: Brussels, Belgium, 2006.
27. EN (European Committee for Standardization). *EN 12190:1998. Products and Systems for the Protection and Repair of Concrete Structures. Test methods. Determination of Compressive Strength of Repair Mortar*; European Committee for Standardization: Brussels, Belgium, 1998.
28. EN (European Committee for Standardization). *EN 1015-11:2007. Methods of Test for Mortar for Masonry. Determination of Flexural and Compressive Strength of Hardened Mortar*; European Committee for Standardization: Brussels, Belgium, 2007.
29. EN (European Committee for Standardization). *EN 196-1:2016. Methods of Testing Cement—Part 1: Determination of Strength*; European Committee for Standardization: Brussels, Belgium, 2016.
30. EN (European Committee for Standardization). *EN 13813:2002. Screed Material and Floor Screeds—Screed Materials—Properties and Requirements*; European Committee for Standardization: Brussels, Belgium, 2002.
31. Cantisani, G.; D'Andrea, A.; Di Mascio, P.; Loprencipe, G. RILEM Book series. In *8th RILEM International Symposium on Testing and Characterization of Sustainable and Innovative Bituminous Materials*; Canestrari, F., Partl, M., Eds.; Springer: Dordrecht, The Netherlands, 2016; Volume 11. [CrossRef]

32. ASTM (American Society for Testing and Materials International). *ASTM D6927. Standard Test Method for Marshall Stability and Flow of Asphalt Mixtures*; ASTM International: West Conshohocken, PA, USA, 2015.
33. EN (European Committee for Standardization). *EN 12390-2:2009. Testing Hardened Concrete. Part 2: Making and Curing Specimens for Strength Tests*; European Committee for Standardization: Brussels, Belgium, 2009.
34. ASTM (American Society for Testing and Materials International). *ASTM D6926. Standard Practice for Preparation of Asphalt Mixture Specimens Using Marshall Apparatus*; ASTM International: West Conshohocken, PA, USA, 2016.
35. Federal Aviation Administration. *AC 150/5320-6F—Airport Pavement Design and Evaluation*; Federal Aviation Administration: Washington, DC, USA, 2016.
36. NATO. *STANAG 7208 Airfield Pavement Design Criteria—Study Draft*; North Atlantic Treaty Organization: Brussels, Belgium, 2016.
37. Tiago Bonucci, P. Airfield rigid pavement structural design—A review of main aspects and methods of analysis. In *AIP Conference Proceedings, Proceedings of the 2nd International Symposium on Computational Mechanics, ISCM II, and the 12th International Conference on the Enhancement and Promotion of Computational Methods in Engineering and Science, EPMESC XII, Hong Kong–Macau, China; 30 November–3 December 2009*; American Institute of Physics: College Park, MD, USA, 2010; Volume 1233, pp. 1339–1344.
38. Chordá Ayela, A. Flexible pavements in the Barcelona airport enlargement (Pavimentos flexibles en la ampliación del aeropuerto de Barcelona). *Carreteras* **2010**, *4*, 29–42.
39. Di Mascio, P.; Loprencipe, G.; Maggioni, F. Visco-elastic modeling for railway track structure layers [Modellazione del comportamento visco-elastico degli strati della sede ferroviaria]. *Ing. Ferrov.* **2014**, *3*, 207–222.
40. *Elmod Software Manual*; Dynatest: Søborg, Denmark, 2015.
41. *Dynatest FWD/HWD Test Systems*; Owners Manual; Dynatest: Søborg, Denmark, 2014.
42. Boussinesq, J. *Application des Potentials a L'Etude de L'Equilibre et du Mouvement des Solides Elastiques*; Gauthier-Villars: Paris, France, 1885.
43. Norman, J.A.; Mumayiz, S.; Wright, P.H. *Airport Engineering: Planning, Design and Development of 21st Century Airports*, 4th ed.; John Wiley & Sons: Hoboken, NJ, USA, 2011; ISBN 978-0-470-39855-5.
44. Federal Aviation Administration. *AC 150/5335-5C, Standardized Method of Reporting Airport Pavement Strength—PCN*; Federal Aviation Administration: Washington, WA, USA, 2014.
45. ICAO. *Annex 14, Volume I Aerodromes*; International Civil Aviation Organization: Montreal, QC, Canada, 2013.
46. Di Mascio, P.; Moretti, L.; Panunzi, F. Economic Sustainability of Concrete Pavements. *Procedia Soc. Behav. Sci.* **2012**, *53*, 125–133.
47. Moretti, L. Technical and economic sustainability of concrete pavements. *Mod. Appl. Sci.* **2014**, *8*, 1–9. [CrossRef]

Article

Tire-Pavement Friction Characteristics with Elastic Properties of Asphalt Pavements

Miao Yu [1,2,*], Guoxiong Wu [3,*], Lingyun Kong [1] and Yu Tang [1]

1. National and Regional Engineering Lab for Transportation Construction Materials, College of Civil Engineering, Chongqing Jiaotong University, 66 Xuefu Ave, Nanan Qu, Chongqing 400074, China; klyyqr2002@163.com (L.K.); faye-yu@163.com (Y.T.)
2. Highway School, Chang'an University, Middle-Section of Nan'er Huan Road, Xi'an 710064, China
3. Chongqing Jianzhu College, 857 Lihua Ave, Nanan Qu, Chongqing 400072, China

* Correspondence: yumiaoym@126.com (M.Y.); wgx_ph.d@163.com (G.W.); Tel.: +86-023-62789023 (M.Y.); +86-023-61849988 (G.W.)

Received: 16 September 2017; Accepted: 26 October 2017; Published: 1 November 2017

Abstract: The skid-resisting performance of pavement is a critical factor in traffic safety. Recent studies primarily analyze this behavior by examining the macro or micro texture of the pavement. It is inevitable that skid-resistance declines with time because the texture of pavement deteriorates throughout its service life. The primary objective of this paper is to evaluate the use of different asphalt pavements, varying in resilience, to optimize braking performance on pavement. Based on the systematic dynamics of tire-pavement contact, and analysis of the tire-road coupled friction mechanism and the effect of enlarging the tire-pavement contact area, road skid resistance was investigated by altering the elastic modulus of asphalt pavement. First, this research constructed the kinetic contact model to simulate tire-pavement friction. Next, the following aspects of contact behaviors were studied when braking: tread deformation in the tangential pavement interface, actual tire-pavement contact in the course, and the frictional braking force transmitted from the pavement to the tires. It was observed that with improvements in pavement elasticity, the actual tire-pavement contact area increased, which gives us the ability to effectively strengthen the frictional adhesion of the tire to the pavement. It should not be overlooked that the improvement in skid resistance was caused by an increase in pavement elasticity. This research approach provides a theoretical basis and design reference for the anti-skid research of asphalt pavements.

Keywords: skid-resistance; asphalt pavement; tire; friction; contact area; braking force coefficient

---

1. Introduction

Skid resisting performance is the primary factor affecting traffic safety, specifically during vehicle braking where the anti-skid deficiencies found in pavement will prolong braking time and increase braking distance [1]. Due to this concern, highway authorities and researchers have been exploring the anti-skid properties of pavement through the frictional contact analyses of three key factors.

The first factor is water. In the early 1960s, Horne and Dreher 1963 [2] put forward the famous National Aeronautics and Space Administration (NASA) hydroplaning equation, which is used to calculate the hydroplaning speed of a tire. Later, Martin 1966, Eshel 1967, and Tsakonas et al. 1968 [3–5] also simulated the hydroplaning of a tire, but in only two dimensions. In the following thirty years, both the NASA equation and the two-dimensional contact model were promoted by Horne et al. 1986 [6] on the applicable conditions and accuracy. In the 21st century, a three-dimensional frictional model of inflated tire and rigid pavement contact, simulating moist conditions, was developed by G.P. Ong et al. 2007, 2008, 2010, 2012 [7–10]. With this model, the degradation mechanism of wet-pavement skid resistance and the subsequent effects of tire slip velocity on rigid pavement under different degrees of wetness were examined.

The second factor is the frictional property of tire rubber. The friction between tire and pavement is mainly determined by adhesive and hysteretic friction. Hysteric friction is enhanced with rise in temperature. For this analysis, K.A. Grosch 2001 [11] chose to use the rubber friction model based upon the tire rubber's viscoelasticity. Coupling thermos-mechanics with the finite element method, Srirangam and Anupam 2013 [12,13] estimated the variations in hysteretic friction during tire-pavement contact due to differences in temperature. This estimation was calculated by simulating a tire rolling on rigid pavement in CAPA-3D (developed and dated by the group of Mechanics of Infrastructure Materials at Delft University of Technology, The Netherlands) Finite Element system.

The third final component in exploring the anti-skid properties of pavement is the surface texture of the pavement. Both the macro-texture and micro-texture are typically used to represent the skid resistance of pavement. The correlation between pavement texture and skid resisting ability are often evaluated via various test methods (Forster, S. W. 1990, Burak Sengoz 2014, Malal Kane 2015 [14–16]). For instance, Srirangam and Anupam 2013 [12,13] captured morphological data reflecting the surface of asphalt pavement through X-ray tomography. Furthermore, they reconstructed the texture of the pavement in CAPA-3D and analyzed their simulation of the kinetic contact of tires rolling on the rough, rigid-pavement [17].

After analyzing the aforementioned literature, we found that research on skid resistance primarily focuses on the macro or micro-texture of the pavement, by which the anti-skid property could be achieved. In addition, there are two methods to model asphalt pavement in the finite element analysis. One approach is modeling the pavement as a flat rigid surface, while the other one is an analytical rigid body with a macro-texture surface. It is apparent that both means of modeling consider the asphalt pavement as a rigid body, disregarding its elasticity.

In the field of terrain vehicle mechanics, the fundamental source of vehicle braking is the friction between the tires and pavement. As illustrated in Figure 1a, the pedal force F_p is delivered to the wheel brake the moment the driver steps on the pedal, forming the frictional moment T_μ opposite the momentum of the tire. This moment could be regarded as the circumferential component F_μ, which is applied to the tire-pavement contact region to prohibit the rotation of the tire. Simultaneously, the pavement generates a reactive force on the vehicle, namely, the ground braking force F_b, which ultimately slows down vehicles or even allows them come to a complete halt. In other words, F_b is equal to F_μ, which is also comparable to T_μ divided by the tire radius r. Nevertheless, restrained by the upper limit of the adhesive force F_ϕ (the maximum braking force $F_{b\max}$ is equal to F_ϕ), F_b will not rise as soon as it runs up to F_ϕ (Figure 1b). Furthermore, from the aspect of tribology, it can be concluded that physical properties of the interface, such as variations in material stiffness, will result in the change in contact area, which will then have an influence on the frictional force (Chengtao Wang 2002 [18]). Eventually, the actual tire-pavement area will vary as a result of pavement characteristics such as the diversification of recoverable resilient. Accordingly, the force of adhesion F_ϕ will also change, accounting for differentiation in the tire-oriented pavement braking function (Jide Zhuang 1986 [19], Liangxi Wang 2008 [20]).

Data produced from the uniaxial compressive strength testing of pavement with normally structured the asphalt specimens is generally above 1000 MPa (You et al. 2009 [21]; Goh et al. 2011 [22]); a value seemingly much higher than the longitudinal tire stiffness. Therefore, in consideration of time expended in a simulated calculation, the deformation of asphalt pavements is usually ignored in modeling (Srirangam 2013 [13], Hao Wang 2014 [23]). In other words, the pavement is defined as a rigid body in simulations. Nonetheless, along with the diversification of paving materials, the elastic modulus of asphalt mixtures also varies significantly. For instance, dry process crumb rubber modified (CRM) asphalt mixtures, which plays an indispensable role in noise reduction and the de-icing (Chunxiu Zhou 2006 [24], Xudong Wang 2008 [25]), is characterized by its remarkable elasticity. In addition, a high content of crumb rubber, which is added to asphalt mixtures, usually reduces the mixture modulus to half of that of the original (Lili Yao 2012 [26]). Such elasticity has the capacity to facilitate the resilience of CRM asphalt mixtures, which are characterized by a large elastic deformation under external loading conditions and a dramatic return to its original shape immediately after the removal of the load in comparison to that of ordinary asphalt mixtures (Miao Yu, Guoxiong Wu 2014 [27–29]). In view of the distinctly narrowed gap of moduli between tires and resilient pavements such as CRM asphalt pavement, the precision of the simulation will inevitably decrease if the resilient pavement still needs to be modeled as a rigid body. Furthermore, according to past literature, the actual stiffness of asphalt pavement is not used in tire-pavement coupled modeling, meaning that the impact of pavement stiffness on tire-pavement friction has not been examined yet (Hao Wang 2014 [23], Reginald B. Kogbara [30], Shahriar Najafi [31]).

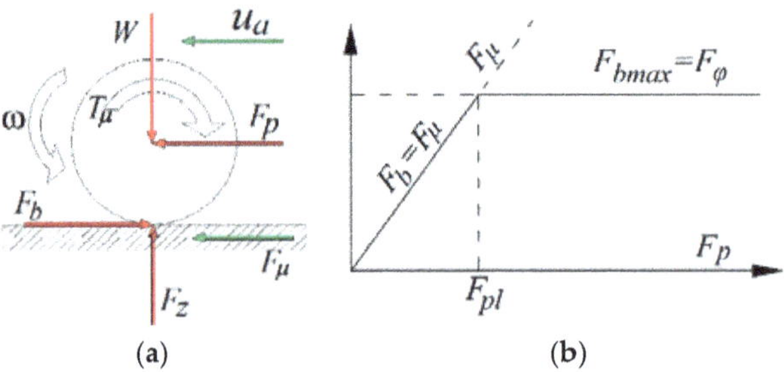

Figure 1. Principles of braking force. (**a**) Principles of braking force; (**b**) Relationship of F_μ, F_b and F_ϕ during braking.

Based on the above background research, the primary objective of this paper is to evaluate the effect of different asphalt pavements of variable resilience on braking performance on pavement. This goal was initially achieved using a 3-D finite tire-asphalt pavement interaction model. By adjusting parameters such as tire pressure and the elastic modulus, behaviors of the dynamic tire-pavement contraction were studied by discussing the varying features of tire-pavement contact such as tread deformation in the tangential interface, actual contact area between the tire and the pavement surface course, and the braking force supplied by the pavement to the tires.

2. Establishment of Tire-Pavement Frictional Contact Simulation Model

To begin, the traffic load and vehicle braking were simulated by establishing 3D tire models. Next, a road model was built by adopting the pavement structure of "asphalt surface layers + cement stabilized macadam base + lime-ash cushion + subgrade." Meanwhile, frictional contact laws were defined in ABAQUS (a commercial program of Finite Element Analysis, produced by Dassault Systèmes® Johnston, RI, USA, founded in 1978) , and subsequently, the contrastive analysis of kinetic friction between the tire and the flexible pavement surface was conducted on specimens created with various combinations of paving material parameters.

The simulation model of the dynamic friction due to tire-pavement contact was created in a three-step process shown in Figure 2. Specifically, a 3D model was built, then the steady state of the kinetic contact was analyzed in ABAQUS/Standard solver based upon static contact.

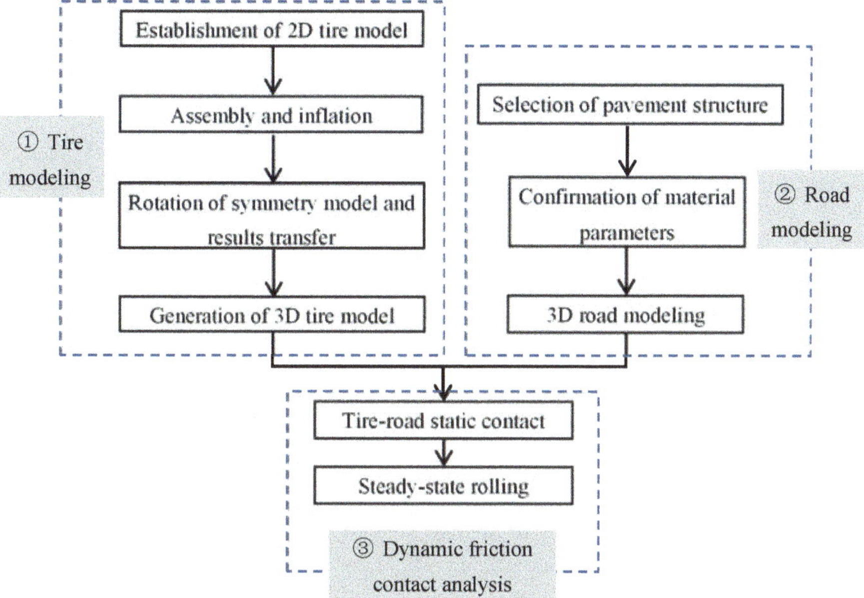

Figure 2. 3D Tire-road contact modeling steps in ABAQUS.

2.1. Establishment of 3D Model for Grooving Radial Tires

Referring to 'Steady-state rolling analysis of a tire' of ABAQUS 6.10 Example Problem Manual [32], three steps were used to create the 3D models of grooving radial tires: (1) simplification before modeling; (2) importing a 2D tire cross-sectional model into ABAQUS; and (3) 3D model generation of pneumatic tires:

(1) Simplification before Modeling. In order to improve the computational efficiency, the total cross section of a 2D model was drawn using CAD2014 software (developed by Autodesk corporation in San Rafael, CA, USA, founded in 1982), which was simplified according to the following:

1. Reduce the poor shape of the cross-sectional element;
2. Simplify the contact constraint between the bead chafer and the wheel hub (see Figure 3 for details. Next, simplify the wheel hub as a rigid constraint element, which shares a node with the tire bead);
3. Use the rebar elements to simulate the tire's steel cord and inner liners.

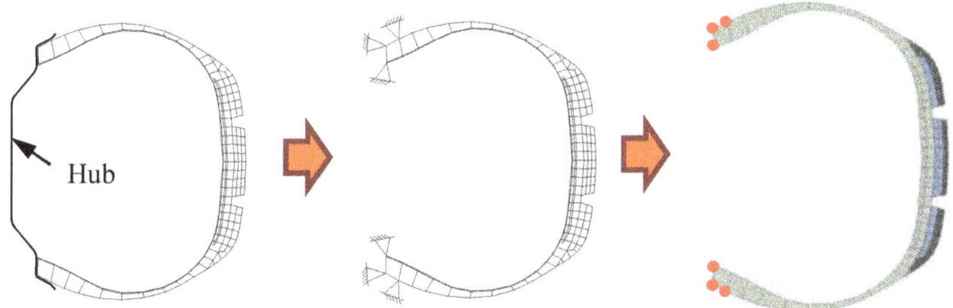

Figure 3. Simplification of contact conditions between tire bead and wheel hub.

(2) Importing a 2D Tire Cross-sectional Model into ABAQUS. To start, the simplifed 2D tire cross-sectional model was imported into ABAQUS. Subsequently, the steel cord of the tire and inner liners were defined as the primary tire materials.

1 Rubber materials

This study aims to establish a 3D model for a 175SR14 groove tire. The Neo-Hooken model was selected to describe the constitutive relationship of the super elasticity (Yanfeng Zhu 2006 [33], Yongguan Wang 2007 [34]). A Prony series is adopted to define the viscoelasticity of the rubber in ABAQUS. Both the above parameters were calibrated to acquire deflection values close to experimental measurements. Please see Table 1 for the relative tire rubber parameters (the simulated and tested datum).

Table 1. Tire rubber material parameters.

Material Parameters	Neo-Hooken Model Parameters in ABAQUS			Tensile Modulus/MPa		Tensile Strength/MPa	Elongation Ratio/%	Permanent Deformation/%
	C_{10} (kPa)	C_{01} (kPa)	D_1 (kPa)	*M100*	*M300*	*Tb*	*Eb*	*Ps*
Tread	835	0	0	2.5	9.9	19.1	544	15
Sidewall	1000	0	0	2.3	9.3	15.1	450	10

Note: In this table, M100 and M300, respectively, represent the tension the specimens bear under a 100% and 300% tensile ratio per unit area.

2 Steel cord-rubber composite materials

Steel cord-rubber composite materials are usually composed of cords in different orientations. The nonlinearity of rubber results in anisotropy and nonlinearity of the steel cord-rubber composite material in tires. Therefore, a rebar layer was applied to simulate the composite materials layer. First, the steel cord and inner liner were defined in ABAQUS and secondly, the membrane element that represents the steel cord in the defined steel cord layer was embedded. Meanwhile, the membrane element representing the inner liner in the inner rubber layer was also embedded to accomplish the simulation of the steel cord and inner liner. The parameters of the steel cord (Belt-1, Belt-2) and inner liner (Carcass) were defined successively (see Table 2).

Table 2. Description of material properties used in the model development.

Tire Section	Young's Modulus (GPa)	Poisson's Ratio (−)	Density (g/cm^3)	Area per Bar (mm^2)	Spacing (mm)	Orientation Angle (°)
Tread Sidewall	From Lab test	0.45	1.12 1.15	-	-	-
Belt-1 Belt-2	172.2	0.3	5900	0.212	1.16	70 110
Carcass	9.9	0.3	1500	0.421	1.00	0

(3) 3D model generation of pneumatic tires.

Definitions the section unit type, mesh partition, and 2D tire inflation (tire pressure of 250 kPa) were accomplished in this phase. Furthermore, the 2D tire section was used to generate a 3D tire model. Therefore, The final step in tire modeling was to transform the 2D section property into a 3D tire model.

2.2. Pavement Structure Modeling

2.2.1. Pavement Structure Setting

The road asphalt pavement structure was adopted as 4 cm + 6 cm + 8 cm, as illustrated in Table 3, which is commonly used in China's major highways. The thickness of the base and the bed course was 18 and 31 cm respectively. A displacement restriction is imposed on both longitudinal ends in the x direction and transverse ends in the y direction of the pavement; the consolidation constraint was applied to the bottom surface separately. Moreover, the entirety of the 3D pavement model was divided according to the principle that meshing density should be cut from the region of tire-pavement contact to the boundaries in every direction.

2.2.2. Design of Road Pavement Materials

In attempt to ameliorate the enhance of asphalt pavement, several authors found that adding crumb rubber into the asphalt mixture by dry process can improve the overall elastic deformation capacity, thus increasing the tire grip (Xudong Wang 2008 [25]) and strengthening braking performance of the pavement (Chengtao Wang 2002 [18], Jide Zhuang 1986 [19], Liangxi Wang 2008 [20]). Therefore, based on the research findings of Miao Yu and Guoxiong Wu 2014 [27–29], the asphalt specimens with 0%, 4% and 5.5% crumb rubber content were fabricated (below, the corresponding pavements are functionally named as low-E, med-E, and high-E in sequence). The crumb rubber content herein represents the percentage of crumb rubber volume accounts for total aggregate volume of asphalt specimens. Furthermore, the volumetric parameters were measured for the purpose of comparison between the effects of different pavement properties on tire brake performance. Table 3 outlines the pavement structure and relevant material parameters. Besides, data of elasticity modulus in this table represents uniaxial compressive modulus of resilience, originating from uniaxial compression test at 20 °C.

Table 3. Parameters of pavement structure modeling.

Pavement Structure	Layer Thickness (cm)	Elasticity Modulus (MPa)	Poisson's Ratio	Density (g/cm^3)
Surface course	4	1480 (without crumb rubber, low-E)	0.35	2.474
		1265 (with 4% crumb rubber, med-E)	0.40	2.420
		886 (with 5.5% crumb rubber, high-E)	0.40	2.387
Leveling course	6	1100	0.35	2.432
Asphalt base course	8	1000	0.35	2.471
Cement-stabilized Macadam Base Course	18	1500	0.25	2.056
Lime-fly ash Soil Cushion	31	750	0.25	1.924
Subgrade	500	35	0.35	1.918

Note: In this paper, low-E, med-E and high-E respectively represent the elasticity of pavement with 0%, 4% and 5.5% crumb rubber.

2.3. Tire-Flexible Pavement Frictional Contact Simulation

2.3.1. Definition of the Tire-Pavement Frictional Contact in ABAQUS

The following parameters facilitating frictional contact and the subsequent tire-pavement interaction analysis (Y Liu et al. 2015 [35]) were defined.

① Slip ratio σ_x (Equation (1)) is a representation of the slip status during the tire rolling phase,

$$\sigma_x = \frac{u_a - r_{eff} \cdot \omega_w}{u_a} \times 100\% \tag{1}$$

where

ω_w—the rotational angular velocity of the wheel hub; rad/s
r_{eff}—the effective radius of the wheel; m
u_a—the actual longitudinal velocity of the tire; m/s

② The static friction coefficient μ (Equation (2)) is defined as the friction between the tire and pavement in the initial contact phase (the tire is about to slide), in which the tire velocity is zero

$$\mu = F/N \tag{2}$$

where

F—the tangential force between tire and pavement contact
N—the normal force, perpendicular to the contact between tire and pavement

③ The braking force coefficient ϕ_b (Equation (3)) describes the braking ability of asphalt pavements. In addition, ϕ_b is the Skid Number (*SN*), representing tires-pavement contact when braking; a larger ϕ_b will account for a higher braking efficiency of the tires.

$$\phi_b = 100 \times F_x/F_z \tag{3}$$

where

F_x—the horizontal braking force of the pavement against the tire at the contact region
F_z—the upper load borne by the tires

④ In order to facilitate comparative analysis, the actual longitudinal velocity of the center of the tire was set at 80 km/h, and the static friction coefficient μ was 0.7 (Jide Zhuang 1986 [19], Judge, A. W. 2008 [36]).

⑤ Tire-asphalt pavement contact definition

Static contact between the tire and asphalt pavement could be modeled via two methods; displacement control and load control. In comparison to the latter, the former usually decreases calculation time and in comparison, produces more reliable results. Thus, the displacement control of the tire and pavement contact was established by applying 2 cm displacement on the rigid reference node of the tire. Secondly, the load control of the tire and pavement contact was set up, that is, causing us to remove the pre-assigned displacement and apply vertical displacement on the tire reference node with respect to the pavement. The tire-pavement static contact was determined via the two steps above. Furthermore, the tire-pavement contact was set at "hard contact" of "finite sliding". Meanwhile, the method of combining isotropic Coulomb friction combined with the penalty function algorithm was adopted to realize the tire-pavement frictional contact simulation model in Figure 4a.

2.3.2. Validation of the 3D Model of Pneumatic Tire-Asphalt Pavement Contact

The complicated nonlinear characteristic of the kinetic contact between tire and pavement will have a major effect not only on the convergence of both the implicit and explicit analyses, but also on the accuracy and computation time of the above analyses. Therefore, before carrying out the dynamic analysis of the tire-pavement contact, it is necessary to successively validate the simulation of the pneumatic tire and its static contact with the pavement.

① Tire inflation

Figure 4b shows the Mises equivalent stress nephogram of the 2D tire cross section. It can be seen that the steel cord bears the primary load of the inner tire after inflation. While the whole tire is under tension, the maximum stress lies in the middle of the steel cord found in the tread. To facilitate investigation of the stress-strain characteristics found in the tire rubber materials, the inner liner of the tire was removed.

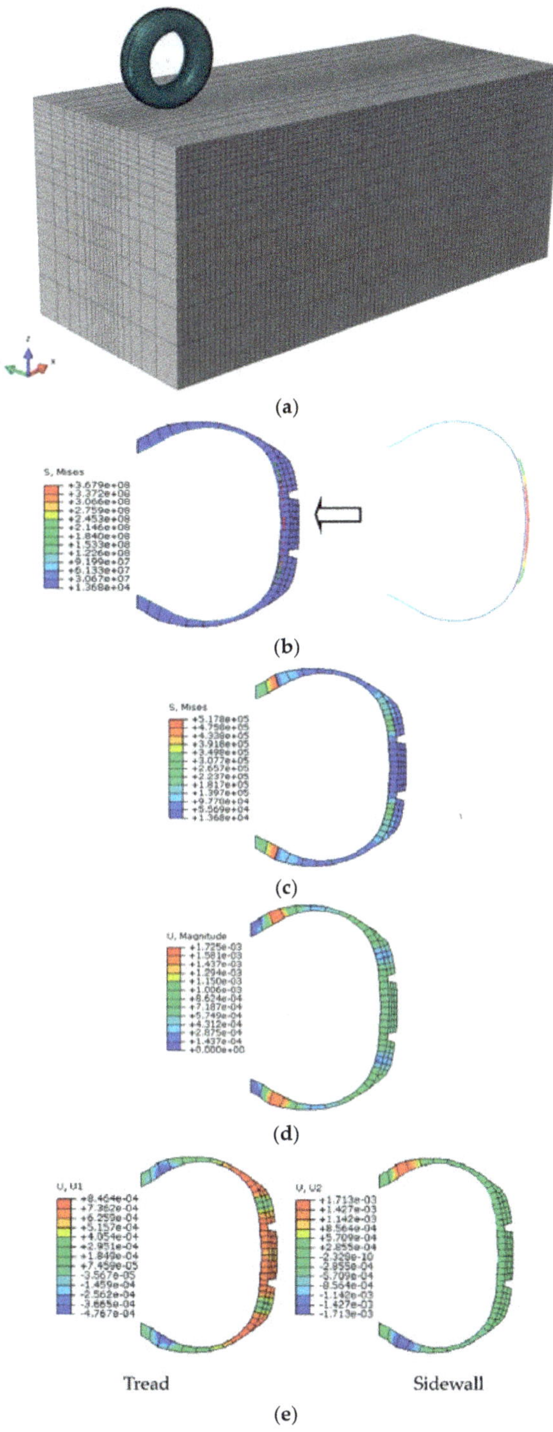

Figure 4. *Cont.*

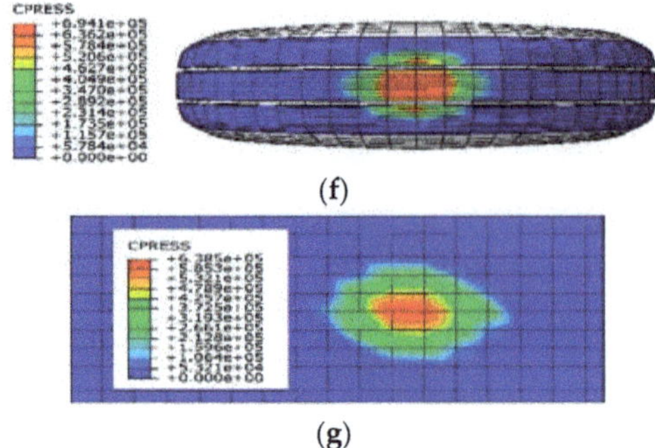

Figure 4. Stress and strain nephogram of the 3D model of pneumatic tire-asphalt pavement contact. (a) 3D Tire-pavement contact model; (b) Mises equivalent stress nephogram of the inflated tire; (c) Mises equivalent stress nephogram of the inflated tire rubber materials; (d) Deformation nephogram of the inflated tire rubber materials; (e) Deformation nephogram of the inflated tire; (f) Pressure distribution of tire tread on the contact region; (g) Pressure distribution of Pavement surface on the contact region.

It can be observed in Figure 4c,d that both the maximum values of the Mises stress and the total strain appeared near the hub constraints when assembled with a wheel hub and bearing inflation pressure. In contrast, when the steel cord bore the maximum tensile stress, the tread rubber presented relatively smaller stress without the self-weight load. As depicted in Figure 4d, it was found that the consolidation constraints contributed to zero displacement around the wheel hub, whereas the aligned regions showed higher stress. In addition, the tread showed lower Mises stress, accounting for a larger deformation of the rubber materials near the wheel hub of the sidewall. Figure 4e depicts the deformation nephogram of the inflated tire along its tread and sidewall directions. It is noted that the deformation in 1 orientation (U1) was primarily created in the connecting region of the sidewall and shoulder. On the other hand, the deformation in 2 directions (U2) was always concentrated on the sidewall near the hub. It can be concluded that, from the above analysis of Figure 4c–e, the principle deformation region lied in the sidewall of tires, in accordance with practical conditions.

② Static tire contact conditions

Under the tire-pavement contact conditions shown in Figure 4f,g, tire grooves accounted for the concentration of the tire-pavement ground pressure, which lied mainly within the longitudinal pattern rather than out of the grooves. Accordingly, the contact pressure on the pavement surface was also gathered in the middle of the contact region.

A photograph of the tire's static loaded test performed by UTTM Stiffness Tester (Testing Service GmbH, Aachen, North Rhine-Westphalia, Germany), is presented in Figure 5a. Comparison of the test data of load-tire vertical deformation, size of contact region as well as the actual contact area with the relative numerically predicted results are shown in Figure 5b and Table 4. It can be seen that the predicted results seem very close to the results of the lab experiments.

Table 4. Contact area comparison of the numerically simulated results and experiment measurements.

Tire Pressure (kPa)	Applied Load (N)	Experimentally Measured Data			Numerically Simulated Results		
		Length (L) (mm)	Width (W) (mm)	Area (cm^2)	Length (L) (mm)	Width (W) (mm)	Area (cm^2)
200	2200	125.0	102.5	104.0	119.3	99.7	98.5
220	3200	147.6	106.1	135.0	142.1	103.6	128.6

Note: L and W is the relative maximum value of the length and width in the contact zone.

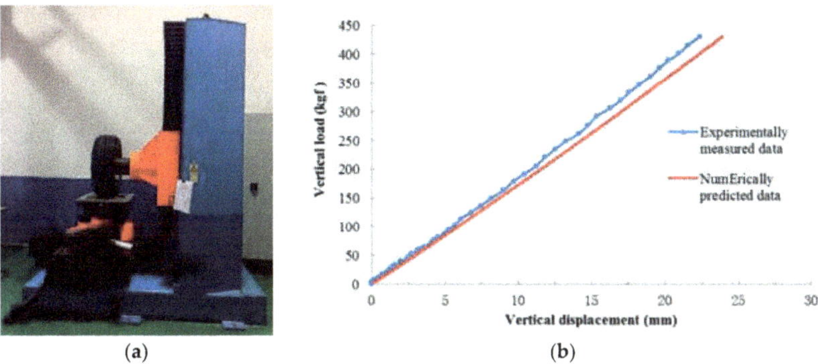

(a) (b)

Figure 5. Verification of simulation model against experimental data on tire static test. (**a**) Tire static loaded test; (**b**) Load-tire vertical deformation comparison of numerically simulated results and experiment measurements.

3. Simulation Analysis of Tire-Asphalt Pavement Frictional Contact

3.1. Simulation Conditions Design of the Tire-Asphalt Pavement Frictional Contact

It is well known that the braking force (coefficient) changes with variations in slip ratio. At the early stage of the tire braking procedure, the adhesive force, F_ϕ, is proportional to the slip ratio. However, because the pavement braking force, F_b, is restrained by the adhesive force, F_ϕ, F_b will stop rising upon reaching the maximum F_ϕ value, and the slip ratio will continue to increase. Therefore, it is relevant to note that if pavement could provide tires with a greater adhesive force, F_ϕ, tires will receive a greater braking force, F_b, from the pavement, facilitating the vehicle braking efficiency. Thus, this study analyzes elasticity's contribute to tire braking efficiency on pavement using various slip ratios. The three slip ratios listed in Table 5 were used to analyze the frictional contact of tire-pavement.

Table 5. Conditions of brake efficiency in different slip ratios.

Pavement Types	Slip Ratio (%)		
	7	10	14
low-E	low-SR	med-SR	high-SR
med-E	low-SR	med-SR	high-SR
high-E	low-SR	med-SR	high-SR

Note: In this table, E represents pavement Elasticity with or without crumb rubber addition, namely, in accordance with the pavement type in Table 3. Besides, SR is the abbreviation of Slip Ratio.

3.2. Tread Deformation

Tire braking behavior has a distinct impact on tangential tire deformation. Therefore, the tangential deformation was examined to investigate tire-pavement frictional behaviors.

Figure 6 depicts tangential deformation of the tire tread along vehicle's direction of motion. It can be inferred from the nephogram that:

When the slip ratio is only 7%, the tread's tangential deformation is insignificant. Nevertheless, as the pavement's elasticity is improved, the deformed area has become longer and wider, extending to the marginal region of sidewall in the direction of the tire's motion. These phenomena account for the enlargement of the tire-pavement contact area; the effect tire adhesion on highly elastic pavement is a superior braking agent in comparison to that of common pavement with no crumbed rubber additives.

Secondly, under each of the nine conditions, the direction of tread deformation is opposite to the tire's motion path. Under conditions of high-E, the tread deformation area is significantly larger in comparison to conditions of low-E and med-E. This pattern is also found in the slip direction of the tire head's mesh. The above findings explain that while stiffness decreased and elasticity improvements were made in the paving materials, the deformity and braking force applied to the tires increases. Therefore, the tire-pavement contact area enlarges with elasticity. This trend indicates that appropriate decreases in pavement stiffness could facilitate enough tire-pavement contact to enhance tire adhesion, improving brake efficiency.

By examining the tread deformation along the direction of tire movement under 10% and 14% slip ratio, we see that when the slip ratio is 14%, the tread deformation slightly decreases, indicating that tire adhesion does not always improve with the increased slip ratio. This change agree with the law that as the slip ratio is between 10% and 15%, the tire braking force coefficient will reach the maximum value [18].

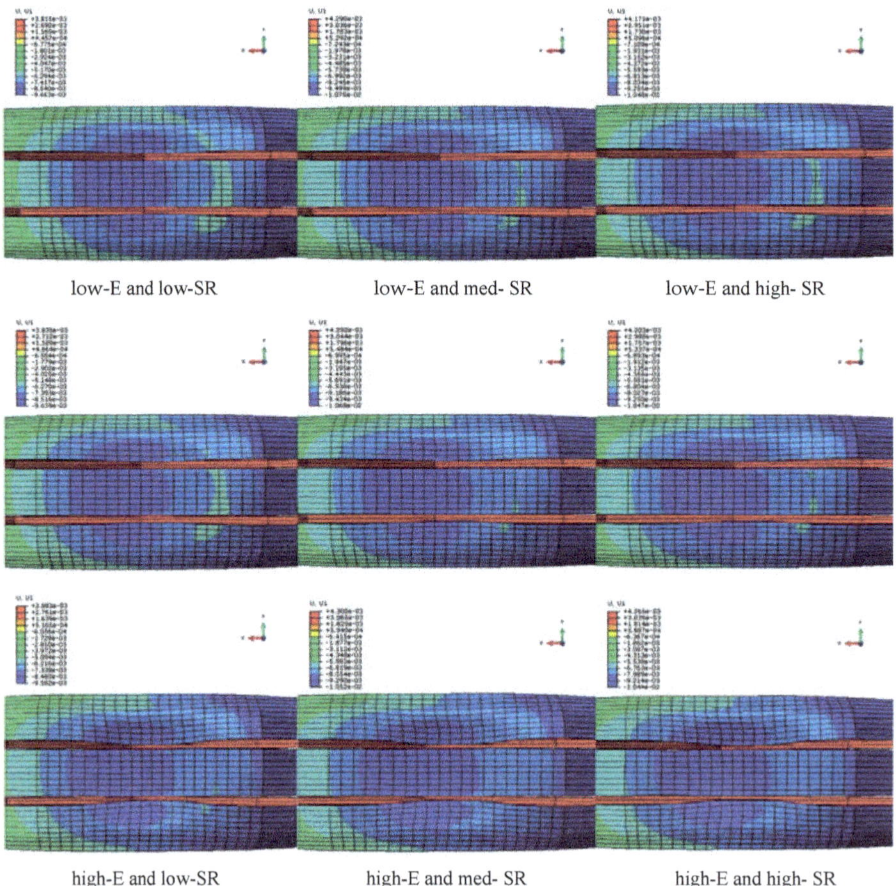

Figure 6. Tangential deformation nephogram of tire tread under different slip ratios.

3.3. Pavement Braking Effect

Tire slip ratio is directly correlated to the tangential deformation found in the tire-pavement contact zone. Therefore, the tangential displacement of the pavement surface, along the direction of tire movement (Figure 7), is primarily used to investigate pavement deformation under various slip ratios. Figure 8 illustrates the pavement's tangential deformation response under three types of tire slip ratios. The comparison shows that pavement tangential deformation varies most significantly with the alternation of slip ratios. For example, consider a 7% slip ratio; the tangential deformation is generally no more than 0.1mm, whereas the deformation could reach 1mm as the slip ratio reaches 10% or higher. The figure also depicts the impact of pavement modulus on tangential interaction, which should not be ignored. It can be observed that in highly elastic pavement (high-E), the tangential resistance to tire rolling is more significant than that of other pavements (low-E and med-E) under the same slip ratio. Therefore, it can be concluded that changes in tire-pavement tangential behavior caused by slight alterations in pavement resilience can be neglected due to their negligibility in comparison to slip ratio.

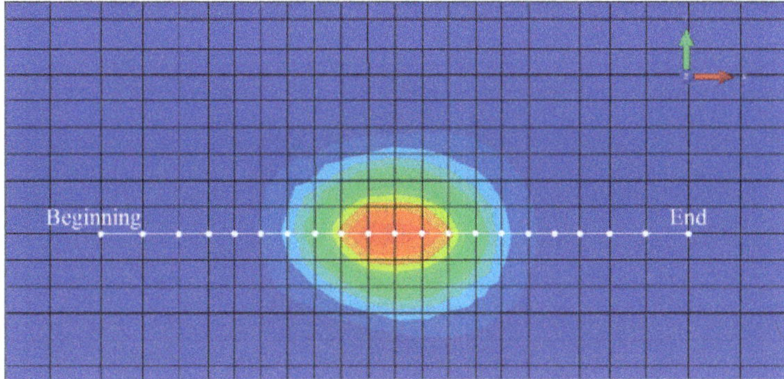

Figure 7. Vertical route of the tire movement.

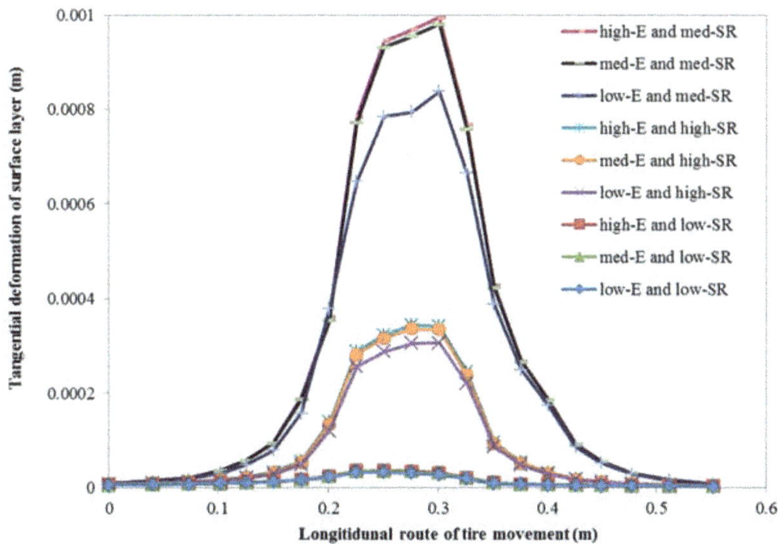

Figure 8. Tangential deformation of pavement surface layer under different slip ratios of tires.

Figures 9 and 10 respectively present the correlation of contact area and pavement braking force under different slip ratios. As the slip ratio increases from 7 to 10%, the contact area enlarged, intensifying the tire-pavement friction, causing an upward trend in the braking force applied by the pavement. When the slip ratio raised from 10 to 14%, the contact area began to reduce, accounting for degradation of the tire adhesion effect. According to Figure 6, the braking force also declined along with increasing slip ratio. The two figures illustrate that tire-pavement contact area is related to pavement braking force. In other words, tire adhesion has a substantial impact on the braking force applied by asphalt pavements.

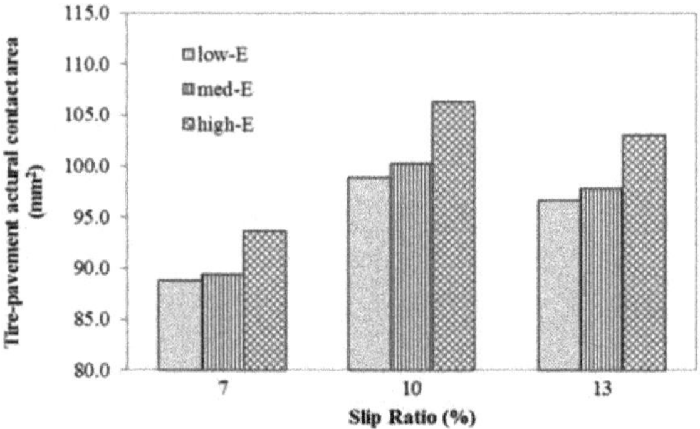

Figure 9. Tire-pavement contact area under different slip ratios.

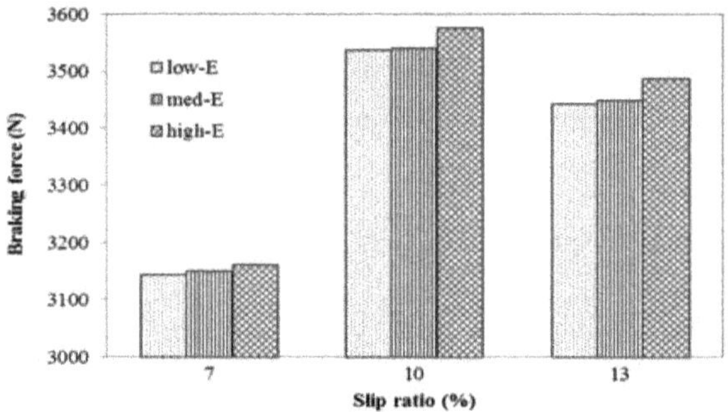

Figure 10. Braking force of pavement under different slip ratios.

Figure 11 also shows that the braking force coefficient of pavement does not always enlarge with increases in slip ratio; this coefficient will decrease after reaching its peak value. Using the tire modeled in this research, it was found that pavement type has the most significant braking impact at a slip ratio value near 10%. Furthermore, with reduction in contact area, the effect of pavement type on braking diminished with slip ratios greater than 10%. However, it can be observed that for the most elastic pavement (high-E), the corresponding braking force coefficient is greater than the other two types of less elastic pavement. As the slip ratio was further improved, the braking effect of the high elastic pavement began to degrade, nonetheless, this pavement's braking effect is superior to that of the other pavements. Therefore, it can be concluded that the enhancement of pavement elasticity will give rise to the braking efficiency of the tires under different slip ratios.

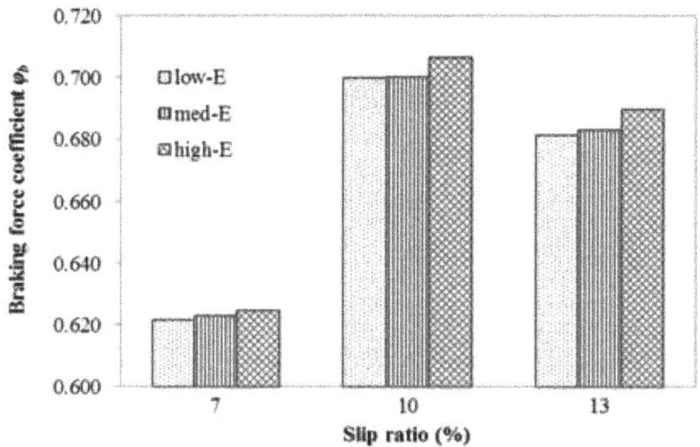

Figure 11. Pavement braking force coefficient under different slip ratios.

4. Conclusions and Discussion

To increase brake efficiency and enhance driving safety, the 3D dynamic frictional contact model was created to conduct a reliable analysis of tire-pavement friction. Through this method, the following conclusions were determined.

First: increasing pavement elasticity could enlarge the tire-pavement actual contact area, providing tires with stronger grip.

Second: slip ratio has the most significant impact on frictional behaviors during tire-pavement contact. However, under the same slip ratio, high elasticity significantly interferes with tire rolling in comparison to pavements of ordinary elasticity.

Third: the braking force coefficient can be slightly raised by the distinct improvement of pavement elasticity. However, braking distance is sensitive to tiny variations in this coefficient, allowing anti-skidding improvements to be made in asphalt by remarkably increasing the pavement's elasticity.

Pavement skid resistance, coupled with vehicle, tire and pavement surface layer, is a problem that includes factors such as aggregate hardness, pavement surface texture and stiffening of paving materials. Moreover, analyzed from the aspect of the tire, this issue also investigated the effect of tire inflation pressure; tread pattern and the physical and mechanical properties of tread rubber materials. In addition, investigated from the angle of tire-pavement coupling, it is a systemic problem that incorporates tire slip ratio, tire-pavement contact behavior, and even the appropriate friction law. Conventional methods focus upon macro and micro texture effects on the frictional properties of the tire, to decrease the skid resistance of asphalt pavement. Investigating the changes in skid resistance caused by altering the recoverable resilient deformation property of pavement is a new approach to increasing driver safety. This method can be used to facilitate tire grip and enhance braking performance. This paper investigates pavement skid resistance from the aspect of paving materials' moduli; tire-pavement friction coupled with pavement texture, heat transfer by coupled frictional behaviors, stiffness modulus, and water film effects will be discussed in future research.

Acknowledgments: This work is financially supported by the 2014 Chongqing Higher Education Institution's Excellent Achievements Program (Major Project) (Theme: Warm mix, semi-flexible compound pavement Technology (KJZH14104)), 2015 Chongqing Higher Education Institution's scientific and technological research program (Theme: Research on the skid Resistance Performance of CRM asphalt pavement based on the tire-pavement kinetic friction test system (KJ1500508)), and the 2016 National Natural Science Foundation of China (Tire braking-oriented anti-skid mechanism and estimating model of friction coefficient of asphalt pavement, No. 51608085).

Author Contributions: Miao Yu was in charge of the whole research program; Guoxiong Wu built the trial protocol; Lingyun Kong and Yu Tang performed the experiments and data analysis. All authors discussed and contributed to the preparation and revision of the manuscript.

Conflicts of Interest: The authors declare no conflict of interest.

References

1. Henry, J.J. *Evaluation of Pavement Friction Characteristics (A Synthesis of Highway Practice)*; NCHRP Synthesis No. 291; National Cooperative Highway Research Program; Transportation Research Board: Washington, DC, USA, 2000.
2. Horne, W.B.; Dreher, R.C. *Phenomena of Pneumatic Tire Hydroplaning*; NASA TN D-2056; National Aeronautics and Space Administration: Washington, DC, USA, 1963.
3. Martin, C.S. *Hydrodynamics of Tire Hydroplaning*; Final Rep. Project B-608; Georgia Institute of Technology: Atlanta, GA, USA, 1966.
4. Eshel, A.A. *Study of Tires on a Wet Runway*; Rep. No. RR67-24; Ampex Corp.: Redwood City, CA, USA, 1967.
5. Tsakonas, S.; Henry, C.J.; Jacobs, W.R. *Hydrodynamics of Aircraft Tire Hydroplaning*; NASA CR-1125; National Aeronautics and Space Administration: Washington, DC, USA, 1968.
6. Horne, W.B.; Yager, T.J.; Ivey, D.L. *Recent Studies to Investigate Effects of Tire Footprint Aspect Ratio on Dynamic Hydroplaning Speed*; Pottinger, M.G., Yager, T.J., Eds.; The Tire Pavement Interface, ASTM STP 929; ASTM: Philadelphia, PA, USA, 1986; pp. 26–46.
7. Ong, G.P.; Fwa, T.F. Wet-pavement hydroplaning risk and skid resistance: Modeling. *J. Transp. Eng.* **2007**, *133*, 590–598. [CrossRef]
8. Fwa, T.F.; Ong, G.P. Wet-pavement hydroplaning risk and skid resistance: Analysis. *J. Transp. Eng.* **2008**, *134*, 182–190. [CrossRef]
9. Ong, G.P.; Fwa, T.F. Modeling skid resistance of commercial trucks on highways. *J. Transp. Eng.* **2010**, *136*, 510–517. [CrossRef]
10. Fwa, T.F.; Pasindu, H.R.; Ong, G.P. Critical rut depth for pavement maintenance based on vehicle skidding and hydroplaning consideration. *J. Transp. Eng.* **2012**, *138*, 423–429. [CrossRef]
11. Grosch, K.A. The relation between the friction and viscoelastic properties of rubber. *Proc. R. Soc. Lond. Ser. A* **1962**, *274*, 21–39. [CrossRef]
12. Srirangam, S.K.; Anupam, K.; Scarpas, A.; Kösters, A. Influence of tire temperature increase on friction measurements-I: Laboratory tests and finite element modeling aspects. In Proceedings of the Transportation Research Board Annual Meeting, Washington, DC, USA, 13–17 January 2013.
13. Anupam, K.; Srirangam, S.K.; Scarpas, A.; Kasbergen, C. Influence of temperature on the tire-pavement friction-II: Analyses. In Proceedings of the Transportation Research Board Annual Meeting, Washington, DC, USA, 13–17 January 2013.
14. Forster, S.W. *Pavement Microtexture and Its Relation to Skid Resistance*; 151B164; Transportation Research Record; Transportation Research Board: Washington, DC, USA, 1990.
15. Sengoz, B.; Onsori, A.; Topal, A. Effect of aggregate shape on the surface properties of flexible pavement. *KSCE J. Civ. Eng.* **2014**, *18*, 1364–1371. [CrossRef]
16. Kane, M.; Rado, Z.; Timmons, A. Exploring the texture–friction relationship: From texture empirical decomposition to pavement friction. *Int. J. Pavement Eng.* **2015**, *16*, 919–928. [CrossRef]
17. Srirangam, S.K.; Anupam, K.; Scarpas, A.; Kasbergen, C. Development of a thermomechanical tyre–pavement interaction model. *Int. J. Pavement Eng.* **2015**, *16*, 721–729. [CrossRef]
18. Wang, C.T.; Yao, Z.Q.; Chen, M. *Vehicle Tribology*; Shanghai Jiaotong University Press: Shanghai, China, 2002; pp. 400–423.
19. Zhuang, J. *Vehicle-Terramechanics*; China Machine Press: Beijing, China, 1986.
20. Wang, L.X.; Wang, H.Y. *Vehicle Dynamics*; National Defence Industry Press: Beijing, China, 2008.
21. You, Z.; Adhikari, S.; Kutay, M.E. Dynamic modulus simulation of the asphalt concrete using the X-ray computed tomography images. *Mater. Struct.* **2009**, *42*, 617–630. [CrossRef]
22. Goh, S.W.; You, Z.; Williams, R.C.; Li, X. Preliminary dynamic modulus criteria of HMA for field rutting of asphalt pavements: Michigan's experience. *J. Transp. Eng.* **2011**, *137*, 37–45. [CrossRef]
23. Wang, H.; Al-Qadi, I.L.; Stanciulescu, I. Effect of surface friction on tire-pavement contact stress during vehicle maneuvering. *J. Eng. Mech.* **2014**, *140*, 04014001. [CrossRef]
24. Zhou, C. Study on Application Technology of Rubber Particle Asphalt Mixture in Ice and Snow Region. Ph.D. Thesis, Harbin University of Industry, Harbin, China, 2006.
25. Wang, X. *The Apply Technology of the Crumb Rubber in the Asphalt and Mixture*; China Communications Press: Beijing, China, 2008; pp. 167–197.

26. Yao, L. Research on Key Technology of Granulated Crumb Rubber Anti-ice Asphalt Pavement. Ph.D. Thesis, Chang'an Universtity, Xi'an, China, 2012.
27. Yu, M.; Wu, G. Research on the mix design of dry process crumb rubber modified asphalt mixture. *J. Build. Mater.* **2014**, *17*, 100–105.
28. Yu, M.; Wu, G.; Zhou, J.; Easa, S. Proposed compaction procedure for dry process crumb rubber modified asphalt mixtures using air void content and expansion ratio. *J. Test. Eval.* **2014**, *42*, 328–338. [CrossRef]
29. Yu, M. Study on the Dry Process Crumb Rubber Modified Anti-Skid Layer Based on the Tire-Road Coupling. Ph.D. Thesis, Chongqing Jiaotong Universtity, Chongqing, China, 2014.
30. Kogbara, R.B.; Masad, E.A.; Kassem, E.; Scarpas, A.T.; Anupam, K. A state-of-the-art review of parameters influencing measurement and modeling of skid resistance of asphalt pavements. *Constr. Build. Mater.* **2017**, *133*, 330–339. [CrossRef]
31. Najafi, S.; Flintsch, G.W.; Medina, A. Linking roadway crashes and tire-pavement friction: A case study. *Int. J. Pavement Eng.* **2017**, *18*, 119–127. [CrossRef]
32. Koishi, M.; Kabe, K.; Shiratori, M. Tire cornering simulation using explicit finite element analysis code. *J. Appl. Polym. Sci.* **2000**, *78*, 1566–1572. [CrossRef]
33. Zhu, Y.; Liu, F.; Huang, X.; Li, L. Constitutive model of rubber materials. *Rubber Ind.* **2006**, *53*, 119–125.
34. Wang, Y.; Li, X.; Huang, Y. Selection for constitutive model in rubber calculation. The chemical industry and engineering society of China. In Proceedings of the 4th Seminar China Rubber Products, Zhuzhou, Hunan, China, 17 November 2007; pp. 443–449.
35. Liu, Y.; You, Z.P.; Yao, H. An idealized discrete element model for pavement-wheel interaction. *J. Mar. Sci. Technol.* **2015**, *23*, 339–343.
36. Judge, A.W. Automobile engines. In *Theory, Design, Construction, Operation, Testing and Maintenance*; Hervey Press: Hervey, Australia, 2008.

Article

Technologies and Principles of Hot Recycling and Investigation of Preheated Reclaimed Asphalt Pavement Batching Process in an Asphalt Mixing Plant

Henrikas Sivilevičius, Justas Bražiūnas *and Olegas Prentkovskis

Department of Mobile Machinery and Railway Transport, Faculty of Transport Engineering, Vilnius Gediminas Technical University, Plytinės g. 27, LT-10105 Vilnius, Lithuania; henrikas.sivilevicius@vgtu.lt (H.S.); olegas.prentkovskis@vgtu.lt (O.P.)

* Correspondence: justas.braziunas@vgtu.lt; Tel.: +370-612-18487

Received: 18 September 2017; Accepted: 23 October 2017; Published: 25 October 2017

Abstract: More and more recycled asphalt mixtures with high reclaimed asphalt pavement (RAP) content are used in road pavement. Having determined and evaluated RAP composition (aged bitumen content and aggregate gradation) and properties, a suitable recycling agent and virgin materials are selected in the design process. The gradation of hot mix asphalt (HMA) mixture that is recycled in an asphalt mixing plant (AMP) shall correspond to its optimal gradation set out in its job-mix formula (JMF). When RAP is recycled in an AMP, inevitable systematic and random errors of performed technological operations and inhomogeneity of virgin materials and RAP have a significant influence. These factors influence the variation of components quantities of recycled hot mix asphalt (RHMA) mixture and deviations from JMF. In this study, the principles of asphalt pavement hot recycling are systematized, which allows analysis of the factors of components' interaction influencing the results of the recycling process. The paper also presents and analyses asphalt recycling technologies in AMP and their comparative analysis. During the season of asphalt mixture production in 2014, statistical parameters were calculated according to the data obtained from one of the companies, which collected and systematized RAP batch masses, when before batching it was pre-dried and pre-heated in an additional dryer. These parameters of batch mass and RAP content in RHMA position and variation were used when evaluating the accuracy and precision of the recycling process in AMP. The obtained data showed that when RHMA mixtures are produced in a modern batch-type AMP, RAP is batched accurately, but not precisely enough.

Keywords: asphalt pavement; recycling principles; asphalt mixing plant (AMP); reclaimed asphalt pavement (RAP); batching errors

---

1. Introduction

Production of asphalt can incorporate reclaimed materials from the deconstruction of road surfaces. This aids in reducing production costs and saves natural resources, bitumen, and aggregates. However, recycling can only be justified if the performance and longevity of the produced pavement is equal or better than that of traditional mixtures. Mixtures containing RAP have not always demonstrated such performance. Premature cracking due to the aged bitumen is one of the main reasons that agencies are reluctant to increase allowed RAP content in the mixtures [1,2]. Another reported problem is the variability of RAP material, which does not allow to securely assume that the produced mixture properties will reflect laboratory design [3]. Finally, problems with production technology have been reported, including excessive emissions and issues with consistency of mixture that has resulted in poor pavement performance [4,5].

These reported problems, however, can be accounted for and the desired performance for mixes containing RAP can be ensured by taking comprehensive measures in three broad areas:

- Materials. It must be ensured that the reclaimed materials correspond to the requirements, they are homogeneous, with a low moisture content, and without contamination.
- Mix design should be performed according to best practices, depending on recycled asphalt content and its properties to ensure the reduction of stiffness of the aged binder, as well as the desired performance properties for the entire life-cycle of the pavement [6].
- The production technology must allow for incorporating the desired RAP content by heating it to the necessary temperature and ensuring homogeneous blending with virgin materials without creating emissions [7].

Most laboratory studies have demonstrated evidence that high RAP pavements can achieve the desired pavement performance and longevity that are equal to conventional asphalt [8–10]. Adequate mix design procedures can even allow for creating mixtures containing 100% recycled asphalt [11,12]. It has also been confirmed that adequate full-scale RAP management, production, and mix design procedures can lead to excellent field results. The research by [13,14] provides evidence that adequate management and processing of RAP can ensure homogeneity of RAP equal or better than that of virgin aggregates. A comprehensive study by National Center for Asphalt Technology (NCAT) [15] evaluated the long-term performance of 18 sites across North America with pavements containing 30% RAP and concluded that the performance is similar to pavement constructed from virgin materials with no RAP. Thus, it becomes evident that both adequate design procedures and full-scale production technologies are currently available and allow for high RAP production [16]. So, why is it not being done routinely?

One of the reasons is that, although a lot of research is being done in laboratory, independent scientific research on production technologies and other full-scale operations and the means to improve them is scarce. This is likely because of the high research costs and many available (and constantly changing) technologies. At the same time, as reported by [17], a large portion of the failures of mixtures containing RAP have been caused by the "use of unprocessed RAP and hot-mix plants that were not designed to handle high RAP contents." Thus, a clear, full-scale research is necessary. The pavement performance is equally dependent on design procedures and consistent production and only adequate accounting for both of these factors can ensure satisfactory pavement performance [18].

This article aims to link different features of a full-scale production to allow for a theoretical evaluation of the entire production process chain. Only a comprehensive approach like this can permit analysing the "bottle necks", which preclude increasing the reclaimed asphalt content in the mixture or to find the defective link in the chain in the case of unsatisfactory performance of the pavement. Such an approach is also necessary when aiming at a cleaner production. The entire chain of operations must be analysed considering the long term consequences. This encompasses undertaking daily, weekly, and yearly analyses. This article will present the theoretical basis of the asphalt production chain and then focus on the asphalt production with regards to ensuring the desired mixture homogeneity.

The aim of the study is to present the analysis of mandatory actions of asphalt pavement hot recycling based on scientific principles and to investigate the parameters of technological operations, performed in batch AMP with additional RAP drying and heating.

2. Theoretical Formulation of the Problem

2.1. Principles of Asphalt Pavement Hot Recycling

When developing asphalt pavement recycling methods and technologies, the most important principles of this relatively new and promising method of recycling road pavement by reusing its asphalt materials shall be known and applied. We present these principles based on our personal theoretical knowledge and practical experience, as well the analysis of the findings of the investigations conducted by other researchers. The scheme of twelve principles is presented in Figure 1.

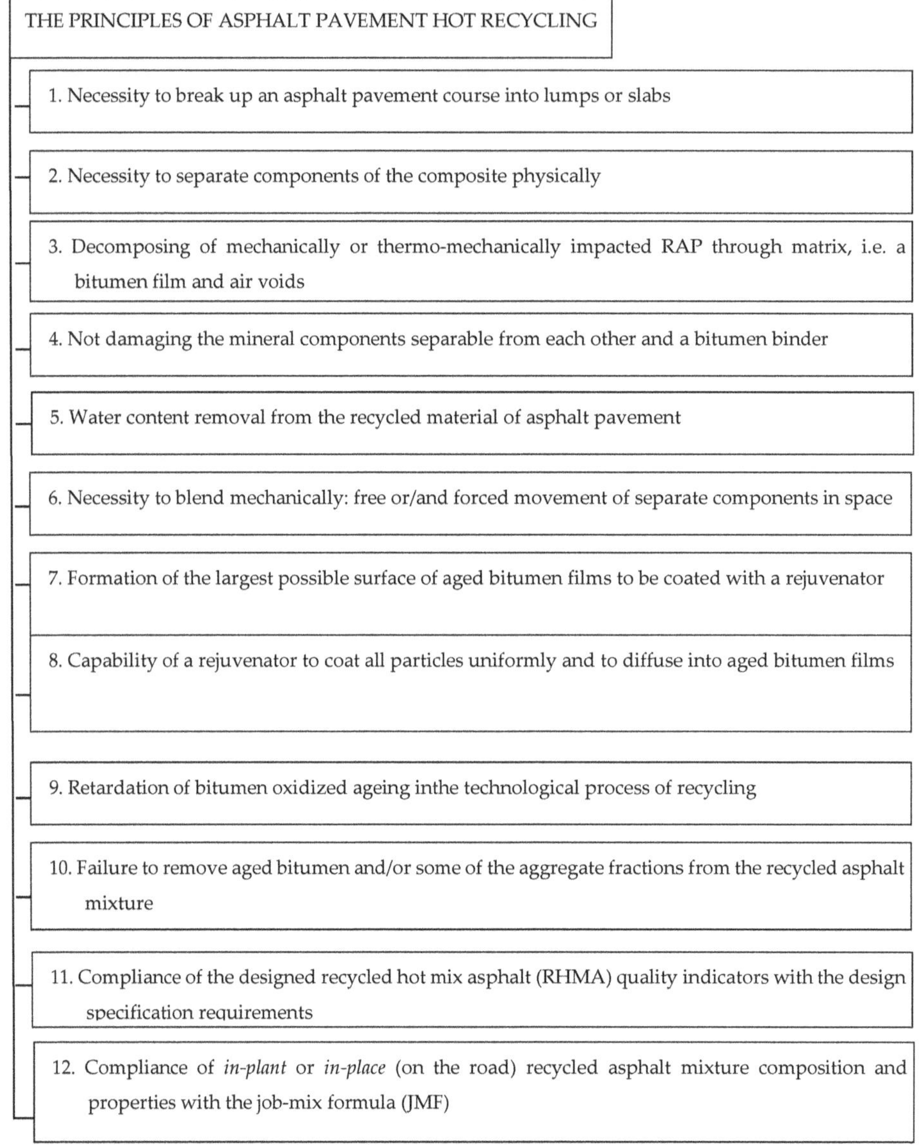

Figure 1. The system of scientific principles of asphalt pavement hot recycling.

The first principle (necessity to break up an asphalt pavement course into lumps, or slabs). The asphalt of the reclaimed pavement containing the matrix and inserts of the composite shall be mechanically, thermally, or hydrothermally crushed (broken-up) into separate components. This enables to improve its whole structure and properties by incorporating some materials, not only those in the surface course of the road pavement [19,20]. It is rather complicated and ineffective to incorporate rejuvenating additives into a compacted and dense pavement through air voids from the top. A ravelled aged asphalt course of the road pavement, which is not crushed into small particles, would not be uniformly recycledthroughout the whole thickness of the course.

The second principle (necessity to separate physically the components of the composite). The asphalt of the recycled pavement shall be downsized to its granular particles (RAP). Aggregate particles of different sizes (solid phase) are separated by breaking-up the films of the bitumen bonding them (liquid phase). As adhesion is stronger than cohesion, thermo-mechanically impacted RAP frequently decompose not only at the border of bitumen and particles, but in the course of volumetric bitumen and air voids. Particles are separated from each other in order that when they are changing their position in space, they allow rejuvenating agents (binder or/and mineral aggregates) to permeate through open gaps, which would distribute uniformly among the obtained particles of the recycled asphalt mixture [21].

The third principle (decomposing of mechanically or thermo-mechanically impacted RAP through matrix, i.e., a bitumen film and air voids). The RAP of aged pavement shall be broken up in order that the material that binds asphalt (the variation system of a bitumen binder and mineral fillers or volumetric bitumen films, which cement the composite material into a conglomerate) decomposed. During its recycling process, surfaces may merge with microcracks occurring in the aged road pavement and air voids. It would be desirable that the surfaces were formed in a matrix, not in inserts (coarse and fine aggregate), when asphalt decomposes (is milled, broken-up or crushed). In the contact zone of a milling machine cutter or a stone breaker (crusher) with reclaimed asphalt, new uncoated surfaces form when stone and fine aggregate particles split, which increases the area of all particles of the mixture and the necessary summative amount of the recycled bitumen. When asphalt pavement is cold-milled or its broken lumps or slabs are crushed in a stonebreaker, some coarse and fine aggregate particles inevitably split up (break up). When heated road pavement is milled with a mobile recycler, the probability of particle splitting decreases. If the degree of aged mineral aggregate fractioning is unknown and it is not taken into consideration when batching additional virgin mineral aggregates or an asphalt mixture, RHMA with a higher content of fine particles is obtained.

The fourth principle (not damaging the mineral components separable from each other and a bitumen binder). The methods and technologies of separating solidified and liquefied phases of asphalt pavement from each other shall ensure that aged bitumen and mineral particles are suitable for the use in a recycled asphalt mixture. Thermal weakening of the matrix (bitumen or asphalt binding material) by reducing the viscosity of aged bitumen before decomposing to the strength that is lower than that of inserts (particles of coarse and fine aggregate) decomposes reclaimed asphalt through matrix, not inserts, due to decreasing adhesion and cohesion. Splitting and crumbling of aggregate particles shall be avoided. When reclaimed asphalt is heated, the properties of aged bitumen should not worsen due to overheating. Usually, RAP that has been heated in the RAP channel goes to a hot RAP storage bin when it has reached its desired temperature (100–120 °C) at the end of circulation line. When it is stored in the bin, it will maintain a warm temperature. Chemical solvents that are used for bitumen recovery, but not for the improvement of its properties, are not suitable for separating the components of RAP from each other.

The fifth principle (water content removal from the reclaimed material of asphalt pavement). The RAP that was milled or crushed and sieved in a stonebreaker almost always contains a certain amount of water. It is likely to become a wet hot-in-place recycled asphalt pavement. When stockpiled, precipitation water increases the water content of this material. Therefore, it is recommended to store RAP that is prepared for use in a stockhouse or in a shelter. This water shall be removed by applying the laws of heat transfer in some equipment [22,23]. Wet and cold RAP, usually sized up to 32 mm, can be dried and heated up in an AMP in the following two ways: first, cold RAP is mixed with superheated virgin mineral materials, takes over part of their heat, dries and heats up; second, RAP is heated up in a parallel additional dryer (in-dryer) using the heat that is radiated by the burner. When the road pavement is recycled with a mobile recycling equipment (hot-in-place recycling), it is heated up by infrared burners from the top [24]. When water films are removed from the surface of RAP particles and microcracks, the adhesion of the recycled summative (aged and virgin bitumen) bituminous binder and mineral aggregates increases, which influences the strength of the recycled asphalt.

The sixth principle (Necessity to blend mechanically: free or/and forced movement of separated components in space). All or most of the particles of different sizes of the reclaimed asphalt shall be separated from each other so that when mixed, they could change their position in relation to each other and their location in a mix or a flow. When during initial decomposition road pavement structure courses are cold-milled or flat asphalt lumps or their slabs (pieces) are crushed (pulverized), and particles are obtained. When granular asphalt is heated up during the second decomposition, viscosity, adhesion, and cohesion of the old bitumen are reduced. Therefore, when exposed to mixing forces weaker than those of milling, particles partially break up and mineral particles redistribute in space, thereby opening new unbound surfaces that could be exposed to rejuvenators (additives). When the heated-up asphalt pavement is milled or broken, and when the films of the reclaimed old liquefied bitumen binder melt sufficiently, adhesion (bonding of bitumen molecules with mineral particles) and cohesion (interaction of bitumen molecules) are reduced. This makes the separation of particles from each other easier and takes less mechanical power, but requires more heat.

The seventh principle (formation of the largest possible surface of aged bitumen films to be coated with a rejuvenator). Particles of granular mixture obtained from the heated-up, and the mixed RAP is coated with extremely viscous oriented inner bitumen films of a certain thickness with a higher content of asphaltenes [25]. Due to adsorption into open voids of particles and microcracks (splits), and due to stronger adhesion, they barely melt and decompose. These particles are also coated with outer bitumen films, which weakly interact with the surface of particles as they do not directly contact with them. Volumetric bitumen films make up an outer layer, in which the content of liquid tars and oils reduces during the road use. At the temperature lower than that of the oriented bitumen they melt and become a viscous liquid that can flow in separate drops and at the molecular level mix with the rejuvenators recycling the old bitumen or virgin more liquefied bitumen [26,27]. Volumetric bitumen films oil the surfaces of mobile particles and facilitate the mixing process of the recycled asphalt mixture. When mixing the recycled asphalt mixture, more and more new surfaces of aged particles, which can absorb the batched rejuvenators or a virgin liquefied bitumen, open [28,29]. It takes more energy for finer particles to separate from each other. If they are not separated, rejuvenators, which improve aged bitumen, cannot pass and part of the mixture remains unrecycled.

The eighth principle (capability of a rejuvenator to coat all particles uniformly and to diffuse into aged bitumen films). The rejuvenators improving the properties of the aged bitumen in the recycled asphalt mixture shall uniformly distribute on the surface of each mineral particle, coated with oriented and volumetric bitumen films [7,30]. The thickness of a newly formed film of the rejuvenator shall ensure that upon completion of its diffusion into the double coating of aged bitumen, the properties of its summative bitumen are restored to the appropriate structure and properties of the recycled asphalt mixture. The deeper that a rejuvenator or a virgin bitumen penetrates into the double-coating system of reclaimed bitumen films due to diffusion after the technological process of recycling, the more complete this process is and the less asphalt properties change in the beginning of its exploitation [31–33]. The long duration of the diffusion of the rejuvenator into aged bitumen does not allow the consideration of the process of asphalt recycling as complete [34]. The duration of diffusive mixing of reclaimed old and virgin bitumen in films should be as short as possible, as only upon completion of this process the new structure of the recycled asphalt structure forms and its required (rated) resilience is obtained. Classically approach, Fick's law and Arrhenius equation describes this theory [33].

The ninth principle (retardation of bitumen oxidized ageing in the technological process of recycling). When the aged structure decomposes at high temperature of asphalt recycling process, unbound surfaces of aged volumetric bitumen contact with ambient air. The components of aged bitumen intensively oxidize and vapour even more due to ageing, thereby worsening the properties of summative bitumen [35,36]. Therefore, the temperature of the recycled asphalt mixture should not be too high and the duration of works should not be too long. When stored in a silo of an AMP and transported in a truck for a considerable period of time, laid, and compacted slowly, the structure and properties of the hot-recycled asphalt mixture deteriorate.

The tenth principle (failure to remove aged bitumen and/or some of the aggregate's fractions from the recycled asphalt mixture). The gradation of aggregates in an asphalt mixture complies with a certain optimal content of bituminous binder, which ensures its best properties [37]. When reclaimed asphalt is being recycled, the properties of aged bitumen are improved by admixing a rejuvenator or virgin bitumen, which always increases the content of summative bitumen and makes it not optimal. In order to optimize the content of summative bitumen, aggregates of appropriate fractions shall be added [38]. It is practically impossible to remove aged bitumen and (or) some aggregate fraction from recycled asphalt. It is possible only to add rejuvenators. Virgin mineral aggregates, which improve the gradation absorb part of the aged volumetric bitumen, transferring from the surface of the granular particles of reclaimed asphalt (during hot-in-plant recycling). When asphalt pavement is hot-in-place recycled with a mobile Remixer, the particles of additional HMA mixture are coated with virgin bitumen and transfer part of it to aged particles. It is likely that during the mass exchange process, aged bitumen is transferred to new particles of an additional HMA mixture. This way, a reversible double coating of new particles is obtained: exogenous coating of aged bitumen and indigenous coating of new bitumen films. These two different hot recycling technologies impact on the different formation of the structure of double-coating of a bitumen film, which coats aged and virgin particles, the thickness of the coatings, diffusion processes, and the properties of the recycled asphalt.

The eleventh principle (compliance of the designed RHMA quality indicators with the design specification requirements). RHMA is designed using the RAP, virgin mineral aggregates, virgin bitumen, or any other rejuvenator, when their factual properties and composition are identified and evaluated. The proportion of virgin bituminous binding materials and aggregates is selected taking into account the composition of RAP, its homogeneity and the properties of aged bitumen [39–41]. The aim is to obtain such physical and mechanical properties of RHMA, which are not worse than those of the mixtures made of purely virgin materials (for mixtures of the mix group) and the type of bituminous mixture and which comply with the specifications set out in European Standart (norm) EN 13108-1, ..., EN 13108-7 and EN 13108-20. The lower the homogeneity of RAP is, the less amount of it can be batched in the manufactured recycled HMA mixture [42]. RAP homogeneity is evaluated according to EN 13108-20 and TRA ASFALTAS 08.

The twelfth principle (compliance of in-plant or in-place recycled asphalt mixture composition and properties with the JMF). The recycled hot bituminous mixture manufactured from RAP and virgin materials in an AMP or in a mobile Remixer shall meet Factory Production Control requirements, as specified in EN 13108-21. Therefore, the errors of technological parameters are minimized during the recycling process, which enables to deviate the factual composition of the recycled HMA mixture, temperature, physical, and mechanical indicators from the JMF within permitted (tolerance) limits [43].

2.2. *Model of Technologies of RAP Recycling*

Such asphalt pavement recycling principles are entirely or partially applied in hot-mix recycling process of RAP in an AMP. The recycling process may be divided into sequential technological operations (Figure 2).

In conventional AMP constructed to recycle aged RAP, additional technological equipment to feed, store, supply (transport), batch, dry, heat up, and mix RAP with virgin materials be used (Figure 3). Recently, the increasing content of RAP has been observed. The structure of the equipment designed to recycle RAP depends on the maximum permitted RAP percentage content, the methods of drying and heating up, the place of batching in AMP, as well as mixing with virgin materials. When a small percentage (up to 10% or 20% of RAP content from the total mass of recycled asphalt mixture) is used, a batch-type AMP without additional heating of RAP is the most suitable one (Figure 3D). AMP of this type enables the delivery RAP into its four types of equipment: 1—hot elevator; 2—hot aggregate bin (compartment of bypass); 3—hot aggregate scale bin; and, 4—mixer. To batch the increased RAP percentage content (from 20% to 50%), it is dried and heated up in an additional cylinder dryer, incorporated into a batch-type AMP (Figure 3C).

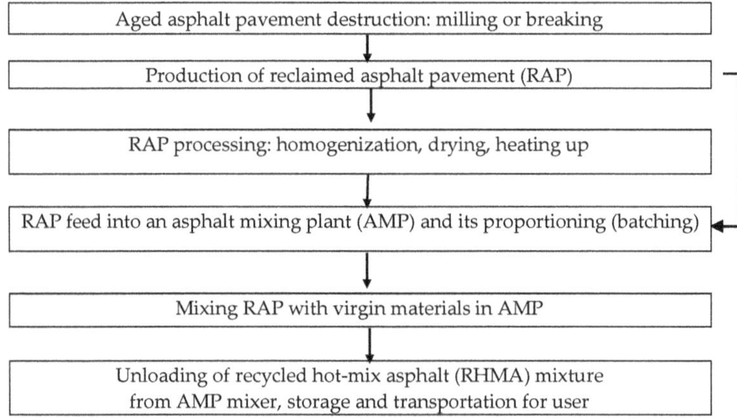

Figure 2. Production operations of the in-plant recycling process of aged asphalt pavement.

RAP can also be recycled in a continuous AMP with integrated drum-mix (Figure 3A) or double barrel dryers-mixers (Figure 3B). When using this equipment, 50% or more of wet and cold particles of RAP could be batched. All of the asphalt recycling technologies presented in Figure 3 have both advantages and disadvantages.

Recent theoretical and practical achievements enable the recycling of up to 100% of RAP, of course, we need to evaluate that the studies were conducted in ideal and lab-controlled conditions [44,45]. RAP inhomogeneity has a huge impact on its maximum permitted percentage in a recycled RHMA. It is only in rare cases that RAP is sieved into two or more different size fractions. A lack of an efficient practical method, which enables to homogenize RAP prepared for use, does not allow the use of high RAP content in a recycled hot-mix asphalt (RHMA) mixture. As a rule, according to the standards it is allowed use less RAP content in a recycled asphalt mixture of the pavement surface course than in a binder or a base course [42].

When RHMA mixture is produced in an AMP, cold and wet RAP is dried and heated up to break them down into separate component particles, which could absorb thin films of a rejuvenator on their surface. RAP may dry out and heat up in two technological ways of heat transfer: first, (Figure 3A,B,D technologies) without preheating the reclaimed material [46], when RAP take over some of the heat of aggregates (superheated virgin materials, superheating temperature) in contact with superheated aggregates; second, (Figure 3C technology) the flow of RAP is dried and heated up (up to 100–110 °C) in a separate parallel dryer, where a mixture of heated up gas and air is emanated from a burner [47].

The most important technological requirements for the production of RHMA mixtures are as follows:

- the conformity of its gradation (content of mineral filler, fine aggregate and coarse aggregate and bituminous binder content with JMF; and,
- the conformity of the RHMA mixture temperature to the type of used bituminous binder grade.

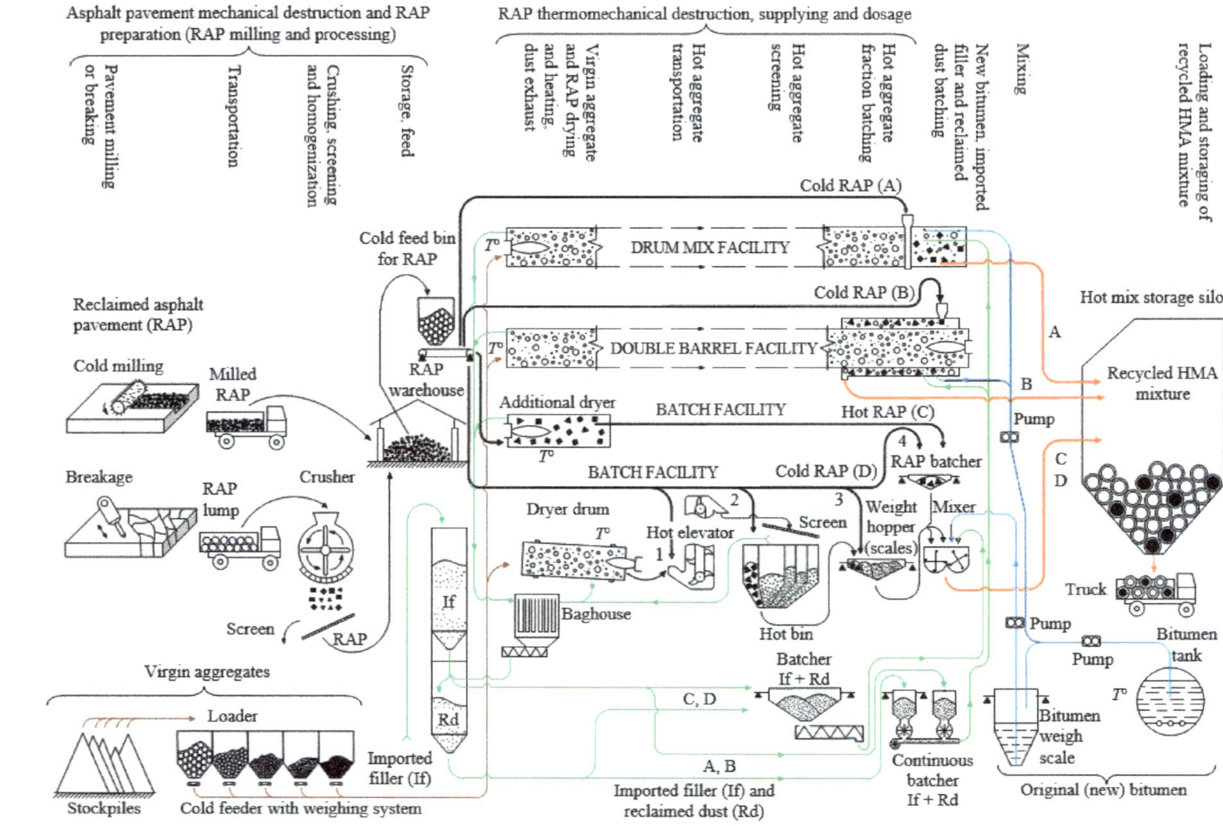

Figure 3. The model of asphalt pavement recycling operations and equipment used in batch plant (**C**,**D**), drum mix (**A**) and a double barrel-drum mix (**B**) when reclaimed asphalt pavement (RAP) is supplied and dosed to different places.

Technical requirements TRA ASFALTAS 08 drafted in compliance with the above-mentioned standards requires that the max allowed content of RAP K_i should be calculated according to Equations (1) and (2) taking into account the homogeneity of RAP. Such homogeneity is determined by the range indicators for individual properties (a_i). K_i shall be computed by the general permissible deviations $N_{adm,i}$ depending on the property of RAP and the type of recycled HMA mixture:

$$K_i = \frac{0.50 N_{adm,i}}{a_i} \times 100 \tag{1}$$

or

$$K_i = \frac{0.33 N_{adm,i}}{a_i} \times 100, \tag{2}$$

where a_i is the range between the max $x_{i,max}$ and the min $x_{i,min}$ values of RAP i-property indicator estimated for the sample (min sample size $n = 5$).

With respect to all of the properties of base and surface-base course mixtures, Equation (1) shall apply. It shall be also applied for the softening point of binder course and surface course mixtures. With respect to all other properties, Equation (2) shall apply [42].

RAP is not only homogeneous, but it is also dosed with systemic and random errors of a certain magnitude in the AMP that increase the deviations of the RHMA component content from JMF and variation [38].

Derivation of the procedure for calculation of the maximum allowable RAP content is as follows. ASTM D 4460 gives equations for calculating standard deviation values for quantities determined from calculations involving two other values. From these equations, the following formula for calculating the standard deviation of a blend of two materials can be derived:

$$\sigma_m = \sqrt{\alpha^2 \sigma_a^2 + (1 - \alpha)^2 \sigma_b^2 + \left(\overline{X}_a^2 + \overline{X}_b^2\right) \sigma_\alpha^2}, \tag{3}$$

where σ_m—standard deviation of the mixture; σ_a—standard deviation of component "a"; σ_b—standard deviation of component "b"; α—proportion of component "a" in the mixture; $\overline{X}_a^2$— mean value for component "a"; $\overline{X}_b^2$—mean value for component "b"; and, σ_α—standard deviation of the proportions.

We can rewrite this for percent passing for a selected sieve for HMA mixtures consisting of a blend of new HMA materials with RAP [48]:

$$\sigma_{PM} = \sqrt{w_R^2 \sigma_{PR}^2 + w_N^2 \sigma_{PN}^2 + \left(\overline{P}_R^2 + \overline{P}_N^2\right) \sigma_W^2}, \tag{4}$$

where σ_{PM}—standard deviation of percent passing for a selected sieve for the mixture with RAP (i.e., RHMA); w_R—weight fraction of RAP in the mixture; σ_{PR}—standard deviation of percent passing for the selected sieve for the RAP; w_N—weight fraction of new materials (new HMA) in the mixture $= (1 - w_R)$; σ_{PN}—standard deviation of percent passing for the selected sieve for the new HMA; $\overline{P}_R$—mean value for RAP% passing for the selected sieve; $\overline{P}_N$—mean value for new HMA% passing for the selected sieve; and, σ_w—standard deviation of the weight fractions, also called "batching variability".

Equation (3) can be solved for the maximum amount of RAP that can be added to new material without increasing the standard deviation for percent passing on the selected sieve above a selected maximum value by application of the quadratic equation:

$$Max.RAP\% = \frac{-b + \sqrt{b^2 - 4ac}}{2a} 100\%, \tag{5}$$

where $Max.RAP\%$—maximum amount of RAP that can be added to the mix, weight%; $a = \sigma_{PR}^2 + \sigma_{PN}^2$; $b = -2\sigma_{PN}^2$; $c = \sigma_{PN}^2 + \left(\overline{P_{PN}^2} + P_N^2\right)\sigma_w^2 - \sigma_{PM/Max}^2$; and, $\sigma_{PM/Max}^2$ is the maximum allowable standard deviation for percent passing for the selected sieve.

Because asphalt is temperature sensitive, good preheating of RAP is helpful to eliminate the agglomerates of RAP to promote the dispersion of RAP during the hot recycling process [49]. When applying the first heat transfer method, superheating temperature of virgin mineral materials (aggregate) is calculated according to the following empirical formula applied in practice [50]:

$$T_{v.a.} = \frac{T_{RHMA} - S \cdot T_{RAP}}{1 - S} + \frac{4 \cdot S \cdot W_{RAP}}{1 - S}(637 - T_{RAP}), \quad (6)$$

where $T_{v.a.}$—superheating temperature of virgin mineral materials (aggregates), °C; T_{RHMA}—required temperature of recycled hot mix asphalt, which depends on the grade of the bitumen binder, °C; S—ratio of RAP batching mass in the recycled hot mix asphalt mixture, in units; T_{RAP}—temperature of RAP, °C; and, W_{RAP}—moisture (water content) of RAP, in units.

This formula (1) contains four variables, two of which (T_{RHMA}, S) show the temperature of RAP and component composition, whereas other two (T_{RAP}, W_{RAP})—show RAP temperature and moisture. Experimental investigations showed that RAP may absorb from 0.7 to 6.5% of moisture. The average made up 3.40%. Therefore, it is recommended to store it in warehouses. It is likely that not all humidity is removed from RAP in a parallel drum. The quantity and temperature of a batched.

RAP, which depends on its transfer type impacted by the production technology of RHMA in AMP, allows to calculate the temperature at which virgin aggregates shall be heated. According to formula (6), it was identified how the temperature required for heating virgin aggregates varies when the moisture of RAP varies from 0 to 8%, and its quantity in RHMA varies from 5 to 50% (Figure 4). When making calculations, it was assumed that the initial temperature of RAP is equal to ambient temperature of 15 °C (Figure 4a). RAP was heated in a parallel drum to the working temperature of 100°C (Figure 4b), and it also contained some residual temperature.

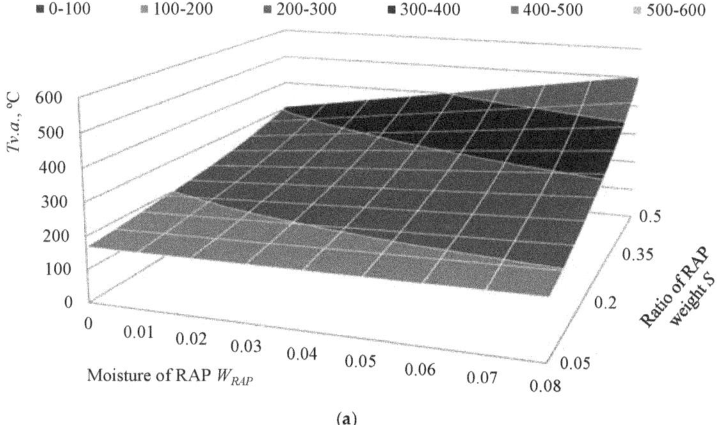

(a)

Figure 4. *Cont.*

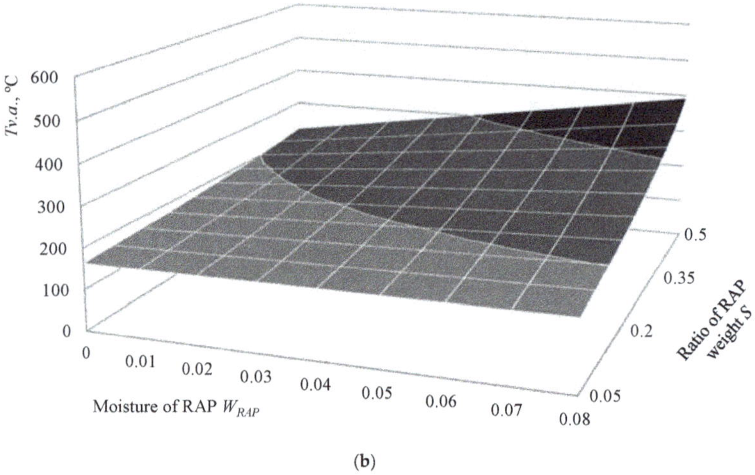

(b)

Figure 4. Dependence of superheating temperature of virgin mineral materials on the ratio of RAP weight and moisture of RAP when T_{RAP} is as follows: (**a**)—15 °C; (**b**)—100 °C.

It was estimated that if RAP content in RHMA mixture is 50% and RAP moisture is 2%, then the temperature required to heat virgin aggregates when producing 160 °C RHMA with RAP heated in a parallel drum is 262 °C, and with non-heated RAP-even 354 °C. The difference made up 92 °C.

3. RAP Batching Accuracy and Precision

RHMA mixture of different RAP content was produced in one asphalt producing company in a batch-type asphalt mixing plant (AMP), Ammann Euro A 240, the highest capability of which was Q_{max} = 240 t/h, and the highest mass of one mix batch was 3000 kg. RAP feeding, transporting, drying, and heating, as well as batching devices were mounted in addition (Figure 5).

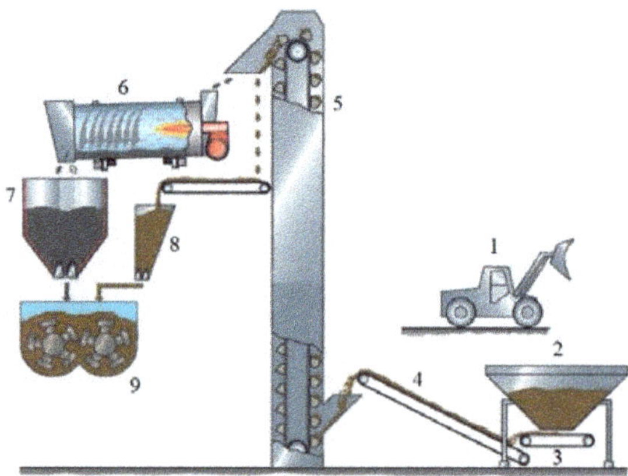

Figure 5. The technological scheme of RAP hot and cold recycling in batch AMP: 1—wheel loader; 2—RAP feeder; 3—belt scale; 4—collecting conveyor; 5—RAP elevator; 6—parallel drum; 7—recycling buffer silo with weighing appliance and recycling scale; 8—cold recycling addition via buffer silo and belt scale; 9—mixer.

3.1. Methods

The following mixture for a road pavement surface and base courses of two types was produced in a batchAMP with preheated RAP: with 25% and 50% of RAP, AC 16 PD and AC 22 PS, respectively. The grade of AC 16 DP according to the national standard TRA ASFALTAS 09 is asphalt concrete (AC); maximum size of particles—16mm; purpose—for pavement single ply surface course (PD). The grade of AC 22 PS is also asphalt concrete (AC); maximum size of particles—22 mm; purpose—for pavement base course (P); also, purpose—for the heavy load (S). To investigate the accuracy and precision of RAP batching, technological data about RHMA mixtures produced in 2014, which were stored in an APM handling computer, were used. AMP handling software presents rated (target) and factual data on the name and number of job-mix formula (JMF), weight, mixing time, and temperature of each RHMA mix batch. Masses of all mineral materials, bitumen, additives, and RAP weighed batches are presented as well.

For statistical purposes, only the data when most of RHMA mixture was produced were used: 10-day AC 16 PD mixture production data and 12-day AC 22 PS grade hot-mix asphalt (HMA) mixture production data. Quite large sample sizes, presented in Table 1, allowed the study to obtain rather dependable investigation data. RHMA mix batches of less than 2800 kg (outliers) were eliminated from each sample size. Statistical investigation was conducted on the data of 3041 mix batches, for the production of which approximately 3400 t of RAP was used and approximately 9000 t of RHMA mixture were produced.

Table 1. Sample sizes of the data used in the investigation

Type of RHMA	Days	Total Sample Size	Sample Size Without Outliers
AC 16 PD	10	1424	1324
AC 22 PS	12	1617	1569

Statistical data were calculated through the use of Statgraphics Centurion XVII software package (Statpoint Technologies, Inc., Warrenton, VA, USA) for statistical processing, which allowed to identify the accuracy and precision of RAP batching in a batch-type AMP. AMP handling computer data of each working day's lot were grouped and classified by days and grades of RHMA mixture. Total weight Q_{batchi} of RHMA mix batch and weighed RAP batch mass q_{RAPi} were calculated. RAP percentage content in each asphalt mix batch was calculated according to this data:

$$x_{RAPi} = \frac{q_{RAPi}}{Q_{batchi}} \times 100 \tag{7}$$

where x_{RAPi}—RAP content in i RHMA mix batch, %; q_{RAPi}—RAP batch mass, kg, Q_{batchi}—RHMA mix batch mass, kg; i—i RHMA mix batch ($i = 1, 2, \ldots N$, N—RHMA mix batches' number).

Statistical indices of the produced RHMA mixture of both grades RAP batch mass position and variation were calculated, and diagrams and histograms were drawn (Figures 6 and 7).

The accuracy of RAP batching was identified by comparing average RAP content deviation from JMF in the produced RHMA mixture:

$$\Delta q_{RAP} = \frac{\overline{q}_{RAP} - q_{RAP,JMF}}{q_{RAP,JMF}} \times 100, \tag{8}$$

where $\overline{q}_{RAP}$—average of RAP batches mass, kg; $q_{RAP,JMF}$—target RAP batch mass according to JMF, kg.

Standard deviation was used to evaluate the precision of RAP batching (Figure 7).

Normal distribution of both sample sizes of RAP batch mass was tested, which enabled the calculation of the average precision of RAP batching process. Normal distribution of experimental data was investigated by applying skewness and kurtosis. The values of the coefficient of skewness *sk* and coefficient of kurtosis ku calculated from data of sample size were compared with their respective standard deviations s_{sk} and s_{ku}:

$$s_{sk} = \sqrt{\frac{6n(n-1)}{(n-2)(n+1)(n+3)}}, \tag{9}$$

$$s_{ku} = \sqrt{\frac{24n(n-1)^2}{(n-3)(n-2)(n+3)(n+5)}}, \tag{10}$$

where n—the sample size (number of measurements). When $|sk| < 3s_{sk}$ and $|ku| < 5s_{ku}$, it can be considered that the normality hypothesis of empirical data is accepted. Otherwise, the raised hypothesis is rejected or accepted as doubtful.

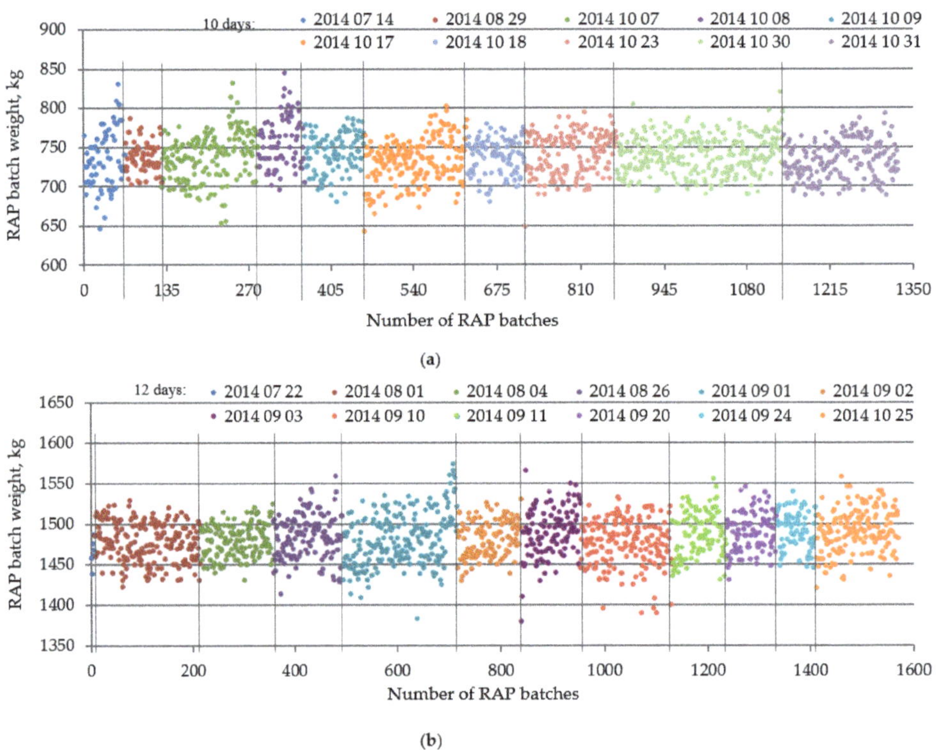

Figure 6. Distribution of RAP separate batch masses. When its job mix formula (JMF) is as follows: (**a**)—750 kg; (**b**)—1500 kg.

3.2. Materials

RHMA mixtures were produced according to various JMF's. To produce AC 16 PD grade mixture, imported fillers (IF), crushed gravel sifting, reclaimed dust (RD), natural sand, various fraction crushed gravel, and 100/150 grade road bitumen were used. AC 22 PS grade RHMA mixture was produced from 0/2 mm and 0/5 mm fraction natural sand, 5/8 mm, 8/11 mm and 11/22 mm fraction gravel, and 50/70 grade road bitumen. Additives were not used for RHMA mixture production. The batch masses of all materials' finite batching in one recycled HMA mix batch, the mass of which according to JMF is 3000 kg, are presented in Table 2.

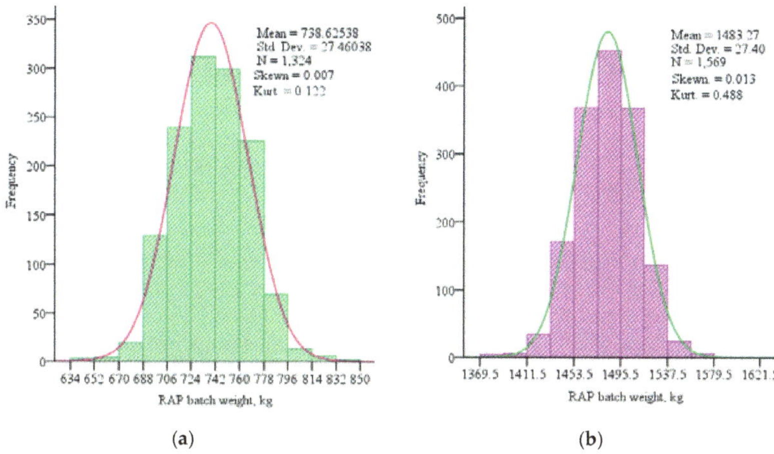

Figure 7. Distribution histograms, theoretical curves of normal distribution and statistical parameters of RAP batch masses in HMA batches of mixtures: (**a**)—AC 16 PD (750 kg); (**b**)—AC 22 PS (1500 kg).

Table 2. RAP composition, %.

Particle Size, mm	Subsample Number					Average
	I	II	III	IV	V	
>11.2	6.3	7.5	8.4	8.7	7.5	7.7
>2	69.4	73.4	71.2	68.5	68.7	70.6
0.063–2	30.4	26.5	28.6	31.3	31.2	29.2
<0.125	0.5	0.3	0.2	0.3	0.3	0.3
<0.063	0.2	0.1	0.2	0.2	0.1	0.2
Bitumen	4.78	4.73	4.86	4.81	4.43	4.72

To produce RHMA mixture, crushed and sieved RAP, the size of which was not larger than 32 mm, were used. RAP was prepared according to the third principle and technological scheme (Figure 3). When designing the composition of RHMA mixture and determining JMF, RAP grading and hot fraction were evaluated. Several sample units were taken from the adjacent places of the sheltered stock of RAP and prepared for use. The material of several sample units was put into a subsample. In total, five subsamples were taken this way, each of which was tested separately in a laboratory. RAP investigation data are presented in Table 3.

Table 3. RHMA mixture batch composition according to JMF (finite batching materials' batch mass in kg in one RHMA mix batch of 3000 kg).

RHMA Mixture Group	RAP	IF	RD	Hot Aggregate Fractions, mm						Virgin Bitumen
				0/2	2/5	5/8	8/11	11/16	11/32	
AC 16 PD	750	99	99	612	219	411	384	306	-	120
AC 22 PS	1500	-	-	141	87	297	144	-	780	51

The mean of five subsamples was taken as RAP average composition. RAP homogeneity was estimated by standard deviation.

4. Results and Discussion

The aim of the study was to identify how accurately and precisely RAP is batched when producing an asphalt mixture. The experiment data show that when producing RHMA mixture AC 16 PD (with 25% RAP) for pavement single ply surface course, the batched quantity of RAP into an AMP mixer was on average $\overline{q}_{RAP} = 738.63$ kg. According to Equation (8), batching accuracy $\Delta q_{RAP} = \frac{738.63-750}{750} \times 100 = -1.52\%$. RAP batch mass varied from 643 kg to 845 kg. Range $R = q_{max} - q_{min} = 202$ kg (Figure 6a). Standard deviation showing batch mass variation was $s_{q_{RAP}} = 27.46$ kg and coefficient of variation was $V_q = 3.72\%$. The specifications do not regulate AMP batching tolerances of bitumen, hot and cold mineral materials. They are specified in ASTM Standard Specification for mixing plants for hot-mixed, hot-laid bituminous paving mixture. According to the Specifications above, the automatic proportioning system shall be capable of consistently delivering materials within the full range of a batch size of the following tolerances (Table 4).

Table 4. Tolerances of batch masses of AMP batchers' materials according to ASTM D 995-95b.

Materials	Total Batch Mass of a Paving Mix, %
Batch aggregate component	±1.5
Mineral filler	±0.5
Bituminous material	±0.1
Zero return (aggregate)	±0.5
Zero return (bituminous material)	±0.1

The electric circuits for the above delivery tolerances of each cut-off interlock shall be capable of providing the total span for the full allowable tolerance for maximum batch size. Tolerance controls shall be automatically or manually adjustable to provide spans that are suitable for less than full-size batches. The automatic controls and interlock cut-off circuits shall be capable of being consistently coordinated with the batching scale or meter within an accuracy of 0.2% of the nominal capacity of the said scale or meter throughout the full range of the batch sizes according to ASTM D995-95b.

Having investigated all 1569 asphalt mix batches of grade AC 22 PS RHMA mixture (with 50% RAP content) produced in 12 days for the base course, it was identified that RAP is batched rather accurately. When the mean of RAP of all the batches is $\overline{q}_{RAP} = 1483.27$ kg, and $q_{RAP,JMF} = 1500$ kg, then according to the mean RAP is batched rather accurately: $\Delta q_{RAP} = \frac{1483.27-1500}{1500} \times 100 = -1.11\%$ and error is less than 1.5%, i.e., lower than tolerances. The value of batching preciseness is standard deviation $s_{q_{RAP}} = 27.40$ kg, and variation coefficient $V_q = 1.85\%$.The maximum mass of a RAP batch is 1631 kg and the minimum mass is 1380 kg. Range $R = q_{max} - q_{min} = 251$ kg (Figure 6b).

Taking into consideration the specification that not only RAP batch mass mean is $\overline{q}_{RAP}$, but its weighed batch mass q_{RAP} differed from rated value $q_{RAP,JMF}$ does not exceed the permitted tolerances (±1.5%), when $q_{RAP,JMF} = 750$ kg, the amount of weighed RAP shall be not less than 738.75 kg and not more than 761.25 kg. When $q_{RAP,JMF} = 1500$ kg, the minimum permitted batch mass of RAP may be 1477.5 kg, and the maximum batch mass 1522.5 kg. In fact, the diagrams (Figure 7) above show that the mass of quite a lot of RAP separate batches does not meet tolerance specifications. Only 29% of them (392 out of 1324) are within the range of permitted tolerances, when the estimated $q_{RAP,JMF} = 750$ kg and 51% (803 out of 1569) when it was $q_{RAP,JMF} = 1500$ kg.

The normality of data distribution was verified according to skewness and kurtosis. Empirical coefficient of skewness of RAP batch masses *sk* were 0.007 and 0.013, and the coefficient of kurtosis *ku* were 0.122 and 0.488, respectively (Figure 7). They were compared to the critical values which depend only on sample size *n*. When RAP batch sample sizes *n* were 1324 and 1569, standard deviations of skewness and kurtosis were calculated as follows: $s_{sk} = 0.067$, $s_{sk} = 0.062$ and $s_{ku} = 0.134$, $s_{ku} = 0.123$ (AC 16PD and AC 22PS mixtures, respectively). The estimated s_{sk} values were multiplied by 3, which resulted in $3s_{sk} = 0.202$ and $3s_{sk} = 0.185$, respectively. Standard deviation of kurtosis s_{ku} was multiplied by 5, which resulted in $5s_{ku} = 0.672$ and $5s_{ku} = 0.617$, respectively.

When $|sk| < 3s_{sk}$ and $|ku| < 5s_{ku}$, it could be stated that the hypothesis of normality of the empirical data is confirmed. The normality of sample size data distribution can be verified using more complex methods, such as Kolmogorov and Pearson. Null hypothesis can be verified according to the Bartlett criterion, where the available data is close to normal distribution.

Thus, Bartlett's criterion was used to compare the variances of various sample sizes according to normal distribution. This criterion is convenient to compare two or more variances of the normal distribution sample sizes when these sample sizes are unequal. If statistics B is less than $\chi^2_{\alpha,\nu}$, RAP batch mass variation is statistically equal. Therefore, the average calculated value of sample variance s_q^{-2} = 752.52 kg and general standard deviation $\bar{s}_q$ = 27.43 kg. Statistics B = 0.0094 is less than critical $\chi^2_{\alpha,\nu}$ = 3.84 with the selected $\alpha = 0.05$ significance level and degrees of freedom $\nu = l-1$ (where l—the number of comparable variances (samples). Therefore, it could be concluded that both (with 25% and 50% RAP content) batch mass variation (standard deviation s_q = 27.46 kg and s_q = 27.40 kg) are statistically equal.

The physical and mechanical properties of a recycled HMA mixture are influenced by RAP percentage content in each mix batch of RHMA mixture. This parameter allows to determine the proportion of RAP in the whole mix batch. When hot fractions, IF and RD batches are weighed in AMP, their mass like that of RAP has a certain variation, which influences the variation of RHMA mixture batch masses (Figure 8). Therefore, the mass of RHMA mixture batches always differs from JMF, i.e., from 3000 kg and is of stochastic type.

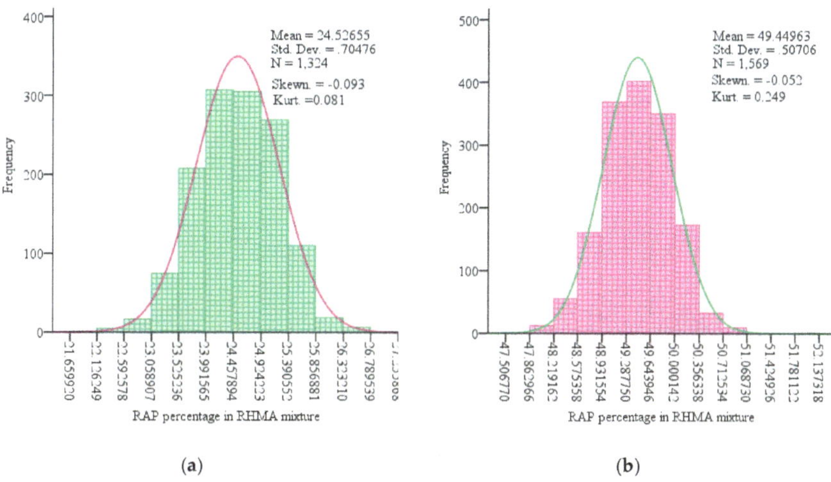

(a) (b)

Figure 8. Normal distribution curves and statistical indices of histogram of RAP percentage content in recycled hot mix asphalt (RHMA) mixture, when its content according to JMF had to be: (**a**)—25%; (**b**)—50%.

Average values $\overline{x}_{RAP}$ of RAP percentage Equation (7) of both sample sizes out of 1324 (with 25% RAP) and 1569 (with 50% RAP) were approximate to JMF (Figure 8). $\overline{x}_{RAP_{25}}$ = 24.53%, $\overline{x}_{RAP_{50}}$ = 49.45%. Standard deviation s_x of RAP percentage content was rather low: in mix batches with 25% of RAP content—0.705%; in mix batches with 50% of RAP content—0.507%, and their variation coefficients were vc_{x25} = 2.87% and vc_{x50} = 1.03%, respectively.

Average values $\overline{x}_{RAP}$ of RAP percentage content Equation (7) of both sample sizes out of 1324 (with RAP percentage content in RHMA mix batches were approximate to those designed in JMF; however, histograms show that most of the data is grouped below the value of JMF (Figure 8). There was a tendency of lack of RAP content in RHMA mixture.

When asphalt pavement is hot-mix in-plant recycled, the sequence of technological operations depends on the structure of an AMP and additional equipment for feeding, transporting, and proportioning or batching of RAP. The model constructed by the authors contains two types of technological operations. First, hot-mix recycling in a batch plant, when RAP is dried and heated up due to the contact with superheated virgin mineral materials, i.e., conductive heat transfer occurs. Second, when RAP is dried and heated up in an additional parallel dryer, i.e., convective heat transfer occurs. These two recycling technologies are popular in the European Union countries. The model also contains hot-mix recycling in drum-mix and double-barrel plant technologies, when RAP dry out and heat up due to the heat exchange between them and superheated virgin mineral materials: conductive heat transfer occurs. These technologies are more popular in the USA.

When up to 10% or 20% of RAP is used, such aged asphalt pavement material may be supplied to four batch AMP places: the bottom part of a bucket hot elevator with aggregate, hot fractions' unsieved mineral materials bypass hot bin, hot mineral material fractions batcher's weight hopper, forced mixing two-shaft mixer. Such four RAP feeding places into AMP, which were selected in advance, have both advantages and disadvantages.

The aim is to consume less heat energy (gas or liquid fuel), required to remove water content in RAP and to reduce the pollution of ambient air. Ready-to-use RAP is stored in a closed warehouse, which protects the material from precipitation water. When the conductive heat transfer method is applied to vaporize the water content in RAP and to reduce the viscosity of aged bitumen, the superheated temperature of virgin mineral materials mostly depends on RAP moisture (water content) and its percentage in RHMA. When the value of these parameters increases, the required temperature of heating and drying of virgin aggregates in a drum increases, which is calculated according to the presented empirical formula.

The composition of HMA mixture with less than 100% of RAP developed in a laboratory as accurately and precisely as possible shall be guaranteed when producing it in an AMP. Inevitable systematic and random errors of batching RAP increase the deviations of RHMA mixture component content from JMF and their variation. The size of errors depends on AMP structure, technical condition, RAP percentage in RHMA and AMP operator's actions when setting the technological parameters of the recycling process and handling them.

5. Conclusions

The application of these twelve road asphalt pavement hot recycling principles allows us to expect the best possible properties and pavement of the course with reclaimed asphalt pavement. These principles may be used for asphalt pavement hot-in-place surface recycling and hot-mix in-plant recycling.

According to JMF, RAP content was 25% (750 kg batch mass, 10 days, $n = 1324$ batches) and during the investigation, it was 50% (1500 kg batch mass, 12 days, $n = 1569$ batches). The mean of 738.6 kg of all the AMP batches with its required content of 25% mass was less than 750 kg, i.e., by 1.52%, and with the required RAP content of 50%, the mean of 1483.3 kg was less than 1500 kg, i.e., by 1.1%. The accuracy of RAP batching are almost met or completely met the tolerances of the mean specified in ASTM D995-95b on the batch aggregate component ($\pm$1.5%) content. Due to the high variation of the batch mass of separate RAP, the tolerance of $\pm$1.5% was met only by 29.6% and 51.5%, respectively, of all RAP batches. The accuracy and compliance of batch masses with tolerances increases when a higher percentage of RAP is batched.

The standard deviation mean calculated according to two standard deviations ($s_q = 27.46$ kg and $s_q = 27.40$ kg) was 27.4 kg. The variation coefficient of RAP batch mass was 3.72% (when RAP batch mass according to JMF was 750 kg) and approximately two times less (1.85%), when according to JMF RAP batch mass was 1500 kg.

RAP batch mass deviations from JMF and their variation influence RAP percentage content in a recycled HMA mixture. When the mass of an asphalt mix batch inevitably varies, the variation of

RAP percentage in RHMA mixture increases even more. When RAP content according to JMF was 25% in RHMA, the mean of its percentage content was 24.53%, and when according to JMF it was 50%, in fact, it was 49.45%. In both cases, RAP percentage content in RHMA mixture varies according to normal distribution. Standard deviations of RAP percentage content in RHMA mixture were 0.705% and 0.507%, and statistically did not diverge.

Acknowledgments: The authors would like to acknowledge Fegda JSC for providing part of materials and laboratory equipment required for the research work.

Author Contributions: Henrikas Sivilevičius systematized hot regeneration principles, submitted the model of asphalt pavement recycling operations (Figure 3) and wrote the introduction. Justas Bražiūnas conceived and designed the experiments, wrote and edited the paper. Olegas Prenkovskis processed and analyzed the data. All authors read and approved the final manuscript.

Conflicts of Interest: The authors declare no conflict of interest.

References

1. Tran, N.H.; Taylor, A.; Willis, R. *Effect of Rejuvenator on Performance Properties of HMA Mixtures with High RAP and RAS Contents*; Report Number 12-05; National Center for Asphalt Technology: Auburn, AL, USA, 2012.
2. Zaumanis, M.; Mallick, R.B. Review of very high-content reclaimed asphalt use in plant-produced pavement: State of the art. *Int. J. Pavement Eng.* **2015**, *16*, 39–55. [CrossRef]
3. Solaimanian, M.; Savory, E. Variability Analysis of Hot-Mix Asphalt Concrete Containing High Percentage of Reclaimed Asphalt Pavement. *Transp. Res. Rec.* **1996**, *1543*, 89–96. [CrossRef]
4. Howard, I.L.; Cooley, L.A., Jr.; Doyle, J.D. *Laboratory Testing and Economic Analysis of High RAP Warm Mixed Asphalt*; Mississippi Department of Transportation: Jackson, MS, USA, 2009.
5. Bloomquist, D.; Diamond, G.; Oden, M.; Ruth, B.; Tia, M. *Engineering and Environmental Aspects of Recycled Materials for Highway Construction*; Western Research Institute: Laramie, WY, USA, 1993.
6. Apeagyei, A.K.; Clark, T.M.; Rorrer, T.M. Stiffnes of high-RAP asphalt mixtures: Virginia's experience. *J. Mater. Civ. Eng.* **2013**, *25*, 747–754. [CrossRef]
7. Rad, F.Y.; Sefidmazgi, N.R.; Bahia, H. Application of diffusion mechanism. Degree of Blending between fresh and recycled asphalt pavement binder in dynamic shear rheometer. *Transp. Res. Rec. J. Transp. Res. Board.* **2014**, *2444*, 71–77. [CrossRef]
8. Miró, R.; Valdés, G.; Martínez, A.; Segura, P.; Rodriguez, C. Evaluation of high modulus mixture behaviour with high reclaimed asphalt pavement (RAP) percentages for sustainable road construction. *Const. Build. Mater.* **2011**, *25*, 3854–3862. [CrossRef]
9. Mogawer, W.; Bennert, T.; Daniel, J.S.; Bonaquist, R.; Austerman, A.; Booshehrian, A. Performance characteristics of plant produced high RAP mixtures. *Road Mater. Pavement Des.* **2012**, *13*, 183–208. [CrossRef]
10. Bražiūnas, J.; Sivilevičius, H. The bitumen batching system's modernization and its effective at the asphalt mixing plant. *Transport* **2010**, *25*, 325–335. [CrossRef]
11. Zaumanis, M.; Mallick, R.B.; Frank, R. 100% recycled hot mix asphalt: A review and analysis. *Resour. Conserv. Recycl.* **2014**, *92*, 230–245. [CrossRef]
12. Santos, L.G.D.P.; Baptista, A.M.D.C.; Capitão, S.D. Assessment of the Use of Hot-Mix Recycled Asphalt Concrete in Plant. *J. Transp. Eng.* **2010**, *136*, 1159–1164. [CrossRef]
13. Estakhri, C.; Spiegelman, C.; Gajewski, B.; Yang, G.; Little, D. *Recycled Hot-Mix Asphalt Concrete in Florida: A Variability Study*; International Center for Aggregates Research: Austin, TX, USA, 1999.
14. West, R.; Willis, J.R.; Marasteanu, M. *Improved Mix Design, Evaluation, and Materials Management Practices for Hot Mix Asphalt with High Reclaimed Asphalt Pavement Content*; NCHRP Report 752; NCHRP: Washington, DC, USA, 2013.
15. National Center for Asphalt Technology (NCAT). LTPP Data Shows RAP Mixes Perform as Well as Virgin Mixes. In *Asphalt Technology News*; NCAT: Auburn, AL, USA, 2009; Volume 21.
16. Diefenderfer, S.; Nair, H. Evaluation of Production, Construction, and Properties of High Reclaimed Asphalt Pavement Mixture. *Transp. Res. Rec. J. Transp. Res. Board.* **2014**, *2445*, 75–82. [CrossRef]
17. Bonaquist, R. Can I run more RAP? *HMAT Hot Mix Asph. Technol.* **2007**, *12*, 11–13.

18. Mohammad, L.; Wu, Z.; Zhang, C.; Khattak, M.; Abadie, C. Variability of Air Voids and Mechanistic Properties of Plant-Produced Asphalt Mixtures. *Transp. Res. Rec. J. Transp. Res. Board.* **2004**, *1891*, 85–97. [CrossRef]
19. Ali, H.; Grzybowski, K. Life cycle of hot in-place pavement recycling. Case study. *Transp. Res. Rec. J. Transp. Res. Board.* **2012**, *2292*, 29–35. [CrossRef]
20. Tahmoorian, F.; Samali, B.; Tam, V.W.Y.; Yeaman, J. Evaluation of mechanical properties of recycled material for utilization in the asphalt mixtures. *Appl. Sci.* **2017**, *7*, 763. [CrossRef]
21. Kriz, P.; Grant, D.L.; Veloza, B.A.; Gale, M.J.; Blahey, A.G.; Brownie, J.H.; Shirts, R.D. Blending and diffusion of reclaimed asphalt pavement and virgin asphalt binders. *Road Mater. Pavement Des.* **2014**, *15*, 78–112. [CrossRef]
22. Le Guen, L.; Huchet, F.; Tamagny, P. Drying and heating modelling of granular flow: Application to the mix-asphalt process. *J. Appl. Fluid Mech.* **2011**, *4*, 71–80.
23. Peinado, D.; de Vega, M.; Garcia-Hernando, N.; Marugan-Cruz, C. Energy and exergy analysis in an asphalt's rotary dryer. *Appl. Therm. Eng.* **2011**, *31*, 1039–1049. [CrossRef]
24. Wang, H.; Hao, P.; Xue, L. Laboratory evaluation of microwave heating method for hot in-place recycling. *J. Test. Eval.* **2011**, *39*, 1070–1077.
25. Carpenter, S.H.; Wolosick, J.R. Modifier influence in the characterization of hot-mix recycled material. *Transp. Res. Rec. J. Transp. Res. Board.* **1980**, *777*, 15–22.
26. Nahar, S.N.; Mohajeri, M.; Schmets, A.J.M.; Scarpas, A.; van de Ven, M.F.C.; Schitter, G. First observation of blending-zone morphology at interface of reclaimed asphalt binder and virgin bitumen. *Transp. Res. Rec. J. Transp. Res. Board.* **2013**, *2370*, 1–9. [CrossRef]
27. Willis, J.R.; Turner, O.; de Goes Padula, F.; Tran, N.; Julian, G. Effects of changing virgin binder grade and content on high reclaimed asphalt pavement mixture properties. *Transp. Res. Rec. J. Transp. Res. Board.* **2013**, *2371*, 66–73. [CrossRef]
28. Dony, A.; Colin, J.; Bruneau, D.; Drouadainc, I.; Navaro, J. Reclaimed asphalt concretes with high recycling rates: Changes in reclaimed binder properties according to rejuvenating agent. *Constr. Build. Mater.* **2013**, *41*, 175–181. [CrossRef]
29. Lin, J.; Guo, P.; Xie, J.; Wu, S.; Chen, M. Effect of rejuvenator scaler materials on the properties of aged asphalt binder. *J. Mater. Civ. Eng.* **2013**, *25*, 829–835. [CrossRef]
30. Čygas, D.; Mučinis, D.; Sivilevičius, H.; Abukauskas, N. Dependence of the recycled asphalt mixture physical and mechanical properties on the grade and amount of rejuvenating bitumen. *Balt. J. Road Bridge Eng.* **2011**, *6*, 124–134. [CrossRef]
31. Huang, S.-C.; Qin, Q.; Grimes, W.R.; Pauli, A.T.; Galer, R. Influence of Rejuvenators on the Physical Properties of RAP Binders. *J. Test. Eval.* **2015**, *43*, 594–603. [CrossRef]
32. Shirodkar, P.; Mehta, Y.; Nolan, A.; Sonpal, K.; Norton, A.; Tomlison, C.; Dubois, E.; Sullivan, P.; Sauber, R. A study to determine the degree of partial blending of reclaimed asphalt pavement (RAP) binder for high RAP hot mix asphalt. *Constr. Build. Mater.* **2011**, *25*, 150–155. [CrossRef]
33. Karlsson, R.; Isacsson, U. Material-related aspects of asphalt recycling—State-of-art. *J. Mater. Civ. Eng.* **2006**, *18*, 81–92. [CrossRef]
34. Wang, F.; Wang, Z.; Li, C.; Xiao, Y.; Wu, S.; Pan, P. The rejuvenating effect in hot asphalt recycling by mortar transfer ratio and image analysis. *Materials* **2017**, *10*, 574. [CrossRef] [PubMed]
35. Poulikakos, L.D.; dos Santos, S.; Bueno, M.; Kuentzel, S.; Hugener, M.; Partl, M.N. Influence of short and long-term aging on chemical, microstructural and macro-mechanical properties of recycled asphalt mixtures. *Constr. Build. Mater.* **2014**, *51*, 414–423. [CrossRef]
36. Huang, B.; Li, G.; Vukosavljevic, D.; Shu, X.; Egan, B. Laboratory investigation of mixing hot-mix asphalt with reclaimed asphalt pavement. *Transp. Res. Rec. J. Transp. Res. Board.* **2005**, *1929*, 37–45. [CrossRef]
37. Bražiūnas, J.; Sivilevičius, H.; Virbickas, R. Dependences of SMA mixture and its bituminous binder properties on bitumen batching system, mixing time and temperature on asphalt mixing plant. *J. Civ. Eng. Manag.* **2013**, *19*, 862–872. [CrossRef]
38. Vislavičius, K.; Sivilevičius, H. Effect of reclaimed asphalt pavement gradation variation on the homogeneity of recycled hot-mix asphalt. *Arch. Civ. Mech. Eng.* **2013**, *13*, 345–353. [CrossRef]

39. Aurangzeb, Q.; Al-Qadi, I.L.; Abuawad, I.M.; Pine, W.J.; Trepanier, J.S. Achieving desired volumetrics and performance for mixtures with high percentage of reclaimed asphalt pavement. *Transp. Res. Rec. J. Transp. Res. Board.* **2012**, *2294*, 34–42. [CrossRef]
40. Sivilevičius, H.; Vislavičius, K.; Bražiūnas, J. Technological and economic design of asphalt mixture composition based on optimization methods. *Technol. Econ. Dev. Econ.* **2017**, *23*, 627–648. [CrossRef]
41. Valdés, G.; Pérez-Jiménez, F.; Miró, R.; Martínez, A.; Botella, R. Experimental study of recycled asphalt mixtures with high percentages of reclaimed asphalt pavement (RAP). *Constr. Build. Mater.* **2011**, *25*, 1289–1297. [CrossRef]
42. Mučinis, D.; Sivilevičius, H.; Oginskas, R. Factors determining the inhomogeneity of reclaimed asphalt pavement and estimation of its components content variation parameters. *Balt. J. Road Bridge Eng.* **2009**, *4*, 69–79. [CrossRef]
43. Mogawer, W.S.; Austerman, A.J.; Bonaquist, R. Determining the influence of plant type and production parameters on performance of plant-produced reclaimed asphalt pavement mixtures. *Transp. Res. Rec. J. Transp. Res. Board.* **2012**, *2268*, 71–81. [CrossRef]
44. Silva, H.M.R.D.; Oliveira, J.R.M.; Jesus, C.M.G. Are totally recycled hot mix asphalts a sustainable alternative for road paving? *Resour. Conserv. Recycl.* **2012**, *60*, 38–48. [CrossRef]
45. Zhao, S.; Huang, B.; Shu, X.; Woods, M.E. Quantitative Characterization of Binder Blending. *Transp. Res. Rec. J. Transp. Res. Board.* **2015**, *2506*, 72–80. [CrossRef]
46. Babtista, A.M.; Picado-Santos, L.G.; Capitão, S.D. Design of hot-mix recycled asphalt concrete produced in plant without preheating the reclaimed material. *Int. J. Pavement Eng.* **2013**, *14*, 95–102. [CrossRef]
47. DeDene, C.D.; Voller, V.R.; Marasteanu, M.O.; Dave, E.V. Calculation of particle heating times of reclaimed asphalt pavement material. *Road Mater. Pavement Des.* **2014**, *15*, 721–732. [CrossRef]
48. National Cooperative Highway Research Program (NCHRP). *A Manual for Design of Hot Mix Asphalt with Commentary. National Cooperative Highway Research Program. Transportation Research Board of the National Academics*; NCHRP Report 673; NCHRP: Washington, DC, USA, 2011; p. 273.
49. Ma, T.; Huang, X.; Zhao, Y.; Zhang, Y.; Wang, H. Influences of preheating temperature of RAP on properties of hot-mix recycled asphalt mixture. *J. Test. Eval.* **2016**, *44*, 762–769. [CrossRef]
50. Baroux, R. Recyclage des enrobés bitumineux en centrale d'enrobage. Les problémes de matériel. *Bulletin de Liaison des Ratoires des Ponts et Chaussées* **1980**, *105*, 98–102.

Article

Evaluation of Adhesion and Hysteresis Friction of Rubber–Pavement System

Mohammad Al-Assi * and Emad Kassem

Department of Civil and Environmental Engineering, University of Idaho, Moscow, ID 83844, USA; ekassem@uidaho.edu

* Correspondence: alas9935@vandals.uidaho.edu; Tel.: +1-208-310-4850

Received: 1 September 2017; Accepted: 30 September 2017; Published: 7 October 2017

Abstract: Tire-pavement friction is a key component in road safety. Adhesion and hysteresis are the two main mechanisms that affect the friction between rubber tires and pavements. This study experimentally examined the relationship between rubber–pavement adhesion and friction. The adhesive bond energy between rubber and pavement surfaces was calculated by measuring the surface energy components of rubber and aggregates. The friction was measured in the laboratory using a dynamic friction tester. The results revealed that there is a fair correlation between the adhesive bond energy and measured coefficient of friction. A rubber–pavement system with higher adhesion provided higher friction at low speed. In addition, the results demonstrated that there is a strong correlation between rubber–pavement friction and rubber properties. Softer rubber provided higher friction and vice versa. The results of this study provide an experimental verification of the relationship between adhesion and pavement surface friction. The adhesive bond energy and rubber rheological properties could be incorporated in computational models to study tire-pavement friction in different conditions (e.g., speed and temperature).

Keywords: adhesion friction; hysteresis friction; adhesive bond energy; hot mix asphalt

---

1. Introduction

Tire-pavement friction is one of the main factors that contribute to road safety. An adequate level of friction between the vehicle tire and pavement surface reduces the number of crashes, especially in wet pavement conditions [1]; therefore, an appropriate level of traction between rubber tires and pavements is essential for safe driving. Rubber is the principal component of vehicle tires. It is an elastomer that exhibits unique physical and chemical properties. It has low Young's modulus and high yield strain, as compared with other materials.

It is documented that rubber generates three different forms of friction: adhesion, deformation, and wear. Previous studies have shown that the adhesion force is the most influential component in rubber friction in dry conditions and low speed [2,3]. It is postulated that the adhesion between rubber tires and pavements is highly influenced by the surface free energy, which depends on the chemical composition of rubbers. Yet, to the best of the authors' knowledge, no studies have been carried out to investigate the effect of surface free energy of rubber on traction with pavements.

The theory of rubber friction recognizes three major friction forces: adhesion force, bulk deformations (hysteresis), and cohesion losses due to the wearing of rubber [4]. The adhesion friction force is formed due to rubber–pavement interaction at a microtexture level and highly dependent on the true area of contact, and thus the adhesion friction is dominant for smooth contact at low speeds and dry conditions [3]. The hysteresis component of friction is caused by the bulk deformation of rubber material as it comes into contact with pavement asperities. Such deformation causes energy losses as rubber slides over the pavement surface. These energy losses are caused by the hysteretic losses in the rubber material due to loading and unloading [3]. The third major component of friction (cohesion losses) is caused by the wearing of rubber as it slides over the pavement surface. Pavement texture has an important role in the friction components—a recent study showed that higher pavement microtexture and macrotexture result in higher frictional properties [5]. The study also showed that pavement surfaces initially had lower frictional properties due to the thin film of asphalt coating the aggregates at the surface. The friction increased as the binder film was removed by traffic, exposing the microtexture. Then, the friction deceased due to polishing and abrasion of the aggregates at the surface.

2. Objectives

The main goal of this study was to investigate the effect of adhesion between rubber and pavement surface on friction at low speed. In addition, we evaluated the effect of rubber rheological properties on friction with pavements. These objectives were achieved by conducting the following tasks:

1. Measure the surface energy components of different rubber and aggregate materials.
2. Calculate the adhesive bond energy between rubber materials and aggregates.
3. Measure the rheological properties of test rubber materials at different temperatures and loading frequencies.
4. Measure the coefficient of friction between rubber and hot mix asphalt (HMA) substrates.
5. Investigate the correlation between the calculated adhesive bond energy and the measured coefficient of friction between rubber and pavement surfaces.
6. Investigate the correlation between the rubber dynamic modulus and the coefficient of friction.

We measured the surface free energy of test rubber materials using a Sessile drop device, while the surface energy components of aggregates were measured using a universal sorption device (USD). The surface free energy of the rubber materials and aggregates were used to calculate the adhesion between the rubber materials and aggregates used in preparing the HMA test substrates. The friction between rubber sliders prepared using test rubber materials and HMA substrates was measured using the dynamic friction tester (DFT) in dry conditions at low speed. In addition, we measured the dynamic modulus of test rubber materials at different temperatures and loading frequencies using a dynamic mechanical analyzer (DMA).

3. Theory of Surface Energy

There are two major components contributing to pavement friction: adhesion and hysteresis. The adhesion component of pavement friction is dominant at low speeds in dry conditions, while the hysteresis component is dominant at higher speeds in wet conditions [6]. The adhesion between any materials is a function of their surface free energy components, where the adhesion force is developed at the molecular level [7,8]. The atoms at surface have lower level of bond energy compared to the atoms in bulk. As molecules in the material bulk are surrounded by other molecules from all sides, an external work must be applied to create a new surface area [7]. This work is known as the total surface free energy of the material and is denoted with Greek letter gamma (γ), the surface free energy is measured in units of ergs/cm^2 or mJ/m^2. The two most recognized theories to explain the surface free energy are: the two-component theory, and the acid-base theory. According to the two-component theory, the total surface free energy is caused by dispersion forces (e.g., Lifshitz-van der Waals forces) and specific forces (e.g., H-bonding). The total surface free energy is the sum of these two forces as shown in the Equation (1) [9,10].

$$\gamma = \gamma^{\text{Dispersive}} + \gamma^{\text{Specific}} \tag{1}$$

According to the acid-base theory, the total surface free energy for any material has three components; nonpolar component also known as Lifshitz-van der Waals (γ^{LW}), and two polar components: Lewis acid (γ^{+}), and Lewis base (γ^{-}) [11,12]. The total surface energy can be calculated from these three components as given in Equation (2).

$$\gamma^{\mathrm{Total}} = \gamma^{\mathrm{LW}} + 2\sqrt{\gamma^{+}\gamma^{-}} \quad (2)$$

Several theories explain the adhesion between two materials by studying the interfacial forces between the materials [13]. The adhesion between rubber tires and pavement surface is one mechanism leading to friction and is dominant at low speeds. The adhesion between rubber and aggregate can be measured if the surface free energies of both materials are known. The adhesion between two materials (A and B) is a function of the surface free energy components of these two materials, as given in Equation (3) [7].

$$W_{AB} = 2\sqrt{\gamma_A^{LW}\gamma_B^{LW}} + 2\sqrt{\gamma_A^{+}\gamma_B^{-}} + 2\sqrt{\gamma_A^{-}\gamma_B^{+}} \quad (3)$$

The amount of work required to separate the two materials at the interface in vacuum is referred as the adhesive bond energy (W_{AB}). The adhesive bond energy between rubber sample (subscript A) and an aggregate (subscript B) can be calculated using Equation (3), while the cohesive bond energy of a single material (e.g., rubber) (W_{AA}) can be calculated according to Equation (4) [7].

$$W_{AA} = \gamma_A^{LW} + 2\sqrt{\gamma_A^{LW}\gamma_B^{LW}} \quad (4)$$

3.1. Methods for Measuring Surface Energy

3.1.1. Sessile Drop Method

The Sessile drop method is commonly used to measure the surface free energy between a probe liquid and solid surface by means of measuring the contact angle between a droplet of a selected probe liquid and material surface [14]. The Sessile drop test measures the contact angle between a drop of liquid of known surface energy and a solid material, to calculate the surface free energy (SFE) of solid material surfaces. The Young's equation (Equation (5)), defines the equilibrium at the three-phase contact of solid-liquid and gas [15].

$$\gamma_{SV} = \gamma_{Sl} + \gamma_{LV}\cos(\theta) \quad (5)$$

where θ is the contact angle between a solid surface and a drop of probe liquid. The contact angle is measured from a static image using a charged-coupled device (CCD) camera as shown in Figure 1. The Young equation assumes that the surface is chemically homogeneous and topographically smooth.

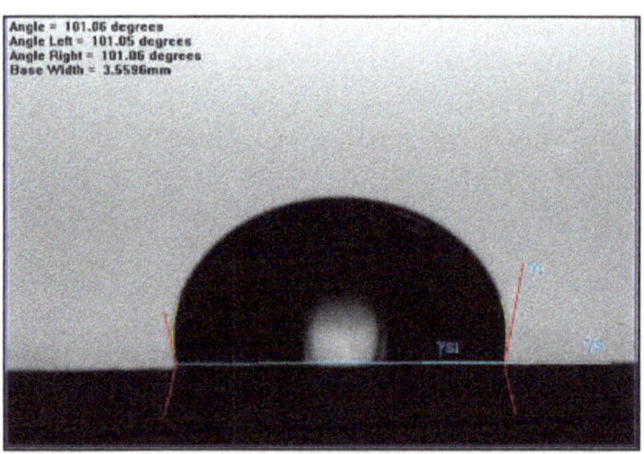

Figure 1. Contact angle between liquid and solid surfaces.

In the Sessile drop technique, a droplet of probe liquid is placed on the test surface using a micro-syringe. The Sessile drop device (Figure 2a) uses a CCD camera static image to measure the contact angle between the used probe liquid of known surface energy and the rubber surface. The contact angle is the average value of the measured right and left contact angles (Figure 2b).

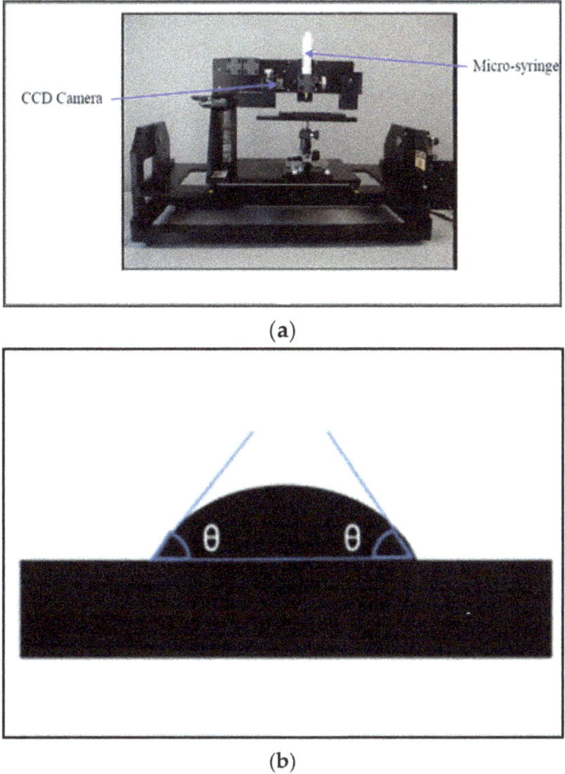

(a)

(b)

Figure 2. (a) Sessile drop device; (b) Sketch of Sessile drop contact angle measurement.

To determine the surface energy components of a material, three different probe liquids are used in the test. However, in order to minimize experimental error in calculating the surface energy components, Little and Bhsain [7] recommended using five different probe liquids (water, glycerol, ethylene glycol, formamide, and diiodomethane) with known surface energy components. Table 1 summarizes the surface energy components of the five probe liquids used in this study.

Table 1. Surface free energy components for selected probe liquids (ergs/cm^2).

Probe Liquid	γ^{LW}	γ^{+}	γ^{-}	γ^{Total}
Water	21.80	25.50	25.50	72.80
Glycerol	34.00	3.92	57.40	64.00
Formamide	39.00	2.28	39.60	58.00
Methylene Iodide	50.80	0.00	0.00	50.80
Ethylene Glycol	29.00	1.92	47.00	47.99

3.1.2. Universal Sorption Device

The surface free energy components of aggregates are often measured using the universal sorption device (USD). In this method, the aggregate particles are placed in a sealed cell under vacuum and controlled temperature. The USD uses a magnetic suspension balance to measure the mass of the aggregate particle and probe vapor adsorbed on the aggregate surface when aggregates are subjected to different pressures of probe vapor. The relationship between the amount of probe vapor adsorbed on aggregate surface and the partial pressure is called the adsorption isotherm which is used to determine the spreading pressure (π_e) [7]. The spreading pressure (π_e) is a function of the surface energy components of aggregate particles (S) and probe vapor (V), as given in Equation (6).

$$\pi_e + 2\gamma_V^{Tot} = 2\sqrt{\gamma_S^{LW}\gamma_V^{LW}} + 2\sqrt{\gamma_S^{+}\gamma_V^{-}} + 2\sqrt{\gamma_S^{-}\gamma_V^{+}} \quad (6)$$

Several researchers used the USD to measure the surface free energy components of aggregates [16–18]. In this study, the USD test was used to determine the surface free energy of the gabbro and limestone aggregates. The surface energy components of rubber and aggregates are used to calculate the adhesive bond energy or work of adhesion (W_{AB}), (Equation (3)) between rubber sliders and HMA surfaces.

4. Test Materials and Laboratory Experiments

In this study, six different rubber materials were used to investigate the frictional properties between laboratory-prepared rubber sliders and HMA slabs prepared from limestone and gabbro aggregates. The rubber materials used in this study included pure gum, Styrene Butadiene Rubber (SBR), Nitrile, Ethylene Propylene Diene Monomer (EPDM), Neoprene, and Butyl. These materials were selected in testing as they cover a wide range of rubber properties in terms of dynamic modulus, elongation, and durability. Table 2 summarizes the rubber properties as provided by the manufacturer.

Table 2. Rubber properties as provided by the manufacturer.

Rubber Type	Tensile Strength (PSI)	Elongation (%)	Composition	Durometer
Pure Gum	3000	600	Organic Gum	40 A Nominal
SBR	800	250	Styrene Butadiene	65 Shore A
EPDM	800	400	Synthetic	60 A Nominal
Nitrile	950	250	Synthetic	50 A Nominal
Neoprene	1000	220	Synthetic	60 A Nominal
Butyl	1000	350	Isobutylene Isoprene	55 Shore A

Two sources of aggregates were evaluated in this study (gabbro and limestone). The gabbro and limestone aggregates were used to prepare HMA slabs. The gabbro is an igneous rock while the limestone is a sedimentary rock. Gabbro has rough surface compared to limestone [19]. These aggregates are used in road construction in the State of Qatar. Table 3 presents the aggregate gradation used to prepare both gabbro and limestone substrates. It should be noted that the same aggregate gradation and binder type (Pen 60–70) were used in preparing the asphalt mixture substrates. The optimum binder content was found to be 4.3% for gabbro and 4.7% for limestone. The mixing and compaction temperatures were 143 °C and 135 °C, respectively.

Table 3. Aggregate gradation used for asphalt mixture slabs.

Sieve Size	% Passing
1.5″	100.0
1″	98.6
3/4″	88.2
1/2″	76.9
3/8″	68.9
N4	47.1
N8	26.5
N16	15.8
N30	10.5
N50	7.9
N100	6.1
N200	4.2
Pan	0.0

The asphalt mixtures were prepared following the AASHTO T-312, and the test slabs were compacted using a small vibratory compactor as shown in Figure 3.

Figure 3. Preparing hot mix asphalt (HMA) slabs.

5. Measuring Frictional Properties

We used a dynamic friction tester (DFT) to measure the coefficient of friction between rubber sliders (Figure 4), prepared using the test rubber materials, and HMA substrate at low speed (20 km/h) in dry conditions. The DFT uses three rubber sliders attached to a rotating desk (Figure 5). The rotating desk is lowered on the surface once the specified speed is reached and the coefficient of friction is measured with speed until the rotating desk comes to a complete stop.

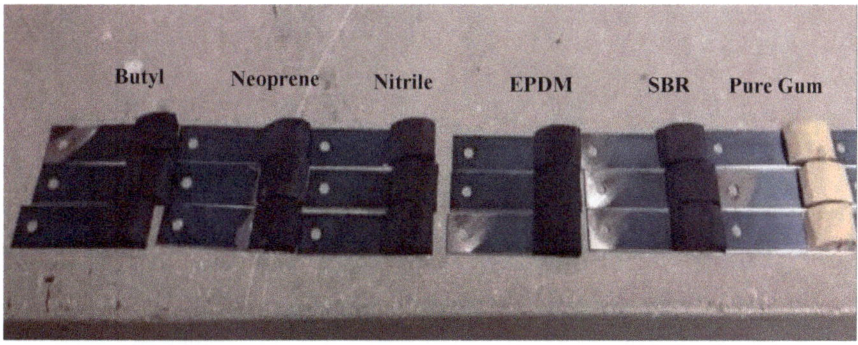

Figure 4. Prepared rubber sliders from test rubber materials.

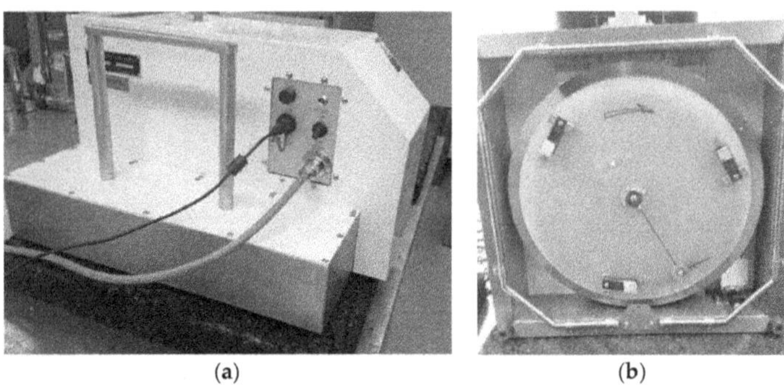

(a) (b)

Figure 5. (**a**) Dynamic friction tester (DFT) device; (**b**) Bottom of the DFT with three rubber sliders.

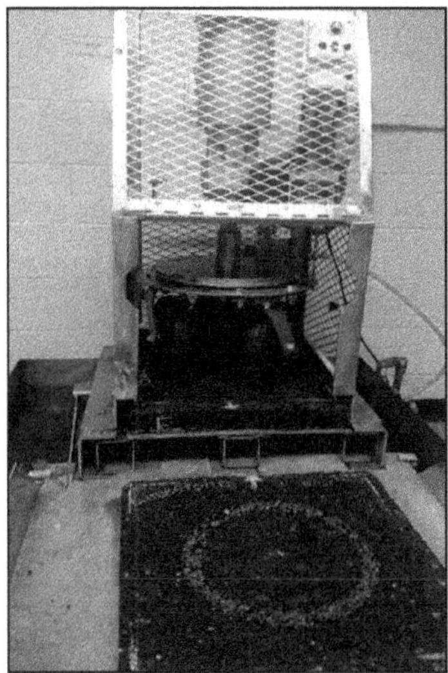

Figure 6. Three-wheel polisher.

For the prepared HMA substrates, the aggregates on the surface were coated with a thin asphalt binder. We used a three-wheel polisher to remove this thin film of binder to ensure direct contact between rubber sliders and aggregates. The surface energy components of asphalt binder are different than those of aggregate samples and thus it may affect the adhesive bond energy between rubber and surface aggregates. In the field, this thin layer of asphalt binder is removed under traffic in a short period of time. The three-wheel polisher consists of pneumatic rubber wheels that rotate over the test HMA substrate as shown in Figure 6. The DFT device was used to measure the coefficient of friction between the prepared rubber sliders (Figure 4) and HMA substrates.

6. Measuring Rubber Properties

The dynamic mechanical analyzer (DMA) device was used to measure the dynamic modulus of rubber samples tested in this study (Figure 7). The rubber test sample for the DMA testing was 50 mm in length, 16.8 mm in width, and 6.4 mm thick. The test was performed at different test temperatures (0 to 70 °C) and frequencies (0.1–70 Hz). The temperature was controlled during the test by placing the DMA inside an environmental chamber. The dynamic modulus is calculated by dividing the maximum applied cyclic tensile stress by the resulting strain.

Figure 7. Testing rubber specimen in the dynamic mechanical analyzer (DMA) device inside a temperature-control chamber.

7. Test Results and Discussion

7.1. Adhesion Friction

The average contact angle between rubber samples and the test probe liquids was recorded using the Sessile drop device. Each measurement represents the average contact angle measured from right and left. The results for the measured contact angles in degrees are presented in Table 4. Based on the contact angle measurements, the surface energy components were calculated for each rubber material using Equation (7), which is discussed in detail by Little and Bhasin [7].

$$W_{LS} = \gamma_L (1 + \cos\theta) = 2\sqrt{\gamma_S^{LW}\gamma_V^{LW}} + 2\sqrt{\gamma_S^{+}\gamma_V^{-}} + 2\sqrt{\gamma_S^{-}\gamma_V^{+}} \quad (7)$$

The surface free energy components are summarized in Table 5. The surface free energy components of the aggregates were measured using a Universal Sorption Device (USD) from a previous study [20]. The surface free energy components for the gabbro and limestone aggregates are shown in Table 6. It was found that gabbro rock has the higher total surface free energy compared to limestone.

Table 4. Average contact angle between rubbers and probe liquids using Sessile drop device (degrees).

Probe Liquid	Water		Glycerol		Ethylene Glycol		Formamide		Diidomethane	
Rubber Type	Contact Angle	Std. Dev	Contact Angle	Std. Dev	Contact Angle	Std. Dev	Contact Angle	Std. Dev	Contact Angle	Std. Dev
Pure gum	89.47	0.69	120.18	0.86	87.14	0.74	83.03	1.21	67.43	0.76
SBR	100.98	0.86	101.28	1.47	87.16	1.05	84.19	0.66	71.74	0.55
Nitrile	89.04	0.74	94.77	0.82	84.32	0.89	90.35	0.99	63.03	1.04
EPDM	108.36	0.74	112.45	0.67	89.61	0.75	93.29	0.87	72.22	13.12
Neoprene	124.58	0.89	112.02	0.74	90.23	0.95	87.54	1.75	69.63	1.64
Butyl	111.32	0.96	117.02	1.21	92.49	0.71	107.16	0.93	64.43	0.98

Table 5. Measured surface energy of rubber materials using the Sessile drop device.

Rubber	Surface Energy Components (ergs/cm^2)				Standard Deviation (ergs/cm^2)		
	γ^{LW}	γ^{+}	γ^{-}	γ^{Total}	γ^{LW}	γ^{+}	γ^{-}
Pure gum	21.21	0.00	2.26	21.21	0.44	0.00	0.42
SBR	22.11	0.00	2.24	22.11	0.46	0.00	0.40
Nitrile	17.08	0.00	10.33	17.08	0.54	0.00	0.67
EPDM	16.77	0.00	0.35	16.77	1.53	0.00	0.17
Neoprene	15.30	0.00	0.00	15.30	0.69	0.00	0.00
Butyl	18.79	0.00	0.00	18.79	0.51	0.00	0.00

Table 6. Measured surface energy of aggregates using universal sorption device (USD) device, (adapted from [20]).

Material	Surface Energy Components (ergs/cm^2)			
	γ^{LW}	γ^{+}	γ^{-}	γ^{Total}
Limestone	69.35	0.28	1075.40	104.18
Gabbro	57.37	3.34	6277.96	346.85

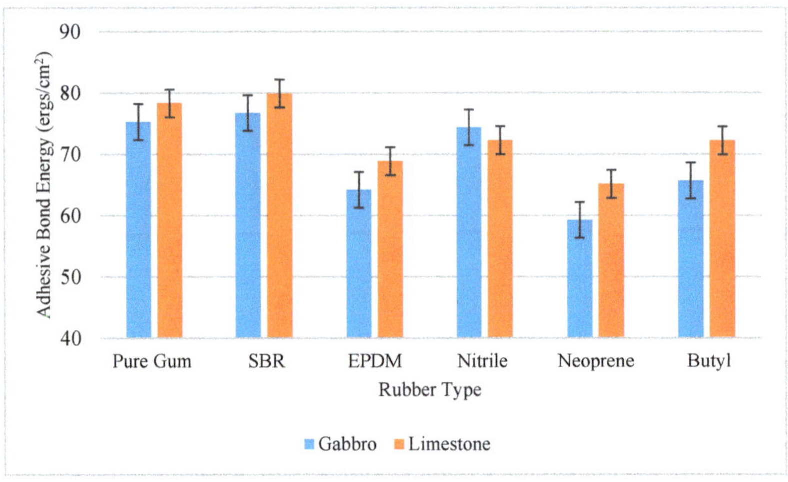

Figure 8. Adhesive bond energy between rubbers and aggregates materials.

The adhesive bond energy between different rubber and aggregate samples was calculated from their surface free energy components according to Equation (3). Figure 8 shows the calculated adhesive bond energies between the rubber and aggregates. The SBR and Pure Gum rubber materials were found to provide the highest adhesion with test substrates made with gabbro and limestone. The adhesive bond energy between limestone and SBR rubber was 79.91 ergs/cm^2. The Neoprene was found to provide the lowest adhesion with HMA substrates compared to other rubber materials. Also, the limestone was found to provide higher adhesion with rubber materials compared to gabbro in most cases.

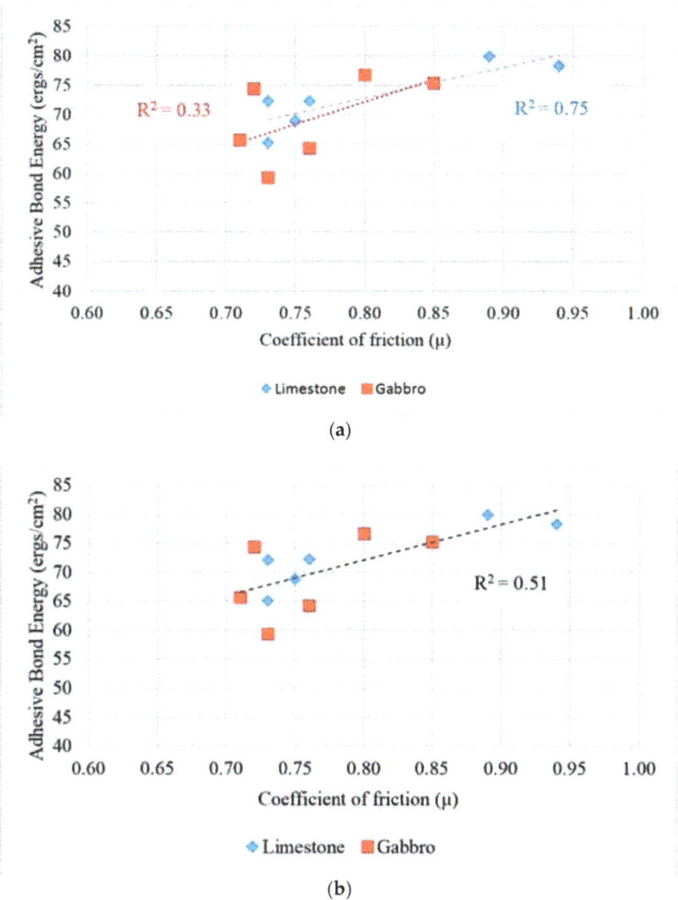

Figure 9. Measured coefficient of friction vs. the adhesive bond energy for (**a**) limestone and gabbro slabs separately (**b**) both gabbro and limestone slabs.

The coefficient of friction between rubber sliders and HMA substrates was measured using the DFT at low speed (20 km/h) in dry conditions. The adhesion component of the friction is dominant at low speed. Figure 9a shows the relationship between the adhesive bond energy and the coefficient of friction for gabbro and limestone separately, while Figure 9b shows the relationship between the coefficient of friction versus the adhesive bond energy between test rubber materials and all the test HMA substrates (both gabbro and limestone). It was found that there is a fair correlation between the adhesive bond energy and measured coefficient of friction. This relationship demonstrates that higher adhesion between rubber and pavement surface increases the coefficient of friction at low speed as one expects; however, this study provided an experimental verification to such relationship. In the meantime, we believe that this relationship is affected by two experimental limitations. First, the asphalt binder film was not fully removed from the aggregates on the surface. We attempted to minimize this effect be removing most of the film thickness by polishing the slab using the three-wheel polisher for 5000 cycles until the aggregate particles surface is exposed. The application of more cycles was found to polish the aggregates which should be avoided since the surface energy components of aggregate will change with change in the texture of the surface. Second, the surface energy components of rubber were calculated based on the contact angle between smooth rubber samples and probe liquids.

Meanwhile, we observed that the surface of rubber sliders, used in DFT, was not completely smooth due to cutting the rubber sheet to prepare the sliders. These two limitations may contribute to this fair correlation between adhesive bond energy and measured coefficient of friction at low speed.

7.2. Hysteresis Friction

The DMA results demonstrated that the dynamic modulus (E*) of rubber material increased with the loading frequency and decreased with temperature as expected. Figure 10 shows the E* master curves of the test rubber materials. The Pure Gum rubber provided the lowest E* value compared to other rubber materials, while the Nitrile rubber had the highest E*. From Figure 10, it can be clearly seen that the Pure Gum rubber is the softest followed by the SBR rubber while the Nitrile rubber is the stiffest. The rubbers dynamic modulus at room temperature (20 °C) were correlated with the measured coefficient of friction.

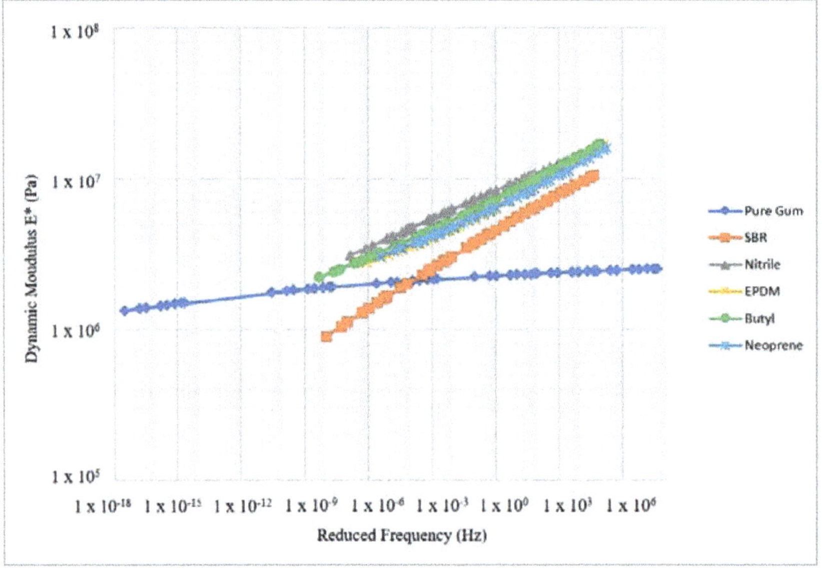

Figure 10. E* Master curves of the test rubber materials.

Figure 11 shows the relationship between the dynamic modulus values of rubber samples and the coefficient of friction between rubber sliders and HMA substrates. There is a strong correlation between the complex modulus at room temperature (20 °C) and the coefficient of friction at 30 km/h. The researchers observed that there was no significant change in the coefficient of friction measured using DFT with speed after 30 km/h. Figure 11 demonstrates that softer rubbers provide higher coefficient of friction while stiffer rubbers provided lower coefficient of friction regardless the aggregate type.

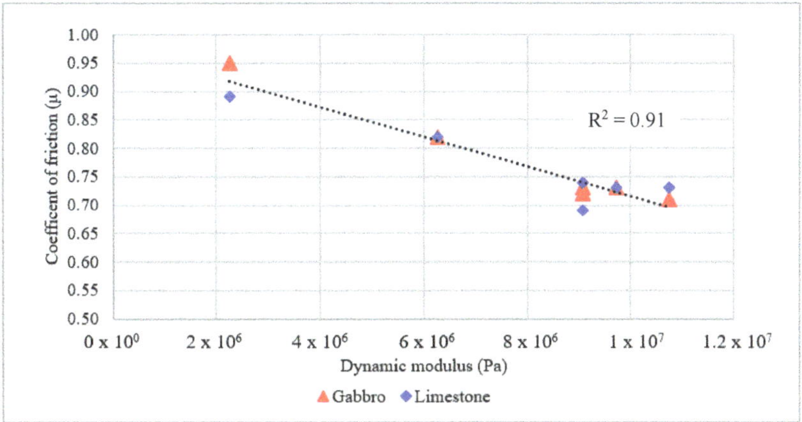

Figure 11. Relationship between the rubber dynamic modulus and the coefficient of friction.

8. Summary and Conclusions

This study experimentally examined the relationship between the rubber–pavement adhesion and friction. We calculated the surface energy components for rubber materials by measuring the contact angle between rubber materials and different probe liquids. The contact angle was measured using a Sessile drop device. Similarly, the surface energy components of aggregate samples were measured using a USD device. The adhesive bond energy between rubber and aggregates was calculated based on the surface energy components of both materials. In addition, we measured the coefficient of friction between different rubber materials and HMA substrates prepared using different rock types. The results showed that there is a fair correlation between adhesive bond energy and friction between rubber and pavement surfaces. In addition, the results demonstrated that there is a strong relationship between friction and rubber elastic modulus. Softer rubber provided higher friction. The results of this study provide an experimental verification of the relationship between adhesion and friction of rubber and pavement surface. The adhesive bond energy and rubber rheological properties could be incorporated in computational models to study tire-pavement friction in different conditions (e.g., speed and temperature). We recommend testing more aggregate types and rubber materials to evaluate the adhesion friction.

Acknowledgments: This publication was made possible by an NPRP award (NPRP No. 7-482-2-184: Thermo-Mechanical Tire-Pavement Interaction: Computational Modelling and Field Measurements) from the Qatar National Research Fund (QNRF – a member of The Qatar Foundation). The statements made herein are solely the responsibility of the authors.

Author Contributions: Mohammad Al-Assi and Emad Kassem conceived, designed and performed the experiments, analyzed the data, and wrote the paper.

Conflicts of Interest: The authors declare no conflict of interest.

References

1. Noyce, D.; Bahia, H.; Yambo, J.; Kim, G. *Incorporating Road Safety into Pavement Management: Maximizing Asphalt Pavement Surface Friction for Road Safety Improvements*; Draft Literature Review and State Surveys; Midwest Regional University Transportation Center (UMTRI): Madison, WI, USA, 2005.
2. Roberts, A.D. Rubber Adhesion at High Rolling Speeds. *J. Nat. Rubber Res.* **1988**, 3, 4.
3. Hall, J.W.; Smith, K.L.; Titus-Glover, L. *Guide for Pavement Friction*; Contractors final report for NCHRP Project 01-43; National Cooperative Highway Research Program: Washington, DC, USA, 2009. Available online: https://www.nap.edu/download/23038# (accessed on 12 August 2017).

4. Kummer, H. *Unified Theory of Rubber and Tire Friction*, 1st ed.; College of Engineering, Pennsylvania State University: University Park, PA, USA, 1966.
5. Woodward, D.; Millar, P.; Lantieri, C.; Sangiorgi, C.; Vignali, V. The wear of Stone Mastic Asphalt due to slow speed high stress simulated laboratory trafficking. *Constr. Build. Mater.* **2016**, *110*, 270–277. [CrossRef]
6. Dunford, A. Friction and the Texture of Aggregate Particles Used in the Road Surface Course. Ph.D. Thesis, University of Nottingham, Nottingham, UK, 2013.
7. Little, D.; Bhasin, A. *NCHRP Report W104: Using Surface Energy Measurements to Select Materials for Asphalt Pavement*; Transportation Research Board: Washington, DC, USA, 2006.
8. Berg, J. *Interfaces and Colloids*; World Scientific: Hackensack, NJ, USA, 2010; pp. 250–268.
9. Fowkes, F. Determination of Interfacial Tensions, Contact Angles, and Dispersion Forces in Surfaces by Assuming Additivity of Intermolecular Interactions in Surfaces. *J. Phys. Chem.* **1962**, *66*, 382. [CrossRef]
10. Fowkes, F. Attractive Forces at Interfaces. *Ind. Eng. Chem.* **1964**, *56*, 40–52. [CrossRef]
11. Van Oss, C.; Chaudhury, M.; Good, R. Interfacial Lifshitz-Van Der Waals and Polar Interactions in Macroscopic Systems. *Chem. Rev.* **1988**, *88*, 927–941. [CrossRef]
12. Van Oss, C. Use of The Combined Lifshitz–Van Der Waals and Lewis Acid–Base Approaches in Determining the Apolar and Polar Contributions to Surface and Interfacial Tensions and Free Energies. *J. Adhes. Sci. Technol.* **2002**, *16*, 669–677. [CrossRef]
13. Adamson, A.; Gast, A. *Physical Chemistry of Surfaces*, 1st ed.; Wiley: New York, NY, USA, 1997.
14. Hejda, F.; Solar, P.; Kousal, J. Surface Free Energy Determination by Contact Angle. In Proceedings of the 19th Annual Conference of Doctoral Students—WDS 2010, Prague, Czech, 1–4 June 2010; Part III. pp. 25–30.
15. Bracco, G.; Holst, B. *Surface Science Techniques*, 1st ed.; Springer: Berlin, Germany, 2013.
16. Cheng, D. Surface Free Energy of Asphalt-Aggregate Systems and PerformanceAnalysis of Asphalt Concrete Based on Surface Free Energy. Ph.D. Thesis, Texas A&M University, College Station, TX, USA, 2002.
17. Hefer, A. Adhesion in Bitumen-Aggregate Systems and Quantification of the Effects of Water on the Adhesive Bond. Ph.D. Thesis, Texas A&M University, College Station, TX, USA, 2004.
18. Zollinger, C. Application of Surface Energy Measurements to Evaluate Moisture Susceptibility of Asphalt and Aggregates. Master's Thesis, Texas A&M University, College Station, TX, USA, 2005.
19. Masad, E.; Rezaei, A.; Chowdhury, A.; Harris, P. Predicting Asphalt Mixture skId Resistance Based on Aggregate Characteristics. Technical Report 0-5627-1. 2009. Available online: https://static.tti.tamu.edu/tti.tamu.edu/documents/0-5627-1.pdf (accessed on 17 August 2017).
20. Kassem, E.; Garcia Cucalon, L.; Masad, E.; Little, D. Effect of warm mix additives on the interfacial bonding characteristics of asphalt binders. *Int. J. Pavement Eng.* **2016**, 1–14. [CrossRef]

Article

Research on Performance of a Dense Graded Ultra-Thin Wearing Course Mixture

Lei Geng [1], Tao Ma [2,*], Junhui Zhang [3], Xiaoming Huang [2] and Pengsen Hu [2]

1 Jiangsu Sinoroad Engineering Research Institute Co. LTD., 19 Lanhua Road, Nanjing 211800, China; gl@sinoroad.com
2 School of Transportation, Southeast University, 2 Sipailou, Nanjing 210096, China; huangxm@seu.edu.cn (X.H.); 220163555@seu.edu.cn (P.H.)
3 State Engineering Laboratory of Highway Maintenance Technology, Changsha University of Science and Technology, Changsha 410114, China; zjhseu@csust.edu.cn
* Correspondence: matao@seu.edu.cn; Tel.: +86-15805160021

Received: 7 July 2017; Accepted: 2 August 2017; Published: 7 August 2017

Abstract: This paper focused on the design and performance characterization of a modified ultra-thin wearing course mixture (M-UWM). A dense graded ultra-thin wearing course mixture with nominal maximum aggregate size of 10 mm was designed and named as UWM10. A multi-chain polyolefin modifier was used to modify the performance of UWM10 to get M-UWM10. Based on different laboratory performance tests including wheel tracking tests, low-temperature bending beam tests, immersion Marshall tests and freeze-thaw splitting tests, the high temperature rutting resistance, low-temperature cracking resistance and moisture resistance of the designed M-UWM10 were evaluated. The texture depth tests and wheel tracking tests were combined to characterize the degradation behaviour of the surface texture depth of M-UWM10. Based on test roads, the bonding conditions between the wearing course layer that consisted of M-UWM10 and its sublayer were evaluated by computed tomography (CT) scanning test and pull out test. Filed texture depth tests were also conducted on the test roads. It is proved that the designed wearing course mixture M-UWM10 shows excellent pavement performance as well as better wearing resistance and interlayer bonding than the traditional wearing course mixture.

Keywords: ultra-thin wearing course; road performance; texture depth; interlayer bonding

---

1. Introduction

An ultra-thin paving technique, usually with 1.5 to 2.5 centimetres pavement thickness, namely Novachip, originated from France in the late 1970s [1]. It has several advantages like short construction time, good anti-skid performance and good water permeability [1–4]. These distinctive features have been gradually gaining popularity, as ultra-thin wearing layers are being frequently applied in old highway maintenance as well as serving as wearing layers in new constructed highways [2,3,5,6]. Apart from the excellent functional performance, the ultra-thin paving technique is also cost-effective, environmental-friendly and a sustainable construction method [7–9], which is attracting more and more attention from researchers.

Kandhal and Lockett [8] examined the road performance of two Novachip projects in Alabama, USA. After 4.5 years of service, both two Novachip projects showed no significant ravelling, concluding that the cohesion between the ultra-thin wearing layer and underlying asphalt layer is good. However, the ultra-thin wearing layer presents more surface friction than the hot mixture asphalt (HMA) wearing layer.

In order to identify appropriate evaluation indexes for ultra-thin wearing layers, Tan, Yao, Wang, Bian and Yu-Xiang [5] conducted several experiments (thermal stress restrained sample tests, permeability tests, skid resistance tests, and indoor abrasion tests) on asphalt mixture with three different surface layers—stone mastic asphalt mixture with a nominal maximum aggregate size of 10 mm (SMA10), ultra-thin asphalt concrete with a nominal maximum aggregate size of 10 mm (UTAC10) and Novachip Type C. The result shows that frozen broken temperature and frozen broken strength could be regarded as appropriate evaluation indexes for low temperature performance of ultra-thin surface layers. Moreover, sliding attenuation degree and initial British Pendulum Number (BPN) value are proposed to measure the anti-skid performance of the ultra-thin surface layer. Furthermore, the author indicates that the freeze-thaw splitting strength ratio and freeze-thaw splitting strength should be combined to evaluate resistance of water damage performance.

Zeng et al. [10] studied the anti-skid performance, permeability and texture depth of three different asphalt mixtures—dense-graded asphalt mixture with a nominal maximum aggregate size of 13.2 mm (AC13), open graded friction course with a nominal maximum aggregate size of 13.2 mm (OGFC13) and semi-open gradation Novachip Type C. Results show that Novachip Type C as an ultra-thin wearing layer has better overall performance in terms of skid resistance, permeability and texture depth. Furthermore, Novachip Type C was proven to have better shear capability than other types of asphalt mixture, under the same tack coat material and dosage. Moreover, Yang, et al. [11] investigated the moisture stability, high temperature stability and low temperature cracking resistance of the same three types of asphalt mixture (AC-13, OGFC-13 and semi-open gradation Novachip Type C). The result shows that Novachip Type C has better high temperature stability than the other two types of asphalt mixture. However, the low temperature cracking resistance and moisture stability performance of Novachip Type C is worse than the AC-13 asphalt mixture but better than the OGFC-13 asphalt mixture.

Yang, Shen and Gao [4] argue that the current design specification in China [12] does not take temperature into consideration and the interlayer contact is presumed as completely continuous. The actual environmental condition subjected to pavement structure is not sufficiently considered in the specification. Thus, the author conducted an experiment to investigate the shear stress between the ultra-thin wearing layer and the underlying asphalt layer under different temperatures and interlayer contact situations. The result shows that maximum interlayer shear stress is negatively correlated with temperature and interlayer sliding coefficient.

The cohesiveness between the thin surface layer and underlying asphalt layer is proven to be an important factor that influences the performance and durability of the thin surface layer [13]. Thus, Wu [14] proposed a modified pull test method to test cohesiveness between the ultra-thin wearing layer and underlying asphalt layer. Three types of ultra-thin wearing layers (Epoxy modified bitumen, Resin and Cement) were selected. The experiment was conducted under three different temperatures (0, 10 and 20 °C) with a loading speed of 0.025 MPa/s. The result demonstrates the feasibility of the proposed modified test method, provided that the interface of the two layers is the weakest area. Moreover, the author also demonstrated that the Epoxy modified bitumen-based surface layer has better cohesiveness than the other two types of layers.

Pavement performance decreases over time, and thus needs to be rehabilitated once the existing pavement is unable to satisfy the traffic demand. Several researchers also focused on ultra-thin wearing layer recycling since high quality materials are used in the ultra-thin wearing layer. Both environment protection and cost reduction could be achieved by recycling high-quality used materials. Rahaman et al. [15] conducted research to investigate if the reclaimed ultra-thin wearing layer material could be used in the chip seal or Superpave mixtures. Sweep test based on the American Society for Testing and Materials (ASTM) D7000-04 was conducted to observe the chip retention of ultra-thin wearing layer millings. Different percentages (0%, 10%, and 20%) of ultra-thin wearing layer millings were added to the Superpave mixture. Rutting, stripping and moisture sensitivity tests were conducted to evaluate the performance of the Superpave mixture. The sweep test indicates that reclaimed ultra-thin wearing layer millings do not contribute to chip retention. However, an ultra-thin wearing layer milling addition was proven to have a positive effect on the performance of Superpave mixture.

The ultra-thin paving technique has been extensively used in China [16], since its first adoption in the 1990s [17]. However, several drawbacks of the ultra-thin paving technique occurred in engineering practices such as low degree of compaction, moisture damage and rapid attenuation of sliding resistance at a later stage. As to open-graded ultra-thin paving material, the situation could be even worse. Because of the high void ratio of open-graded ultra-thin paving material, water could easily permeate into the pavement structure. It would lower the bonding strength between pavement layers, causing raveling and reducing the durability of ultra-thin paving materials.

In view of the existing problems of the current ultra-thin technique, a dense-graded ultra-thin wearing mixture was designed and then further modified by a multi-chain polyolefin modifier to get the modified ultra-thin wearing mixture in this study. Both laboratory and field tests were conducted to fully evaluate the road performances, wearing resistance and bonding conditions with sub-layers of the modified ultra-thin wearing mixture.

2. Experimental

2.1. Mix Design for Different Asphalt Mixtures

Styrene-butadiene-styrene (SBS) modified asphalt with penetration grade of PG-70, basalt aggregate and limestone powders were used for all of the asphalt mixtures prepared in this study. According to the recommendation by Chinese specification and previous studies, the polyester fibre was used for SMA mixtures. To improve the rutting resistance and the anti-wearing performance of ultra-thin wearing course mixture (UWM), a multi-chain polyolefin modifier was added to the SBS modified asphalt to prepare modified ultrathin wearing course mixtures (M-UWM).

Five different asphalt mixtures were studied including an SMA mixture with nominal maximum aggregate size of 13.2 mm (SMA13), dense-graded asphalt mixture with nominal maximum aggregate size of 13.2 mm (AC13), SMA mixture with nominal maximum aggregate size of 9.5 mm (SMA10), ultrathin wearing course mixture with nominal maximum aggregate size of 9.5 mm (UWM10), and a modified ultrathin wearing course mixture with a nominal maximum aggregate size of 9.5 mm (M-UWM10). The SMA13 and AC13 are commonly used in the pavement surface layer while SMA10, UWM10 and M-UWM10 are mainly used in the wearing course above the pavement surface layer [18]. While the SMA10 is a traditional wearing course mixture, the UWM10 and M-UWM10 are newly developed wearing course mixtures. Based on Marshall mix design, the gradations for different asphalt mixtures are shown in Figure 1 and the Marshall design parameters for different asphalt mixtures are listed in Table 1.

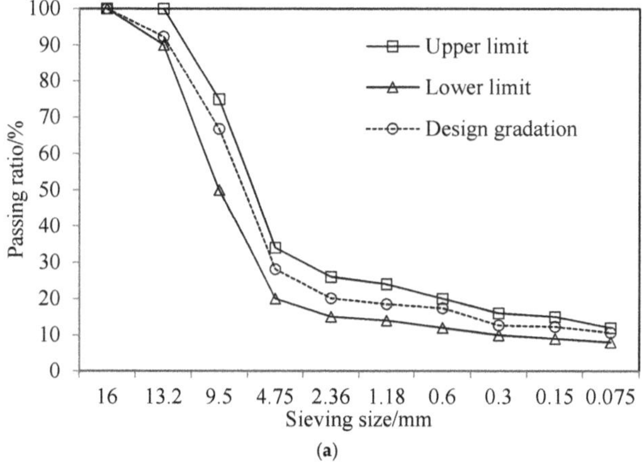

(a)

Figure 1. *Cont.*

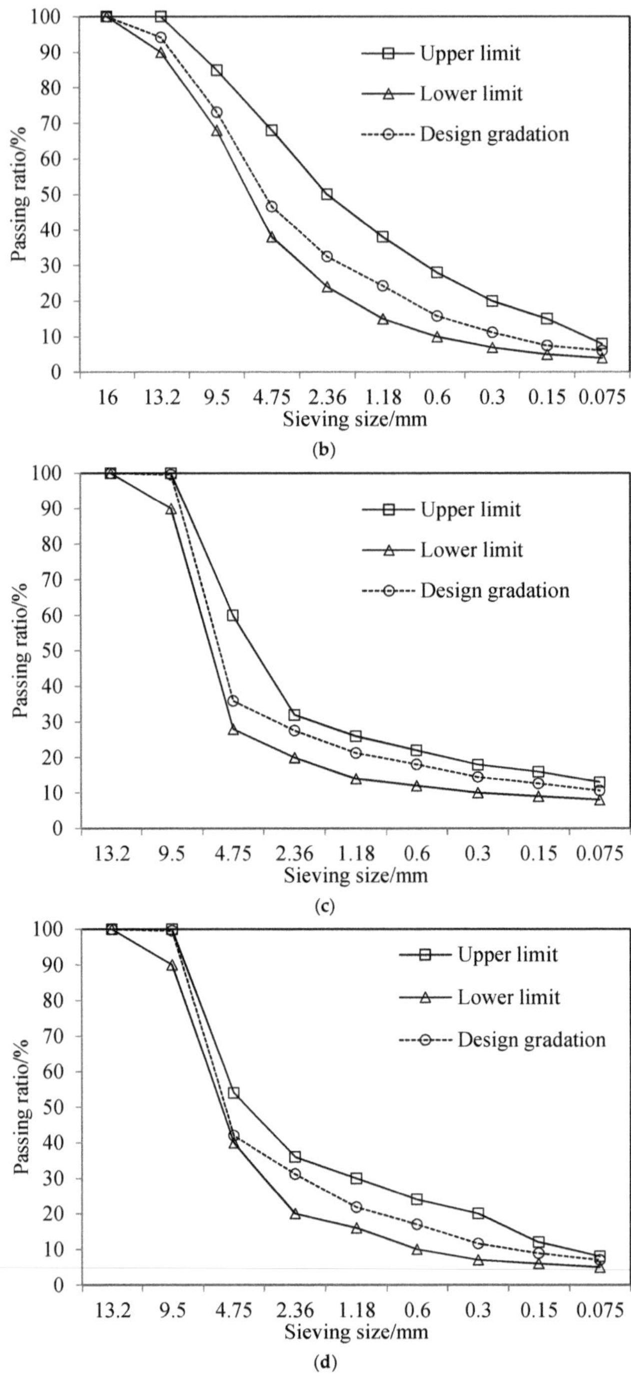

Figure 1. The designed gradation for different asphalt mixtures: (**a**) SMA13; (**b**) AC13; (**c**) SMA10; (**d**) UWM10 and M-UWM10.

Table 1. Marshall design parameters for different asphalt mixtures.

Mixture Type	Asphalt Content (%)	VV (%)	VMA (%)	VFA (%)	Marshall Stability (kN)	Flow Value (0.1 mm)
SMA13	6.0	4.3	18.28	76.7	8.9	26.8
AC13	5.1	4.8	19.90	75.9	8.52	27.4
SMA10	6.3	4.2	18.53	77.5	10.54	28.9
UWM10	5.2	5.1	16.0	68.0	9.83	26
M-UWM10	5.2	4.7	15.6	70.1	12.00	24

VV: percent air voids in bituminous mixtures; VMA: percent voids in mineral aggregate in bituminous mixtures; VFA: percent voids in mineral aggregate that are filled with asphalt in bituminous mixtures.

Based on the designed asphalt mixtures, two test roads were paved. One was paved with AC13 used in the surface layer and M-UWM10 used in the wearing course above the AC13 layer. The other was paved with SMA13 used in the surface layer and SMA10 used in the wearing course above the SMA13 layer.

2.2. Test Procedures

2.2.1. Road Performance Evaluation

Since the designed ultra-thin wearing course mixtures are mainly used in the top surface layer or the wearing course layer of pavement to improve the surface functions instead of structure capacity of asphalt pavement, they should have good road performance, especially excellent rutting resistance and water stability to bear the environment conditioning and wheel loading. Thus, based on the standard test specifications of China, the road performances including permeability, high-temperature rutting resistance, low-temperature cracking resistance, moisture resistance were evaluated by permeability tests, wheel tracking tests, low-temperature bending beam tests, immersion Marshall tests and freeze-thaw splitting tests [19,20]. The wheel tracking test was conducted at 60 °C to get the dynamic stability to describe and compare the high-temperature rutting resistance of different asphalt mixtures. The low-temperature bending beam test was conducted at −10 °C to get the failure strain to describe and compare the low-temperature cracking resistance of different asphalt mixtures. Both immersion Marshall tests and freeze-thaw splitting tests were conducted to get the Marshall strength ratio and tensile strength ratio, separately, in order to describe and compare the moisture resistance of different asphalt mixtures.

2.2.2. Texture Depth Evaluation

To guarantee the driving safety, high-quality skidding resistance is an important characteristic for asphalt mixtures used in the top surface layer or the wearing course layer of pavement. Thus, the surface texture depth (TD) of asphalt mixture and pavement is usually used to describe the skidding resistance. However, the texture depth (TD) test that followed the standard protocol of Chinese specification cannot reveal the wearing resistance of asphalt mixtures, which is another important characteristic for asphalt mixtures used in the top surface layer or the wearing course layer of pavement. Thus, to reveal the degradation behaviour of surface texture of different asphalt mixtures that can represent their wearing resistance, the texture depth test was combined with the wheel tracking test in this study. The designed test processes were summarized as follows:

1. After the rectangular specimen with dimensions of 300 × 300 × 50 mm was prepared for the wheel tracking test, its original texture depth (TD) named K_0 was measured, as shown in Figure 2a.
2. After the silica sands on the specimen surface were washed away, the specimen was submitted to the wheel tracking test for 180 minutes.
3. After the wheel tracking test, the rutting area right underneath the wheel loading positions was cut from the tested specimen and then weighed to get its mass m_1, as shown in Figure 2b.

4. According to the texture depth test, sands were laid on the surface of the cut strip specimen, as shown in Figure 2c, and then the total mass m^2 for the sands and strip specimen was weighed, as shown in Figure 2d.
5. The texture depth K_1 for the cut strip specimen, which represents the rutting area under the wheel loading positions after the wheel tracking test, was determined by Equation (1):

$$K_1 = \frac{(m_2 - m_1)}{\rho_G \cdot s}, \quad (1)$$

where s is the top surface area of the cut strip specimen, and ρ_G is the density of the silica sands.

6. By comparing K_0 and K_1, the degradation degree of the surface texture depth caused by wheel loading can be obtained.

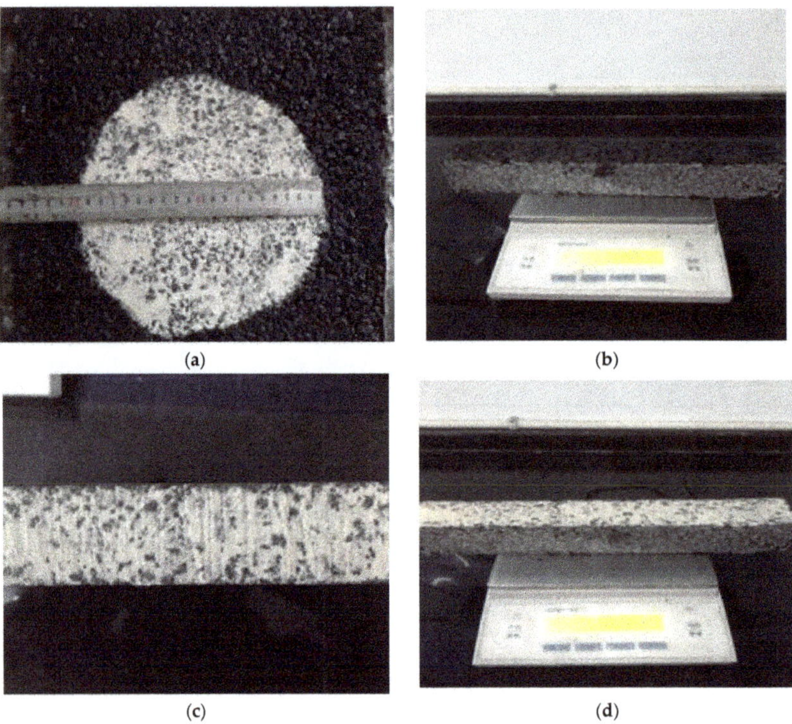

Figure 2. Test processes for combined texture depth test and wheel loading test: (**a**) measuring K_0 for the specimen of wheel loading test; (**b**) measuring m_1 for the cut strip specimen; (**c**) laying sands on the top surface of the strip specimen; (**d**) measuring m_2 for the strip specimen with sands.

2.2.3. Interlay Bonding Evaluation

The interlayer bonding between the wearing course layer and its sublayer has important influences on its long-term durability. Thus, both the CT scanning test and the pull-out test were used to evaluate the interlayer bonding conditions.

The CT scanning test was conducted to evaluate the air voids within the wearing course mixture and between the wearing course layer and its sublayer. The basic principle for the 3D CT scanning technique was explained in Figure 3a and the basic parameters were shown in Table 2. Firstly, the specimen was scanned by transmitted X-rays from multiple directions, and then attenuated X-rays after transmission were gathered by the detector to rebuild 2D or 3D grey level images. The scanned specimen was a standard cylindrical sample cored from pavement and was cut flat at each side by the cutting machine. The test conditions were: voltage 200 kV, current 0.43 mA, and integration time of 300 ms.

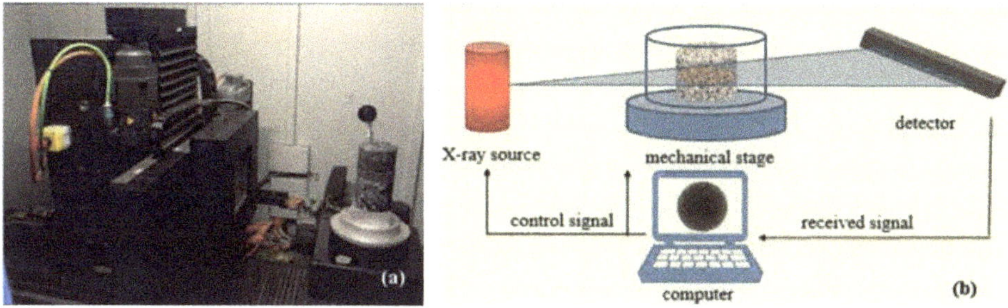

Figure 3. The basic principle and equipment for a computed tomography (CT) scanning test.

Table 2. Basic parameters of the computed tomography (CT) technique.

Type	x-axis (mm)	y-axis (mm)	z-axis (mm)	3D-xy-Pixel Size, 3D-z-Pixel Size (mm)
Specification	200	420	630	0.113

As shown in Figure 4, the pull-out tests were conducted on the field pavements with M-UWM10 wearing course and SMA10 wearing course. The bonding strength between the wearing course layer and its sublayer can be measured.

Figure 4. Field pull out test.

3. Result and Discussions

3.1. Road Performance Evaluation

Road performance evaluation were conducted for all of the five designed asphalt mixtures including M-UWM10, UWM10, SMA10, SMA13 and AC13. The test results of dynamic stability from the wheel tracking test, failure strain from the low-temperature bending beam test, the Marshall stability ratio from the immersion Marshall test and the tensile strength ratio from a freeze-thaw splitting test were shown in Figure 5, respectively. The requirements for a different road performance index from the Chinese specifications [21,22] were shown in Table 3.

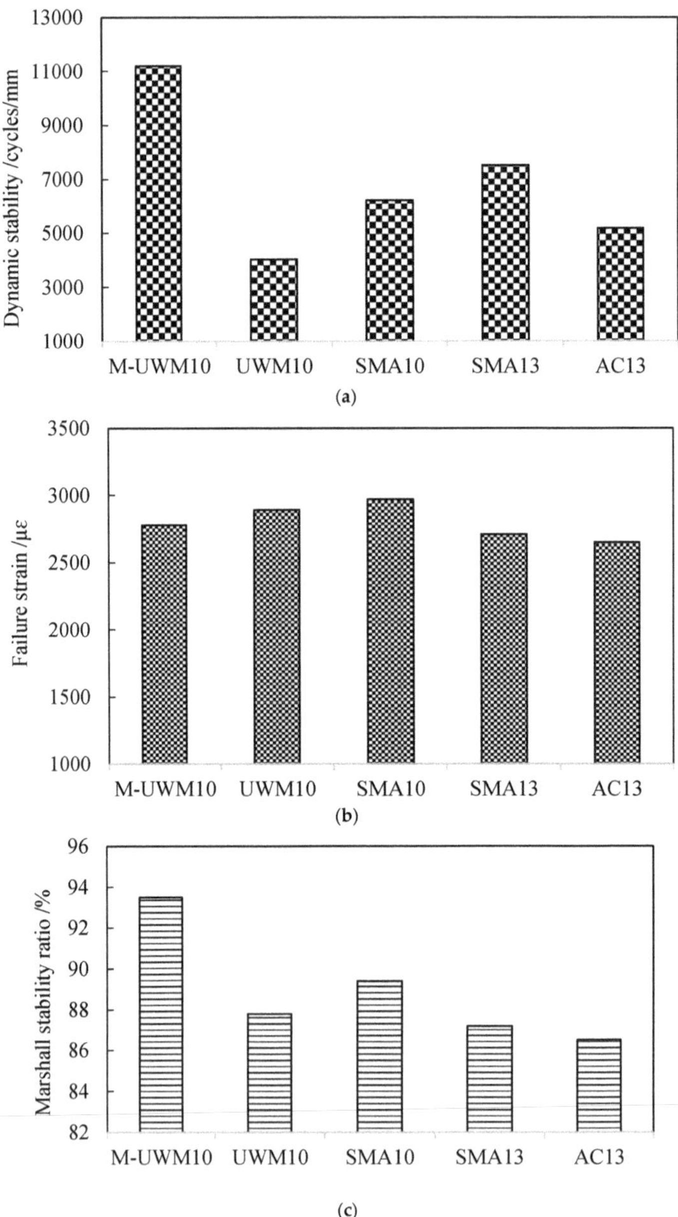

Figure 5. *Cont.*

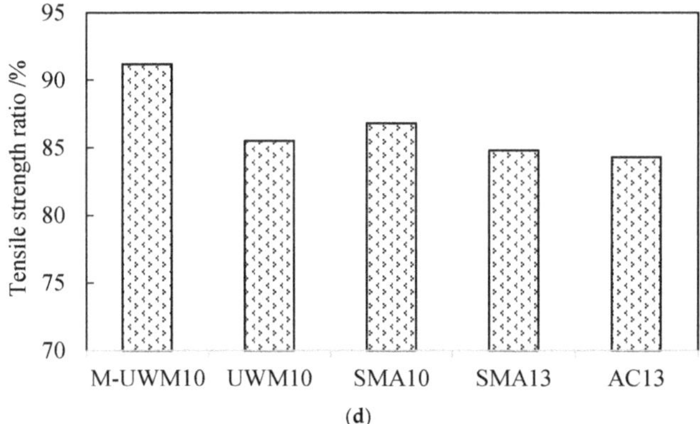

(d)

Figure 5. Test results for road performance evaluation: (**a**) dynamic stability; (**b**) failure strain; (**c**) Marshall stability ratio; (**d**) tensile strength ratio.

Table 3. Pavement performance verification for the M-UWM10 mixture

Performance Index	Dynamic Stability (cycles/mm)	Failure Strain (με)	Marshall Stability Ratio (%)	Tensile Strength Ratio (%)
Requirements	≥3000	≥2500	≥85	≥80
Test method	T0719-2011	T0715-2011	T0709-2011	T0729-2000

First of all, from Figure 5 and Table 3, it can be seen that all of the designed asphalt mixtures can well meet the basic specification requirements. Based on Figure 5a, it can be seen that the M-UWM10 has much higher dynamic stability than the other asphalt mixtures. It proves that the multi-chain polyolefin modifier can well improve the high-temperature rutting resistance of the wearing course. From Figure 5b, it can be seen that the M-UWM10 has a similar failure strain to the other asphalt mixtures. It indicates that the multi-chain polyolefin modifier shows few influences on the low-temperature cracking resistance of the wearing course. From Figure 5c,d, it is clearly seen that the M-UWM10 has the highest Marshall stability ratio and tensile strength ratio. It means that the multi-chain polyolefin modifier can well improve the moisture resistance of the wearing course. Thus, the M-UWM10 has a better high-temperature rutting resistance and moisture resistance than other asphalt mixtures, which are beneficial for the wearing course to bear heavy traffic loading.

3.2. Texture Depth Evaluation

According to the previous test plan, all of the designed asphalt mixtures including M-UWM10, UWM10, SMA10, SMA13, and AC13 were submitted to the combination of the texture depth test and the wheel tracking test. The measured texture depth for different asphalt mixtures before and after the wheel tracking test are shown in Figure 6a, and the loss ratios defined as the texture depth after the wheel tracking test compared to the texture depth before the wheel tracking test are shown in Figure 6b. The rutting depth, for all of the five asphalt mixtures after the wheel tracking test are also shown in Figure 6b, and the correlation between the reduced ratio of texture depth and the rutting depth after the wheel tracking test is shown in Figure 6c. As shown in Figure 7, field texture depth tests were also conducted for the two test roads. Table 4 shows the field measured values of the original texture depth and the texture depth after one year of traffic loading for the two test roads with the M-UWM10 wearing course and the SMA10 wearing course, respectively.

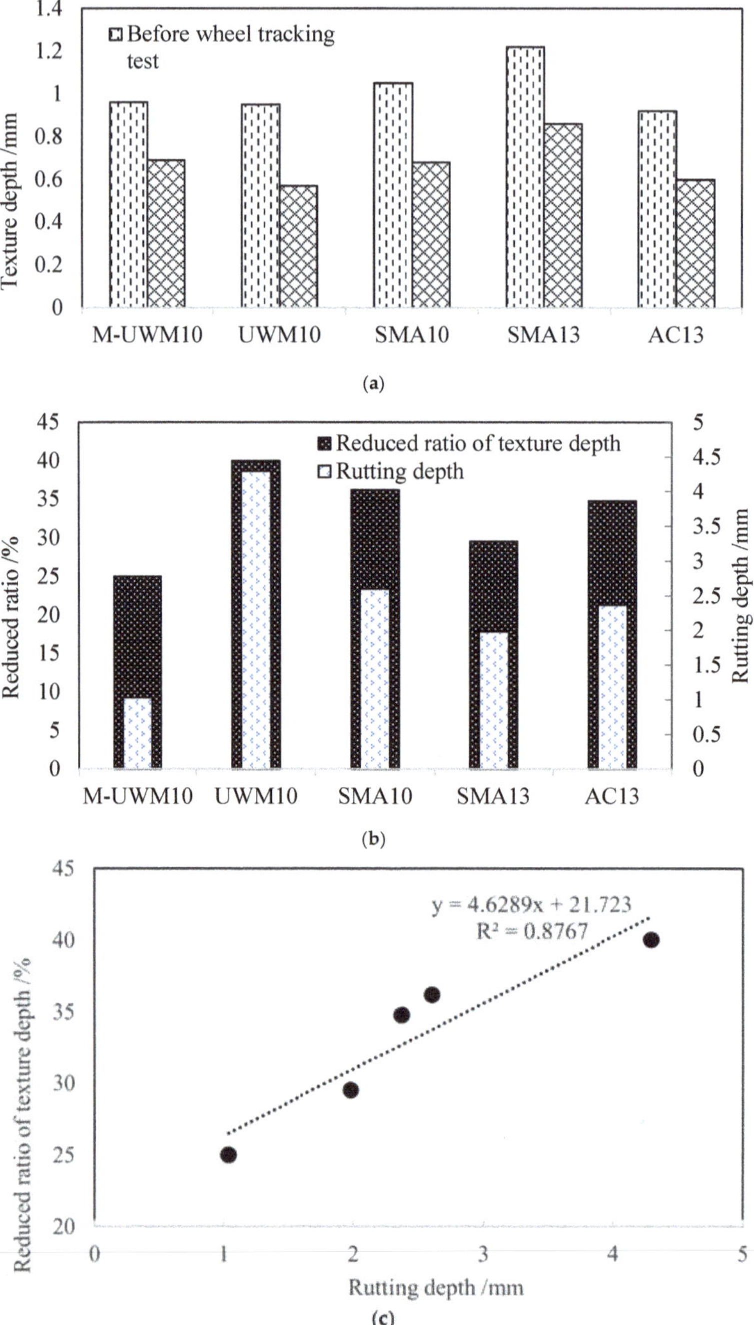

Figure 6. Test results from the combination of the texture depth test and the wheel tracking test: (**a**) texture depth before and after wheel tracking test; (**b**) loss ratio of texture depth after wheel loading; (**c**) correlation between the reduced ratio of texture depth and the rutting depth after the wheel tracking test.

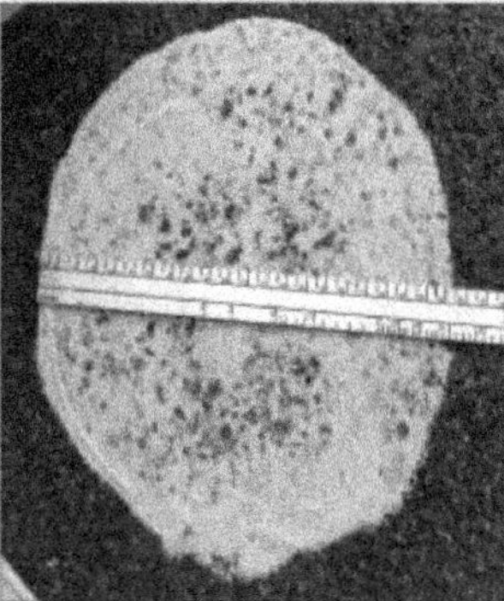

Figure 7. Field texture depth test.

Table 4. Field test results of texture depth test on different test roads.

Type of Wearing Course Mixtures	Original Texture Depth (mm)	Texture Depth after One Year (mm)	Reduced Ratio (%)
M-UWM10	0.85	0.78	8.2
SMA10	0.92	0.8	13.0

From Figure 6a, it can be seen that the SMA mixtures have higher texture depth than the other three asphalt mixtures, which have similar texture depth. It is mainly because the SMA mixtures use gap gradation while the other three asphalt mixtures use dense gradation. From the data after the wheel tracking test shown in Figure 6a, it can be seen that, due to its original high texture depth, SMA13 still have the highest texture depth, the M-UWM10 and SMA10 have similar texture depth, and the UWM10 and AC13 show a similar texture depth, which are the lowest. From Figure 6b, it can be further seen that the UWM10 has the biggest reduced ratio of texture depth, SMA10 and AC13 show a similar reduced ratio, which are lower than that of UMW10, while M-UWM10 and SMA13 have a similar reduced ratio, which are lower than the other three mixtures. Furthermore, the M-UWM10 shows a lower reduced ratio than SMA13.

From Figure 6c, it could be observed that the reduced ratio of texture depth increases with the growing of rutting depth. It indicates that the degradation behaviour of texture depth has a good correlation with the rutting resistance of asphalt mixture, and higher rutting resistance leads to less degradation of texture depth during the wheel tracking test.

The field test results confirm well with the laboratory test analysis. From Table 4, it can be seen that, although the SMA10 has a bigger texture depth than the M-UWM10 before traffic loading, after one year of traffic loading, the SMA10 shows a higher reduced ratio of texture depth than the M-UWM10, and the texture depth of the two mixtures becomes similar to each other.

Thus, the M- UWM10 has similar skidding resistance with other dense graded asphalt mixtures such as AC13 but better wearing resistance than other asphalt mixtures including both the dense graded asphalt mixture and the gap graded asphalt mixture.

3.3. Interlay Bonding Evaluation

Samples consisted of the wearing course mixture and the top surface layer mixture was cored from the two test roads and submitted to CT scanning tests. Examples of the cored samples and the 3D rebuilt images for the two wearing course mixtures including M-UWM10 and SMA10 based on the CT scanning were shown in Figure 8. The CT scanning results for the M-UWM10 sample and the SMA10 sample are shown in Figures 9 and 10, respectively.

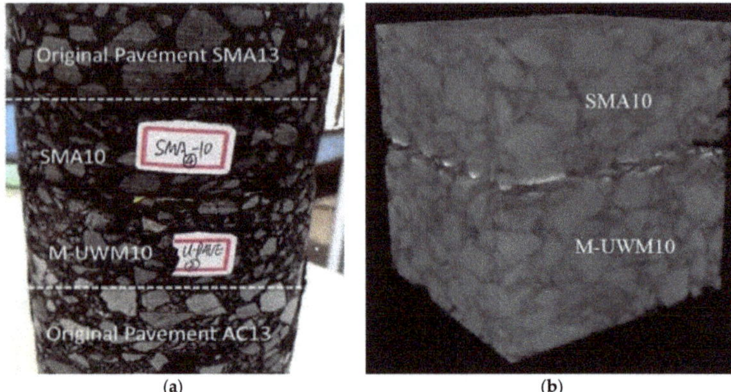

Figure 8. Examples of the cored samples from test roads and the restructuring graphs of wearing course mixtures: (**a**) cored samples; (**b**) restructuring graphs for M-UWM10 and SMA10.

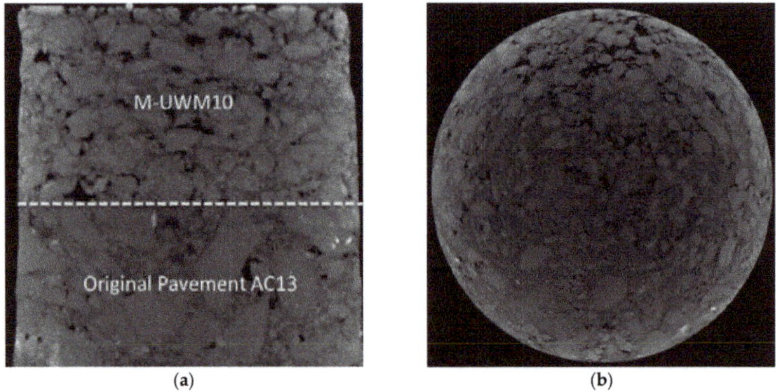

Figure 9. The scanning images for cored sample from M-UWM10 pavement: (**a**) front view for cored sample; (**b**) the interface between wearing course with M-UWM10 and sub-layer with AC13.

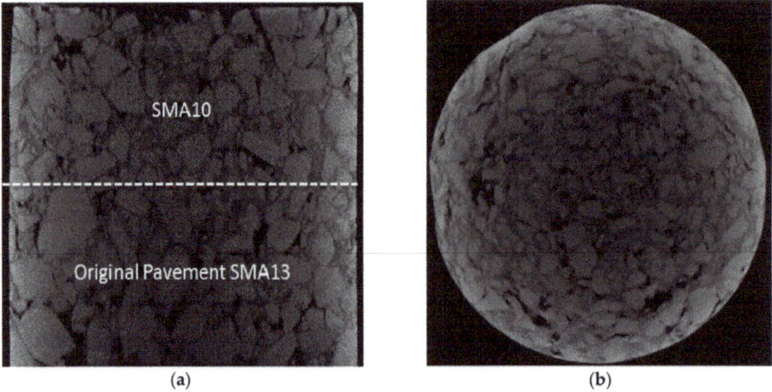

Figure 10. The scanning images for the cored sample from SMA10 pavement: (**a**) front view for cored sample; (**b**) the interface between wearing course with SMA10 and sub-layer with SMA13.

Based on the CT scanning images of the cored samples, grey processing of the images was conducted and Origin software (version, Manufacturer, City, US State abbrev. if applicable, Country) was used to calculate the air void ratios. The grey level versus probability density distribution curve was depicted in Figure 11. Since the grey level value for air voids was usually less than 30, the air void ratio was calculated by integrating from 0 to 30. The distribution of the air void ratios along the scanning height of the cored samples for the two different wearing course mixtures are shown in Figure 12. The average air void ratio within the wearing course mixture and the average air void ration on the interface between the wearing course mixture and the sub-layer mixture were also calculated. The calculate results for the two different cored samples were shown in Figure 13.

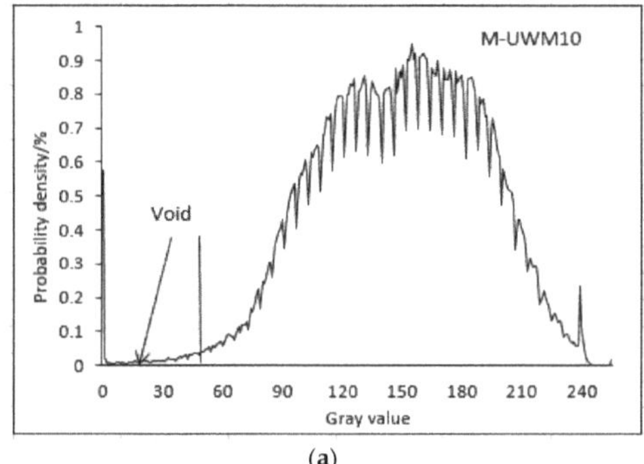

(a)

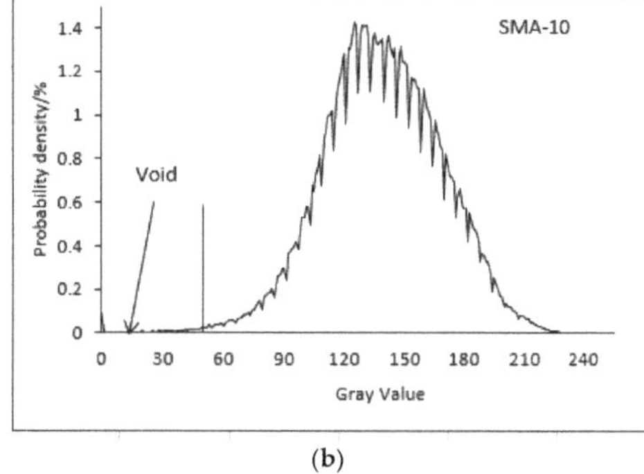

(b)

Figure 11. Grey level- probability density distribution curves: (**a**) M-UWM10; (**b**) SMA10.

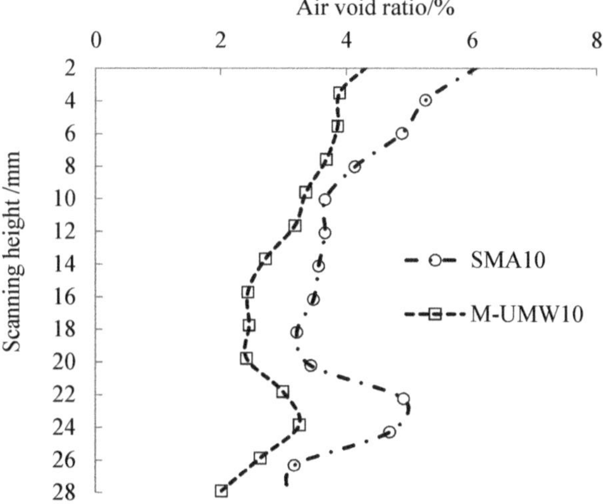

Figure 12. Distribution of air void ratios along the scanning height of the cored samples.

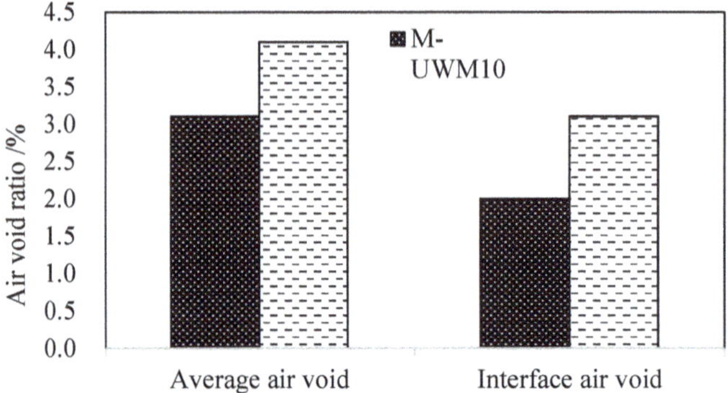

Figure 13. Average air void ratios and interface air void ratios for different cored asphalt mixtures.

From Figures 12 and 13, it can be seen that, no matter for the air void at the same height or for the average air void, the air void ratios of the M-UWM10 are lower than that of the SMA10. It could be attributed to better construction workability of M-UWM10 than SMA10. Meanwhile, it can be seen from Figure 13 that the interface air void ratio of M-UWM10 pavement is lower than that of the SMA10 pavement. Both the smaller average air void ratio and interface air void ratio are helpful to improve the bonding effect between the wearing course and the sub-layer. It is well confirmed by the test results from the field pull out test. Figure 14 shows the pull-out samples from field pavements for M-UWM10 and SMA10. The pull-out force and pull-out stress for the two wearing course mixtures are shown in Figure 15. It can be seen, that the bonding strength between M-UWM10 and its sublayer is much better than that between SMA10 and its sublayer. Thus, the wearing course consisted of M-UWM10 shows better structural integrity with the underneath pavement surface than the wearing course consisted of SMA10.

Figure 14. Field pull out test samples: (**a**) M-UWM10; (**b**) SMA10.

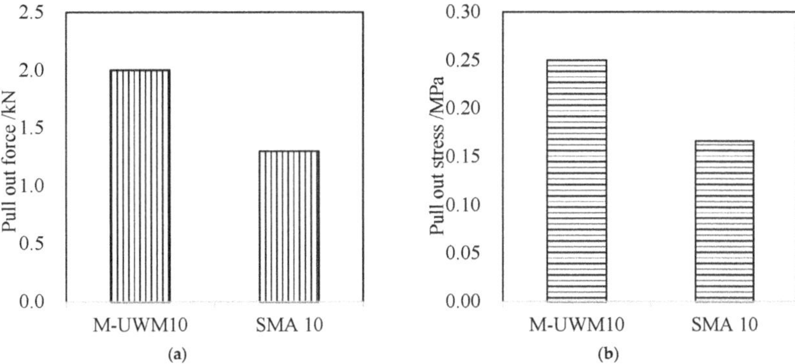

Figure 15. Pull out test results: (**a**) pull out force; (**b**) pull out stress.

4. Conclusions

Based on the laboratory and field tests, the following conclusions can be drawn:

1. A dense-graded ultra-thin wearing course mixture with multi-chain polyolefin modifier and SBS modified asphalt was prepared in this paper. It is proved that the designed asphalt mixture has satisfied high-temperature rutting resistance, low-temperature cracking resistance and moisture resistance to bear the traffic loading and environment effects.
2. The test results from combined tests of the wheel tracking test and the texture depth test proved that the designed wearing course mixture shows satisfied skidding resistance and wearing resistance. Field texture depth test results confirmed that the designed wearing course mixture is promising to keep long-term skidding resistance during traffic loading.
3. Laboratory and field tests based on the test road indicated that the wearing course paved with the designed ultra-thin wearing course mixture can provide satisfied water-proof and interlayer bonding effects, which are helpful to improve the pavement durability.
4. Future work will focus on the fatigue resistance of the ultra-thin wearing course mixture.

Acknowledgments: This work is financially supported by the Natural Science Foundation of China (No. 51378006 and No. 51378123), the Natural Science Foundation of Jiangsu (BK20161421), the Fundamental Research Funds for the Central Universities (No. 2242015R30027), and the Open Fund of the State Engineering Laboratory of Highway Maintenance Technology (Changsha University of Science and Technology, kfj160104). The author would like to thank his collaborators and support from the Jiangsu Sinoroad Engineering Research Institute Co., Ltd and Southeast University.

Author Contributions: Tao Ma conceived and designed the experiments; Pengsen Hu performed the experiments; Junhui Zhang analyzed the data; Xiaoming Huang contributed materials and apparatus; Lei Geng wrote the paper.

Conflicts of Interest: The founding sponsors had no role in the design of the study; in the collection, analyses, or interpretation of data; in the writing of the manuscript, and in the decision to publish the results.

References

1. Luo, Y.L. Brief introduction to application of maintenance technology on ultra-thin wearing course for asphalt pavement of expressway. *Guangdong Highway Commun.* **2011**. [CrossRef]
2. Pei, P.; Peng, G.J. Application of ultra-thin wear layer technology in expressway maintenance. *W. China Commun. Sci. Technol.* **2015**, *12*, 23–26. [CrossRef]
3. Qiao, X.L. Application of novachip ultra-thin wearing layer in highway maintenance. *Transport. Standard.* **2013**, *18*. Available online: http://xueshu.baidu.com/s?wd=paperuri%3A%2832b4e61b2b25b0508cfc06e363ef4551%29&filter=sc_long_sign&tn=SE_xueshusource_2kduw22v&sc_vurl=http%3A%2F%2Fen.cnki.com.cn%2FArticle_en%2FCJFDTOTAL-JTBH201318005.htm&ie=utf-8&sc_us=14205075173698836325 (accessed on 6 August 2017).

4. Yang, Y.H.; Shen, Y.; Gao, X.X. Analysis interlayer shear stress of ultra-thin wearing course considering temperature and different interlayer contact conditions. *Appl. Mech. Mater.* **2014**, *505–506*, 102–105. [CrossRef]
5. Tan, Y.Q.; Yao, L.; Wang, H.P.; Bian, X.; Yu-Xiang, Q.I. Performance evaluation indicator of ultra-thin wearing course asphalt mixture. *J. Harbin Inst. Tech.* **2012**, *44*, 73–77.
6. Zhi-Tao, H.U.; Niu, X.X. Application of novachip ultra-thin wearing layer of asphalt pavement on the preventive maintenance engineering of expressways. *Guangdong Highway Commun.* **2009**, *3*. Available online: http://xueshu.baidu.com/s?wd=paperuri%3A%2817fe1a1403e497837a8ed55c1959682d%29&filter=sc_long_sign&tn=SE_xueshusource_2kduw22v&sc_vurl=http%3A%2F%2Fen.cnki.com.cn%2FArticle_en%2FCJFDTOTAL-GDGT200903003.htm&ie=utf-8&sc_us=6153204051113523490 (accessed on 6 August 2017).
7. Chan, S.; Lane, B.; Kazmierowski, T. Pavement preservation—a solution for sustainability. *Transport. Res. Rec.* **2011**, *47*, 36–42. [CrossRef]
8. Kandhal, P.S.; Lockett, L. *Construction and performance of ultrathin asphalt friction course*; NCAT Report No. 97-5; National Center for Asphalt Technology of Auburn University: Auburn, AL, USA, September 1997.
9. Pretorius, F.J.; Wise, J.C.; Henderson, M. Development of application differentiated ultra-thin asphalt friction courses for southern african application. In Proceedings of the 8th Conference on Asphalt Pavements for Southern Africa, Sun City, South Africa, 12–16 September 2004.
10. Zeng, M.L.; Peng, L.Q.; Chao-Fan, W.U.; Tan, B.Y. Experimental study of the performance of ultrathin asphalt friction course. *J. Wuhan Univ. Technol.* **2012**, *4*. U414.
11. Yang, X.; Ling, J.; Chaofan, W.U.; Zeng, Z. Evaluation of performance of novachip ultrathin asphalt friction course. *Highway Eng.* **2013**, *1*. Available online: http://en.cnki.com.cn/Article_en/CJFDTOTAL-ZNGL201301004.htm (accessed on 6 August 2017).
12. Ministry of Transport of the People's Republic of China. *Specifications for design of highway asphalt*; China Communications Press: Beijing, China, 2006.
13. Xiao, F.; Amirkhanian, S.N. Effects of liquid antistrip additives on rheology and moisture susceptibility of water bearing warm mixtures. *Constr. Build. Mater.* **2010**, *24*, 1649–1655. [CrossRef]
14. Wu, S. Assessment of bonding behaviours between ultrathin surface layer and asphalt mixture layer using modified pull test. *J. Adhes. Sci. Technol.* **2015**, *29*, 1508–1521.
15. Rahaman, F.; Musty, H.Y.; Hossain, M. Evaluation of recycled asphalt pavement materials from ultra-thin bonded bituminous surface. In Proceedings of the GeoCongress 2012, Oakland, CA, USA, 25–29 March 2012. [CrossRef]
16. Yang, Y.H.; Liu, Z.; Gao, J.X.; Zhu, G.S. Bonding characteristics analysis of ultra-thin overlay of asphalt pavement layer. *Appl. Mech. Mater.* **2012**, *178–181*, 1245–1249. [CrossRef]
17. Weidong, C.; Jianrong, S.; Hengchun, H. Introduction of Technology of Ultra-thin Asphalt Friction Course. *Petrol. Asphalt* **2005**, *19*, 56–59.
18. Pasetto, M.; Baldo, N. Influence of the aggregate skeleton design method on the permanent deformation resistance of stone mastic asphalt. *Mater. Res. Innovations* **2014**, *18*, S3-96–S3-101. [CrossRef]
19. Ministry of Transport of the People's Republic of China. *Standard test methods of bitumen and bituminous mixtures for highway engineering*; China Communications Press: Beijing, China, 2011.
20. Ma, T.; Geng, L.; Ding, X.; Zhang, D.; Huang, X. Experimental study of deicing asphalt mixture with anti-icing additives. *Constr. Build. Mater.* **2016**, *127*, 653–662. [CrossRef]
21. Ministry of Transport of the People's Republic of China. *Standard specification for construction of highway asphalt pavements*; China Communications Press: Beijing, China, 2004.
22. Ma, T.; Wang, H.; He, L.; Zhao, Y.; Huang, X.; Chen, J. Property characterization of asphalt binders and mixtures modified by different crumb rubbers. *J. Mater. Civil Eng.* **2017**, *29*. [CrossRef]

Article

Improving Asphalt Mixture Performance by Partially Replacing Bitumen with Waste Motor Oil and Elastomer Modifiers

Sara Fernandes [1], Joana Peralta [2], Joel R. M. Oliveira [1] , R. Christopher Williams [3] and Hugo M. R. D. Silva [1,*]

1. CTAC, Centre for Territory, Environment and Construction, University of Minho, 4800 058 Guimarães, Portugal; id4966@alunos.uminho.pt (S.F.); joliveira@civil.uminho.pt (J.R.M.O.)
2. Wacker Chemical Corporation, 6870 Tilghman Street, Allentown, PA 18106-9346, USA; Joana.Peralta@wacker.com
3. Department of Civil, Construction and Environmental Engineering, Iowa State University, Ames, IA 50011-3232, USA; rwilliam@iastate.edu

* Correspondence: hugo@civil.uminho.pt; Tel.: +351-253-5102-00

Received: 21 July 2017; Accepted: 2 August 2017; Published: 5 August 2017

Featured Application: This work deals with the partial substitution of asphalt binder with waste motor oil and elastomer modifiers in order to obtain improved mixtures for pavement application purposes, which is also compared to an alternative mixture using bio-oil. The featured application of this solution is the production of new asphalt binders and mixtures for paving works in general.

Abstract: The environmental concern about waste generation and the gradual decrease of oil reserves has led the way to finding new waste materials that may partially replace the bitumens used in the road paving industry. Used motor oil from vehicles is a waste product that could answer that demand, but it can also drastically reduce the viscosity, increasing the asphalt mixture's rutting potential. Therefore, polymer modification should be used in order to avoid compromising the required performance of asphalt mixtures when higher amounts of waste motor oil are used. Thus, this study was aimed at assessing the performance of an asphalt binder/mixture obtained by replacing part of a paving grade bitumen (35/50) with 10% waste motor oil and 5% styrene-butadiene-styrene (SBS) as an elastomer modifier. A comparison was also made with the results of a previous study using a blend of bio-oil from fast pyrolysis and ground tire rubber modifier as a partial substitute for usual PG64-22 bitumen. The asphalt binders were tested by means of Fourier infrared spectra and dynamic shear rheology, namely by assessing their continuous high-performance grade. Later, the water sensitivity, fatigue cracking resistance, dynamic modulus and rut resistance performance of the resulting asphalt mixtures was evaluated. It was concluded that the new binder studied in this work improves the asphalt mixture's performance, making it an excellent solution for paving works.

Keywords: waste motor oil; styrene-butadiene-styrene; bitumen modification; bio-binder; asphalt mixtures; performance

---

1. Introduction

Currently, there is a growing concern to reuse waste materials and to conserve or minimize the use of natural resources in road paving. Thus, many studies have recently emerged in which these issues, or part of them, are discussed [1–3]. Under this assumption, the study of new asphalt binders that partially incorporate wastes is essential to reduce the use of bitumen directly obtained from oil sources, which is important for the sustainable development of road paving construction. Some studies related to this matter refer to non-petroleum binders [4] and synthetic binders made up of used oils, resins and polymers [5].

One example of a bitumen substitute or asphalt rejuvenator that has been studied lately is bio-oil, mainly produced from fast pyrolysis of biomass materials [6,7]; i.e., renewable organic matter, namely from agriculture, forestry or urban wastes. Bio-oil behavior at intermediate and high temperatures is similar to that of a conventional bitumen [8], but it is too stiff, and is quite brittle at low temperatures. However, that problem could be solved by adding ground rubber from used tires (GTR) in order to obtain a bio-binder with a very good performance [9].

The incorporation of waste motor oil in asphalt mixtures is also being tested to prevent aging, or as a bitumen rejuvenator [10,11], due to its lower viscosity, which results in lower mixing and compaction temperatures [12,13]. This solution is also environmentally friendly, because waste motor oil is not totally recyclable [14]. However, the addition of waste motor oil could bring some disadvantages for asphalt mixture performance, namely a reduced elastic recovery and rutting resistance [15]. Thus, polymer modification should be used to minimize some of these problems, and the elastomer styrene-butadiene-styrene (SBS) could be considered the most suitable modifier [16] of bitumen after addition of waste motor oil. In fact, SBS increases the asphalt mixtures rutting resistance at high temperatures and cracking resistance at low temperatures, and it also improves the tensile strength and elastic recovery properties of bitumen [17].

Taking into account that bio-oil from fast pyrolysis was not available in the necessary amounts for road paving works in several countries because this technology is not used globally at an industrial scale, this work focused on studying an alternative solution using waste motor oil to partially replace the amount of bitumen used in asphalt mixtures. Therefore, an asphalt mixture was produced with a bitumen modified with waste motor oil and SBS, and its performance was evaluated and compared with a conventional control mixture produced with 35/50 bitumen, and with the mixture previously studied by Peralta et al. [6] with bio-oil and ground rubber.

This new binder can be seen as a sustainable solution for road paving works, because both the economic and environmental criteria are being considered, as suggested by other authors [18,19]. In fact, by reducing the amount of bitumen used in asphalt mixtures, a high amount of waste materials can be used, with a consequent reduction in the global cost of the mixtures. However, the asphalt mixture produced with this new binder must perform as well as a conventional mixture produced with a current bitumen in order to ensure real advantages in a comprehensive way, and this factor is assessed in this work.

2. Materials and Methods

The materials and methods used in this study were selected to compare the results with those previously obtained by Peralta et al. [6] using bio-oil and ground rubber, with the necessary adjustments for the European region.

2.1. Materials

The materials used for production of the new asphalt binder in this study are those presented below:

- Conventional bitumen (B35/50) supplied by Cepsa Portugal (located in Matosinhos, Portugal);
- Styrene-butadiene-styrene (SBS) elastomer, supplied by Indústrias Invicta S.A. (located in Porto, Portugal), with a maximum size of 4 mm; and
- Waste motor oil (MotorOil) from heavy vehicles supplied by Correia & Correia (located in Sertã, Portugal), without any kind of treatment.

Additionally, the following aggregate fractions were selected from Bezzeras quarries (located in Guimarães, Portugal), in order to obtain an asphalt concrete mixture (AC 14 surf) for surface layers of road pavements:

- Fraction 0/4 of crushed granite with a nominal maximum aggregate size (NMAS) of 4.0 mm;
- Fraction 4/6 of crushed granite with a NMAS of 6.3 mm;
- Fraction 6/14 of crushed granite with a NMAS of 14.0 mm; and
- Limestone filler with a NMAS of 0.125 mm.

It should be noted that the final aggregate gradation of the AC 14 surf mixture is quite similar to that used by Peralta et al. [6] (with 9.5 mm NMAS), which allows their direct comparison. Primarily, they had been used as the main type of mixtures applied in surface layers in both regions where the studies were developed.

2.2. Experimental Procedure

The experimental work developed in this study was divided into two main phases of production and characterization:

- The initial phase of production and testing of the asphalt binders (conventional bitumen B35/50 and new binder with waste motor oil and SBS), and
- The final phase of production and testing of the asphalt mixtures (conventional mixture with B35/50 bitumen and asphalt mixture with the new binder).

2.2.1. Production and Testing of Asphalt Binders

The production of the binder partially replaced with waste motor oil and SBS was carried out by adding 10 wt % waste motor oil and 5 wt % SBS to the base bitumen (B35/50), as concluded in a previous work [20]. A low shear mixer (IKA RW 20 equipment, IKA®-Werke GmbH & Co. KG, Staufen, Germany) was used to obtain an initial blend of the materials at a temperature of 180 °C. Then, the binder was placed in a high shear mixer (IKA T 65 D ULTRA-TURRAX equipment, IKA®-Werke GmbH & Co. KG, Staufen, Germany), at a speed of 7200 rpm for 20 min at 180 °C in order to obtain a homogeneous blend.

After production, Fourier transform infrared (FTIR) spectroscopy tests were performed to assess the differences and the interaction between the materials used to produce the new binder developed in this work. FTIR spectra can be a very useful tool for the analysis of the macromolecules of the studied materials, because various organic chemical components [15] can be identified. The equipment used was a Jasco FT/IR-6600 (Jasco, Easton, MD, USA) with a range of wave numbers from 4000 cm^{-1} to 400 cm^{-1} with 32 scans.

Then, rheology tests were performed in a dynamic shear rheometer (DSR) model 81-PV6002 (Controls, Milan, Italy), according to EN 14770 [21], in order to obtain the stiffness master curve and the high temperature performance grade of the new binder, which was compared to the conventional bitumen B35/50 applied in road paving.

Finally, the viscosity of the binders at very high production temperatures was evaluated with a Brookfield viscometer model DV-II+ (AMETEK Brookfield, Middleboro, MA, USA), according to EN 13302 [22], in order to define the mixing and compaction temperatures to be used in the production of asphalt mixtures in the following phase of this work.

It should be noted that the binder properties obtained in these tests will also be compared to those obtained in the previous study using the asphalt binder with bio-oil and ground tire rubber (GTR) from cryogenic milling (Lehigh Technologies, Tucker, GA, USA) [6]. To produce this material, the bio-oil was initially blended with 15 wt % cryogenic GTR at a speed of 3000 rpm for 1 h at 110 °C, resulting in a cryogenic rubber modified bio-oil (cryoMBO). This blend (20 wt %) was later added to a PG64-22 bitumen (Seneca Petroleum, Des Moines, IA, USA), at a speed of 3000 rpm for 20 min at

120 °C, so as to obtain the modified asphalt binder. The low temperatures used in these processes were due to the sensitivity of bio-oil to higher production temperatures.

Some additional basic characteristics of the bitumen modified with waste motor oil and SBS, like the penetration and ring and ball values, can also be found in a previous study [20].

2.2.2. Production and Testing of Asphalt Mixtures

Concerning the asphalt mixtures studied in this work, with conventional B35/50 bitumen and with the new binder (combining motor oil and SBS), the same optimum binder content of 5.0% resulted from using the European Marshall mix design method [23,24]. These mixtures were designed for a heavy traffic level, higher than 10 million equivalent standard axle loads (ESAL), and thus Marshall specimens were compacted with 75 blows per side.

As previously mentioned, the production and compaction temperatures of both asphalt mixtures were based on Brookfield viscosity results (as discussed in Section 3.1). Specimens prepared for performance tests were produced to meet the target value of 4.0% air voids.

However, each mixture was designed according to the current practice of each country where the studies were developed. Thus, the asphalt mixture with bio-oil and GTR described in the previous study [6], which will be used for comparison purposes, was manufactured with a binder content of 5.5%, using the US SuperPave mix design methodology [25]. The production and compaction temperatures of that mixture with bio-oil and GTR were 150 °C and 140 °C, respectively, in order to meet the same air void target value of 4.0% for that mixture.

After producing the asphalt mixtures, their performance was assessed with the following tests, selected to allow a direct comparison between all materials studied:

- Water/moisture sensitivity according to the standard AASHTO T 283-07 [26];
- Fatigue cracking resistance according to the standard AASHTO T 321-07 [27]; and
- Stiffness modulus in accordance with the standard AASHTO TP 79-10 [28].

Finally, the rutting resistance of the asphalt mixtures studied in this work was evaluated with the wheel tracking test, according to EN 12697-22 [29] standard. The triaxial repeated loading test [28] was used in the previous study of the asphalt mixture with bio-oil and GTR [6] in order to assess its rutting resistance, namely by assessing the accumulated strain values. Despite the different standards used, which do not allow a direct comparison, they evaluate the same property and have the same base concepts. Thus, an indirect comparison will be carried out concerning this property.

3. Results and Discussion

3.1. Asphalt Binder Test Results

First, the differences between the binder materials were analyzed by using FTIR tests, and the corresponding results can be observed in Figure 1.

The FTIR spectra show the chemical transformations/interactions that may or may not have occurred during asphalt binder production. The main absorbance peaks of the spectra of all asphalt binder materials occur in the same wavenumber ranges (700–1800 cm^{-1} and 2800–3050 cm^{-1}) because the results are mainly controlled by the conventional base bitumens. This means that the main functional groups of the conventional base bitumens (B35/50 and PG 64-22) are also present in the modified binders with waste motor oil and SBS (B35/50 & MotorOil & SBS) and with bio-oil and ground rubber (PG64-22 & cryoMBO).

The spectrum of rubber modified bio-oil (cryoMBO) is slightly different from those of asphalt binder materials, mainly due to the presence of some moisture in the bio-oil [6].

Nevertheless, it was concluded that the new modified binders can be a good alternative to conventional bitumens, because they are chemically similar and/or compatible.

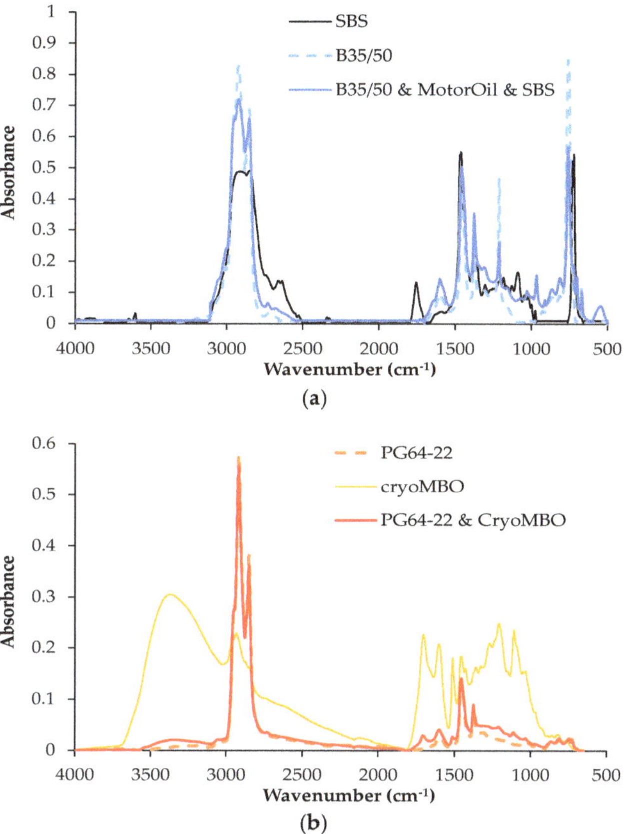

Figure 1. Fourier transform infrared (FTIR) spectra results of: (**a**) styrene-butadiene-styrene (SBS) elastomer, conventional bitumen (B35/50) and modified asphalt binder B35/50 & MotorOil & SBS; (**b**) Rubber modified bio-oil (cryoMBO), bitumen PG64-22 and modified asphalt binder PG64-22 & cryoMBO (adapted from Peralta et al. [6], @ AAPT, 2014).

The rheology of the asphalt binder materials was then evaluated with a dynamic shear rheometer (DSR), by carrying out frequency sweep tests at different temperatures. The stiffness master curves obtained with those tests are presented in Figure 2.

The master curve of the modified binder with motor oil and SBS (B35/50 & MotorOil & SBS) shows a lower slope than that of conventional bitumen (B35/50), which may be associated with a low susceptibility of this binder to frequency and temperature changes. On the other hand, the rheological behavior obtained in the previous study with bio-binder (CryoMBO) [6] showed that it is comparable to conventional bitumen (PG64-22) and the modified binder (PG64-22 & cryoMBO), because all these materials present master curves with very similar shapes. The different behavior of the new binder developed in this study (B35/50 & MotorOil & SBS) may be related to the higher amount of elastomer (SBS) used, in comparison with the ground rubber amount (3%) used in the bio-oil study.

The rheological results were also used to calculate the high temperature continuous performance grades of all asphalt binders, which are presented in Table 1. The addition of 10% waste motor oil and 5% SBS clearly increased the performance grade of the base bitumen B35/50, mainly due to the high amount of elastomer used. In turn, in the past study with bio-oil [6], 20% bio-binder (cryoMBO) was added to the base bitumen (PG 64-22), almost without changing its high temperature performance grade. Thus, the rheological results indicate that the new binder developed in this work (B35/50 & MotorOil & SBS) is rheologically or mechanically superior to the conventional bitumens and the modified binder PG64-22 & cryoMBO, particularly at high service temperatures, indicating its suitability for asphalt road pavements applied in hot climate zones.

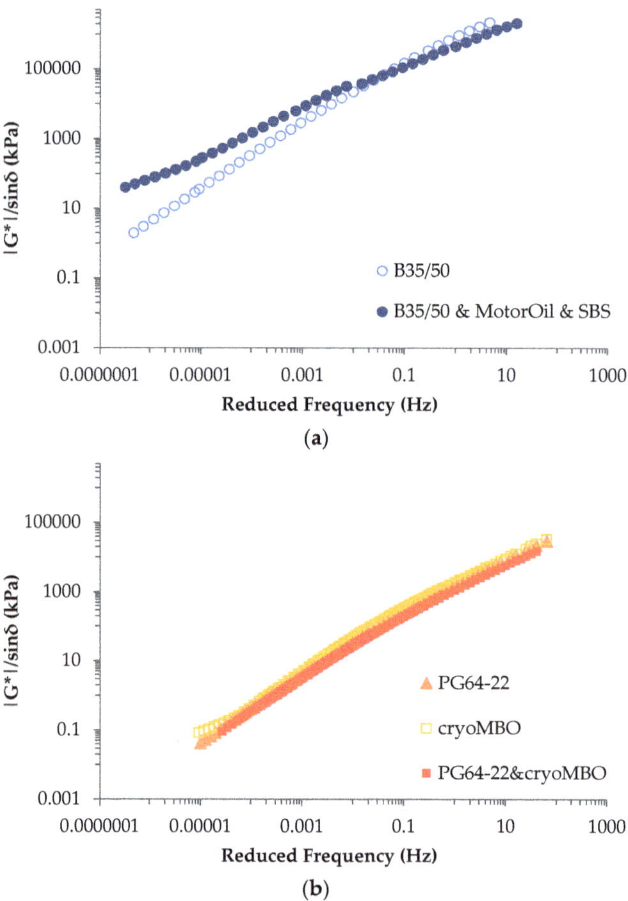

Figure 2. Stiffness master curves (T_{ref} = 20 °C) results of: (**a**) bitumen B35/50 and modified asphalt binder B35/50 & MotorOil & SBS; (**b**) CryoMBO, bitumen PG64-22 and modified asphalt binder PG64-22 & cryoMBO (adapted from Peralta et al. [6], @ AAPT, 2014).

Table 1. High temperature continuous performance grade of conventional bitumens (B35/50 and PG 64-22), rubber modified bio-oil (CryoMBO) and modified asphalt binders (B35/50 & MotorOil & SBS and PG64-22 & cryoMBO).

Materials	High Temperature Continuous Performance Grade (°C)
Conventional bitumen (B35/50)	65.1
Modified asphalt binder (B35/50 & MotorOil & SBS)	76.8
Conventional bitumen (PG 64-22)	66.6
Rubber modified bio-oil (CryoMBO)	61.3
Modified asphalt binder (PG64-22 & cryoMBO)	65.5

Finally, taking the ideal viscosity values for mixing and compaction [30] into account, it was possible to define the optimum temperatures for asphalt mixture production in the next phase of the work. In the case of the asphalt binder with motor oil and SBS, the mixing and compaction temperatures for asphalt mixture production were determined to be 180 °C and 160 °C, respectively, while those temperatures were 160 °C and 140 °C for the conventional asphalt mixture with B35/50 bitumen. The new binder presented mixing and compaction temperatures higher than expected. This may be caused by the SBS particle effect, which greatly increases the viscosity of this binder, although waste motor oil has decreased the viscosity before interaction with SBS modifier.

3.2. Asphalt Mixtures Tests Results

In relation to the water or moisture sensitivity of the studied asphalt mixtures, both in Europe and the US, a minimum tensile strength ratio (TSR) of 70% is recommended [31], although some conservative agencies refer to a TSR value of 80% [32–34]. The novel asphalt mixture studied in this work, with waste motor oil and SBS, showed a TSR value of 80%, which narrowly meets the recommended values to assure the needed durability of this mixture. In the previous study with bio-oil and ground rubber [6], the asphalt mixture presented a TSR value of 84%, which is also greater than 80%, and thus that mixture also fulfilled the requirements.

Subsequently, the results of the fatigue cracking resistance test are summarized in Figure 3, namely for the new asphalt mixture with waste motor oil and SBS (B35/50 & MotorOil & SBS), the conventional mixture with B35/50 bitumen, and also for the asphalt mixture with bio-oil and ground rubber (PG64-22 & cryoMBO). Those results show the variation in the number of loading cycles (N_f) before fatigue failure of each mixture (defined through the reduction of the stiffness to half of its initial value) as a function of the tensile micro strain ($\mu\varepsilon$) applied in the fatigue test.

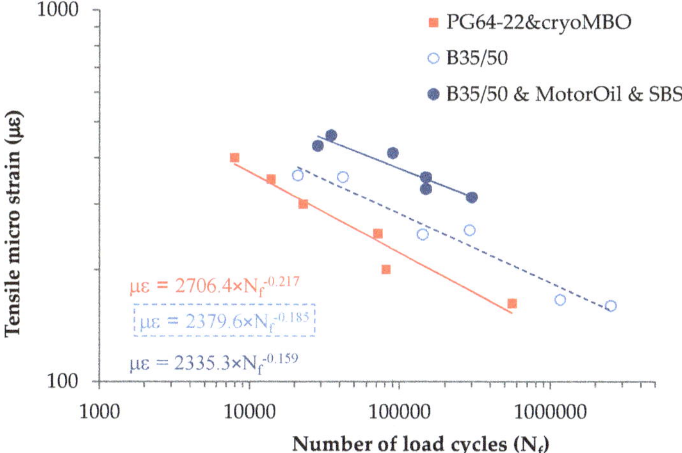

Figure 3. Variation in the number of loading cycles (N_f) before fatigue failure with the tensile micro strain ($\mu\varepsilon$) applied in the fatigue test for the studied mixtures.

Those results demonstrate that the new asphalt mixture with waste motor oil and SBS (B35/50 & MotorOil & SBS) is more resistant to fatigue than the conventional asphalt mixture (B35/50), at least in the range of the tensile strain values used in this test. The low viscosity of the waste motor oil and the high amount of SBS elastomer used may justify the very good fatigue cracking resistance of this new mixture. The asphalt mixture with bio-oil and ground rubber from the previous study [6] presented a lower fatigue resistance, probably due to the higher stiffness of the bio-binder or the lower amount of ground rubber used.

One of the main models used to predict the fatigue life of an asphalt mixture in road pavements was that proposed by Shell [35]. The fatigue laws obtained for both modified asphalt mixtures (with waste motor oil and SBS or with bio-binder and GTR) indicate a fatigue life higher than that computed by the Shell model for an asphalt mixture with 4000 MPa and an asphalt volume content of 12%, which confirms the suitable fatigue performance of those mixtures. Moreover, the experimental fatigue coefficients K_1 and K_2 mentioned by other authors [36] were determined to be 5×10^{-14} and 5.171 for the conventional asphalt mixture with B35/50 bitumen. The new mixture with 10% motor oil and 5% SBS presented fatigue coefficients of $K_1 = 1 \times 10^{-14}$ and $K_2 = 5.508$, while the mixture with bio-binder and GTR presented fatigue coefficients of $K_1 = 1 \times 10^{-11}$ and $K_2 = 4.368$. In conclusion, both mixtures with modified binders (B35/50 & MotorOil & SBS and PG64-22 & cryoMBO) showed an adequate fatigue performance, at least similar to that that of the conventional asphalt mixture.

The stiffness modulus of asphalt mixtures is a very important parameter for road pavement design, and it also influences the mixture fatigue performance. Figure 4 shows the master curves of the dynamic stiffness modulus of the several asphalt mixtures studied in this work for a reference temperature of 21 °C. In order to obtain these master curves, the conventional mixture with B35/50 bitumen and the asphalt mixture with waste motor oil and SBS were tested at 0, 10, 20 and 30 °C, according to the current practice [3], while the asphalt mixture with bio-oil and ground rubber was tested at 4, 21 and 37 °C, following another procedure [37].

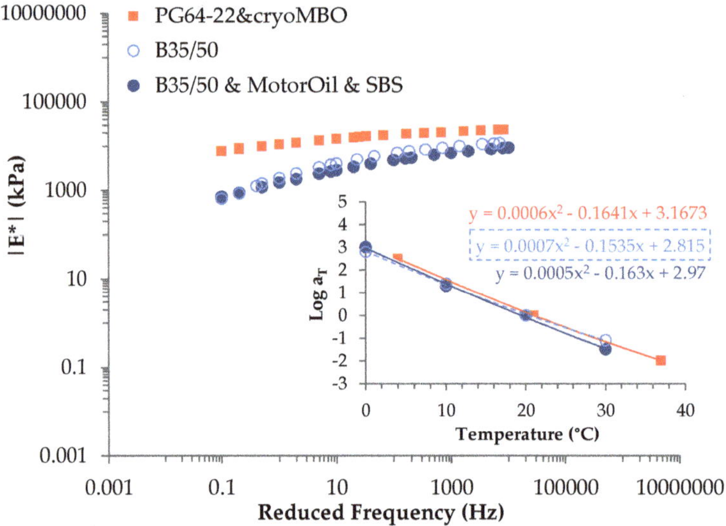

Figure 4. Master curves of the shifted dynamic stiffness modulus for the different asphalt mixtures (T_{ref} = 21 °C) studied in this work (the insert represents the quadratic fitting of the shift factors obtained using the Arrhenius time-temperature superposition principle).

The analysis of the dynamic modulus master curves shows that the asphalt mixture with the waste motor oil and SBS (B35/50 & MotorOil & SBS) is softer than that with bio-oil and ground rubber (PG64-22 & cryoMBO). This result can partially explain the very good fatigue performance previously observed for the new asphalt mixture with waste motor oil and SBS. Furthermore, the stiffness of that mixture (B35/50 & MotorOil & SBS) is similar to that of the conventional mixture with B35/50 bitumen, being slightly higher at very low frequencies and high temperatures. Based on these results, it can be assumed that the asphalt mixture with bio-oil and GTR should perform better at high temperatures, followed by the mixture with waste motor oil and SBS. The high stiffness modulus at high temperatures (or low frequencies) points towards a high rutting resistance of the asphalt mixtures with modified binders, which is in agreement with the results obtained by other authors [38,39].

Finally, different rutting resistance tests were applied to evaluate the performance of the asphalt mixtures with modified binders, as previously mentioned in Section 2.2.2. In this particular study, the rutting resistance was obtained by measuring the rut depth formed by successive passages of a wheel, at a temperature of 50 °C. This test ends after 10,000 load cycles. The wheel-tracking slope in air (WTS_{AIR}) is the main result of this test used to rank the rutting resistance performance of asphalt mixtures, which is calculated based on the evolution of the permanent deformation of the asphalt mixture with the number of cycles during the last 5000 load cycles in the wheel-tracking test. The results of this test demonstrated a higher rutting resistance for the asphalt mixture with waste motor oil and SBS (B35/50 & MotorOil & SBS) in comparison to the conventional asphalt mixture (B35/50), as observed from their final rut values (2.4 mm and 5.0 mm, respectively). Moreover, the new asphalt mixture with waste motor oil and SBS exhibited a very low WTS_{AIR} value (0.05 mm/10^3), which confirms the good rutting resistance performance of this mixture in comparison to the conventional asphalt mixture (WTS_{AIR} value of 0.20 mm/10^3). The rutting resistance of the asphalt mixture with bio-oil and ground rubber (PG64-22 & cryoMBO), obtained in the previous study [6], was evaluated through the accumulated strain values of the asphalt mixture in a triaxial repeated loading test. The accumulated strain values obtained in that test at 37 °C and 54 °C were 0.14% and 1.00%, respectively. When compared to the conventional asphalt mixture (PG64-22), the results obtained for this mixture with bio-binder are exceptionally good at both temperatures, demonstrating its excellent rutting resistance performance. In conclusion, the very high rutting resistance performance of the asphalt mixtures with modified binders, which is not impaired by the low viscosity of the waste motor oil, confirms the conclusions drawn from the dynamic modulus results.

4. Conclusions

The new asphalt binder and mixture evaluated in this work showed that the blend of bitumen, waste motor oil and SBS has better properties than those of conventional bitumen (B35/50), and results in an asphalt mixture with improved performance in comparison to the conventional asphalt mixture.

Actually, the asphalt mixtures with the new modified binders, namely with waste motor oil and SBS (B35/50 & MotorOil & SBS) and with bio-oil and ground rubber (PG64-22 & cryoMBO), showed very good performance in all tests carried out during this study. Thus, a good rutting and fatigue cracking resistance can be expected from these mixtures. In addition, these mixtures are not sensitive to moisture, and their durability can be assured. In particular, the asphalt mixture with waste motor oil and SBS showed a higher fatigue cracking resistance, while the mixture with bio-oil and ground rubber presented higher stiffness and rutting resistance, but both mixtures generally performed well during the laboratorial study. It also becomes clear that the new mixture with waste motor oil and SBS performs better than the conventional mixture B35/50.

The very good performance of the new binders and mixtures with waste-derived oils and elastomer modifiers made the potential of these economical and environmentally friendly solutions to be used in real road pavements even more evident. It should be noted that these solutions should now be applied and evaluated in a real pavement trial, in order to have these alternative binders and mixtures validated by road administrations and ready to be applied by the paving industry.

Acknowledgments: The authors gratefully acknowledge the funding provided by the Portuguese Government (Portuguese Foundation for Science and Technology) and by the Competitiveness Factors Operational Programme (EU/ERDF funds), through the FCT project PLASTIROADS (PTDC/ECM/119179/2010) and the PhD grant SFRH/BD98379/2013. Authors also want to acknowledge all companies mentioned in this document for supplying all the materials, which were essential to obtain the final results presented in this paper.

Author Contributions: Sara Fernandes (S.F.), Joana Peralta (J.P.), Joel Oliveira (J.O.), Christopher Williams (C.W.) and Hugo Silva (H.S.) conceived and designed the experimental work; S.F and J.P. performed the experiments mentioned in the paper; S.F., J.P., J.O., C.W. and H.S. analyzed the data and contributed to the completion of the manuscript.

Conflicts of Interest: The authors declare no conflict of interest.

References

1. Costa, L.M.B.; Silva, H.M.R.D.; Oliveira, J.R.M.; Fernandes, S.R.M. Incorporation of waste plastic in asphalt binders to improve their performance in the pavement. *Int. J. Pavement Res. Technol.* **2013**, *6*, 457–464.
2. Anastasiou, E.K.; Liapis, A.; Papayianni, I. Comparative life cycle assessment of concrete road pavements using industrial by-products as alternative materials. *Resour. Conserv. Recycl.* **2015**, *101*, 1–8. [CrossRef]
3. Costa, L.; Peralta, J.; Oliveira, J.; Silva, H. A New Life for Cross-Linked Plastic Waste as Aggregates and Binder Modifier for Asphalt Mixtures. *Appl. Sci.* **2017**, *7*, 603. [CrossRef]
4. Metwally, M.A.R.M.; Williams, R.C. *Development of Non-Petroleum Based Binders for Use in Flexible Pavements (Final Report)*; Institute for Transportation Iowa State University: Ames, IA, USA, 2010.
5. Fuentes-Audén, C.; Martínez-Boza, F.J.; Navarro, F.J.; Partal, P.; Gallegos, C. Formulation of new synthetic binders: Thermo-mechanical properties of recycled polymer/oil blends. *Polym. Test.* **2007**, *26*, 323–332. [CrossRef]
6. Peralta, J.; Williams, R.C.; Silva, H.M.R.D.; Machado, A.V.A. Recombination of asphalt with bio-asphalt: Binder formulation and asphalt mixes application. *Asph. Paving Technol. Assoc. Asph. Paving Technol.* **2014**, *83*, 1–36.
7. Kowalski, K.; Król, J.; Bańkowski, W.; Radziszewski, P.; Sarnowski, M. Thermal and Fatigue Evaluation of Asphalt Mixtures Containing RAP Treated with a Bio-Agent. *Appl. Sci.* **2017**, *7*, 216. [CrossRef]
8. Raouf, M.A.; Williams, R.C. General rheological properties of fractionated switchgrass bio-oil as a pavement material. *Road Mater. Pavement Des.* **2010**, *11*, 325–353. [CrossRef]
9. Peralta, J.; Williams, R.C.; Rover, M.; Silva, H.M.R.D. Development of a rubber-modified fractionated bio-oil for use as noncrude petroleum binder in flexible pavements. *Transp. Res. Board* **2012**.
10. Dedene, C.; Mills-Beale, J.; You, Z. Properties of recovered asphalt binder blended with waste engine oil: A preliminary study. In *ICCTP 2011: Towards Sustainable Transportation Systems, Proceedings of the 11th International Conference of Chinese Transportation Professionals, Nanjing, China, 14–17 August 2011*; ASCE Libarary: Reston, VA, USA, 2011; pp. 4399–4406.
11. DeDene, C.D.; You, Z. The performance of aged asphalt materials rejuvenated with waste engine oil. *Int. J. Pavement Res. Technol.* **2014**, *7*, 145–152.
12. Silva, H.M.R.D.; Oliveira, J.R.M.; Jesus, C.M.G. Are totally recycled hot mix asphalts a sustainable alternative for road paving? *Resour. Conserv. Recycl.* **2012**, *60*, 38–48. [CrossRef]
13. Lesueur, D. The colloidal structure of bitumen: Consequences on the rheology and on the mechanisms of bitumen modification. *Adv. Colloid Interface Sci.* **2009**, *145*, 42–82. [CrossRef] [PubMed]
14. Kuczenski, B.; Geyer, R.; Zink, T.; Henderson, A. Material flow analysis of lubricating oil use in California. *Resour. Conserv. Recycl.* **2014**, *93*, 59–66. [CrossRef]
15. Jia, X.; Huang, B.; Bowers, B.F.; Zhao, S. Infrared spectra and rheological properties of asphalt cement containing waste engine oil residues. *Constr. Build. Mater.* **2014**, *50*, 683–691. [CrossRef]
16. Yildirim, Y. Polymer modified asphalt binders. *Constr. Build. Mater.* **2007**, *21*, 66–72. [CrossRef]
17. Ahmedzade, P. The investigation and comparison effects of SBS and SBS with new reactive terpolymer on the rheological properties of bitumen. *Constr. Build. Mater.* **2013**, *38*, 285–291. [CrossRef]
18. Moretti, L.; Di Mascio, P.; D'Andrea, A. Environmental Impact Assessment of Road Asphalt Pavements. *Mod. Appl. Sci.* **2013**, *7*. [CrossRef]
19. Moretti, L.; Mandrone, V.; D'Andrea, A.; Caro, S. Comparative "from Cradle to Gate" Life Cycle Assessments of Hot Mix Asphalt (HMA) Materials. *Sustainability* **2017**, *9*, 400. [CrossRef]
20. Fernandes, S.; Costa, L.; Silva, H.; Oliveira, J. Effect of incorporating different waste materials in bitumen. *Ciênc. Tecnol. Mater.* **2017**, *29*, e204–e209. [CrossRef]
21. *Bitumen and Bituminous Binders. Determination of Complex Shear Modulus and Phase Angle. Dynamic Shear Rheometer (DSR)*; EN 14770; European Commitee for Standardization: Brussels, Belgium, 2012.

22. *Bitumen and Bituminous Binders. Determination of Dynamic Viscosity of Bituminous Binder Using a Rotating Spindle Apparatus*; EN 13302; European Commitee for Standardization: Brussels, Belgium, 2010.
23. *Bituminous Mixtures. Material Specifications. Asphalt Concrete*; EN 13108-1; European Commitee for Standardization: Brussels, Belgium, 2006.
24. *Bituminous Mixtures—Test Methods for Hot Mix Asphalt—Part 34: Marshall Test*; EN 12697-34; European Commitee for Standardization: Brussels, Belgium, 2012.
25. Asphalt Institute. *Superpave®: Performance Graded Asphalt Binder Specification and Testing*, 3rd ed.; Asphalt Institute: Lexington, KY, USA, 2003.
26. *Resistance of Compacted Bituminous Mixture to Moisture Induced Damage*; AASHTO T 283-07; American Association of State Highway & Transportation Officials: Washington, DC, USA, 2007.
27. *Standard Method of Test for Determining the Fatigue Life of Compacted Hot-Mix Asphalt (HMA) Subjected to Repeated Flexural Bending*; AASHTO T 321-07; American Association of State Highway & Transportation Officials: Washington, DC, USA, 2007.
28. *Standard Method of Test for Determining the Dynamic Modulus and Flow Number for Hot Mix Asphalt (HMA) Using the Asphalt Mixture Performance Tester (AMPT)*; AASHTO TP 79-10; American Association of State Highway & Transportation Officials: Washington, DC, USA, 2010.
29. *Bituminous Mixtures—Test Methods for Hot Mix Asphalt—Part 22: Wheel Tracking*; EN 12697-22; European Commitee for Standardization: Brussels, Belgium, 2012.
30. Read, J.; Whiteoak, D.; Bitumen, S. *The Shell Bitumen Handbook*; Thomas Telford: London, UK, 2003.
31. Lottman, R.P. *Predicting Moisture-Induced Damage to Asphaltic Concrete Field Evaluation (No. 246)*; Transportation Research Board: Washington, DC, USA, 1982.
32. Asphalt Institute. *Superpave® Mix Design*, 3rd ed.; Asphalt Institute: Lexington, KY, USA, 2001.
33. Roberts, F.L.; Kandhal, P.S.; Brown, E.R.; Lee, D.-Y.; Kennedy, T.W. *Hot Mix Asphalt Materials, Mixture Design and Construction*; Transportation Research Board: Lanham, MD, USA, 1996; p. 603.
34. Brown, E.R.; Kandhal, P.S.; Roberts, F.L.; Kim, Y.R.; Lee, D.-Y.; Kennedy, T.W. *Hot Mix Asphalt Materials, Mixture Design, and Construction*; National Asphalt Pavement Association Research and Education Foundation: Lanham, MD, USA, 2009.
35. Shell. *Shell Pavement Design Manual: Asphalt Pavements and Overlays for Road Traffic*; Shell International Petroleum Co. Ltd.: London, UK, 1978.
36. Pais, J.C.; Pereira, P.A.A.; Minhoto, M.J.C.; Fontes, L.P.T.L.; Kumar, D.S.N.V.A.; Silva, B.T.A. The prediction of fatigue life using the K_1-K_2 relationship. In *2nd Workshop on Four Point Bending*; Pais, J.C., Ed.; University of Minho: Guimarães, Portugal, 2009.
37. Garcia, G.; Thompson, M. *HMA Dynamic Modulus Predictive Models—A Review*; National Technical Information Service: Alexandria, VA, USA, 2007.
38. Williams, R.C.; Cascione, A.; Haugen, D.S.; Buttlar, W.G.; Bentsen, R.A.; Behnke, J. *Characterization of Hot Mix Asphalt Containing Post-Consumer Recycled Asphalt Shingles and Fractionated Reclaimed Asphalt Pavement*; Final Report; Illinois State Toll Highway Authority: Downers Grove, IL, USA, 2011.
39. West, R.; Willis, J.R.; Marasteanu, M.O. *Improved Mix Design, Evaluation, and Materials Management Practices for Hot Mix Asphalt with High Reclaimed Asphalt Pavement Content*; Report No. 752; National Cooperative Highway Research Program, Transportation Research Board of The National Academies: Washington, DC, USA, 2013.

Article

Mechanical Resilience of Modified Bitumen at Different Cooling Rates: A Rheological and Atomic Force Microscopy Investigation

Cesare Oliviero Rossi [1,*], Saltanat Ashimova [1,2], Pietro Calandra [3], Maria Penelope De Santo [4] and Ruggero Angelico [5,6,*]

1 Department of Chemistry and Chemical Technologies, University of Calabria, 87036 Arcavacata di Rende (CS), Italy; salta_32@mail.ru
2 Kazakhstan Highway Research Institute, Nurpeisova Str., 2A, Almaty 050061, Kazakhstan
3 CNR-ISMN, National Council of Research, Via Salaria km 29.300, 00015 Monterotondo Stazione (RM), Italy; pietro.calandra@ismn.cnr.it
4 Department of Physics and CNR-Nanotec, University of Calabria, 87036 Rende (CS), Italy; maria.desanto@fis.unical.it
5 Department of Agricultural, Environmental and Food Sciences (DIAAA), University of Molise, Via De Sanctis, 86100 Campobasso (CB), Italy
6 CSGI (Center for Colloid and Surface Science), Via della Lastruccia 3, I-50019 Sesto Fiorentino (FI), Italy
* Correspondence: cesare.oliviero@unical.it (C.O.R.); angelico@unimol.it (R.A.); Tel./Fax: +39-0984-492045 (C.O.R.); +39-0874-404649 (R.A.)

Received: 8 July 2017; Accepted: 28 July 2017; Published: 31 July 2017

Abstract: Due to the wide variation in geographic and climatic conditions, the search for high-performance bituminous materials is becoming more and more urgent to increase the useful life of pavements and reduce the enormous cost of road maintenance. Extensive research has been done by testing various bitumen modifiers, although most of them are petroleum-derived additives, such as polymers, rubbers and plastic, which in turn do not prevent oxidative aging of the binder. Thus, as an alternative to the most common polymeric rheological modifiers, selected binder additives falling in the categories of organosilane (P2KA), polyphosphoric acid (PPA) and food grade phospholipids (LCS) were homogeneously mixed to a base bitumen. The goal was to analyse the micro-morphology of the bitumens (neat and modified) subjected to different cooling rates and to find the corresponding correlations in the mechanical response domain. Therefore, microstructural investigations carried out by Atomic Force Microscopy (AFM) and fundamental rheological tests based on oscillatory dynamic rheology, were used to evaluate the effect of additives on the bitumen structure and compared with pristine binder as a reference. The tested bitumen additives have been shown to elicit different mechanical behaviours by varying the cooling rate. By comparing rheological data, analysed in the framework of the "weak gel" model, and AFM images, it was found that both P2KA and PPA altered the material structure in a different manner whereas LCS revealed superior performances, acting as "mechanical buffer" in the whole explored range of cooling rates.

Keywords: modified bitumen; Atomic Force Microscopy; Dynamic Oscillatory Rheology; complex modulus; "weak gel" model

1. Introduction

In today's road construction technology, the demand for bitumen characterised by high mechanical properties becomes increasingly insistent, even if it represents only a minor component of asphalt (5–8% by weight of binder) [1].

Bitumen, which is a complex solid or semisolid colloidal dispersion of asphaltenes into a continuous oily phase constituted by saturated paraffins, aromatics and resins [2–4], is a viscoelastic material whose mechanical response is both time and temperature dependent [5]. However, because of the wide variation in geographical and climatic conditions, a careful selection of bituminous materials is required to increase the useful life of the pavement and reduce the huge cost of road maintenance. For instance, for a good road performance, it would be highly desirable that the deformation properties of bitumen remain unchanged under the effect of different cooling rates, and mitigate the susceptibility of asphalt concrete to several drawbacks such as thermal cracking and thermal stress accumulation [6]. Indeed, depending on the geographical areas, the bitumen used in hot-mix asphalt concretes for road construction may suffer severe thermal shocks during placement of the asphalt top layer onto pavements, characterised by a wide range of surface temperatures [7,8]. Therefore, to facilitate the evaluation of cooling rate sensitivity in controlled lab-scale conditions, the temperature of both pristine bitumen and bitumen modified with three additives was varied from 105 °C to 25 °C using cooling rates of 1, 5 and 10 °C/min, respectively. Being aware that the selected cooling rates might not exactly reproduce the realistic pavement conditions met in a field study, the purpose of this analysis was to gain a preliminary understanding of the impact that additives distinct from the most common polymers [9–11] may have on the bitumen rheological response to changes in the cooling rate. Therefore, a rheological investigation was performed to compare the mechanical behaviour manifested by a base bitumen modified with three additives, namely, organosilane (P2KA), polyphosphoric acid (PPA) and food grade phospholipids (LCS), whose properties as rheological/adhesion bitumen modifiers have been widely investigated [12–15]. Specifically, oscillatory rheological tests were carried out to monitor the dependence of complex mechanical modulus on the type of additive at various cooling rates. According to the present research project, we seek the most efficient additive that would make the mechanical modulus of the correspondent modified bitumen less temperature-sensitive in the explored cooling ramp range. A parallel structural investigation by using Atomic Force Microscopy (AFM) has been also undertaken at 25 °C to monitor changes in the micro-morphology of both virgin and modified bitumens once subjected to different cooling rates. Indeed, previous AFM investigations were found successful in studying the bitumen microstructure at nanoscale level, including its surface morphology dependence on various physico-chemical parameters [16–18]. Therefore, the results illustrated in the present study and obtained from a combination of oscillatory rheology and AFM measurements have been found very useful in the identification of the best additive able to leave the bituminous structure nearly unaffected by the action of different cooling rates applied in range of 1–10 °C/min.

2. Materials and Methods

2.1. Materials

The asphalt binder was kindly supplied by Loprete Costruzioni Stradali (Terranova Sappo Minulio, Calabria, Italy) and was used as base bitumen. It was produced in Italy and the crude oil was from Saudi Arabia. The neat bitumen was modified by adding commercial additives, namely, (a) phospholipids in the form of light yellow powder (hereafter LCS) provided by Somercom srl (Catania, Italy); (b) organosilane P2KA provided by KimiCal s.r.l. (Rende, Italy) and (c) polyphosphoric acid (PPA) provided by Sigma Aldrich (Milano, Italy).

2.2. Sample Preparations and Setup of Cooling Ramps

The additives were mixed separately to hot bitumen (140–160 °C) at fixed content of 2% wt/wt [12–14] by using a mechanical stirrer (IKA RW20, Königswinter, Germany). First, 100 g of bitumen was heated up to 140–160 °C until it flowed fully, then a given amount of additive was added to the melted bitumen under a high-speed shear mixer at 500–700 rpm. Furthermore, the mixtures were stirred again at 140–160 °C for 30 min. After mixing, three different cooling ramp rates were applied to bitumen samples modified with P2KA, PPA and LCS, respectively. Parallel temperature ramp tests were carried out on unmodified bitumen as a reference system. The experimental conditions were isothermal annealing for 10 min at 160 °C then cooling at 1, 5 and 10 °C/min until room temperature was reached.

2.3. SARA Determination

The Iatroscan MK 5 Thin Layer Chromatography (TLC) was used for the chemical characterisation of bitumen by separating it into four fractions: Saturates, Aromatics, Resins and Asphaltenes (SARA) [19]. During the measurement, the separation took place on the surface of silica-coated rods. The detection of the amount of different groups was according to the flame ionisation. The sample was dissolved in peroxide-free tetrahydrofurane solvent to reach a 2% (w/v) solution. Saturated components of the sample were developed in n-heptane solvent while the aromatics were developed in a 4:1 mixture of toluene and n-heptane. Afterwards, the rods had to be dipped into a third tank, which was a 95 to 5% mixture of dichloromethane and methanol. That organic medium proved suitable to develop the resin fraction whereas the asphaltene fraction was left on the lower end of the rods. Details of bitumen composition are listed in Table 1.

Table 1. Group composition of the tested neat bitumen.

SAMPLE	SARA Fraction in Weight % (±0.1)
Saturated	4.2
Aromatics	51.6
Resins	21.3
Asphaltenes	22.9

2.4. Empirical Characterisation

Penetration tests for bitumens were performed according to the standard procedure (ASTM D946) [20]. The bitumen consistency was evaluated by measuring the penetration depth (531/2-T101, Tecnotest, Castelfranco, Treviso, Italy) of a stainless steel needle of standard dimensions under determinate charge conditions (100 g), time (5 s) and temperature (25 °C).

2.5. AFM Microstructure Analysis

Atomic Force Microscopy equipment (Multimode VIII with a Nanoscope V controller, Bruker, Karlsruhe, Germany) was used to analyse the samples. The AFM was used in tapping mode, where the cantilever oscillates up and down close to its resonance frequency so that the tip contacts the sample surface intermittently. When the tip is brought close to the surface, the vibration of the cantilever is influenced by the tip–sample interaction. In particular, shifts in the phase angle of vibration of the cantilever are due to the energy dissipation in the tip–sample ensemble. The phase shift provides information on surface properties such as stiffness, viscoelasticity and adhesion. For measurements, Antimony-doped silicon probes (TAP150A, Bruker) with resonance frequency 150 kHz and nominal tip radius of curvature 10 nm were used. All the measurements were performed at room temperature. Phase images were acquired simultaneously with the topographic mode. Materials with different viscoelasticity were clearly distinguishable. The softer domains appeared dark while the stiffer ones appeared bright in the phase images (see Figure 1A–H).

2.6. Isothermal Rheological Tests after Different Cooling Ramps

After each cooling ramp, samples were subjected to oscillatory rheological tests at constant temperature t = 25 °C, controlled by a Peltier element (±0.1 °C), using a dynamic stress-controlled rheometer (SR5, Rheometric Scientific, Piscataway, NJ, USA) equipped with a parallel plate geometry (gap 2.0 ± 0.1 mm, diameter 25 mm). The linear viscoelastic regime of both neat and modified bitumens was checked through the determination of the complex shear modulus $G^*(\omega)$ in the regime of small-amplitude oscillatory shear [21,22]. $G^*(\omega)$ can be considered the sample's total resistance to deformation when repeatedly sheared. For viscoelastic materials, $G^*(\omega)$ is split into a real and an imaginary part, respectively, [23]:

$$G^*(\omega) = G'(\omega) + i\,G''(\omega) \tag{1}$$

The frequency-dependent functions $G'(\omega)$ and $G''(\omega)$ define the in-phase (storage) and the out-of-phase (loss) moduli, respectively, i being the imaginary unit of the complex number. $G'(\omega)$ is a measure of the reversible, elastic energy, while $G''(\omega)$ represents the irreversible viscous dissipation of the mechanical energy [24]. Both the storage and loss moduli are related to each other through the phase angle δ defined by:

$$\tan\delta = G''(\omega)/G'(\omega) \tag{2}$$

Aimed at investigating the material structure, frequency sweep tests were performed at 25 °C and proper stress values were applied to guarantee linear viscoelastic conditions.

3. Results and Discussion

Bitumen was characterised by the SARA method and four different groups were individuated: Saturates, Aromatics, Resins and Asphaltenes (SARA). We recall here that according to the current accepted colloidal model for bitumen, asphaltene molecules rich in resins as peptizing agents self-assembly into micellar-like structures dispersed into the continuous phase composed mainly by the saturated and aromatic oil fractions (maltene) [25–27]. The SARA content of the pristine bitumen was determined (see Section 2) and the results are shown in Table 1.

3.1. AFM Results

Both neat and modified bitumens have been studied in micro-scale at room temperature (RT) after being subjected to different cooling rates. Figure 1 collects AFM phase images acquired after the tested specimens had been slowly cooled at 1 °C/min from 105 °C to RT (Figure 1A,C,E,G) and compared to an analogous series of images from the same samples subjected to a faster cooling ramp at 10 °C/min (Figure 1B,D,F,H). A first effect of cooling rate can be observed on the unmodified bitumen where coarse and isolated aggregates (which form the catana-phase [16,28]), with irregular or ellipsoid shaped domains dispersed into maltene matrix, are replaced by smaller oblong shaped structures with a rippled interior (1A vs. 1B). The occurrence of those discrete domains has been attributed to the crystallisation process of the paraffin wax fraction [16–18,29] where the formation of small crystalline nuclei may be kinetically favoured with respect to the particle growth process if bitumen undergoes too fast cooling rates [30]. A more dramatic micro-morphology change can be observed when the organosilane additive (P2KA) is added to the base bitumen (1C vs. 1D). After a slow cooling ramp temperature (1 °C/min, 1C), a rough surface morphology has been imaged at RT, constituted by irregular often-interlocking domains. As the sample is cooled at a faster rate (10 °C/min, 1D), the phase-contrast images reveal several small submicrometer crystalline structures resembling the 'ant-like' spots observed by Ramm et al. [31], considered as metastable aggregates subjected to a kinetic rather than thermodynamic control.

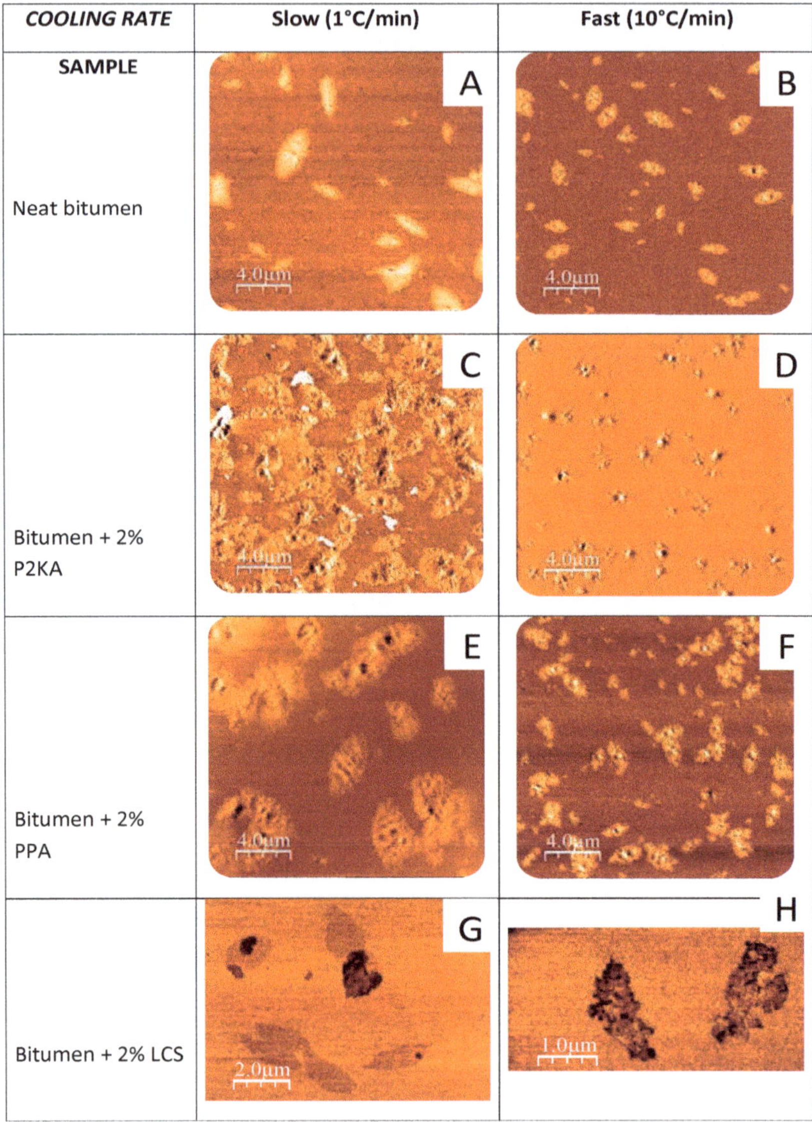

Figure 1. Atomic Force Microscopy (AFM) phase images at room temperature (RT) of both neat bitumen and bitumens modified by addition of 2% P2KA, PPA and LCS, respectively. The microphotographs were acquired after different thermal treatments: slowly cooled at 1 °C/min to RT (**A**,**C**,**E**,**G**) and quickly cooled at 10 °C/min to RT (**B**,**D**,**F**,**H**).

Addition of polyphosphoric acid (PPA) gives rise to surface structuring very similar to pristine bitumen. Indeed, after a slow cooling rate, large pseudo-spherical or lenticular domains have been imaged (1E) characterised by catana-phase with transverse stripes of high and low surface height surrounded by peri-phase regions; likewise, it has been observed in analogous AFM studies [18,32]. However, after subjecting the bitumen to rapid cooling treatments, smaller crystallites with irregular or ellipsoidal shapes are formed, yet retaining the same inner rippled microstructure (1F).

A completely different scenario has been found for bitumen modified with 2 wt % of phospholipids (LCS). Indeed, irregular domains with low phase contrast coexist with fractal-like and more defined particles with similar average dimensions. Those micro-morphological features have been found unaffected by the different cooling ramps applied to LCS-modified bitumen, as can be easily verified by comparing the AFM phase images acquired after slow cooling (1G) and fast cooling (1H) rates, respectively. During the investigation of LCS samples, the contact between the tip and the sample was unstable, due to the large attractive forces caused by the probable presence of LCS even on the sample surface, and this led to difficulties in imaging large areas. The measurements were, then, performed on small areas in order to visualise the single domain's size.

It is worth noting that the apparent aggregates' invariance detected at microstructural level regardless of the chosen ramp is reflected also by a correspondent mechanical resilience manifested by LCS-modified bitumen, as will be described in more details in the next paragraph.

3.2. Oscillatory Shear Experiments

The viscoelastic properties of bitumens were analysed by oscillatory experiments (frequency-sweeps and temperature sweeps) at 25 °C. Preliminary stress-sweep tests were also performed by applying small strain amplitudes in order to define linear viscoelastic conditions. The frequency dependence of the experimental complex modulus $|G^*|$ measured at 25 °C for both unmodified and modified bitumens is illustrated in Figure 2 in correspondence to different thermal cooling gradients of, respectively, 1 °C/min (A), 5 °C/min (B) and 10 °C/min (C).

Clear evidence can first be observed at low rates (1°/min) and fast (10°/min), showing substantial differences of $|G^*|$ among the various samples, attributable to the presence of the additives (see Figure 2A,C). An exception can be found for P2KA and PPA modified bitumens, characterised by $|G^*|$ data, which are almost overlapped for 1 °C/min (Figure 2A). However, the effect of the additives seems to be cancelled out at the intermediate cooling rate of 5 °C/min, where the respective rheological behaviours are hardly distinguishable from those of neat bitumen (see Figure 2B). As observed from AFM images, the development of amorphous/crystalline polydomains in the bitumens is controlled by nucleation and growth rates of disperse catana/peri-phase induced by different cooling rates, which in turn are affected also by the presence of additives. It should be expected that upon rapid cooling small aggregates would engender the formation of a network stabilised by physical interactions with a consequent increase of complex modulus $|G^*|$. To understand whether the added compounds are somewhat engaged in the formation of noncovalent interactions between asphaltene aggregates able to stabilise or even destabilise supramolecular networks upon cooling, the rheological data have been analysed in the framework of the colloidal gel model.

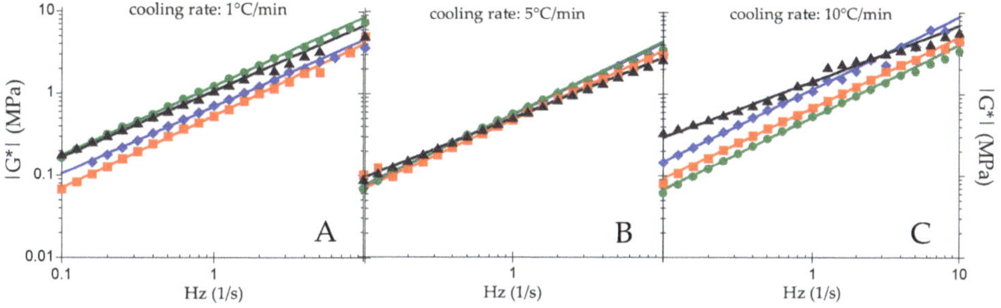

Figure 2. Complex modulus $|G^*|$ vs frequency determined at 25 °C for neat bitumen (diamonds), and bitumens modified by addition of 2% P2KA (circles), LCS (squares) and PPA (triangles), respectively, after the specimens were subjected to different thermal treatment: (**A**) 1 °C/min, (**B**) 5 °C/min and (**C**) 10 °C/min. Solid lines represent the best non-linear fits according to the power law of Equation (3), whose fitting parameters z and A have been listed in Tables 2 and 3, respectively.

According to the theory of Bohlin [33] and Winter [34], which has been applied to several colloidal complex systems [35–37] and widely reported in literature as the "weak-gel model" [38], a weak-gel material is defined as a complex system characterised by a cooperative arrangement of flow units connected by weak physical interactions that cooperatively ensure the stability of the structure. Thus, the weak-gel model provides a direct link between the microstructure of the material and its rheological properties. The most important parameter is the "coordination number", z, which is the number of flow units interacting with each other to give the observed flow response. It was shown in reference [33] that, above the Newtonian region, there exists a regime characterised by the following flow equation:

$$|G^*(\omega)| = \sqrt{G\prime(\omega)^2 + G''(\omega)^2} = A\omega^{\frac{1}{z}} \quad (3)$$

where A is a proper constant related to the overall stiffness or resistance to deformation of the material within the linear viscoelastic region at an angular frequency of 1 rad/s. Clearly, a log–log plots of $|G^*|$ vs. frequency should yield a straight line with slope $1/z$ and intercept A.

In Tables 2 and 3, the parameters z and A, calculated from non-linear fitting of viscoelastic data to Equation (1), are listed for all investigated samples subjected to different cooling rates. Three systems out of four, namely, neat bitumen and bitumens modified with P2KA and LCS, share a similar z variability in the narrow range 1.12–1.22, confirming the presence of interacting asphaltenes that form lightly entangled networks. On the contrary, PPA shows a slightly higher flow coordination number, especially in the range 5–10 °C/min, evidencing its network-promoting effect, according to previous investigations [39,40]. In any case, the calculated z values indicate that the coordination numbers are slightly affected by the thermal history as evidenced by the upper plot in Figure 3. A more useful parameter is represented by the constant prefactor of Equation (1) characterising the "interaction strength" of the three-dimensional structure of a gel, which is very sensitive to the cooling rates. Interestingly, an initial decrement of A is observed for all the tested specimens in correspondence of the increment 1 °C/min → 5 °C/min in the cooling rate. Upon further increase to 10 °C/min, both neat and PPA-modified bitumen show an upturn in A whereas P2KA and LCS give rise to a minor change of A (see the lower plot in Figure 3). What is worthy to remark here is that addition of LCS to bitumen provides the smallest A-variation within the explored range of cooling rates and this indicates that overall LCS is able to mitigate the effects provoked by a wide range of thermal gradients. The addition of LCS, whose adhesion efficiency has been recently ascertained [41], precludes the growth of nuclei formed during the nucleation stage, thus making the bulk bitumen structure almost unperturbed by drastic temperature variations.

Table 2. Dependence of the coordination number z on the cooling rate, calculated as a fitting parameter in Equation (1) adapted to experimental oscillatory rheological data, $|G^*|$ vs. frequency, for neat bitumen and bitumens modified with additives.

Cooling Rate (°C/min)	1	5	10
Sample	z	z	z
Bitumen	1.22 ± 0.01	1.15 ± 0.01	1.13 ± 0.03
Bitumen + P2KA 2%	1.19 ± 0.01	1.13 ± 0.01	1.12 ± 0.01
Bitumen + LCS 2%	1.13 ± 0.01	1.18 ± 0.01	1.15 ± 0.01
Bitumen + PPA 2%	1.25 ± 0.02	1.36 ± 0.01	1.46 ± 0.06

Table 3. Dependence of the "interaction strength" A on the cooling rate, calculated as a fitting parameter in Equation (1) adapted to experimental oscillatory rheological data, $|G^*|$ vs. frequency, for neat bitumen and bitumens modified with additives.

Cooling Rate (°C/min)	1	5	10
Sample	$A \times 10^{-6}$	$A \times 10^{-6}$	$A \times 10^{-6}$
Bitumen	0.68 ± 0.01	$0.558 \pm 1 \times 10^{-3}$	1.14 ± 0.01
Bitumen + P2KA 2%	1.22 ± 0.01	$0.563 \pm 3 \times 10^{-3}$	$0.518 \pm 2 \times 10^{-3}$
Bitumen + LCS 2%	$0.530 \pm 3 \times 10^{-3}$	$0.479 \pm 2 \times 10^{-3}$	$0.675 \pm 2 \times 10^{-3}$
Bitumen + PPA 2%	1.06 ± 0.01	$0.511 \pm 1 \times 10^{-3}$	1.42 ± 0.03

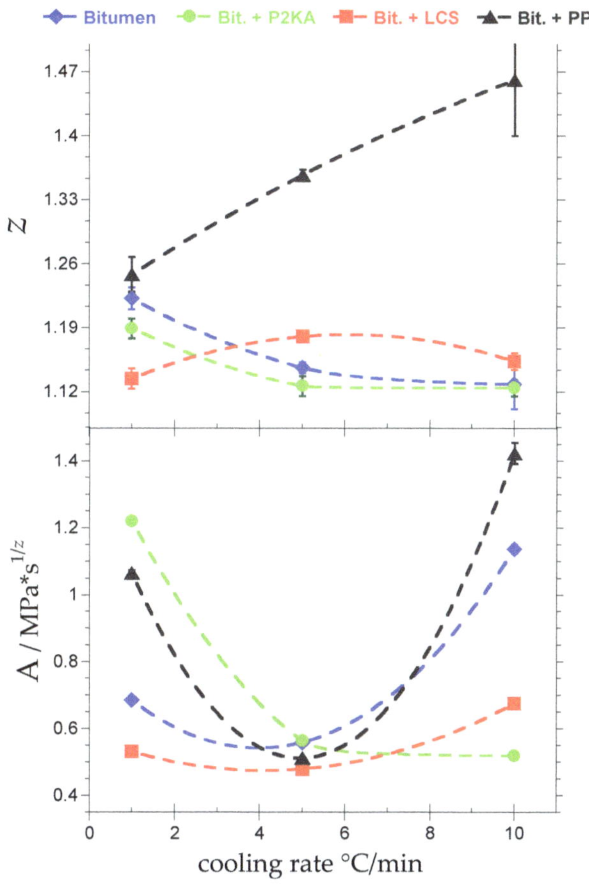

Figure 3. Dependence on the cooling rate of both z coordination number (upper plot) and interaction strength A (lower plot) for neat bitumen (diamonds) and bitumens modified by addition of 2% P2KA (circles), LCS (squares) and PPA (triangles), respectively.

4. Conclusions

A comparative investigation of the mechanical responses manifested by bitumen modified with three non-polymeric additives was performed at 25 °C, after the corresponding specimens were subjected to different cooling rates in the range 1–10 °C/min. The selected range of cooling ramps represented a good compromise between suitable experimental conditions performed at lab-scale and hypothetical cooling events that might occur in realistic environmental conditions.

Aimed at searching for microstructural differences between various types of modified bitumens when they have undergone the action of several thermal ramps, a parallel morphological investigation was also carried out by using Atomic Force Microscopy (AFM). The correspondent rheological response was interpreted under the framework of the "weak gel" model whose analysis revealed the presence of lightly entangled networks with the exception of PPA-modified bitumen. A striking result was recorded from bitumen incorporating a raw mixture of natural phospholipids (2% LCS additive), which was able to leave the asphaltene aggregates fairly unaltered after the hot material was either slowly (1 °C/min) or rapidly (10 °C/min) cooled to the final reference temperature of 25 °C. The complex modulus of the correspondent LCS-modified bitumen was found almost independent of the cooling rate as well. The addition of LCS to bitumen should prove extremely fruitful in increasing the mechanical resistance of bitumen to thermal shocks and provide an attempt to substitute polymer-based rheology modifiers with additives derived from renewable bio-resources.

Acknowledgments: This work was financially supported by KimiCal s.r.l. (Rende, Italy). Authors were very grateful to Loprete SRL for the support offered (supply of bitumen) and Ing. G. Lo Prete who stimulated the need to undertake the present study with strong applicative implications.

Author Contributions: Cesare Oliviero Rossi and Ruggero Angelico conceived and designed the experiments; Maria Penelope De Santo performed the AFM experiments; Saltanat Ashimova prepared the samples and performed the rheological experiments; Pietro Calandra analyzed the data.

Conflicts of Interest: The authors declare no conflict of interest

References

1. Briscoe, O.E. *Asphalt Rheology: Relationship to Mixture*; ASTM Spec. Publ. 941; American Society for Testing and Materials: Philadelphia, PA, USA, 1987.
2. Yen, T.F.; Chilingarian, G.V. *Asphalthenes and Asphalts*; Elsevier: New York, NY, USA, 1994.
3. Rozeveld, S.; Shin, E.; Bhurke, A.; France, L.; Drzal, L. Network morphology of straight and polymer modified asphalt cements. *Microsc. Res. Tech.* **1997**, *38*, 529–543. [CrossRef]
4. Lesueur, D. The colloidal structure of bitumen: Consequences on the rheology and on the mechanisms of bitumen modification. *Adv. Colloid Int. Sci.* **2009**, *145*, 42–82. [CrossRef] [PubMed]
5. Loeber, L.; Muller, G.; Morel, J.; Sutton, O. Bitumen in colloid science: A chemical, structure and rheological approach. *Fuel* **1998**, *77*, 1443–1450. [CrossRef]
6. Apeagyei, A.K.; Dave, E.V.; Buttlar, W.G. Effect of Cooling Rate on Thermal Cracking of Asphalt Concrete Pavements. *J. Assoc. Asph. Paving Technol.* **2008**, *77*, 709–738.
7. Huurman, M.; Mo, L.; Woldekidan, M.F.; Khedoe, R.N.; Moraal, J. Overview of the LOT meso mechanical research into porous asphalt ravelling. In *Advanced Testing and Characterization of Bituminous Materials 1*; CRC Press: Boca Raton, FL, USA, 2009; pp. 507–517.
8. Kluttz, R.; Jellema, E.; Woldekidan, M.; Huurman, M. Highly Modified Bitumen for Prevention of Winter Damage in OGFCs. In Proceedings of the Airfield and Highway Pavement, Los Angeles, CA, USA, 9–12 June 2013; pp. 1075–1087.
9. Oliviero Rossi, C.; Spadafora, A.; Teltayev, B.; Izmailova, G.; Amerbayev, Y.; Bortolotti, V. Polymer modified bitumen: Rheological properties and structural characterization. *Colloids Surf. A Physicochem. Eng. Asp.* **2015**, *480*, 390–397. [CrossRef]
10. Shen, J.; Amirkhanian, S.N.; Xiao, F.; Tang, B. Influence of surface area and size of crumb rubber on high temperature properties of crumb rubber and modified binders. *Constr. Build. Mater.* **2009**, *23*, 304–310. [CrossRef]
11. Mashaan, N.S.; Karim, M.R. Investigating the Rheological Properties of Crumb Rubber Modified Bitumen and its Correlation with Temperature Susceptibility. *Mater. Res.* **2013**, *16*, 116–127. [CrossRef]
12. Baldino, N.; Gabriele, D.; Oliviero Rossi, C.; Seta, L.; Lupi, F.R.; Caputo, P. Low temperature rheology of polyphosphoric acid (PPA) added bitumen. *Constr. Build. Mater.* **2012**, *36*, 592–598. [CrossRef]

13. Rossi, C.O.; Caputo, P.; Baldino, N.; Szerb, E.I.; Teltayev, B. Quantitative evaluation of organosilane-based adhesion promoter effect on bitumen-aggregate bond by contact angle test. *Int. J. Adhes. Adhes.* **2017**, *72*, 117–122. [CrossRef]
14. Rossi, C.O.; Caputo, P.; Baldino, N.; Lupi, F.R.; Miriello, D.; Angelico, R. Effects of adhesion promoters on the contact angle of bitumen-aggregate interface. *Int. J. Adhes. Adhes.* **2016**, *70*, 297–303. [CrossRef]
15. Rossi, C.O.; Taltayev, B.; Angelico, R. Adhesion Promoters in Bituminous Road Materials: A Review. *Appl. Sci.* **2017**, *7*, 524. [CrossRef]
16. Loeber, L.; Sutton, O.; Morel, J.; Valleton, J.M.; Muller, G. New direct observations of asphalts and asphalt binders by scanning electron microscopy and atomic force microscopy. *J. Microsc.* **1996**, *182*, 32–39. [CrossRef]
17. Masson, J.F.; Leblond, V.; Margeson, J. Bitumen morphologies by phase-detection atomic force microscopy. *J. Microsc.* **2006**, *221*, 17–29. [CrossRef] [PubMed]
18. Hung, A.M.; Fini, E.H. AFM study of asphalt binder "bee" structures: Origin, mechanical fracture, topological evolution, and experimental artifacts. *RSC Adv.* **2015**, *5*, 96972–96982. [CrossRef]
19. Yoon, S.; Bhatt, S.D.; Lee, W.; Lee, H.Y.; Jeong, S.Y.; Baeg, J.O.; Lee, C.W. Separation and characterization of bitumen from Athabasca oil sand. *Korean J. Chem. Eng.* **2009**, *26*, 64–71. [CrossRef]
20. ASTM Standard 2005. *Standard for Penetration-Graded Asphalt Cement for Use in Pavement Construction, D946-82*; ASTM International: Montgomery, PA, USA, 2005.
21. Coppola, L.; Gianferri, R.; Rossi, C.O.; Nicotera, I.; Ranieri, G.A. Structural changes in CTAB/H_2O mixtures using a rheological approach. *Phys. Chem. Chem. Phys.* **2004**, *6*, 2364–2372. [CrossRef]
22. Angelico, R.; Carboni, M.; Lampis, S.; Schmidt, J.; Talmon, Y.; Monduzzi, M.; Murgia, S. Physicochemical and rheological properties of a novel monoolein-based vesicle gel. *Soft Matter* **2013**, *9*, 921–928. [CrossRef]
23. Read, J.; Whiteoak, D. *The Shell Bitumen Handbook*, 5th ed.; Hunter, R.N., Ed.; Thomas Telford Publishing: London, UK, 2003.
24. Antunes, F.; Gentile, L.; Rossi, C.O.; Tavano, L.; Ranieri, G.A. Gels of Pluronic F127 and nonionic surfactants from rheological characterization to controlled drug permeation. *Colloids Surf. B Biointerfaces* **2011**, *87*, 42–48. [CrossRef] [PubMed]
25. Andersen, S.I.; Birdi, K.S. Aggregation of asphaltenes as determined by calorimetry. *J. Colloid Interface Sci.* **1991**, *42*, 497–502. [CrossRef]
26. Sheu, E.Y.; Storm, D.A. Colloidal properties of asphaltenes in organic solvents. In *Asphaltenes: Fundamentals and Applications*; Sheu, E.Y., Mullins, O.C., Eds.; Plenum Press: New York, NY, USA, 1995; Chapter 1.
27. Sirota, E.B. Physical structure of asphaltenes. *Energy Fuels* **2005**, *19*, 1290–1296. [CrossRef]
28. Masson, J.F.; Leblond, V.; Margeson, J.; Bundalo-Perc, S. Low—Temperature bitumen stiffness and viscous paraffinic nano-and micro-domains by cryogenic AFM and PDM. *J. Microsc.* **2007**, *227*, 191–202. [CrossRef] [PubMed]
29. Pauli, A.T.; Grimes, R.W.; Beemer, A.G.; Turner, T.F.; Branthaver, J.F. Morphology of asphalts, asphalt fractions and model wax-doped asphalts studied by atomic force microscopy. *Int. J. Pavement Eng.* **2011**, *12*, 291–309. [CrossRef]
30. Fischer, H.R.; Dillingh, B.; Ingenhut, B. Fast solidification kinetics of parts of bituminous binders. *Mater. Struct.* **2016**, *49*, 3335–3340. [CrossRef]
31. Ramm, A.; Sakib, N.; Bhasin, A.; Downer, M.C. Optical characterization of temperature- and composition-dependent microstructure in asphalt binders. *J. Microsc.* **2016**, *262*, 216–225. [CrossRef] [PubMed]
32. Yan, K.; Zhang, H.; Xu, H. Effect of polyphosphoric acid on physical properties, chemical composition and morphology of bitumen. *Constr. Build. Mater.* **2013**, *47*, 92–98. [CrossRef]
33. Bohlin, L. A Theory of Flow as a Cooperative Phenomenon. *J. Colloid Interface Sci.* **1980**, *74*, 423–434. [CrossRef]
34. Winter, H.H. Can the Gel Point of a Cross-linking Polymer Be Detected by the G'—G'' Crossover? *Polym. Eng. Sci.* **1987**, *27*, 1698–1702. [CrossRef]
35. Rossi, C.O.; Coppola, L.; La Mesa, C.; Ranieri, G.A.; Terenzi, M. Gemini surfactant–water mixtures: Some physical–chemical properties. *Colloids Surf. A* **2002**, *201*, 247–260.
36. Coppola, L.; Gianferri, R.; Rossi, C.O.; Nicotera, I. Solution and Liquid-Crystalline microstructures in Sodium Taurodeoxycholate/D_2O mixtures. *Langmuir* **2003**, *19*, 1990–1999. [CrossRef]

37. Coppola, L.; Gentile, L.; Nicotera, I.; Rossi, C.O.; Ranieri, G.A. Evidence of Formation of Ammonium Perfluorononanoate/^2H_2O Multilamellar Vesicles: Morphological Analysis by Rheology and Rheo-^{2}H NMR Experiments. *Langmuir* **2010**, *26*, 19060–19065. [CrossRef] [PubMed]
38. Gabriele, D.; de Cindio, B.; D'Antona, P. A weak gel model for foods. *Rheol. Acta* **2001**, *40*, 120–127. [CrossRef]
39. Gentile, L.; Filippelli, L.; Rossi, C.O.; Baldino, N.; Ranieri, G.A. Rheological and H-NMR Spin-Spin relaxation time for the evaluation of the effects of PPA addition on bitumen. *Mol. Cryst. Liq. Cryst.* **2012**, *558*, 54–63. [CrossRef]
40. Baldino, N.; Gabriele, D.; Lupi, F.R.; Rossi, C.O.; Caputo, P.; Falvo, T. Rheological effects on bitumen of polyphosphoric acid (PPA) addition. *Constr. Build. Mater.* **2013**, *40*, 397–404. [CrossRef]
41. Rossi, C.O.; Caputo, P.; Loise, V.; Miriello, D.; Taltayev, B.; Angelico, R. Role of a food grade additive in the high temperature performance of modified bitumens. *Colloids Surf. A Physicochem. Eng. Asp.* **2017**. [CrossRef]

Article

Using a Molecular Dynamics Simulation to Investigate Asphalt Nano-Cracking under External Loading Conditions

Yue Hou [1], Linbing Wang [2], Dawei Wang [3,*], Xin Qu [3] and Jiangfeng Wu [1]

1 National Center for Materials Service Safety, University of Science and Technology Beijing, Beijing 100083, China; yhou@ustb.edu.cn (Y.H.); b20130489@xs.ustb.edu.cn (J.W.)

2 Joint USTB-Virginia Tech Lab on Multifunctional Materials, USTB, Beijing 100083, China, Virginia Tech, Blacksburg, VA 24061, USA; wangl@vt.edu

3 Institute of Highway Engineering, RWTH Aachen University, D52074 Aachen, Germany; qu@isac.rwth-aachen.de

* Correspondence: wang@isac.rwth-aachen.de

Academic Editor: Zhanping You

Received: 8 June 2017; Accepted: 26 July 2017; Published: 28 July 2017

Abstract: Recent research shows that macro-scale cracking in asphalt binder may originate from its intrinsic defects at the nano-scale. In this paper, a molecular dynamics (MD) simulation was conducted to evaluate the nucleation of natural defects in asphalt. The asphalt microstructure was modeled using an ensemble of three different types of molecules to represent a constituent species: asphaltenes, naphthene aromatics and saturates, where the weight proportion of 20:60:20 was used to create an asphalt-like ensemble of molecules. Tension force was then applied on the molecular boundaries to study the crack initiation and propagation. It was discovered that the natural distribution of atoms at microscale would affect the intrinsic defects in asphalt and further influence crack initiation and propagation in asphalt.

Keywords: MD simulation; asphalt; nano-cracking; low temperature

1. Background and Introduction

Asphalt pavement is one of the main highway forms throughout the world. With the continuous development of the world economy, there have been tremendous numbers of vehicles passing on the pavements every day, causing significant irreversible damage to pavement. It is therefore necessary to study asphalt material performance, since it is directly relevant to its service life, where structure is one of the most important factors that affects its mechanical performance. Recent studies show that the asphalt material microstructure may affect its properties at the macro-scale [1,2].

There have been many experimental approaches on the asphalt material microstructure due to the development of modern testing instruments. Mikhailenko et al. [3], used scanning electron microscopy (SEM) to observe the microstructure of asphalt materials. Fourier transform infrared spectroscopy (FTIR) can be used to study the distribution and contents of asphalt components [4–6]. Gel permeation chromatography (GPC) can determine the molecular weight and distribution of asphalt rapidly and reliably [7]. Differential scanning calorimetry (DSC) can be used to analyze the difference of the thermal stability of asphalt [8,9]. Atomic force microscope (AFM) can be used to observe three-dimensional (3D) spatial images of asphalt [10–12]. These new experimental methods can be used to further investigate the effect of modifiers on the modification of asphalt materials at nanometer scales [13,14]. For example, SEM can be used to observe and compare the microstructures of different nanometer modified asphalts [15], and AFM, FTIR, GPC and DSC can be used to further analyze the modify mechanism of nano-modifier materials.

There also have been many studies on asphalt material properties using the simulation, where molecular dynamics (MD) simulation is one of the effective methods to investigate the microstructure and micro-properties of asphalt materials [16–18]. In the aspect of asphalt oxidation aging, MD simulation can be employed to investigate internal chemical and molecular mechanical property changes [19]. Molecular Dynamics simulation can also be used to determine the thermodynamic properties of asphaltene before and after oxidative aging [20]. In other aspects, molecular dynamics simulation can also be used to study molecular movements and microstructure changes of epoxy modified asphalt in the process of healing [21]. Xu et al. [22], conducted AFM experiments and MD simulation to study the adhesion between asphalt and aggregate. The mechanical properties of asphalt and aggregate can be further investigated by MD simulation at the nano-scale [23], and by investigating mechanical properties of the asphalt-aggregate interface [24].

Although there has been significant progress in asphalt material microstructure and micro-properties, research on the fundamental mechanism of asphalt microstructure cracking using simulations and experiment approaches are very few. In this study, MD simulation was conducted to study micro-crack initiation, and propagation of typical asphalt molecular structures. The ultimate goal was to try to understand the fundamental mechanism of asphalt molecular structure cracking.

2. Molecular Dynamics Simulations

In this study, the MD simulation was employed to investigate the mechanical behavior of asphalt molecules.

2.1. Asphalt Molecular Structure Construction

Asphalt is often characterized chemically by its hydrocarbon class composition, e.g., three main constituent species including asphaltene, naphthene aromatics and saturates. The three-fraction asphalt model was employed to establish the molecular structure following Zhang and Greenfield [25], where three different types of molecules are used to represent the corresponding constituent species, and the three are then formed together to create an asphalt-like ensemble of molecules. In this model, 1,7-dimethylnapthalene was used to represent naphthene aromatics, and n-C_{22} (n-doccosane) molecules were used to represent saturates, with the mass fraction of asphaltenes, naphthene aromatics, and saturates being approximately 20:60:20, respectively.

The condensed-phase optimized molecular potentials for atomistic simulation studies (COMPASS) force field was used for our simulation, considering that this force field could accurately describe the inorganic material properties. The initial molecular structure was constructed by using the commercial molecular computation software Material Studio 7.0 (BIOVIA Company, San Diego, CA, USA). The construction function in amorphous cell tools of Material Studio was used to construct the representative volume element in asphalt binder. The target density of the final configurations was set to 1 g/cm^3. Table 1 shows the detailed configuration of the asphalt-like ensemble. Figure 1 shows the constructed molecular structure, where the edge length (Å) of the lattice is $50.9 \times 50.9 \times 50.9$. The atoms in each molecule in the asphalt-like amorphous cell were then subjected to conjugate gradient energy minimization within 10,000 fs.

Table 1. Configuration of the asphalt-like ensemble.

Molecular Components	Asphaltene	1,7-Dimethylnapthalene	n-Doccosane
Number	18	306	49
Approximate Mass ratio	20	60	20

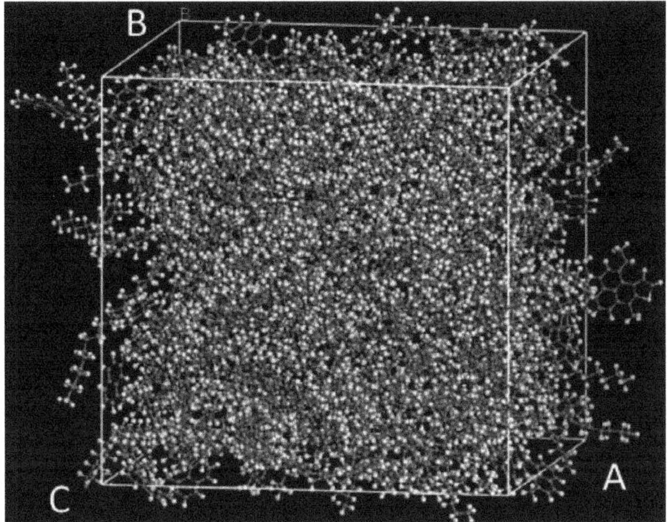

Figure 1. Constructed asphalt in amorphous cell.

Our molecular dynamics simulation was split into two parts: constructing the model in Materials Studio due to convenience, and applying the loading conditions in the Large-scale Atomic/Molecular Massively Parallel Simulator (LAMMPS) software. After construction of the asphalt molecular model in Material Studio, it was exported in a PDB-format file and imported into the VMD (Visual Molecular Dynamics) software (University of Illinois at Urbana-Champaign, Urbana-Champaign, IL, USA). By using the "Topo writelammpsdata asphalt.data full" command, the asphalt data file can be read by the LAMMPS software.

The following potential function was used in our LAMMPS simulation for convenience:

$$E = \varepsilon\left[2\left(\frac{\sigma}{r}\right)^{9} - 3\left(\frac{\sigma}{r}\right)^{6}\right], \; r < r_c \tag{1}$$

where E is the potential, ε is the depth of a potential well that reflects the attraction between two atoms, σ is the distance between atoms when the action potential is equal to 0, r is the distance between two atoms, and r_c is the cutoff.

2.2. Simulation Results

The relaxation process of asphalt molecules is shown in Figure 2. The relaxation was used to reach the energy minimization state of the asphalt molecular structure. It can be seen that the relaxation process would influence the structure of asphalt molecules, where agglomeration and densification occurred.

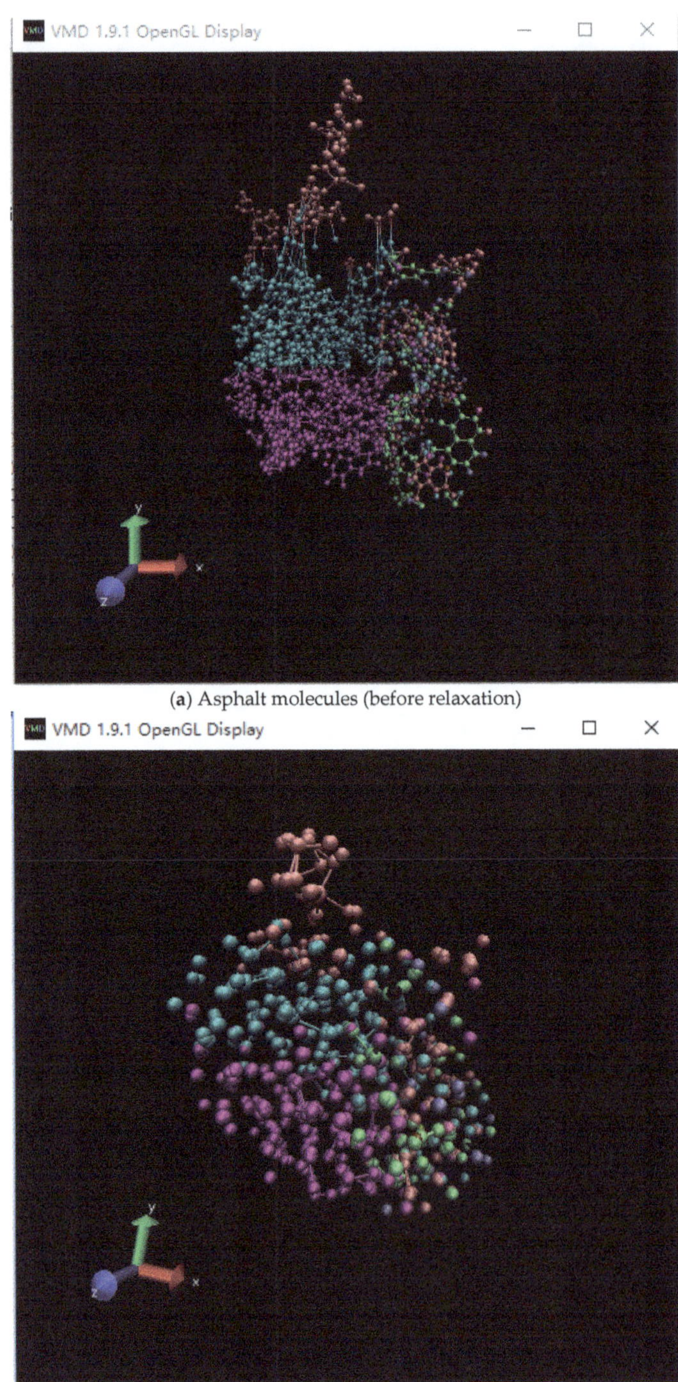

(**a**) Asphalt molecules (before relaxation)

(**b**) Asphalt molecules (after relaxation)

Figure 2. The relaxation process of asphalt molecules.

In the study, molecular dynamics simulation were exerted on the asphalt layer, where "region" and "group" commands were applied to set different regions in the asphalt molecule ensemble. The velocity command was used to set the velocities of different groups at 0.3 Å/fs along the y axis to simulate the state of tension. The ensemble temperature was set to room temperature, which is 298.15 K. For every 100 time steps, the simulation temperature was adjusted. For asphalt molecular ensemble, the energy minimization process was firstly conducted to achieve the stability of the system, and then the simulation of the stretching process was conducted. The time step was set to 0.003 fs, and a total operation time with 6000 steps was conducted. In the OVITO (Open Visualization Tool) software, asphalt molecules were sliced (Figure 3), where the internal development of asphalt molecules under tension could be observed. Figure 3a shows the complete whole microstructure and Figure 3b shows the half-sliced part. It was found that, along with the applying of continuous tension loading, obvious gaps could be found in the internal asphalt molecule structure (Figure 4), where these locations may be the weak regions and prone to easily crack.

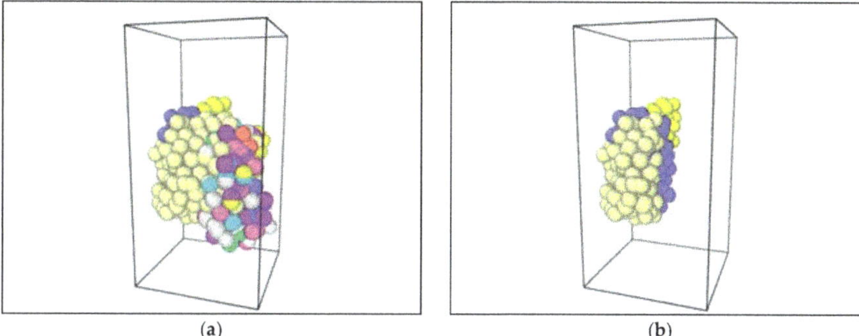

(a) (b)

Figure 3. Slicing of the asphalt molecules.

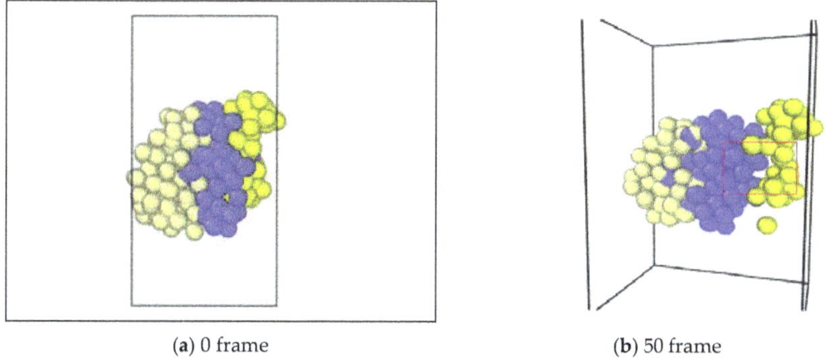

(a) 0 frame (b) 50 frame

Figure 4. Development of gap under tension loading.

The radial distribution function (RDF) is an important parameter to describe molecular structure. The essence of the radial distribution function $g_{ab}(r)$ is the atomic ratio of local density and average density in the system [26]. When A and B atoms are distributed uniformly in the spherical shell, the probability is calculated as follows:

$$\rho_A = \frac{n_A}{V} \tag{2}$$

$$\rho_B = \frac{n_B}{V} \tag{3}$$

$$P_A = 4\rho_A r^2 \delta r \tag{4}$$

$$P_B = 4\pi\rho_B r^2 \delta r \tag{5}$$

where V is the total volume of the simulation system, n_A and n_B are the number of A and B atoms, ρ_A and ρ_B are the average density of atoms in the simulation system, and P_A and P_B are the uniform distribution probabilities of A and B atoms.

By calculating the radial distribution function between different atoms, the local distribution of different atoms could be studied, as shown in Figure 5. It is observed that there existed a small difference in RDF before and after tension loading. For the asphalt microstructure before tension loading, when the distance was 2.375 Å, a maximum value of 16.5250 occurred; for the asphalt microstructure after the tension loading, when the distance was 2.375 Å, maximum value of 17.039 occurred. The trends of the curves of the two curves were completely consistent, indicating that most of the atoms in the asphalt microstructure did not significantly change their locations while a small portion changed the locations. It was further concluded that the applied tension loading would not only result in the whole asphalt microstructure change, but also in some of the specific locations.

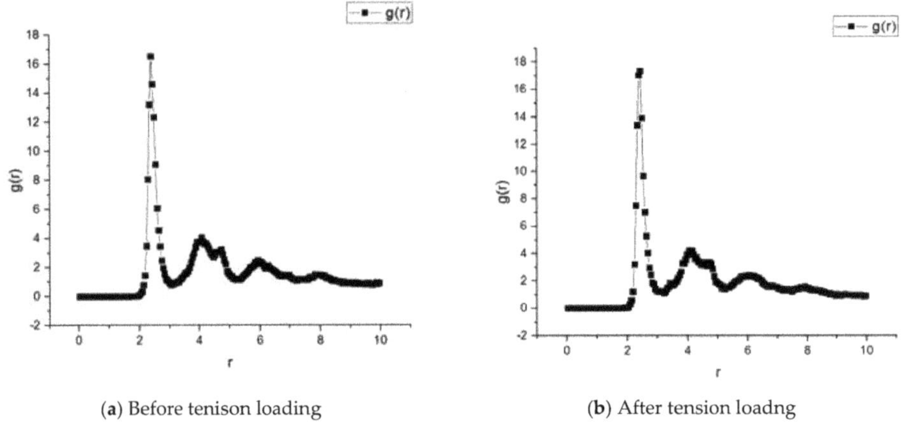

(**a**) Before tenison loading

(**b**) After tension loadng

Figure 5. Radial distribution function (RDF) of asphalt molecule before and after tension loading.

The application of tension loadings on the asphalt molecular structure was done by setting a constant speed in its y direction (vertical direction) of different regions, resulting in the stretching of the asphalt microstructure. Figure 6 shows the morphological changes of asphalt microstructures. Note that different color represents different atom types on the top and three different divisions on the bottoms. It was noticed that by applying the tension force, the original "perfect" structure generally stretched, and a "hole", i.e., natural defect, occurred in the microstructure, which could be seen as the initiation of the microcrack. It was expected that with an increase in the tension loading, the inside small hole propagated and finally resulted in the failure of the asphalt microstructure [27].

In order to analyze the temperature effects on the asphalt molecular, four kinds of different temperatures were chosen, including −40, 0, 25 and 60 °C, to analyze the temperature influence on the morphology and properties of the asphalt molecules during tension. The time step was 0.003 fs, with a total of 5000 steps. The asphalt molecular morphology with different temperture at 5000 timesteps are shown in Figure 7. It was seen that temperature seemed do not have a significant influence on the asphalt microstructure evolution. It might be caused by the small molecule amount in the model. Compared with the internal molecular component distribution, temperature does not have a significant effect on the molecular structure for this computation scale. To further study the temperature effects on asphalt microstructure evolution quantitatively and in more detail, future studies will be conducted on a supercomputer.

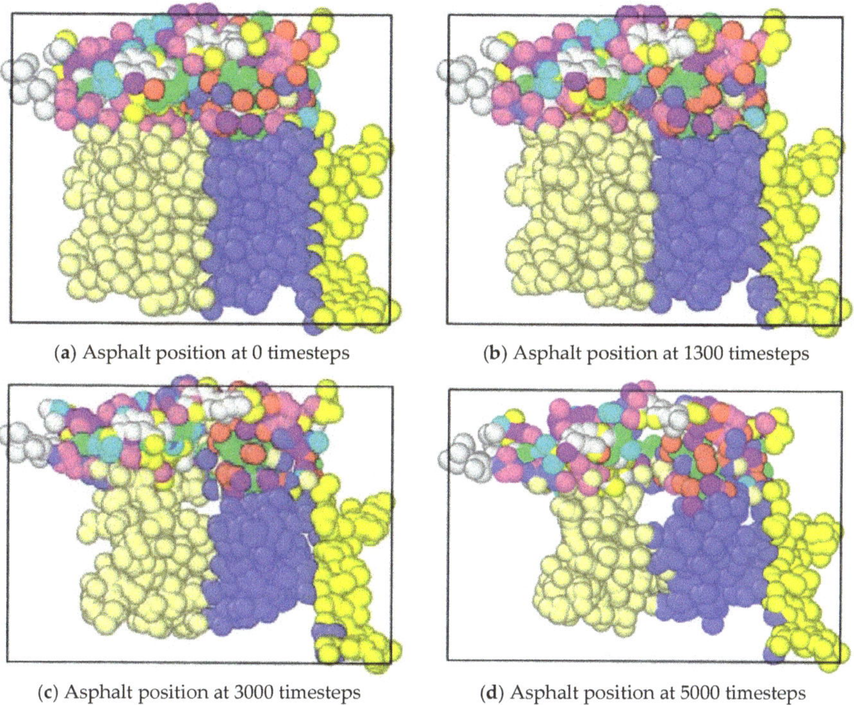

(**a**) Asphalt position at 0 timesteps (**b**) Asphalt position at 1300 timesteps

(**c**) Asphalt position at 3000 timesteps (**d**) Asphalt position at 5000 timesteps

Figure 6. Morphological changes of asphalt molecules under uniaxial tension.

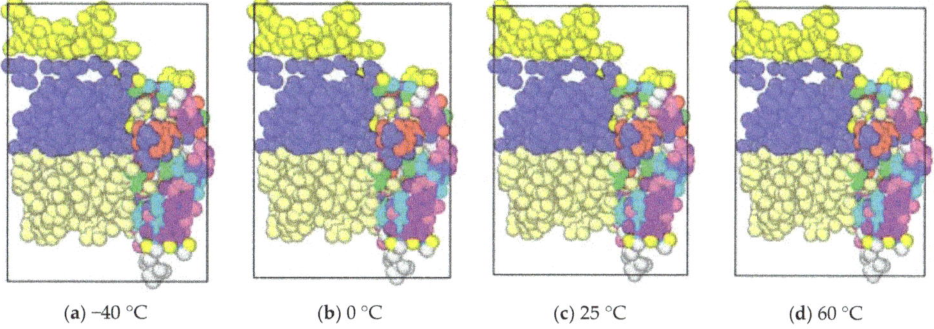

(**a**) −40 °C (**b**) 0 °C (**c**) 25 °C (**d**) 60 °C

Figure 7. Morphological changes of asphalt molecules at different temperature.

Through the analysis, it was found that there were "holes" in the asphalt molecules by stretching, with these holes existing due to the distribution of molecules, creating the original defects. The results showed that the original defects naturally existed in the asphalt molecules, resulting in different cracking performance. Note that since this was a preliminary study on the nucleation of natural defects in asphalt, we did not study this phenomenon quantitatively. In future studies, a parameter such as "total distance moved" by the atoms from some reference points may be proposed to describe this phenomenon.

The strain and stress developments over time for the asphalt microstructure subject to external loading at 25 °C are shown in Figure 8. Figure 8a shows that the strain increased almost linearly, as we applied a constant strain deformation rate. Figure 8b shows that the stress gradually decreased and remained almost constant after 1300 timesteps, indicating that due to the occurrence of the "internal hole", the asphalt microstructure bore less stress, and when the final structure was stabilized, the stress became stable.

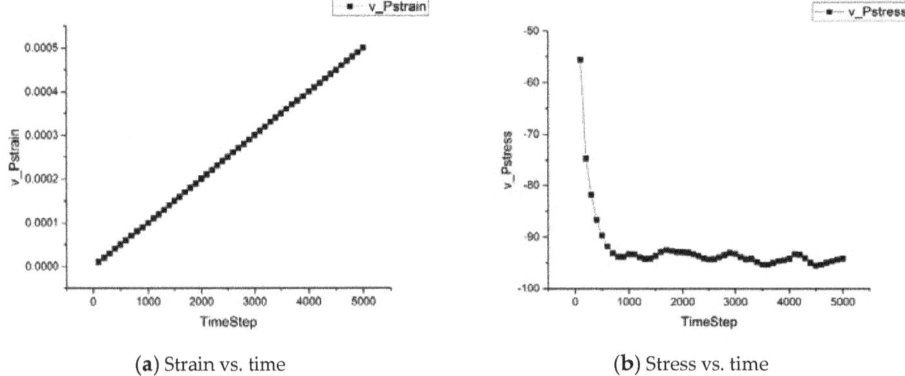

(**a**) Strain vs. time (**b**) Stress vs. time

Figure 8. Strain and stress developments.

It should be mentioned that our current research results seem to be a bit different from the observations in previous studies: 1. Menapace et al. (2016) discovered that temperature will significantly affect the "Bumble-bee" structure in asphalt samples at microscale [28]; 2. Jahangir et al. (2015) claimed that the tension loading would result in the change in asphalt microstructure [11]. There might be two reasons for the differences: 1. We currently only apply a very small tension loading. Therefore, compared with the internal molecule component distribution, the latter will have a larger effect on the molecular structure; 2. The current molecular model is established with relatively small molecule amount due to the limitation of our computer ability. It is expected that a much larger molecular model with millions of molecules may resemble the results by the previous research.

3. Conclusions

In this study, the MD simulation was conducted to model the asphalt molecular structure. The following conclusions were made:

1. It was discovered that by applying the tension force, the original "perfect" structure generally stretched and a "hole", i.e., natural defect, occurred in the microstructure, which can be seen as the initiation of the microcrack. It was expected that with an increase of tension loading, the inside small hole would propagate and finally result in the failure of the asphalt microstructure.
2. It was observed that there existed a difference in RDF before and after tension loading. For the asphalt microstructure before the tension loading, when the distance was 2.375 Å, a maximum value of 16.5250 occurred; for the asphalt microstructure after the tension loading, when the distance was 2.375 Å, maximum value of 17.039 occurred. The trends of the curves for the two curves were completely consistent, indicating that many of the atoms in the asphalt microstructure did not change their locations, while a small portions changed.

Acknowledgments: The first author would like to express his sincerely gratitude to Amit Bhasin at the University of Texas at Austin for sharing the asphalt molecular structure. The research performed in this paper is supported by National Natural Science Foundation of China (No. 41372320), the Natural Science Foundation of Shandong

Province (ZR2015EQ009), the Fundamental Research Funds for the Central Universities (06500036), and the 111 Project (Grant No. B12012).

Author Contributions: Yue Hou, Linbing Wang, Dawei Wang, Xin Qu, and Jiangfeng Wu conducted the numerical simulations; and Yue Hou wrote the paper.

Conflicts of Interest: The authors declare no conflict of interest.

References

1. Guo, M.; Tan, Y.; Wang, L.; Hou, Y. Diffusion of asphaltene, resin, aromatic and saturate components of asphalt on mineral aggregates surface: Molecular dynamics simulation. *Road Mater. Pavement* **2017**, *18*, 149–158. [CrossRef]
2. Hou, Y.; Wang, L.; Wang, D.; Liu, P.; Guo, M.; Yu, J. Characterization of Bitumen Micro-Mechanical Behavior Using AFM, Phase Dynamics Theory and MD Simulation. *Materials* **2017**, *10*, 208. [CrossRef]
3. Mikhailenko, P.; Khadim, H.; Baaj, H. Observation of Asphalt Binder Microstructure with ESEM. In Proceedings of the Transportation Association of Canada Conference, Toronto, ON, Canada, 25–28 September 2016.
4. Sun, D.Q.; Zhang, L.W.; Zhang, X.L. Quantification of SBS content in SBS polymer modified asphalt by FTIR. *Adv. Mater. Res.* **2011**, *287*, 953–960. [CrossRef]
5. Perez-Martinez, M.; Marsac, P.; Gabet, T.; Chailleux, E. Prediction of the Mechanical Properties of Aged Asphalt Mixes from FTIR Measurements. In Proceedings of the 8th RILEM International Symposium on Testing and Characterization of Sustainable and Innovative Bituminous Materials, Ancona, Italy, 7–9 October 2015; Springer: Haarlem, The Netherlands, 2015.
6. Feng, Z.G.; Bian, H.J.; Li, X.J.; Yu, J.Y. FTIR analysis of UV aging on bitumen and its fractions. *Mater. Struct.* **2016**, *49*, 1381–1389. [CrossRef]
7. Zhou, Y.; Zhang, K.; Yin, J.; Zheng, M. Analysis on thermo-stability of modified asphalt using GPC (gel permeation chromatography). *Railway Eng.* **2013**, *8*, U414.
8. Chen, X.L.; Sun, Y.S.; Han, Y.X.; Tian, Z.F. Microstructure and DSC of organic bentonite modified bitumen. *J. Northeast. Univer.* **2012**, *33*, 743–747.
9. Chen, X.L.; Jiang, Y.Z.; Sun, Y.S.; Han, Y.X. Influence of DCCF on the properties of asphalt by DSC. *Adv. Mater. Res.* **2012**, *454*, 35–40. [CrossRef]
10. Nazzal, M.D.; Abu-Qtaish, K.; Powers, D. Using atomic force microscopy to evaluate the nanostructure and nanomechanics of warm mix asphalt. *J. Mater. Civ. Eng.* **2015**, *27*, 1–9. [CrossRef]
11. Jahangir, R.; Little, D.; Bhasin, A. Evolution of asphalt binder microstructure due to tensile loading determined using AFM and image analysis techniques. *Int. J. Pavement Eng.* **2015**, *16*, 337–349. [CrossRef]
12. Tan, Y.; Guo, M. Micro- and nano-characteration of interaction between asphalt and filler. *J. Test. Eval.* **2014**, *42*, 1–9. [CrossRef]
13. Khattak, M.J.; Khattab, A.; Zhang, P.; Rizvi, H.R.; Pesacreta, T. Microstructure and fracture morphology of carbon nano-fiber modified asphalt and hot mix asphalt mixtures. *Mater. Struct.* **2013**, *46*, 2045–2057. [CrossRef]
14. Zareshahabadi, A.; Shokuhfar, A.; Ebrahiminejad, S. Microstructure and properties of nanoclay reinforced asphalt binders. *Defect Diffus. Forum* **2010**, 579–583. [CrossRef]
15. Sun, L.; Wang, H.; Wang, S.; Xiao, D. *Nano-Modified Asphalt and Road Performance*; Science Press: Beijing, China, 2012.
16. Rapaport, D.C.; Blumberg, R.L.; Mckay, S.R.; Christian, W. *The Art of Molecular Dynamics Simulation*; Cambridge University Press: Cambridge, UK, 2004.
17. Hou, Y.; Sun, W.; Huang, Y.; Ayatollahi, M.; Wang, L.; Zhang, J. Diffuse Interface Model to Investigate the Asphalt Concrete Cracking Subjected to Shear Loading at a Low Temperature. *J. Cold Reg. Eng.* **2017**, *31*, 04016009. [CrossRef]
18. Hou, Y.; Guo, M.; Ge, Z.; Wang, L.; Sun, W. Mixed-Mode I-II Cracking Characterization of Mortar using Phase-Field Method. *J. Eng. Mech.* **2017**, *143*, 04017033. [CrossRef]
19. Pan, J.; Tarefder, R.A. Investigation of asphalt aging behaviour due to oxidation using molecular dynamics simulation. *Mol. Simul.* **2016**, *42*, 1–12. [CrossRef]

20. Tarefder, R.A.; Arisa, I.R. Molecular Dynamic Simulation of Oxidative Aging in Asphaltene. In Proceedings of the Pavements and Materials: Characterization and Modeling Symposium at Emi Conference, Los Angeles, CA, USA, 8–11 August 2010.
21. Zhou, X.; Wu, S.; Liu, G.; Pan, P. Molecular simulations and experimental evaluation on the curing of epoxy bitumen. *Mater. Struct.* **2016**, *49*, 1–7. [CrossRef]
22. Xu, M.; Yi, J.; Feng, D.; Huang, Y.; Wang, D. Analysis of adhesive characteristics of asphalt based on atomic force microscopy and molecular dynamics simulation. *ACS Appl. Mater. Interfaces* **2016**, *8*, 12393–12403. [CrossRef] [PubMed]
23. Lu, Y.; Wang, L. Nanoscale Modeling of the Mechanical Properties of Asphalt and Aggregate. In Proceedings of the Pavements and Materials: Characterization and Modeling Symposium at Emi Conference, Los Angeles, CA, USA, 8–11 August 2010.
24. Lu, Y.; Wang, L. Nanoscale modelling of mechanical properties of asphalt–aggregate interface under tensile loading. *Int. J. Pavement Eng.* **2010**, *11*, 393–401. [CrossRef]
25. Zhang, L.; Greenfield, M.L. Analyzing properties of model asphalts using molecular simulation. *Energy Fuels* **2007**, *21*, 1712–1716. [CrossRef]
26. Ding, Y.; Tang, B.; Zhang, Y.; Wei, J.; Cao, X. Molecular Dynamics Simulation to Investigate the Influence of SBS on Molecular Agglomeration Behavior of Asphalt. *J. Mater. Civ. Eng.* **2015**, *27*. [CrossRef]
27. Hou, Y.; Wang, L.; Yue, P.; Pauli, T.; Sun, W. Modeling Mode I: Cracking Failure in Asphalt Binder by Using Nonconserved Phase-Field Model. *J. Mater. Civ. Eng.* **2014**, *26*, 684–691. [CrossRef]
28. Menapace, I.; Masad, E.; Bhasin, A. Effect of treatment temperature on the microstructure of asphalt binders: Insights on the development of dispersed domains. *J. Microsc.* **2016**, *262*, 12–27. [CrossRef] [PubMed]

Article

Evaluation of Mechanical Properties of Recycled Material for Utilization in Asphalt Mixtures

Farzaneh Tahmoorian [1,*], Bijan Samali [1], Vivian W.Y. Tam [2,3] and John Yeaman [4]

1 Centre for Infrastructure Engineering, Western Sydney University, Kingswood, NSW 2751, Australia; B.Samali@westernsydney.edu.au
2 School of Computing, Engineering, and Mathematics, Western Sydney University, Kingswood, NSW 2751, Australia; V.Tam@westernsydney.edu.au
3 College of Civil Engineering, Shenzhen University, Shenzhen 518060, China
4 Faculty of Science, Health, Education and Engineering, University of Sunshine Coast, Sippy Downs, Queensland 4556, Australia; jyeaman@usc.edu.au
* Correspondence: F.Tahmoorian@westernsydney.edu.au; Tel.: +61-2-4736-0866

Received: 28 June 2017; Accepted: 18 July 2017; Published: 27 July 2017

Abstract: With an expanding world, the demand for extensive road networks is increasing. As natural resources become scarce, the necessity of finding alternative resources has led to the idea of applying recycled material to pavement construction including asphalt pavements. Amongst all asphalt components, aggregate constitutes the largest part of asphalt mixtures. Therefore, the utilization of recycled material for aggregate will represent an important opportunity to save virgin material and divert material away from landfills. Because of the large amount of construction waste generation around the world, using recycled construction aggregate (RCA) in asphalt mixtures appears to be an effective utilization of RCA. However, as aggregate plays an important role in the final performance of the asphalt mixture, an understanding of their properties is essential in designing an asphalt mixture. Therefore, in this research, the properties of RCA have been evaluated through laboratory investigations. Based on the test results, it is required that combination of RCA with some other targeted waste materials be considered in asphalt mixture. This paper presents the results of an experimental study to evaluate the RCA properties as an alternative for virgin aggregate in asphalt mixture under different percentages and combination with other aggregates, such as reclaimed asphalt pavement (RAP) and basalt.

Keywords: asphalt; basalt; reclaimed asphalt pavement; recycled construction aggregate; recycled aggregate; pavement

---

1. Introduction

The increasing amount of waste all over the world has shown that effective measures have to be implemented to reduce their negative environmental impact. Landfilling of waste is not a solution, due to danger of leaching and soil impregnation with potential subsequent contamination of underground water.

On the other hand, there are important sustainability benefits associated with the use of recycled material in pavement industry. Recycling helps the environment by reducing resource extraction and the use of virgin material, thereby reducing energy and water use, reducing harmful gas emissions and helping reduce waste to landfills. Buying recycled products also in some cases can reduce cost.

Therefore, the idea of using recycled material in pavement construction has attracted the attention of many researchers (e.g., Karlsson and Isacsson, 2006; Huang et al., 2007; Widyatmoko, 2008; Ahmedzade and Sengoz, 2009; Rafi et al., 2011; Khan and Gundaliya, 2012; Menaria and Sankhla, 2015;

Chandh and Akhila, 2016) in the pavement industry to investigate feasibility of the application of some of waste materials as alternative material in pavement construction [1–8].

Among different layers of flexible pavements, asphalt surface layer plays a fundamental role in flexible pavement structure systems as it should withstand varying traffic loads and constantly changing environmental conditions. Moreover, the asphalt surface layer is critical for safe and comfortable driving. Due to the composition nature of asphalt surface layer, application of solid waste in asphalt layer reduces not only environmental issues associated with waste disposal but also the demand for virgin aggregate which will subsequently result in cost savings and economic advantages, representing a value add application for waste material.

However, the selection of waste material to be used for pavement construction, particularly asphalt surface layer, is of high importance as the application of waste should not adversely influence the structural and functional aspects of the pavements [9–12].

Among different asphalt components, coarse aggregate properties are identified by the researchers (e.g., Vavrik, 2009; Zaniewski and Srinivasan, 2004; Husain, 2014; Al-Mosawe et al., 2015) as the second most important parameter after gradation for the performance of hot-mix asphalt (HMA) because coarse aggregate often forms the skeleton of the asphalt structure and transfers traffic and environmental loads to the underlying base, subbase, and subgrade layers [13–16]. Therefore, the behaviour and performance of asphalt mixture and eventually the asphalt surface layer are directly affected by the material properties and composition of this aggregate skeleton. In fact, the low stiffness of the asphalt mixtures and the excessive rutting in hot-mix asphalt (HMA) pavement surfaces are often attributed to the poor asphalt mixture designs which is primarily controlled by the asphalt binder and aggregate properties [17]. Therefore, except for the fine mixes, the selection of coarse aggregate greatly influences the asphalt layer behaviour.

In addition, since the aggregate represents the major portion of the asphalt mix, from the viewpoint of environmental preservation and effective use of resources, a comprehensive understanding of the engineering properties of the recycled aggregate can provide enormous benefits. Recognizing this fact, the reported studies and research on the utilization of recycled aggregate such as reclaimed asphalt pavement (RAP), recycled construction aggregate (RCA), recycled glass, etc. have increased all over the world over the past two decades [18–29]. Among the recycled aggregates that can be utilized in asphalt mixture, RCA obtained from construction and demolition waste constitute a major part of generated solid waste as a result of renovation and construction projects. Referring to literature survey (e.g., Arulrajah, 2012; Bennert et al., 2000; Blankenagel, 2005; Conceicao et al., 2011; Jayakody et al., 2014; Jimenez et al., 2012; Papp et al., 1998; Nataatmadja and Tan, 2001), although RCA has been used effectively as a base course and subbase course material [30–37], but, few research studies (e.g., Celaura et al., 2010; Hossain et al., 1993; Pereira et al., 2004; Rebbechi and Green, 2005; Berthelot et al., 2010; Wu et al., 2013) have reported the use of RCA in hot-mix asphalt [38–42]. Accordingly, in this research, the properties of RCA have been thoroughly evaluated through the laboratory investigation and tests. The results of these tests have showed that RCA has some shortcomings in satisfying design requirements as asphalt mixtures aggregate, in terms of some properties such as absorption and wet/dry strength variation. Therefore, utilization of RCA in asphalt mixture on its own, can result in less efficient asphalt mixtures and it is, hence, required that combination of RCA with some other targeted and acceptable waste materials and aggregates in certain percentages be considered in designing the asphalt mixture. Accordingly, RCA, RAP, and basalt have been considered as coarse aggregate in this research, and various tests have been conducted on each individual component and in combination. The paper will demonstrate the results of the conducted tests leading to the selection of most acceptable combination of aggregates for designing asphalt mixtures.

It should be noted that because of the diversity in quality and composition of the recycled construction aggregates, this research has been performed on aggregate samples which are collected from a recycling unit in Sydney over a period of one year.

2. Aggregate Properties and Their Relationship to Asphalt Performance

The high proportion of aggregate materials in volumetric design of asphalt mixes inherently links aggregate properties to the strength, stiffness, and generally the performance of the asphalt surface layer.

Because of the important impact of aggregate on the properties of asphalt mixture, a better understanding of the aggregates characteristics is essential in selecting the appropriate materials to optimize the asphalt mixture for strength and durability, and subsequently design a pavement with enough resistance to permanent deformation and cracking.

The most important physical and mechanical characteristics of aggregates include size and gradation, shape and angularity, surface texture, absorption, particle density, durability, toughness and hardness, resistance to polishing, soundness, cleanliness and the deleterious materials contained. Many research studies (e.g., Dahir, 1979; Brown et al., 1989; Brown and Bassett, 1990; Button et al., 1990; Elliot et al., 1991; Krutz and Sebaaly, 1993; Oduroh et al., 2000; Chen and Liao, 2002; Sengoz et al., 2014; Masad et al., 2009; Wu and King, 2011) have been conducted to link the properties of the aggregates to the performance of asphalt concrete pavement [43–53].

The results of these studies have shown that the physical and mechanical properties of the aggregates significantly affect the performance of the asphalt pavements.

Referring to the literature and the research conducted to relate aggregate properties and HMA performance, Figure 1 is generated to illustrate a generalized pattern and a summary of the effects of aggregate properties on the asphalt performance. The figure is the result of extensive literature review during the course of this research study and could be used by the practicing engineers as well as researchers to further improve their understanding of the effects of aggregate constituents on asphalt system performance. The reported relations and correlations shown in Figure 1 exemplify the complexities of mix design issues and considerations involved. This is certainly not unexpected considering the heterogeneity of the asphalt mixes. For example, as shown in this figure, different aggregate properties affect different aspects of asphalt mixture performance, which consequently define pavement service life. Accordingly, in order to design asphalt mixtures with longer service lives and lower production and maintenance costs, the aggregate must have appropriate characteristics.

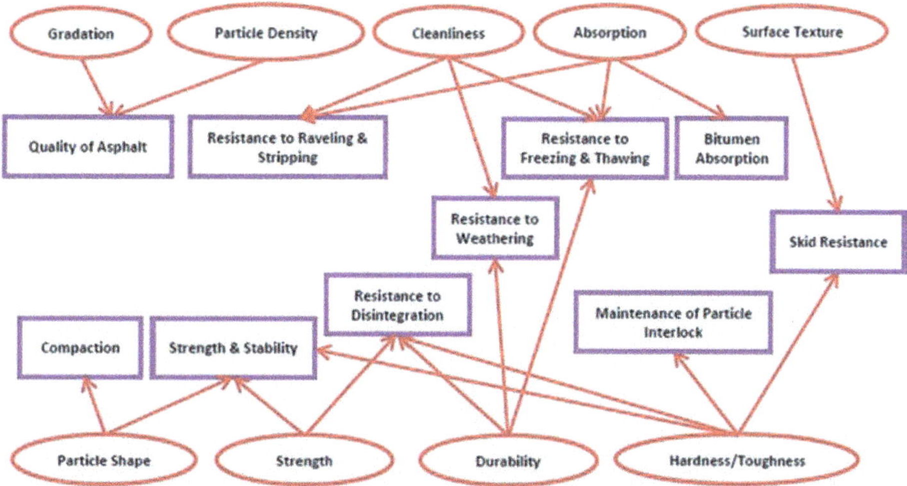

Figure 1. A Summary of the Effects of Aggregate Properties on the Asphalt Performance.

Therefore, the following section describes the experimental work carried out on selected coarse aggregates in order to evaluate the feasibility of using RCA as a part of coarse aggregates in asphalt mixture, and to produce an economical and sustainable asphalt mixture with adequate strength and good workability.

3. Experimental Work

3.1. Materials

In the present study, RCA, RAP, and basalt passing through 20 mm and retained on 4.75 mm I.S sieve have been used throughout the experiments. RAP material used in this research was stockpiled RAP collected from Boral Asphalt Plant (Prospect, NSW, Australia) which is generated from milling and being used in their asphalt projects. It was plant-screened material retained on 19 mm sieve size. The crushed virgin basalt aggregate was obtained from a local supplier. These virgin aggregates were transported from a local quarry (Nepean Quarries) in the vicinity of Sydney. In addition, RCA was collected from a local recycling centre called Revesby Recycling Centre (Revesby, NSW, Australia), a licensed waste facility and transfer station which accepts all construction and demolition waste from both the residential and commercial waste streams. In this centre, RCA is produced through the first sorting process for removing of contaminants such as wood, plastic, metal and glass, then crushing of construction wastes, and finally screening for removal of contaminants such as reinforcement, wood, plastics and gypsum.

3.2. Laboratory Tests

This section reports the laboratory investigation on RCA, RAP and basalt, in order to obtain comprehensive information and data of their properties and to compare these properties with the requirements specified in the standards as well as with the properties of the virgin aggregate. The key properties investigated in this experimental study are presented in Table 1.

Table 1. The Key Properties Investigated in the Experimental Study.

Property	Test Method	Test Name
Gradation and particle size distribution	AS 1141.11.1	Particle Size Distribution (Sieving Method)
Flakiness index of aggregate	AS 1141.15	Flakiness Index
Proportion of misshapen particles	AS 1141.14	Particle Shape by proportional calliper
Water absorption of aggregate	AS1141.6.1	Particle Density and Water Absorption of coarse aggregate
Variation in strength of aggregate in wet/dry condition	AS 1141.22	Wet/Dry Strength Variation
Particle density of aggregate	AS 1141.6.1	Particle Density and Water Absorption of coarse aggregate
Strength and crushing value of aggregate	AS 1141.21	Aggregate Crushing Value
Percentage of weak particles in coarse aggregate	AS 1141.32	Weak particles (including clay lumps, soft and friable particles) in coarse aggregates

In addition, based on the test results on the individual aggregates, necessary tests were conducted on different combinations of these aggregates. The results of these tests are shown in the following sections. It should be noted that three samples were performed for each test and the average of the three samples was reported as the test result.

3.2.1. Particle Size Distribution Test

The gradation of aggregate to be used in asphalt mixtures are evaluated through Particle Size Distribution Test (Figure 2).

Figure 2. Aggregate Gradation by Particle Size Distribution Test.

This test is conducted in accordance with AS 1141.11.1 (2009) and the gradation curves obtained from this test for different coarse aggregates considered in this research, including RCA, RAP and basalt, are shown in Figure 3.

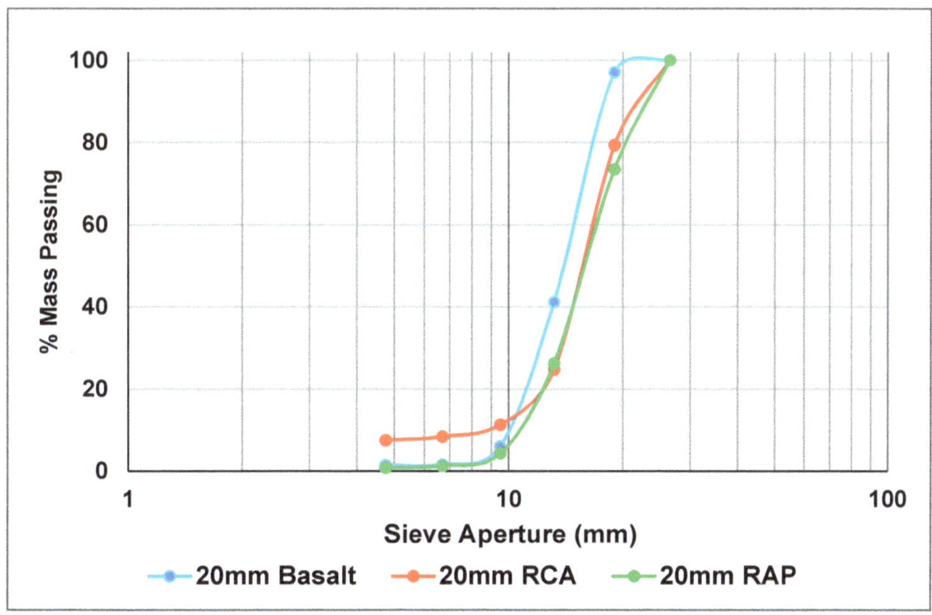

Figure 3. The Results of Particle Size Distribution Test for Coarse Aggregate; RCA: recycled construction aggregate; RAP: reclaimed asphalt pavement.

3.2.2. Particle Shape Test

The results of the studies on aggregate have shown that the aggregate physical shape properties significantly affect both the strength and stability of asphalt mixes [54]. Therefore, in order to design asphalt mixtures with long service lives, the aggregate must have the proper gradation and shape. The particle shape of aggregate substantially influences the mechanical stability of asphalt mix. The presence of excessive flaky and elongated particles is undesirable in asphalt mixtures as they tend to break down during the production and construction, and thus affect the durability of HMAs. Therefore, it is preferable to have rough and angular aggregates rather than smooth and round aggregates.

In this study, the proportion of misshapen aggregates, including the flat particles, elongated particles and, flat and elongated particles found in coarse aggregate is evaluated through the Particle Shape Test (Figure 4). The particle shape test is carried out by proportional caliper, using a 2:1 calliper ratio and based on AS 1141.14 (2007). The results of this test on three samples for each aggregate type (i.e., RCA, RAP and basalt) and the average value are given in Table 2.

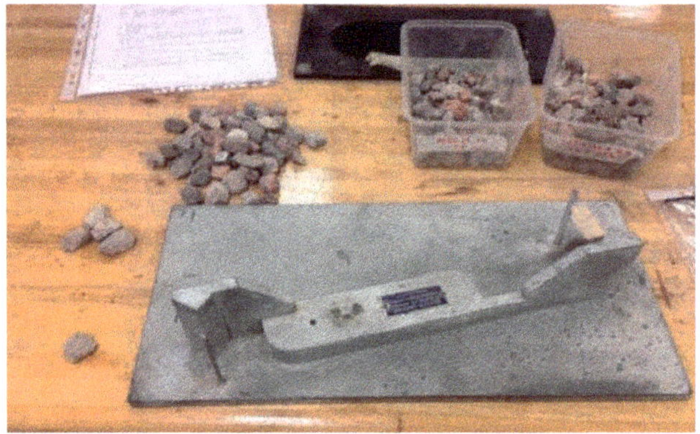

Figure 4. Classification of Aggregate Based on Particle Shape Test.

Table 2. The Results of Particle Shape Test for Coarse Aggregates and Misshapen Percentage Limits for Dense Graded Asphalt Based on Australian Standards.

Sample Number	Misshapen Particles (%)			Australian Standards Misshapen Percentage Limits (%)
	RCA	RAP	Basalt	
Sample 1	5.3	12.0	19.4	35% (max) For heavy and very heavy traffic
Sample 2	6.2	7.7	18.3	
Sample 3	7.0	8.7	17.3	
Average Misshapen Particles Percentage (%)	**6.2**	**9.5**	**18.3**	

As presented in Table 2, basalt materials show more of misshapen particles than RAP and RCA while still below the 35% limit of the Australian standard.

3.2.3. Flakiness Index Test

Some aggregates, on account of their shape, would be unsuitable for asphalt mixture as they would have low potential for developing inter-particle interlock. The percentage by mass of this type of aggregates, namely flaky aggregates is determined by the most commonly used test, called Flakiness Index Test (Figure 5). In this test, the flakiness index is determined by direct measurement using a special slotted sieve, from the ratio of the mass of material passing the slotted sieve to the total mass of the size fraction.

Figure 5. Conducting Flakiness Index Test for Coarse Aggregate.

The flakiness index test is performed based on AS 1141.15 (1999) and the results of this test on three samples for each aggregate type (i.e., RCA, RAP and basalt) and the average flakiness index for each aggregate type are given in Table 3.

Table 3. The Results of Flakiness Index Test for Coarse Aggregates and Flakiness Index Limits for Dense Graded Asphalt Based on Australian Standards.

Sample Number	Flakiness Index (%)			Australian Standards Flakiness Index Limits (%)
	RCA	RAP	Basalt	
Sample 1	5.6	12.8	21.3	25% (max) For heavy and very heavy traffic
Sample 2	8.1	10.1	21.1	
Sample 3	7.1	8.4	14.7	
Average Flakiness Index (%)	**6.9**	**10.4**	**19.0**	

The results of flakiness index test shows that RCA has less flakiness index than basalt and RAP which can positively affect the inter-particle interlock in asphalt mixture.

3.2.4. Particle Density and Water Absorption Test

The absorption is an indication of porosity in aggregate which demonstrates the pore structure of the aggregate. In asphalt mixtures, a porous aggregate increases the binder absorption, resulting in a dry and less cohesive asphalt mixture. In addition, the particle density of the aggregate is an essential property of the aggregate which plays an important role in the whole procedure of asphalt mix design. Therefore, in this research, the particle density and water absorption test is conducted on coarse aggregates (i.e., RCA, RAP and coarse basalt) based on the procedure described in AS 1141.6.1 (2000), as presented in Figure 6.

In this test, the amount of water which a dried sample will absorb is measured. This test is performed on three trials and the related test results on RCA, RAP and basalt are given in Table 4, under apparent, dry, and saturated surface dry (SSD) conditions.

Figure 6. Evaluation of Particle Density of Coarse Aggregate.

Table 4. The Results of Particle Density and Water Absorption Test on Coarse Aggregates and Water Absorption Limits for Dense Graded Asphalt Based on Australian Standards.

Sample Name	Apparent Particle Density (g/cm³)	Particle Density on a Dry Basis (g/cm³)	Particle Density on a SSD Basis (g/cm³)	Water Absorption (%)	Australian Standards Limits for Water Absorption (%)
RCA 1	2.375	2.211	2.352	6.39	2% (max) For heavy and very heavy traffic
RCA 2	2.375	2.211	2.352	6.39	
RCA 3	2.361	2.214	2.349	6.12	
Average Values for RCA	**2.370**	**2.212**	**2.351**	**6.30**	
RAP 1	2.539	2.422	2.468	1.89	
RAP 2	2.544	2.437	2.479	1.72	
RAP 3	2.540	2.433	2.475	1.73	
Average Values for RAP	**2.541**	**2.431**	**2.474**	**1.78**	
Basalt 1	2.635	2.528	2.568	1.60	
Basalt 2	2.630	2.521	2.562	1.64	
Basalt 3	2.654	2.542	2.584	1.67	
Average Values for Basalt	**2.640**	**2.530**	**2.571**	**1.64**	

The results of the particle density and water absorption test on different coarse aggregates (i.e., RCA, RAP and basalt) and their average value, as presented in Table 4, indicate the high absorption of RCA in comparison with RAP and basalt. The RCA water absorption exceeds the limit set by the Australian Standard.

As this research aims to investigate the feasibility of the application of RCA as a recycled material for potential partial replacement of coarse virgin aggregate (basalt) in asphalt mixtures, the particle density and water absorption test is also conducted on the mix of coarse aggregates (i.e., RCA, RAP and coarse basalt) considering different percentages of these materials. Such undertaking was needed in order to get a better understanding of an acceptable range of mix proportions in terms of water absorption.

The results of Particle Density and Water Absorption test on six different mixes of RCA, RAP and basalt and the average water absorption and particle density for each mix are given in Table 5. Despite the fact that above mixes (except when there is no RCA) have water absorption of more than 2%, the use of RCA is still a viable option as discussed in Section 3.3.

Table 5. The Results of Particle Density and Water Absorption Test for the Mix of Coarse Aggregates.

Sample Name		Apparent Particle Density (g/cm³)	Particle Density on a Dry Basis (g/cm³)	Particle Density on a SSD Basis (g/cm³)	Water Absorption (%)
75% Basalt and 25% RCA	1	2.579	2.394	2.466	2.98
	2	2.589	2.406	2.476	2.94
	3	2.601	2.420	2.489	2.88
Average Values for 75% Basalt and 25% RCA		**2.590**	**2.407**	**2.477**	**2.93**
50% Basalt and 50% RCA	1	2.520	2.296	2.385	3.86
	2	2.535	2.322	2.406	3.62
	3	2.527	2.313	2.397	3.65
Average Values for 50% Basalt and 50% RCA		**2.527**	**2.310**	**2.396**	**3.71**
25% Basalt and 75% RCA	1	2.477	2.224	2.326	4.57
	2	2.471	2.207	2.313	4.84
	3	2.480	2.234	2.333	4.44
Average Values for 25% Basalt and 75% RCA		**2.476**	**2.222**	**2.324**	**4.62**
80% Basalt and 20% RAP	1	2.727	2.607	2.651	1.68
	2	2.719	2.600	2.644	1.68
	3	2.724	2.596	2.643	1.80
Average Values for 80% Basalt and 20% RAP		**2.723**	**2.601**	**2.646**	**1.72**
25% Basalt and 25% RCA and 50% RAP	1	2.579	2.411	2.476	2.70
	2	2.581	2.401	2.471	2.89
	3	2.585	2.412	2.479	2.78
Average Values for 25% Basalt and 25% RCA and 50% RAP		**2.582**	**2.408**	**2.475**	**2.79**
25% Basalt and 50% RCA and 25% RAP	1	2.606	2.380	2.466	3.63
	2	2.592	2.356	2.447	3.85
	3	2.596	2.355	2.447	3.94
Average Values for 25% Basalt and 50% RCA and 25% RAP		**2.598**	**2.364**	**2.453**	**3.81**

3.2.5. Crushing Value Test

Aggregates used in road construction should be strong enough to resist crushing under traffic wheel loads [55]. The strength of the coarse aggregates can be evaluated by the Aggregate Crushing Value Test. In this test, the aggregate were crushed by a compression testing machine with a load rate of 40 kN/min to reach the peak load of 400 kN. The percentage of particles produced when the aggregate is crushed under this load and which pass a 2.36 mm sieve is called Aggregate Crushing Value.

The aggregate crushing value provides a relative measure of resistance to crushing under a gradually applied compressive load. To achieve a high quality pavement, it is preferred to utilize the aggregate possessing low crushing value.

In this research, the crushing value of RCA, RAP and basalt is assessed through the Aggregate Crushing Value Test in accordance with AS 1141.21 (1997), as presented in Figure 7. This test was performed in two trials, as required in the standard, and the related test results on RCA, RAP and basalt and the average crushing values for each aggregate type are given in Table 6.

Table 6. The Results of Aggregate Crushing Value Test for Coarse Aggregates and Crushing Value Limits for Dense Graded Asphalt Based on Australian Standards.

Sample Number	Crushing Value (%)			Australian Standards Aggregate Crushing Value Limits (%)
	RCA	RAP	Basalt	
Sample 1	29.53	7.04	9.32	35% (max) For heavy and very heavy traffic
Sample 2	28.88	7.76	8.51	
Average Crushing Value (%)	**29.21**	**7.40**	**8.91**	

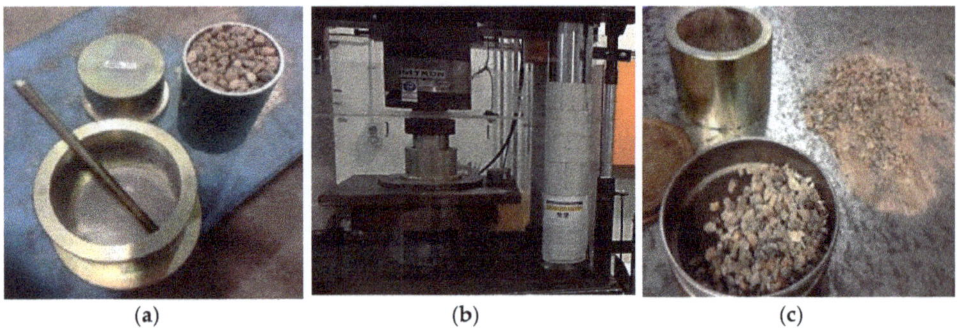

Figure 7. Crushing Value Test for Coarse Aggregates. (a) apparatus to crush aggregate; (b) compression machine ; (c) aggregates after crushing and sieving.

3.2.6. Weak Particle Test

The aggregate cleanliness refers to the presence of foreign or deleterious substances such as soft particles, weak and weathered materials, friable particles, clay lumps, and organic matters. The presence of these materials in the used aggregate can lead to stripping and ravelling in HMAs, as these materials adversely affect the bond between the aggregate and asphalt, and subsequently the stability of the pavement structure. Moreover, these substances disintegrate under traffic loading and wetting and drying cycles.

The cleanliness of aggregate can be evaluated based on the Weak Particles Test. In this test, the percentage of weak particles in coarse aggregate is determined. These particles will deform under finger pressures when wet. In this study, the percentage of weak particles in RCA, RAP and basalt are determined through the Weak Particle Test in accordance with AS 1141.32 (2008).

The weak particle test is conducted on two samples, as specified in the related standard, and the results of this test on RCA, RAP and basalt and the average weak particle percentage for each type of aggregate are presented in Table 7. The test results show that RCA and basalt have higher percentage of weak particles. However, all aggregates still meet the Standard's requirements.

Table 7. The Results of Weak Particle Test for Coarse Aggregates and Weak Particles Percentage Limits for Dense Graded Asphalt Based on Australian Standards.

Sample Number	Weak Particles (%)			Australian Standards Aggregate Crushing Value Limits (%)
	RCA	RAP	Basalt	
Sample 1	0.21	0.05	0.29	1% (max) For heavy and very heavy traffic
Sample 2	0.25	0.03	0.16	
Average Weak Particles Percentage (%)	**0.23**	**0.04**	**0.23**	

3.2.7. Wet/Dry Strength Variation Test

Strength is an important aggregate property which is related to the satisfactory resistance to crushing under the roller during construction, and adequate resistance to surface abrasion under traffic [56]. Therefore, aggregates used in pavement construction should be strong enough to resist crushing during mixing, laying process, compaction, consolidation and during its service life period when they are subjected to various loads applied by traffic [57].

In this research, the variation in strength of aggregate is evaluated by conducting the Wet-Dry Strength Variation Test on RCA, RAP and basalt in accordance with AS 1141.22 (2008), as shown in Figure 8.

(a) (b)

Figure 8. Conducting the Wet/Dry Strength Variation Test on Coarse Aggregate. (**a**) apparatus to crush aggregate; (**b**) compression machine.

This test determines the variation in strength of the aggregates tested after drying in an oven and then saturated yet with a dry surface. Based on the available standards, the wet/dry strength variation of less than 35% indicate a durable material but values as high as 60% could be used in undemanding circumstances.

In this research, the wet/dry strength variation test was conducted on the RCA, RAP and basalt fraction passed through 13.2 mm and retained on 9.5 mm I.S sieve. Different loading was used in order to adjust the applied load for providing the fines within the range of 7.5% and 12.5%. The results of these tests for coarse aggregates are illustrated in Figures 9 and 10 under dry condition and saturated surface dry condition (SSD), respectively. The wet and dry strengths can be inferred from the test results shown in these figures. Based on the obtained data, the wet/dry strength variation was calculated as follows:

$$\text{Wet/dry strength variation} = \frac{D - W}{D} \times 100 \tag{1}$$

where D is the dry strength in kilonewtons, and W is the wet strength in kilonewtons.

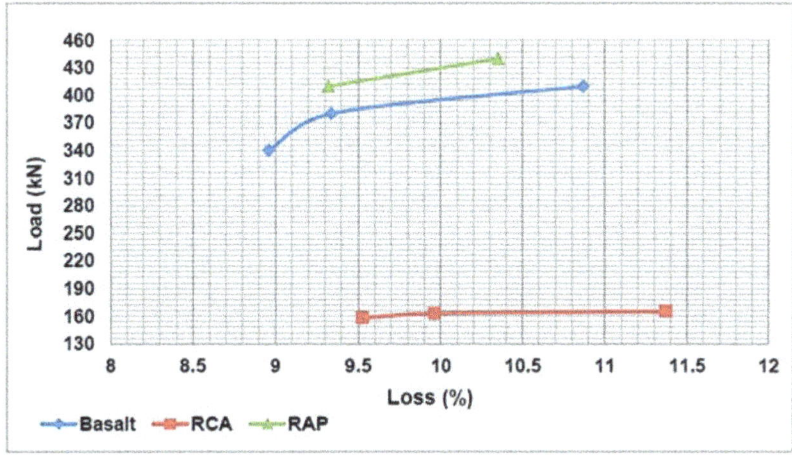

Figure 9. Results of Wet/Dry Strength Test for Coarse Aggregate (Dry Strength).

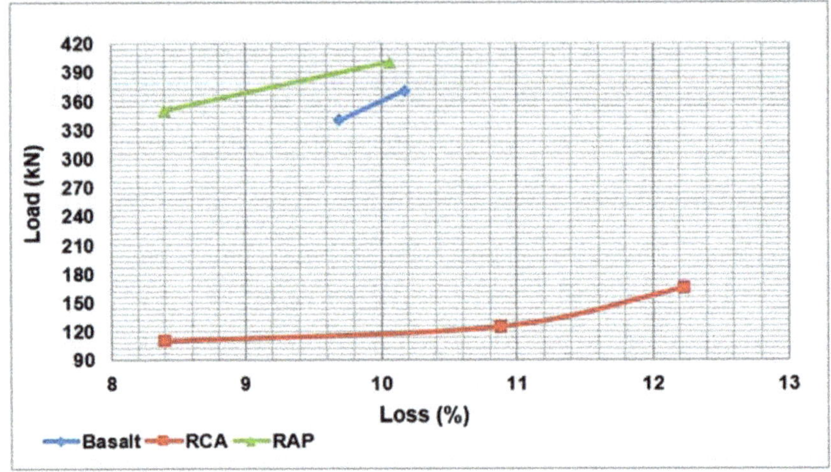

Figure 10. Results of Wet/Dry Strength Test for Coarse Aggregate (Wet Strength).

The results of the calculations for wet strength, dry strength, and wet/dry strength variation for basalt, RAP and RCA are presented in Table 8.

Table 8. The Results of Wet/Dry Strength Variation Test for Aggregates and Strength Limits for Dense Graded Asphalt Based on Australian Standards.

Material	Dry Strength, D (kN)	Wet Strength, W (kN)	Wet/Dry Strength Variation (%)	Australian Standards Wet/Dry Strength Limits (%)
RCA	163.1	119.7	26.6	35% (max) For heavy and very heavy traffic
RAP	429.8	398.2	7.4	
Basalt	392.9	359.4	8.5	

As the results of wet/dry strength test shows, the wet/dry strength variation of RCA is substantially more than the corresponding values for RAP and basalt. Therefore, as mentioned previously, it appears plausible to further investigate the feasibility of the application of RCA for the replacement of part of basalt in asphalt mixtures.

Accordingly, the wet/dry strength variation test was also conducted on different mix of coarse aggregates (i.e., RCA, RAP and coarse basalt) considering different percentages of these materials. Figures 11 and 12 illustrate the results of the wet/dry strength test for several mixes of RCA, RAP and basalt in dry condition and saturated surface dry condition respectively.

Based on the obtained results from these graphs, the wet strength (W) and the dry strength (D) can be determined and subsequently the wet/dry strength variation can be calculated as shown previously. The results of the calculations for wet strength, dry strength, and wet/dry strength variation on different mix of RAP, basalt and RCA are presented in Table 9. The results indicate that all mixes satisfy the maximum 35% limit set by the Australian Standards.

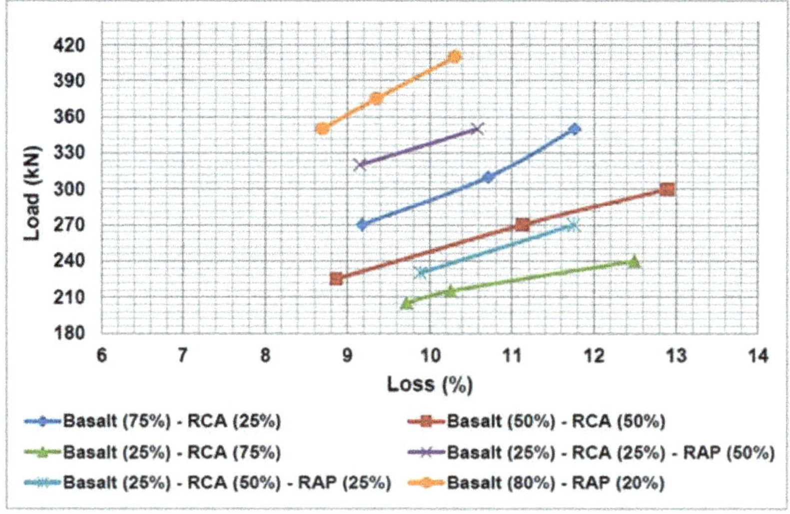

Figure 11. Results of Wet/Dry Strength Test for Mix of Coarse Aggregates (Dry Strength).

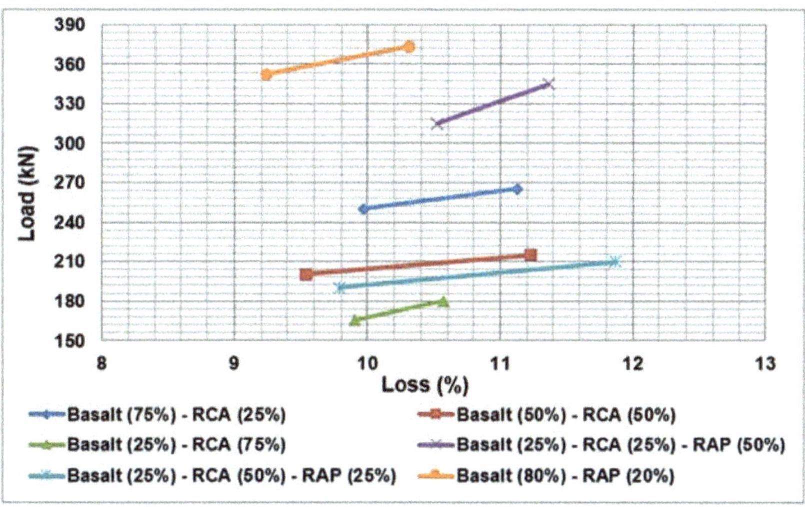

Figure 12. Results of Wet/Dry Strength Test for Mix of Coarse Aggregates (Wet Strength).

Table 9. The Results of Wet/Dry Strength Variation Test for Different Mix of Coarse Aggregates.

Material	Dry Strength, D (kN)	Wet Strength, W (kN)	Wet/Dry Strength Variation (%)	Australian Standards Wet/Dry Strength Limits (%)
Basalt (75%) and RCA (25%)	291.4	250.4	14.1	35% (max) For heavy and very heavy traffic
Basalt (50%) and RCA (50%)	247.6	204.1	17.6	
Basalt (25%) and RCA (75%)	210.4	167.2	20.5	
Basalt (80%) and RAP (20%)	398.9	366.9	8.0	
Basalt (25%), RCA (25%) and RAP (50%)	337.8	296.4	12.3	
Basalt (25%), RCA (50%) and RAP (25%)	232.6	192	17.5	

3.3. Results and Discussions

As presented in the previous sections, in this research, the properties of RCA, RAP, basalt and mix of these aggregates were evaluated by conducting a series of tests. The test results are summarized in Table 10.

Table 10. Summary of the Test Results for the Evaluation of Coarse Aggregate Properties.

Test	Test Method	Aggregate			Typical Limit Based on Australian Standards
		RCA	RAP	Basalt	
Particle Distribution Test	AS 1141.11.1	As presented in relevant Figures and Tables			-
Flakiness Index Test	AS 1141.15	6.91	10.42	19.03	25% (max)
Particle Shape Test	AS 1141.14	6.16	9.47	18.34	35% (max)
Water Absorption	AS 1141.6.1	**6.30**	1.78	1.64	2% (max)
Particle Density	AS 1141.6.1	2.370	2.541	2.640	-
Particle Density on Dry Basis	AS 1141.6.1	2.212	2.431	2.530	-
Particle Density on SSD Basis	AS 1141.6.1	2.351	2.474	2.571	-
Aggregate Crushing Value	AS 1141.21	29.21	7.40	8.91	35% (max)
Weak Particles	AS 1141.32	0.23	0.04	0.23	1% (max)
Wet/Dry Strength Test	AS 1141.22	26.6	7.4	8.5	35% (max)
Wet Strength	AS 1141.22	**119.7**	398.2	359.2	150 kN (min)
Dry Strength	AS 1141.22	163.1	429.8	392.9	-

In addition, to have better comparisons between the aggregate properties, the test results on different aggregates as well as the standard limits are also illustrated in Figure 13.

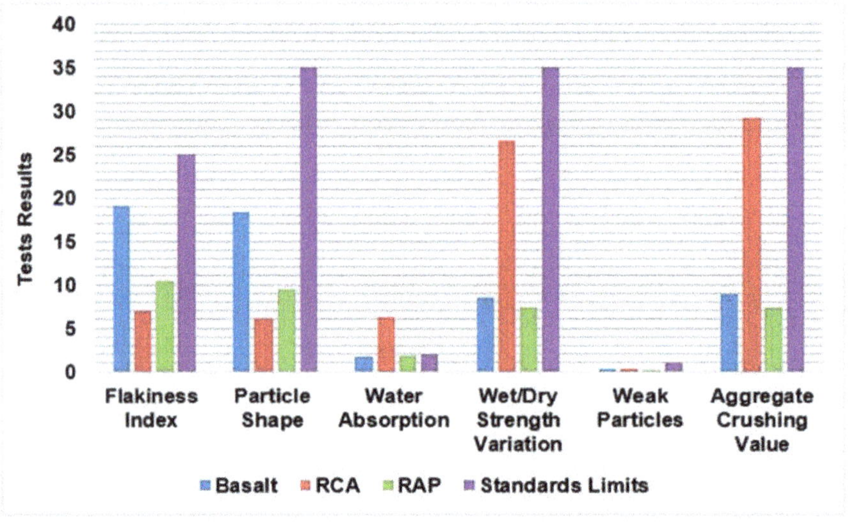

Figure 13. Comparison of Different Aggregate Properties with Standard Limits.

As shown in Figure 13, the results of preliminary tests on coarse aggregates indicate that all properties of RCA, except for water absorption and wet strength (which are shown in bold in Table 10), are within the limits specified by relevant Australian Standards and hence deemed appropriate for use as aggregate in the asphalt mixture. However, for some parameters such as Flakiness Index and Particle Shape which are two dominant characteristics having significant impact on asphalt mixture strength and stability; RCA displays smaller value in comparison with basalt and RAP. This can be one of the strong points of RCA as flakiness index and particle shape are the two important properties for proper compaction, deformation resistance, and workability of asphalt mixture [58].

In addition, as can be observed in Table 10 and Figure 13, the water absorption of RCA is higher than the corresponding value of RAP and basalt and the Australian Standards limit, because it is well known that water absorption requires linked and open cracks in the structure of aggregate and RCA contains cracks due to the crushing processes. Moreover, the great amounts of impurities in RCA can increase the water absorption of RCA. The high water absorption of RCA may result in high bitumen absorption in asphalt mixtures, and hence plays an important role in asphalt mixture design.

Accordingly, since this research aims to investigate the feasibility of the application of RCA for the partial replacement of coarse virgin aggregate (basalt) and in combination with other recycled aggregate (RAP) in asphalt mixtures, the particle density and water absorption tests were conducted on different mix of coarse aggregates while considering different percentages of these materials. The results of these tests are presented in Figure 14.

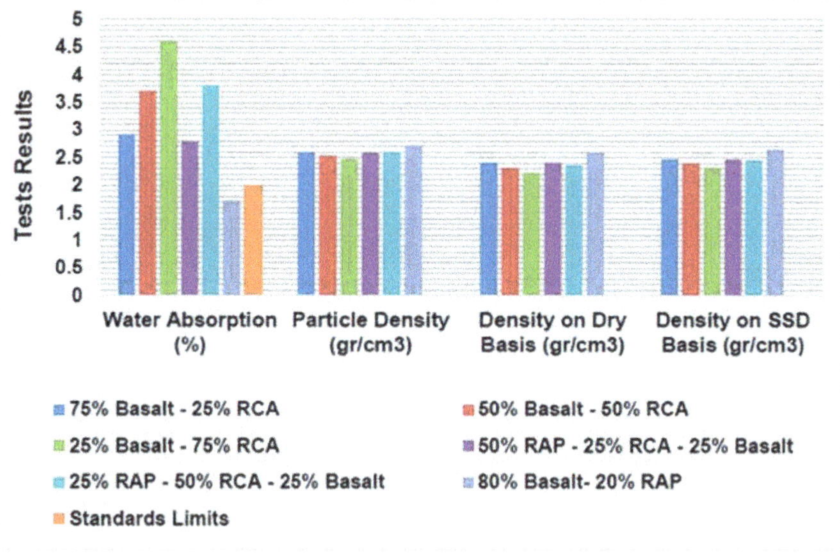

Figure 14. Comparison of Water Absorption and Particle Density of Different Mixes of Coarse Aggregates.

As can be observed in Figure 14, increasing RCA in the mix does not make any substantial change in mix density in comparison with water absorption. In other words, by increasing RCA from 0% to 100% in the mix, the density decreases by 7%, whereas water absorption increases by 74%.

In addition, although wet/dry strength variation of RCA meets the requirements of Australian standards, the test results show that this value is higher than the corresponding value of RAP and basalt. As the wet/dry strength variation is related to the principal mechanical properties which are required for asphalt aggregate, it is of high importance in asphalt mixture design. Therefore, wet/dry strength variation test was also conducted on different mixes of coarse aggregate. Figure 15 shows the comparison of wet strength, dry strength, and wet/dry variation in different mixes of RCA, RAP and basalt.

As illustrated in this figure, the wet/dry strength variation of mix of RCA/basalt increases by increase of the percentage of RCA in the mix, so that the increase of RCA from 0% to 100% will result in 20% increase in wet/dry strength variation. The results of these two tests (i.e., water absorption and particle density test, and wet/dry strength variation test) on mix of coarse aggregates are summarized in Table 11.

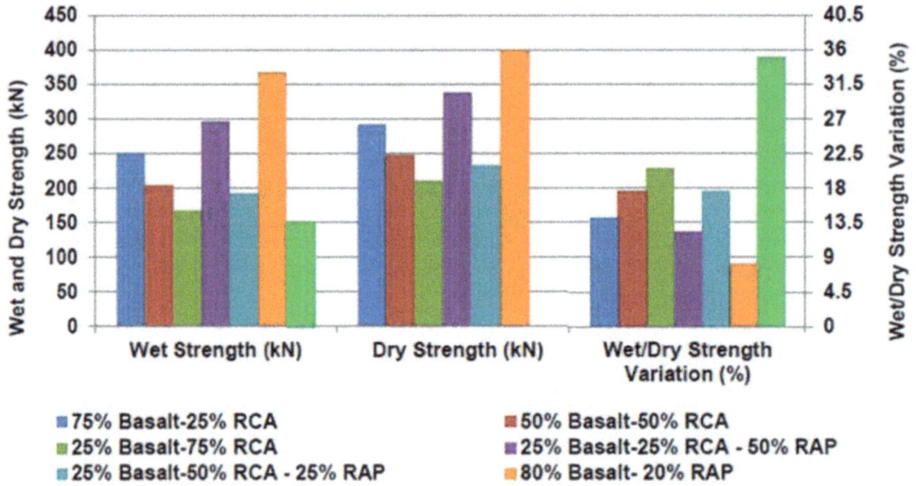

Figure 15. Comparison of the Wet Strength and Dry Strength of Different Mixes of Coarse Aggregates.

Table 11. Summary of Tests Results for Evaluation of Mix of Coarse Aggregates Properties.

Mix	Test Method	Water Absorption	Particle Density	Wet/Dry Strength Variation
Basalt (75%), RCA (25%)	AS 1141.15	2.93	2.590	14.1
Basalt (50%), RCA (50%)	AS 1141.14	3.71	2.527	17.6
Basalt (25%), RCA (75%)	AS 1141.6.1	4.62	2.476	20.5
Basalt (80%), RAP (20%)	AS 1141.6.1	1.72	2.723	8.0
Basalt (25%), RCA (25%), RAP (50%)	AS 1141.21	2.79	2.582	12.3
Basalt (25%), RCA (50%), RAP (25%)	AS 1141.32	3.81	2.598	17.5

The results of tests on mix of coarse aggregate showed that in all cases of RCA ratios, RCA increase causes a decrease in wet and dry strength and an increase in water absorption. This will necessitate the proper selection and optimum combination of RCA and other aggregates.

The coefficient of variation is used as an indication to measure the heterogeneity of test results. The results of calculation of standard deviation (SD) and coefficient of variation (CV) for each set of aggregate mixes are presented in Table 12.

Table 12. Coefficient of Variation and Standard Deviation for Mix of Coarse Aggregates.

Mix	Coefficient of Variation	Standard Deviation
Basalt (100%) + RCA (0%)	2.14	0.035
Basalt (75%) + RCA (25%)	1.71	0.050
Basalt (50%) + RCA (50%)	3.52	0.131
Basalt (25%) + RCA (75%)	4.42	0.204
Basalt (0%) + RCA (100%)	2.47	0.156

As can be observed in Table 12, the coefficient of variation for each data set reveals that the test results dispersion is low and the tests are conducted consistently.

Furthermore, regression analysis is typically applied to the water absorption test results for different combination of RCA and basalt, to show the typical amount of RCA and basalt in a blend to give 2% water absorption which is the standard limit of water absorption based on Australian standards (Figures 16 and 17).

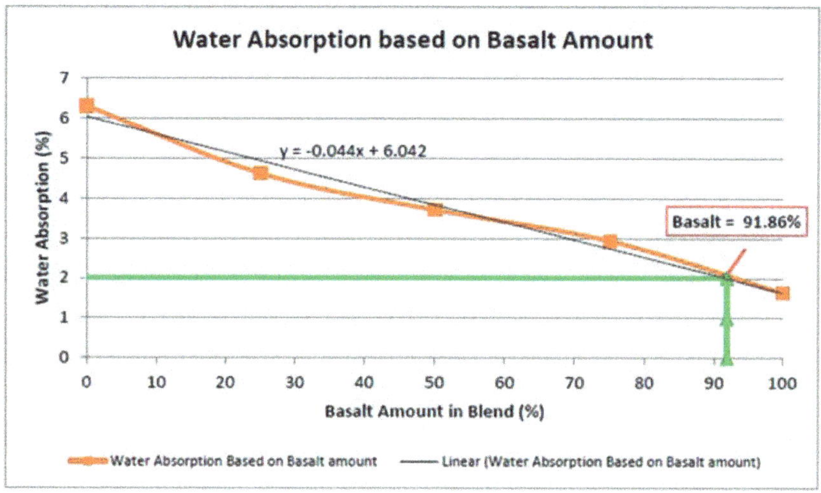

Figure 16. Regression Analysis for Determination of Optimum Basalt Amount.

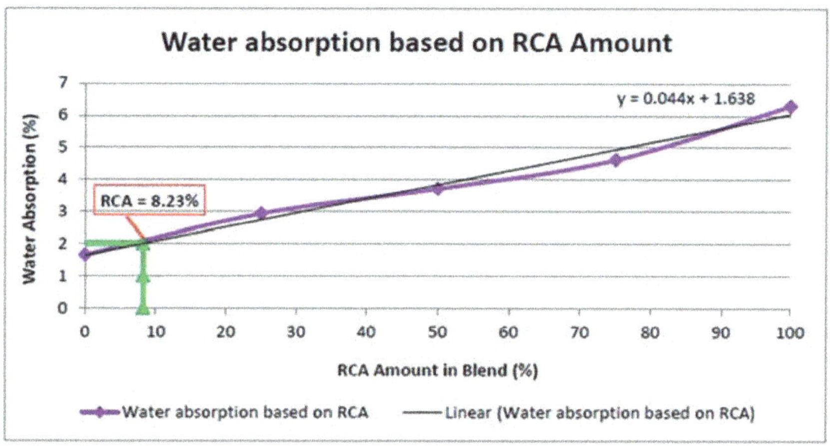

Figure 17. Regression Analysis for Determination of Optimum recycled construction aggregate (RCA) Amount.

As illustrated in Figures 16 and 17, the standard water absorption limit of 2% can be achieved by mixing of almost 8% and 92% of RCA and basalt, respectively. However, based on the available references [59], typically, the amount of binder absorbed by aggregate is 0.3 to 0.7 times the water absorption of the aggregate. In addition, according to this standard, if the sample absorbs between 2% and 4% of its mass, it should be carefully examined by other tests. If the sample absorbs in excess of 4% of its mass, it will rarely prove to be an adequate aggregate for asphalt production. Based on the water absorption results, it can be observed that the combination of 25% RCA and 75% basalt would provide water absorption of 2.93%, and also water absorption of the combination of 50% RCA and 50% basalt would be 3.71%, which are still in the range of aggregate water absorption that suggest further research.

4. Conclusions

Since the coarse aggregate properties are identified by current research as the second most important parameter after gradation for the performance of HMA [49], therefore, in this research, attempts were made to assess the properties of RCA for use in asphalt mixture as coarse aggregate, and this paper presented the summary results of a comprehensive set of preliminary tests on RCA, RAP and basalt as well as different mixes of these aggregates to evaluate their basic mechanical and physical properties. It was argued that information on these fundamental properties were paramount in designing a durable and sustainable asphalt mixtures. To this end, different aggregate and aggregate mixes containing different percentages of RCA, RAP, and basalt were investigated in this research to assess its suitability as coarse aggregate in asphalt. This paper presented the results of this experimental work conducted as a component of a broader research project for designing an asphalt mixture. Based on this research, it was concluded that:

(1) RCA has lower value of flaky and misshapen particles in comparison with RAP and basalt. This implies that asphalt mixtures containing a certain amount of RCA can have better workability, deformation resistance and compaction.
(2) RCA exhibits comparatively more absorption and wet/dry strength variation than conventional aggregate and RAP, while the results of other tests show that RCA still meets the requirements for aggregate in asphalt mixtures. Cracks and adhering mortar and cement paste can be significant reasons for the high water absorbtion of RCA which needs to be compensated for during mix design.
(3) The results of water absorption and particle density test on different mix of coarse aggregates revealed that RCA increase will increase water absorption of the mixture. Therefore, the selection of optimum combination of RCA and other aggregates is required to satisfy the relevant standards requirements.
(4) Regression analysis applied to the results of water absorption test on different combination of RCA and basalt, as illustrated in Figures 16 and 17, indicates that mixing of almost 8% of RCA with natural aggregates will provide the standard water absorption limit of 2%.
(5) Since, according to Austroads (2014), the aggregates with water absorption of between 2% and 4% of their mass should be carefully examined by other tests [59]. This standard limit will allow further investigation of the application of up to 50% of RCA in mixtures because based on the water absorption results, it can be observed that the combination of 25% RCA and 75% basalt would provide water absorption of 2.93%, and the combination of 50% RCA and 50% basalt would provide water absorption of 3.71%, which are still in the range of aggregates that require further research for their water absorption properties.

Acknowledgments: The authors would like to acknowledge Boral Asphalt Company for providing part of materials required for the research work.

Author Contributions: Farzaneh Tahmoorian is the main author and researcher conducting this research as the main component of her PhD studies, Bijan Samali has directed and coordinated the project in a supervisory capacity, Vivian WY Tam and John Yeaman have been consultants to the project providing valuable insights into behavior of aggregates and asphalt.

Conflicts of Interest: The authors declare no conflict of interest.

References

1. Karlsson, R.; Isacsson, U. Material-Related Aspects of Asphalt Recycling-State-of-the-Art. *J. Mater. Civ. Eng.* **2006**, *18*, 81–92. [CrossRef]
2. Huang, Y.; Bird, R.; Heidrich, O. A Review of the Use of Recycled Solid Waste Materials in Asphalt Pavements. *Resour. Conserv. Recycl.* **2007**, *52*, 58–73. [CrossRef]
3. Widyatmoko, I. Mechanistic-Empirical Mixture Design for Hot Mix Asphalt Pavement Recycling. *Constr. Build. Mater.* **2008**, *22*, 77–87. [CrossRef]

4. Ahmedzadeh, P.; Sengoz, B. Evaluation of steel slag coarse aggregate in hot mix asphalt concrete. *J. Hazard. Mater.* **2009**, *165*, 300–305. [CrossRef] [PubMed]
5. Rafi, M.M.; Qadir, A.; Siddiqui, S.H. Experimental Testing of Hot Mix Asphalt Mixture Made of Recycled Aggregates. *Waste Manag. Res.* **2011**, *29*, 1316–1326. [CrossRef] [PubMed]
6. Khan, I.; Gundaliya, P.J. Utilization of Waste Polyethylene Materials in Bituminous Concrete Mix for Improved Performance of Flexible Pavements. *J. Appl. Res.* **2012**, *1*, 85–86.
7. Menaria, Y.; Sankhla, R. Use of Waste Plastic in Flexible Pavements-Green Roads. *Open J. Civ. Eng.* **2015**, *5*, 299–311. [CrossRef]
8. Chandh, K.A.; Akhila, S. A Laboratory Study on Effect of Plastic on Bitumen. *Int. J. Sci. Res.* **2016**, *5*, 1406–1409.
9. Tabsh, S.W.; Abdelfatah, A.S. Influence of Recycled Concrete Aggregates on Strength Properties of Concrete. *Constr. Build. Mater.* **2009**, *23*, 1163–1167. [CrossRef]
10. Mills-Beale, J.; You, Z. The Mechanical Properties of Asphalt Mixtures with Recycled Concrete Aggregates. *Constr. Build. Mater.* **2010**, *24*, 230–235. [CrossRef]
11. Chen, M.; Lin, J.; Wu, S. Potential of Recycled Fine Aggregates Powder as Filler in Asphalt Mixture. *Constr. Build. Mater.* **2011**, *25*, 3909–3914. [CrossRef]
12. Silva, H.M.; Oliveira, J.R.; Jesus, C.M. Are totally recycled hot mix asphalt a sustainable alternative for road paving? *Resour. Conserv. Recycl.* **2012**, *60*, 38–48. [CrossRef]
13. Vavrik, W.R.; Harrell, M.J.; Gillen, S.L. Achieving Perpetual Pavement through Staged Construction. In Proceedings of the International Conference of Perpetual Pavements, Ohio University, Athens, OH, USA, 30 September 2009.
14. Zaniewski, J.P.; Srinivasan, G. Evaluation of Indirect Tensile Strength to Identify Asphalt Concrete Rutting Potential. In *Asphalt Technology Program*; West Virginia University: West Virginia, WV, USA, 2004; Volume 3, pp. 154–196.
15. Husain, N.M.D.; Karim, M.R.; Mahmud, H.B.; Koting, S. Effects of Aggregate Gradation on the Physical Properties of Semiflexible Pavement. *Adv. Mater. Sci. Eng.* **2014**, *2014*, 1–8. [CrossRef]
16. Al-Mosawe, H.; Thom, N.; Airey, G.; Al-Bayati, A. Effect of Aggregate Gradation on the Stiffness of Asphalt Mixtures. *Int. J. Pavement Eng. Asphalt Technol.* **2015**. [CrossRef]
17. Kandhal, P.S.; Cooley, L.A. National Cooperative Highway Research Program Report 464: The Restricted Zone in the Superpave Aggregate Gradation Specification. In *National Cooperative Highway Research Program*; National Research Council: Washington, DC, WA, USA, 2001.
18. Arabani, M.; Azarhoosh, A.R. The Effect of Recycled Concrete Aggregate and Steel Slag on the Dynamic Properties of Asphalt Mixtures. *Constr. Build. Mater.* **2011**, *35*, 1–7. [CrossRef]
19. Jony, H.; Al-Rubaie, M.; Jahad, I. The Effect of Using Glass Powder Filler on Hot Asphalt Concrete Mixtures Properties. *Eng. Technol. J.* **2011**, *29*, 44–57.
20. Pereira, P.; Oliveira, J.; Picado-Santos, L. Mechanical Characterization of Hot Mix Recycled Materials. *Int. J. Pavement Eng.* **2004**, *5*, 211–220. [CrossRef]
21. Fulton, B. Use of Recycled Glass in Pavement Aggregate. In Proceedings of the 23rd Australian Road Research Board International Conference, Adelaide, Australia, 30 July–1 August 2008.
22. Finkle, I.; Ksaibati, K. Recycled Glass Utilization in Highway Construction. In Proceedings of the Transportation Research Board 86th Annual Meeting, University of Wyoming, Laramie, WY, USA, 21–25 January 2007.
23. Wong, Y.D.; Sun, D.D.; Lai, D. Value-Added Utilisation of Recycled Concrete in Hot-Mix Asphalt. *Waste Manag.* **2007**, *27*, 294–301. [CrossRef] [PubMed]
24. Paranavithana, S.; Mohajerani, A. Effects of Recycled Concrete Aggregates on Properties of Asphalt Concrete. *Resour. Conserv. Recycl.* **2006**, *48*, 1–12. [CrossRef]
25. Motter, J.S.; Miranda, L.F.R.; Bernucci, L.L.B. Performance of Hot Mix Asphalt Concrete Produced with Coarse Recycled Concrete Aggregate. *J. Civ. Eng.* **2014**, *27*. [CrossRef]
26. Pérez, I.; Pasandin, A.R.; Medina, L. Hot Mix Asphalt Using C&D Waste as Coarse Aggregates. *Mater. Des.* **2012**, *36*, 840–846.
27. Silva, C.A.R.S. Study of Recycled Aggregate from Civil Construction in Bituminous Mixtures for Urban Roads. Master's Dissertation, Federal University of Ouro Preto, Minas Gerais, Brazil, November 2009.

28. Zhu, J.; Wu, S.; Zhong, J.; Wang, D. Investigation of Asphalt Mixture Containing Demolition Waste Obtained from Earthquake-Damage Buildings. *Constr. Build. Mater.* **2012**, *29*, 466–475. [CrossRef]
29. Zulkati, A.; Wong, Y.D.; Sun, D.D. Mechanistic Performance Of Asphalt-Concrete Mixture Incorporating Coarse Recycled Concrete Aggregate. *J. Mater. Civ. Eng.* **2013**, *10*, 1299–1305. [CrossRef]
30. Arulrajah, A.J.; Piratheepan, M.W.B.; Sivakugan, N. Geotechnical Characteristics of Recycled Crushed Brick Blends for Pavement Subbase Applications. *Can. Geotech. J.* **2012**, *49*, 796–811. [CrossRef]
31. Bennert, T.; Papp, W.J.; Maher, A.; Gucunski, N. Utilization of Construction and Demolition Debris under Traffic-Type Loading in Base and Subbase Applications. *Transp. Res. Rec.* **2000**. [CrossRef]
32. Blankenagel, B.J. Characterization of Recycled Concrete for Use as Pavement Base Material. Master's Thesis, Brigham Young University, Provo, UT, USA, 20 August 2005.
33. Conceicao, F.D.; Motta, R.D.S.; Vasconcelos, K.L.; Bernucci, L. Laboratory Evaluation of Recycled Construction and Demolition Waste for Pavements. *Constr. Build. Mater.* **2011**, *25*, 2972–2979.
34. Jayakody, S.; Gallage, C.; Kumar, A. Assessment of Recycled Concrete Aggregates as a Pavement Material. *Geomech. Eng.* **2014**, *6*, 235–248.
35. Jiménez, J.R.; Ayuso, J.; Agrela, F.; López, M.; Galvín, A.P. Utilization of Unbound Recycled Aggregates from Selected CDW in Unpaved Rural Roads. *J. Resour. Conser. Recycl.* **2012**, *58*, 88–97. [CrossRef]
36. Papp, W.J.; Maher, M.H.; Bennet, T.A.; Gucunski, N. Behavior of construction and demolition debris in base and subbase applications. In *Recycled Materials in Geotechnical Applications*; Vipulanandan, C., Elton, D.J., Eds.; American Society of Civil Engineers: Reston, VA, USA, 1998; pp. 122–136. ISBN 0784403872.
37. Nataatmadja, A.; Tan, Y.L. Resilient Response of Recycled Concrete Road Aggregates. *J. Transp. Eng.* **2001**. [CrossRef]
38. Celauro, C.; Benardo, C.; Gabriele, B. Production of Innovative, Recycled and High-Performance Asphalt for Road Pavements. *Resour. Conser. Recycl. J.* **2010**, *54*, 337–347. [CrossRef]
39. Hossain, M.; Metcalf, D.G.; Scofield, L.A. Performance of Recycled Asphalt Concrete Overlays in Southwestern Arizona. *Transp. Res. Rec.* **1993**, *1427*, 30–37.
40. Rebbechi, J.; Green, M. Going Green: Innovations in Recycling Asphalt. In Proceedings of the AAPA, Queensland, QLD, Australia, 13–16 September 2015.
41. Berthelot, C.; Haichert, R.; Podborochynski, D.; Wandzura, C.; Taylor, B.; Gunther, D. Mechanistic Laboratory Evaluation and Field Construction of Recycled Concrete Materials for Use in Road Substructures. *Transp. Res. Board* **2010**. [CrossRef]
42. Wu, S.; Zhu, J.; Zhong, J.; Wang, D. Influence of Demolition Waste Used as Recycled Aggregate on Performance of Asphalt Mixture. *Road Mater. Pavement Des.* **2013**, *14*, 679–688. [CrossRef]
43. Dahir, S.A. Review of Aggregate Selection Criteria for Improved Wear Resistance and Skid Resistance of Bituminous Surfaces. *J. Test. Eval.* **1979**, *7*, 245–253.
44. Brown, E.R.; McRae, J.L.; Crawley, A.B. Effect of Aggregate on Performance of Bituminous Concrete. In *ASTM STP 1016*; American Society for Testing and Materials: West Conshohocken, PA, USA, 1989; pp. 34–63.
45. Brown, E.R.; Bassett, C.E. Effects of Maximum Aggregate Size on Rutting Potential and Other Properties of Asphalt-Aggregate Mixtures. In *Transportation Research Record*; Transportation Research Board: Washington, DC, USA, 1990; pp. 107–119.
46. Button, J.W.; Perdomo, D.; Lytton, R.L. Influence of Aggregate on Rutting in Asphalt Concrete Pavements. In *Transportation Research Record*; Transportation Research Board: Washington, DC, USA, 1990; pp. 141–152.
47. Elliot, R.P.; Ford, M.C.; Ghanim, M.; Tu, Y.F. Effect of Aggregate Gradation Variation on Asphalt Concrete Mix Properties. In *Transportation Research Record*; Transportation Research Board: Washington, DC, USA, 1991; pp. 52–60.
48. Krutz, N.C.; Sebaaly, P.E. The Effects of Aggregate Gradation on Permanent Deformation of Asphalt Concrete. In Proceedings of the Asphalt Paving Technology, Austin, TX, USA, 22–24 March 1993.
49. Oduroh, P.K.; Mahboub, K.C.; Anderson, R.M. Flat and Elongated Aggregates in Superpave Regime. *J. Mater. Civ. Eng.* **2000**, *12*, 124–130. [CrossRef]
50. Chen, J.S.; Liao, M.S. Evaluation of Internal Resistance in Hot-Mix Asphalt (HMA) Concrete. *Constr. Build. Mater. J.* **2002**, *16*, 313–319. [CrossRef]
51. Sengoz, B.; Onsori, A.; Topal, A. Effect of Aggregate Shape on the Surface Properties of Flexible Pavement, KSCE. *J. Civ. Eng.* **2014**, *18*. [CrossRef]

52. Masad, E.; Rezaei, A.; Chowdhury, A.; Harris, P. Predicting Asphalt Mixture Resistance Based on Aggregate Characteristics. In *Aggregate Resistance to Polishing and Its Relationship to Skid Resistance Project*; National Technical Information Service: Springfield, VA, USA, 2009.
53. Wu, Z.; King, W. Development of Surface Friction Guidelines for LADOTD. In *LTRC Project No. 09-2B*; Louisiana State University: Baton Rouge, LA, USA, April 2011.
54. Marinho, M.N. Analysis of Mechanical Performance of Hot Mixed Asphaltic Concrete Produced with Recycled Coarse Concrete Aggregate. Master's Dissertation, Federal University of Pernambuco, Recife, Brazil, July 2011.
55. Mohajerani, A. A Study of Relationships between Polished Aggregate Friction Value, Aggregate Crushing Value, and Point Load Strength Index. *Aust. Geomech.* **1997**, *1997*, 62–65.
56. Prowell, B.D.; Zhang, J.; Brown, E.R. Aggregate Properties and the Performance of Superpave Designed Hot Mix Asphalt. In *National Cooperative Highway Research Program Report 539*; National Research Council: Washington, DC, USA, 2005. [CrossRef]
57. Dickinson, E.J. *Bituminous Roads in Australia*; Australian Road Research Board: Melbourne, MEL, Australia, 1984.
58. Masad, E.; Al-Rousan, T.; Button, J.; Little, D. Test Methods for Characterizing Aggregate Shape, Texture, and Angularity. In *National Cooperative Highway Research Program Report 555*; National Research Council: Washington, DC, USA, 2007.
59. Guide to Pavement Technology Part 4B: Asphalt, Austroads. 2014. Available online: https://www.onlinepublications.austroads.com.au/items/AGPT04B-1 (accessed on 31 June 2014).

Article

Study of Surfactant Additives for the Manufacture of Warm Mix Asphalt: From Laboratory Design to Asphalt Plant Manufacture

Miguel Sol-Sánchez, Fernando Moreno-Navarro and Mª Carmen Rubio-Gámez *

Construction Engineering Laboratory of the University of Granada (LabIC.UGR), 18071 Granada, Spain; msol@ugr.es (M.S.-S.); fmoreno@ugr.es (F.M.-N.)

* Correspondence: mcrubio@ugr.es; Tel.: +34-958-249-443; Fax: +34-958-246-138

Academic Editors: Zhanping You, Qingli (Barbara) Dai and Feipeng Xiao

Received: 20 June 2017; Accepted: 19 July 2017; Published: 21 July 2017

Abstract: Warm Mix Asphalt (WMA), manufactured at a lower temperature than the traditional Hot Mix Asphalt (HMA), allows for important economic and environmental benefits when considered for application in roads. Nonetheless, despite the benefits, its application in pavement for roads is not as widespread as desired from an environmental point of view; more in-depth studies to investigate its development and wider applicability are required. Thus, the present paper aims to contribute to the implementation of this cleaner technology to produce WMA (based on chemical additives) for its application in pavement for roads, including from the stage of the design of the material in the laboratory (by selecting the most appropriate manufacturing temperature and additive type and dosage) to its production in a conventional industrial plant for its use in a trial section. Results demonstrate that it is possible to reduce the manufacturing temperature of asphalt mixtures by using chemical additives, recording similar mechanical behaviour (or even superior) to conventional hot mixtures when specific studies are developed for the optimal design of the WMA. It was also shown that these mixtures could be produced in a conventional asphalt plant without implementing important changes in equipment, which implies a cost-effective solution that can readily be incorporated into traditional plant procedures.

Keywords: warm mix asphalt; additives; sustainability; water sensitivity; stiffness; plastic deformation

1. Introduction

Asphalt mixtures are the most widely used material in the construction of pavements for roads and highways around the world [1]. However, their use is associated with significant environmental pollution that is generated during manufacturing, placing and maintenance—namely the release of a large volume of greenhouse gases caused by the high manufacturing temperature (around 160 °C). This also leads to high levels of energy consumption, and hence, construction costs [2,3]. Thus, in order to limit these problems [4], a cleaner production of bituminous mixtures is required to decrease the manufacturing temperature without reducing their mechanical behaviour.

One possible alternative is the use of Warm Mix Asphalt (WMA), which is manufactured at 20–40 °C lower than the conventional Hot Mix Asphalt (HMA) [5]. These mixtures were developed to significantly reduce harmful emissions and energy consumption without compromising the mechanical performance of the material [6–8]. In particular, a number of studies [9–11] have demonstrated that the use of WMA in pavements for roads allows for lower emissions, fumes and odors; a reduction of ageing of the bitumen; a decrease in fuel consumption in plant; a more rapid turnover of traffic; and an increase in haulage distances among other economic and environmental advantages, due to the reduction in manufacturing and paving temperature. As a result of these potential benefits, WMA has attracted a considerable amount of attention in the USA as well as in various European countries in recent years [1,9].

The reduction in WMA temperature is the consequence of recent technologies that can be divided into different categories depending on the agent used, and includes organic additives, chemical additives, and foaming (water-based or water-containing) [5]. Despite this qualification, all of these technologies share the same objective—an improvement in mixture workability and aggregate coating at lower temperatures by reducing bitumen viscosity or modifying its surface tension for better aggregate wetting [12]. Among these technologies, the use of chemical additives (which are liquid surfactants that act at the microscopic aggregate/binder interface to reduce internal friction during manufacturing) is one of the most common solutions, since these additives require minimal changes to the manufacturing process. Further, they act as anti-stripping agents that increase the adhesion between aggregates and bitumen, thereby reducing problems associated with WMA such as stripping and raveling [13]. Moreover, in contrast to other techniques such as the use of foam bitumen, their use does not require special investment or modifications in the manufacturing plant. As a result, new WMA surfactant additives that require further investigation have been emerging, and the choice of these materials is often a subjective decision based on the skill of the engineer. Therefore, in-depth studies are required to provide useful knowledge about the suitability of these additives for application in pavement construction as well as their capacity to improve the mechanical performance of asphalt mixtures manufactured at low temperatures [3]. Moreover, most of the work on WMA with chemical additives has focused on laboratory analysis, with relatively few studies conducted in a real life field setting [14–17]. The latter type of study is essential to demonstrate the effectiveness of reproducing these WMAs in real asphalt plants for their application in pavements for roads.

Given this lack of knowledge and its potential to reduce environmental issues associated with asphalt mixtures, the current paper set out to examine the effectiveness of various additives in improving the behaviour of WMA at a range of manufacturing temperatures (145 °C and 120 °C). To do this, the performance of each WMA was compared with that of conventional hot mix asphalt (manufactured at 165 °C) without additives, which was used as a control. Further, the present study evaluates the impact of the dosage of the additive in order to define the optimal design of WMA for its use in pavement for roads. Finally, a study of the reproducibility of the laboratory-designed WMA in a real plant was conducted, along with its application in a trial section, evaluating its mechanical performance in comparison with conventional HMA.

2. Methodology

Materials

For this study, the same type of asphalt mixture was used as that used in the production of both the HMA (used as a reference) and the various WMAs with different chemical additives (applied during manufacturing in order to reduce the temperature of mixing). Such a mixture has a dense-graded mix type AC 22 35/50 S (EN 13108-1) whose mineral skeleton is composed of limestone aggregates (which allows for sufficient contact with the bitumen to achieve a bond between binder and aggregates) [18] for the different fractions (0/6, 6/12, 12/18, and 18/25 mm) with a maximum particle size equal to 22 mm. The main properties of these aggregates are displayed in Table 1, where it is clear that this material presents appropriate characteristics for its application in the manufacturing of asphalt mixtures.

Table 1. Aggregates properties.

Properties		18/25	12/18	6/12	0/6
	Sieves (mm)	% passing	% passing	% passing	% passing
Particle size (UNE-EN 933-1)	25	100	100	100	100
	22	90	100	100	100
	16	4	38	100	100
	8	0	0	23	100
	2	0	0	0	62
	0.5	0	0	0	28
	0.25	0	0	0	16
	0.063	0.1	0.2	0.1	3.0
Coarse aggregate shape. Flakiness index (UNE-EN 933-3)		5.25	5.82	12.01	-
Percentage of fractured face (UNE-EN 933-5)		95.8	95.2	97.1	-
Resistance to fragmentation (UNE-EN 1097-2)		24.7	24.7	24.7	-
Cleaning (organic impurity content) (UNE-EN 146130)		0.26	0.99	2.27	-
Sand equivalent (UNE-EN 933-8)		-	-	-	83.02
Relative density and absorption (UNE-EN 1097-6)	Apparent density (Mg/m^3)	2.73	2.75	2.71	2.80
	ADSS (Mg/m^3)	2.68	2.69	2.64	2.72
	Density after drying (Mg/m^3)	2.70	2.71	2.66	2.75
	Water absorption (%)	0.64	0.81	0.93	1.03

Filler was recovered from the crushing of limestone rocks, which provides fine particles with alkaline properties that facilitate the adhesion with bitumen. This material was less than 0.063 mm in size for more than 96% of the particles, with a density equal to 0.6 Mg/m^3 (EN 1097-3). The binder employed was type B35/50 whose penetration was equal to 44 dmm (EN 1426) with a softening point of approximately 52 °C (EN 1427).

With all of these materials, the asphalt mixture was manufactured using a bitumen dosage equal to 4% over the total mass. The mixture used as a reference to evaluate the influence of the different additives was manufactured at 165 °C, which is a common temperature for this type of mixture. Table 2 lists the main physical and mechanical properties for this material, where it is possible to see that the mixture presents appropriate properties for its application in asphalt pavements for roads.

Table 2. Main properties of the asphalt mixture.

Property	Standard	HMA
Apparent density (g/cm^3)	EN 12697-6	2.477
Mix air void content (%)	EN 12697-8	4.1
Aggregates air void (%)	EN 12697-8	13.7
Marshall stability (kN)	EN 12697-34	17.79
Marshall deformation (mm)	EN 12697-34	4.1

In order to be able to decrease the manufacturing temperature for the WMA, three types of surfactant additives were used, which are referred to in this study as A1 (a traditional surfactant additive with amine composition), A2 (a nano-additive that modifies the alkalinity of the aggregate surface) and A3 (a vegetable additive with active surface properties). Before the manufacturing of the various WMA, the additives were blended with the bitumen by using a rotational blender at 300 revolutions per minute for 10 min. The temperature of blending was around 160 °C, and after this process the bitumen was incorporated directly into the mixer over the aggregates at various temperatures (145 °C or 120 °C, depending on the case studied). The mixing time was similar to that used for the HMA, and the compaction temperature was almost 10 °C lower than the mixing temperature (as was the case for the conventional hot mix asphalt).

3. Testing Plan and Methods

By using the conventional AC 22 S mixture (manufactured at around 165 °C) as a reference, the testing plan developed in this study included three different steps (Table 3) in order to design and apply an asphalt mixture at a lower temperature with appropriate mechanical performance: (i) analysis of the effect of various additives to reduce the manufacturing temperature (evaluating its effectiveness under different manufacturing temperatures); (ii) optimisation of the mixture design (definition of the most appropriate manufacturing temperature and additive dosage) by using the additive selected in the previous stage; and (iii) reproducibility of the WMA in a real asphalt plant, and study of its mechanical behaviour in the laboratory. In addition, the effect of different manufacturing temperatures (145 °C and 120 °C) was analysed in the first two steps, while in the third step an optimal temperature was chosen on the basis of the results obtained in the previous steps.

Table 3. Testing plan.

Study Step	Asphalt Mix	Additive		Temperature °C	Tests
		Type	%		
Effect of different additives on WMA behaviour	HMA165	-	-	165	Workability I. tensile strength Water sensitivity Stiffness Triaxial
	WMA-1-0.5-145 WMA-2-0.5-145 WMA-3-0.5-145	A1 A2 A3	0.5	145	
	WMA-1-0.5-120 WMA-2-0.5-120 WMA-3-0.5-120	A1 A2 A3	0.5	120	
Optimisation of WMA design	HMA165	-	-	165	Workability I. tensile strength Water sensitivity Stiffness Triaxial
	WMA-S-0.5-145 WMA-S-0.05-145	Selected	0.5 0.05	145	
	WMA-S-0.5-120 WMA-S-0.05-120		0.5 0.05	120	
Production of WMA in asphalt plant	HMA165	-	-	165	Density Water sensitivity Stiffness Triaxial
	WMA-S-S-S	Selected	Selected	Selected	

In the first step, the behaviour of 7 asphalt mixes was analysed: the conventional mix used as a reference and manufactured at 165 °C without additives (known as HMA165); and six mixes manufactured at 145 °C and 120 °C, including the three different chemical agents with a dosage of 0.5% over the mass of the bitumen (known as WMA-X-Y-Z, where X refers to type of additive, Y to the dosage, and Z to the manufacturing temperature). This percentage was selected since it is the usual value employed when using chemical additives in the production of WMA [19,20], whilst this quantity was maintained at a constant value for the three additives in order to compare their effect on the mix behaviour. Moreover, the temperature values were selected according to the range commonly used for WMA, which allows for determining the effect of this parameter by using both extreme values of temperatures used in the production of WMA.

To analyse the effect of the different additives on the behaviour of the WMA, the workability and compactibility of the various mixes was studied using a gyratory compactor. In addition, for the different specimens (4 for each mix) obtained from the workability study, its indirect tensile strength was evaluated to analyse the influence of each additive on the coating of the mix at low temperatures. The stiffness modulus test [21], water sensitivity test [22], and triaxial test [23] were also conducted for the different mixes in order to evaluate the impact of the additives on the behaviour of the mixes under the main mechanical properties (bearing capacity, water susceptibility, and resistance to plastic deformations) for their application in pavements for roads.

The second step of the testing plan consisted of studying the effect of the dosage of additive used (for the additive selected as the most appropriate in the previous step) while evaluating the impact of manufacturing temperature. This step was carried out with the aim of both optimising the design of the WMA, and obtaining a mechanical performance that is comparable with the conventional HMA. To do so, the behaviour of the asphalt mix was examined when manufactured at 145 °C and 120 °C by using 0.5% of additive (quantity employed in the previous step) and 0.05%. The percentage was considerably reduced in order to clearly identify the effect of the quantity of this component on the behaviour of the mixture, analysing whether this reduction in additive plays an essential role. Further, the choice of dosage was guided by the manufacturer's suggestion for finding the optimal solution to reduce production costs associated with using the chemical agent. On the other hand, in addition to the WMA mixtures, the performance of a conventional HMA was evaluated for use as a reference for analysing the influence of the additive. The properties evaluated were the workability and compactibility of the different solutions (which included determining the tensile strength), stiffness modulus [21], water sensitivity [22], and triaxial test [23].

The third study step involved the production of the laboratory-designed WMA in a conventional asphalt plant, in accord with the results obtained in the previous stages. In addition, a trial section was paved with both the WMA and the conventional HMA (used as a control) in order to study the long-term behaviour of these mixtures in further studies. During the spreading process, samples of both types of asphalt mix were collected in order to study their behaviour in laboratory, and to then assess the viability of manufacturing this kind of material in conventional asphalt plants. The tests developed to evaluate the response of both mixtures were density [24], stiffness modulus [21], water sensitivity [22], and triaxial test [23].

The workability test was developed by using a gyratory compactor that allows for observing the relationship between the energy transmitted to the specimen and the level of compaction obtained, by analysing the aptitude of the material to be compacted at low temperatures. The total number of cycles applied was 210 gyros, in order to evaluate the evolution of the density of the material. In addition, after manufacturing 4 specimens for each mix, the indirect tensile strength was measured at 20° by using the method developed in [25].

The water sensitivity test [22] involves the manufacture of cylindrical specimens that are compacted with 35 blows on each side by a Marshall hammer in order to reproduce low compaction energy, thereby simulating more unfavorable conditions against water action. The specimens were divided into two sets of three specimens: a dry set and a wet set. The set of dry specimens was stored at room temperature in the laboratory (15 ± 5 °C), whereas a vacuum process was applied to the wet set for 30 ± 5 min until a pressure of 6.7 ± 0.3 kPa was obtained. The wet specimens were then immersed in water at a temperature of 40 °C for a period of 72 h. The next step was to carry out an indirect traction resistance test [25] on each of the cylinders (in both the dry set and the wet set) at a temperature of 15 °C, and to compare their indirect tensile strength through the ITSR (Indirect Tensile Strength Ratio, which is obtained by dividing the strength of the wet set by the strength of the dry set in terms of percentage).

The stiffness modulus was measured according to [21], which consists of manufacturing cylindrical specimens compacted with 75 blows on each side. After conditioning the specimens at 20 °C for at least 2 h, they were placed and secured in a vertical position of one of its diameters, applying 10 load pulses to adjust the magnitude of the load and its duration. In order to measure the deformation of the diameter, 5 additional load pulses were applied to measure and record the load variation and deformation in the time period of each pulse. At the same time, the surface load factor was also determined. As a final result, the stiffness modulus was obtained for two diameters of the specimen (forming an angle of 90 ± 10°).

The triaxial test [23] entails the combination of a confining load of 120 kPa and another cyclic sinusoidal out-of-phase axial loading of 300 kPa at a frequency of 3 Hz for 12,000 load cycles. The creep and permanent deformation parameters for each specimen (cylinders with a diameter of 101.6 mm and a sawn-off height of 60 mm) are calculated, and the results obtained in the test are shown as the mean of the values obtained for three test specimens. During the current study, the triaxial test was carried out at 60 °C in order to assess the resistance of the material to plastic deformation under severe climate conditions.

4. Results

Effect of Different Additives on WMA Behaviour

Figure 1 shows the curves of densification (reduction in air void content) of the asphalt mixtures manufactured at 145 °C (Figure 1a) and 120 °C (Figure 1b) with 0.5% of various additives, as well as the curve measured for the conventional HMA. From the results, it is clear that the use of additives during the manufacturing of mixtures at 145 °C allows for workability that is comparable to that obtained for the conventional HMA, since quite similar air void content (near 4.5–5.5%) was recorded, highlighting the case of the additive A2. In addition, the workability study showed that the use of these chemical additives to reduce the manufacturing temperature to around 20 °C allows for values of density higher than 98%, in reference to the conventional HMA (2.44 Mg/m^3), which indicates its suitability for application in pavements for roads, avoiding the problems often associated with the low level of compaction for this material [26].

For the mixtures manufactured at a lower temperature (around 120 °C), the results revealed that in this case the effectiveness of the additives is slightly lower than in the previous case (manufactured at 145 °C), since higher values of air void content were recorded, indicating lower compaction of the asphalt mixture associated with lower workability due to the increase in bitumen viscosity when the temperature is reduced. Nonetheless, the additives A1 and A2 led to decrease in densification in reference to that presented by the HMA, which allows for density values close to 98%, which was measured for the conventional hot mix asphalt. This shows that the use of these two additives could be appropriate for manufacturing WMA at 120 °C whilst adequate workability and compaction characteristics are obtained in reference to traditional HMA, which is in accord with other studies examining the use of additives for WMA [27,28]. Therefore, these results suggest that the use of these two additives reduces susceptibility to manufacturing temperature with respect to the additive A3, which led to a significant decrease in density when the temperature was reduced from 145 °C to 120 °C.

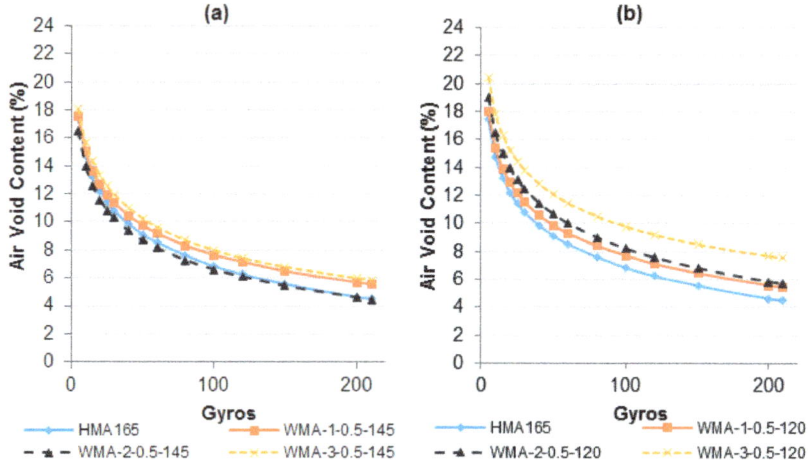

Figure 1. Densification curves of the mixtures with various additives and different manufacturing temperatures. (**a**) 145 °C vs. 165°C; (**b**) 120°C vs. 165°C.

In order to obtain a more in-depth study of the influence of the different additives, Figure 2 displays the values of Indirect Tensile Strength (ITS measured in kPa) measured for the specimens from the workability study (with gyratory compactor) for the different mixtures at low temperatures with the additives A1, A2 and A3, as well as the results obtained for the HMA. In addition, the Indirect Tensile Strength Ratio (ITSR) values obtained in the water sensitivity test [22] are shown in order to evaluate the moisture susceptibility of the various mixtures analysed.

According to the results, it appears that the use of the different additives to manufacture WMA at 145 °C leads to quite similar values of ITS to those measured for the HMA, thereby avoiding the failure of the mixture associated with the reduction in adhesion and cohesion due to the decrease in manufacturing temperature, which could in turn lead to coating problems. Similarly, application of the additives generally improved the ITSR values, which reflects an important reduction in water sensitivity for the WMA manufactured at 145 °C when using the additives analysed. These findings are also in agreement with other studies that employed chemical additives to act as anti-stripping agents [29,30]. However, when the reduction in manufacturing temperature is higher (near 45 °C), it is important to note that a significant decrease in tensile strength could occur, despite the fact that the ITSR are quite similar to those presented by the HMA (with the exception of the additive A3, which showed an important increase in water sensitivity for the mixture manufactured at 120 °C), and therefore, the durability of the material could also be reduced.

With respect to the effectiveness of each of the chemical additives, the present results show that A1 and A2 lead to quite similar WMA performance, obtaining even higher tensile strength (in the case of A2) and lower water sensitivity than the HMA. In reference to the additive A3, this appeared to be less effective in avoiding the reduction in tensile strength and resistance to water due to the decrease in manufacturing temperature, which is probably a consequence of the lower capacity of this chemical additive to improve the adhesiveness between bitumen-aggregates by reducing the surface tension and increasing the wetting of aggregates [13].

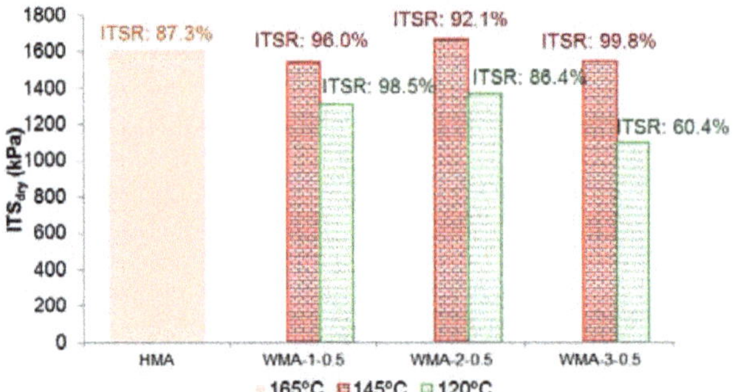

Figure 2. Indirect Tensile Strength (ITS) and Indirect Tensile Strength Ratio (ITSR) results for the mixtures with various additives.

Figure 3 shows the influence of the different additives on the stiffness modulus (at 20 °C) of the asphalt mixture manufactured at low temperatures. The results show that, in general, the WMA presented lower stiffness modulus than the HMA, regardless of the type of additive. In addition, it is clear that the lower the manufacturing temperature, the lower the stiffness modulus. Whilst this could be associated with lower compaction of the specimens (particularly in the case of the mixtures manufactured at 120 °C), the decrease in modulus could also be due to lower ageing of the bitumen during the manufacturing process, since the temperature is reduced by 20–45 °C in reference to the case of the HMA, and therefore, the short-term stiffening of the bitumen is lower [14,19]. Thus, the mixtures manufactured at 120 °C showed higher reduction in stiffness modulus, obtaining quite similar values for the different additives, while in the case of the mixtures manufactured at 145 °C, the additive A2 led to a slight increase in bearing capacity, which could also be influenced by higher density values, according to the results described previously.

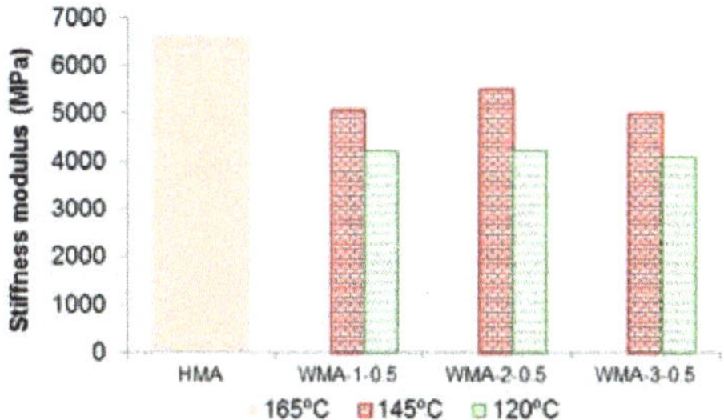

Figure 3. Stiffness modulus at 20 °C for the mixtures with different additives.

As the decrease in stiffness modulus in the WMA could be due to lower stiffening of the bitumen, lower resistance to rutting could be also obtained [14]. Figure 4 displays the resistance to permanent deformation for each mixture in order to show the effectiveness of these chemical additives in reducing the stiffening of the asphalt mixture whilst avoiding higher rutting deformations. In particular, Figure 4 presents the results obtained on the triaxial test, showing the final permanent deformation and the ratio of creep during the last loading cycles. The results indicate that the reduction in manufacturing temperature to 120 °C led to an important increase in plastic deformations, which could be associated with a decreased stiffness of the bitumen along with the lack of compaction. This effect was more marked for additives A1 and A3.

However, in the case of the mixtures manufactured at 145 °C, the decrease in binder stiffening due to the reduction in temperature was balanced with density values close to those shown by the conventional HMA (Figure 1), obtaining fewer permanent deformations, even in the case of additives A2 and A3. This indicates the ability of these chemical additives to reduce the manufacturing temperature of asphalt mixtures, which decreases stiffening whilst allowing the WMA to maintain a bearing capacity and resistance to permanent deformations that is comparable to conventional HMA.

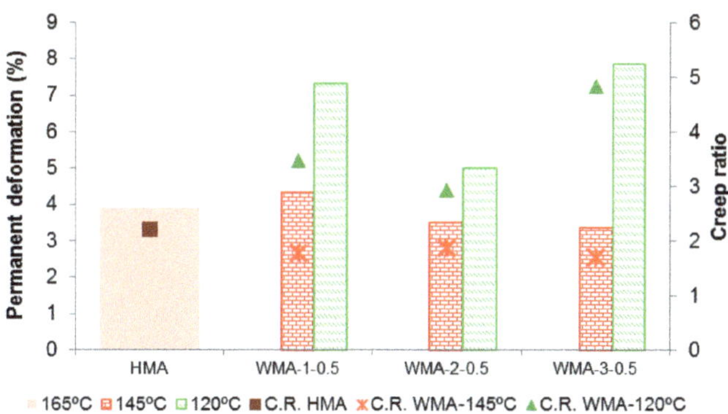

Figure 4. Triaxial results for the Hot Mix Asphalt (HMA) and Warm Mix Asphalt (WMA) with different additives.

5. Optimization of the WMA Design

Based on the previous results, the A2 additive appears be appropriate for use in manufacturing low energy asphalt mixtures, since it shows mechanical performance that is comparable to conventional

HMA, obtaining higher tensile strength, water resistance, bearing capacity, and resistance to plastic deformations than the case of the WMA used with other additives. The additive A2 was therefore selected as the most appropriate for studying the optimization of the WMA design, with a view to its later use in a real asphalt plant. Figure 5 compares the workability of WMA with 0.5% of additive (the quantity commonly used) and 0.05% (the quantity defined to assess the possibility of reducing the amount of additive used), taking the conventional HMA as a reference.

The results show that for both manufacturing temperatures (145 °C and 120 °C, in Figure 5a,b, respectively), the quantity of additive exerts little influence, since both dosages generated rather similar curves of the evolution of the air void content in the asphalt mixtures. Regarding the effect of temperature, it appears that it is possible to obtain similar values of air void content to those measured for the HMA at the manufacturing temperature of 145 °C, which indicates the ability of this additive to improve the workability of the material despite the decrease in temperature. Moreover, despite the reduction in additive effectiveness at lower temperatures (around 120 °C), the air void content for such mixtures was lower than around 6%, obtaining density values higher than 98% over the conventional HMA. Thus, these results indicate that the use of this additive allows for appropriate workability and compaction of the asphalt mixture at low temperatures (within 120–145 °C), presenting little effect of the dosage of additive utilized (between 0.05% and 0.5% in this study).

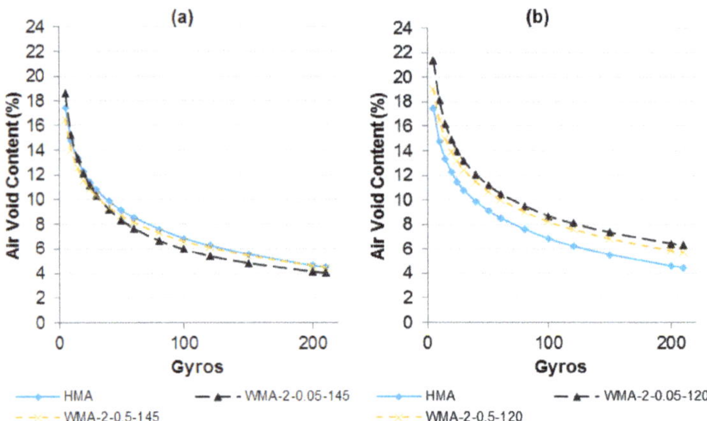

Figure 5. Densification curves for the HMA and WMA with different quantities of additives and manufacturing temperatures. (**a**) 145 °C vs. 165 °C; (**b**) 120 °C vs. 165 °C.

In addition, Figure 6 shows that although the decrease in additive dosage could lead to a slight increase in water sensitivity of the mixture, using a low quantity of this chemical agent (around 0.05% over the bitumen mass) allows for comparable (or even higher) indirect tensile strength to both the conventional HMA and the WMA with a higher amount of additive (0.5%). This fact indicates that despite the lower value of ITSR that is measured in WMA [14], the values of tensile strength under wet conditions make it acceptable for application when compared with the mixtures used as a reference in this study.

In addition, the decrease in additive dosage was shown to be more effective in reducing the effect of manufacturing temperature (within 120–145 °C) on the behaviour of the asphalt mixture under tensile effort and water action, since quite similar values of ITS and ITSR were recorded for both temperatures when 0.05% of additive was used. In contrast, applying a higher amount of additive (0.5%) led to a decrease in tensile strength of the asphalt mixture when the manufacturing temperature was reduced to 120 °C. This fact could be associated with the nano-composition of this additive (A2), which acts at the microscopic aggregate/bitumen interface [13], thereby improving performance when low quantities are used since it is possible to avoid the formation of agglomerations [14].

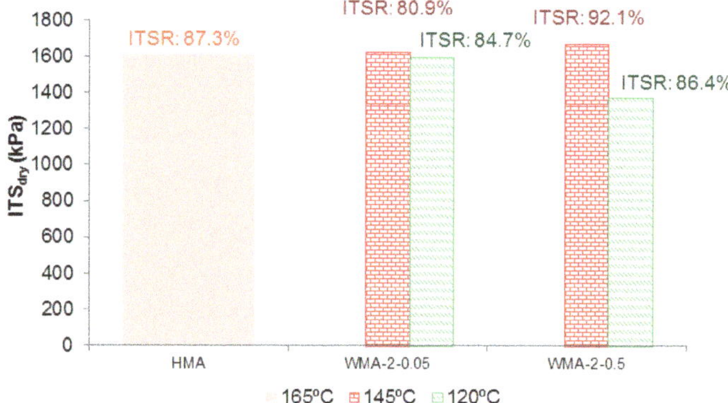

Figure 6. ITS (Indirect Tensile Strength) and water sensitivity results for the HMA and WMA with different quantities of additive under various manufacturing temperatures.

Similarly, Figure 7 shows that reducing the quantity of additive seems to be more effective in yielding a higher stiffness modulus (compared with using 0.5% additive) under the same manufacturing temperatures (producing a similar reduction in bitumen ageing). This is particularly important for a temperature of 120 °C, where a significant increase in bearing capacity was recorded when the dosage of additive was reduced. This could be related to a lower modification of the bitumen properties when the additive percentage is reduced, and thus a lower reduction in stiffness modulus is presented despite the fact that in both cases (0.5% and 0.05% additive) similar density values were recorded.

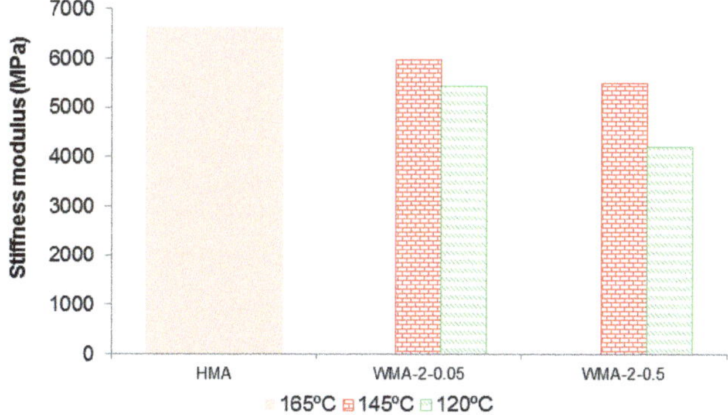

Figure 7. Effect of the quantity of additive and manufacturing temperature on the stiffness modulus of WMA.

Figure 8 displays the results obtained on the triaxial test for the WMA with different quantities of A2 as well as for the conventional HMA, used as a reference. Based on these results, it appears that the use of 0.05% of this additive improves resistance to plastic deformations of the asphalt mixture compared with the HMA (which could be due to improved workability and compaction when the additive is used), showing a similar performance at both of the low temperatures (120–145 °C). However, it should be noted that the increase in additive content could lead to the behaviour of the mixture being more susceptible to the manufacturing temperature, since in the case of using 0.5% additive, the mixture manufactured at 120 °C presented an increase in plastic deformations (both final deformation and ratio of creep), which could lead to the appearance of rutting deformation during its application in pavements for roads in comparison with the other WMA solutions and the conventional HMA.

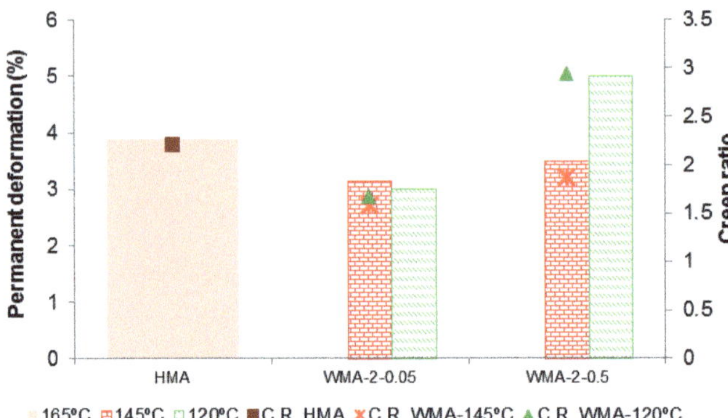

Figure 8. Effect of additive quantity and manufacturing temperature on the resistance to plastic deformations of WMA.

6. Production of WMA in Plant, and Evaluation in Laboratory

In accord with the previous findings, it was observed that the use of 0.05% additive (A2) could lead to appropriate mechanical performance of asphalt mixtures (with adequate workability, bearing capacity, and higher resistance to water action and to permanent deformations than the other dosage analysed) in comparison with HMA. In addition, it appears that the manufacturing temperature (ranging between 120–145 °C) has a relatively lower impact on WMA performance. Therefore, in order to analyse the performance of WMA in the field, an asphalt mixture manufactured at 130 °C including 0.05% of Additive type 2 was produced in a conventional discontinuous plant in order to study its reproducibility without the need to modify plant equipment, evaluating its mechanical performance in reference to HMA manufactured in the same plant at around 165 °C.

As shown in Figure 9A, firstly the additive was mixed directly with the hot bitumen (around 160 °C) for 20 min using a blender in order to obtain a homogenous mix of additive and bitumen. After this, the modified bitumen was mixed with warm aggregates (at 130 °C as shown in Figure 9B), applying a similar mixing time to that used for HMA before starting the spreading and compaction process.

To determine the abilities of WMA manufactured in a conventional discontinuous plant, Table 4 shows the mechanical performance of both WMA and HMA during the spreading process (Figure 9C,D). The results indicate that despite a slight reduction in density of the WMA, both asphalt mixtures showed comparable density and air void content, implying the workability and compactibility of WMA. In addition, these values were quite similar to those measured for the mixtures manufactured in laboratory, which indicates good reproducibility of the WMA in a real asphalt plant.

Regarding the mechanical response of the mixtures, Table 4 shows that the WMA presented an even higher indirect tensile strength ratio under the water action while the tensile strength was quite similar to that measured for the conventional HMA. This could be associated with appropriate coating of aggregates in the WMA despite the decrease in manufacturing temperature, obtaining a good cohesiveness of the mixture as a result of using the chemical additive. Nonetheless, it is important to note that a slight reduction in stiffness and resistance to plastic deformations was recorded for the WMA in reference to the conventional HMA. This effect could be due to lower ageing of the bitumen during the manufacturing process, and therefore less stiffening of the asphalt mixture was observed, which is in accord with other studies [14,19].

Figure 9. Visual appearance of the process for manufacturing the WMA with the chemical additive. (**A**) incorporation of additive to bitumen; (**B**) aggregates temperature; (**C**) WMA temperature; (**D**) spreading and compaction of the WMA.

Table 4. Results recorded for the HMA and WMA manufactured in a real asphalt plant.

Mix	Density (Mg/m^3)	Air Void Content (%)	Water Susceptibility		Stiffness Modulus at 20 °C (MPa)	Triaxial	
			ITS Dry (kPa)	ITSR (%)		P. def. (%)	Creep Ratio
WMA	2.451	5.7	1530.2	96.0	6626.3	2.76	1.1
HMA	2.471	5.0	1569.6	91.4	7371.2	1.71	0.7

7. Conclusions

The present paper aims to analyse the effectiveness of various chemical additives used to manufacture asphalt mixtures at lower temperatures, to determine whether a cleaner technology can be employed for the production of asphalt mixtures with appropriate mechanical behaviour for application in pavements for roads. To this end, this study focused on determining the optimal design parameters of a WMA by examining its use in a real asphalt plant and applying the mixture in a trial section, as well as assessing its mechanical properties through the use of laboratory tests. On the basis of the results obtained in this study, the following conclusions can be drawn:

1. The manufacturing and compaction temperature has a strong impact on the behaviour of the mixture, obtaining better results when the manufacturing temperature is around 145 °C. Nonetheless, the behaviour of mixtures with the A1 and A2 additives appeared to be less susceptible to the effects of temperature than the material produced with A3 additives.
2. For mixtures manufactured at 145 °C, the results showed that the use of certain additives even allowed for an improvement in mixture workability, tensile strength, and water susceptibility in comparison with conventional HMA.
3. The results reveal that the use of certain chemical additives in WMA leads to lower stiffness modulus, without reducing the resistance to plastic deformations in reference to HMA. This

indicates the possibility of obtaining more flexible mixtures without reducing strength, which could result in greater longevity of the mixtures. Nonetheless, there is a need for more in-depth studies on this topic.

4. The A2 additive was identified as the most appropriate in this study on the basis of the laboratory results, and it was shown that using a lower dosage (0.05% over the bitumen mass, instead of 0.5% that is commonly used with chemical additives) allowed for lower susceptibility to the manufacturing temperature, whilst better mechanical performance was recorded. This fact could be associated with the nano-composition of this additive, which acts at the microscopic aggregate/bitumen interface.
5. This study indicates that WMA can be produced in a conventional discontinuous plant without incorporating any significant changes to the equipment and manufacturing process.
6. Moreover, the WMA presented appropriate density and air void values when compared to the HMA, whilst comparable tensile strength and water sensitivity was also recorded. Nonetheless, in this case, the reduction in stiffness modulus of the WMA produced a decrease in resistance to plastic deformations, which should be considered in further studies focusing on its application in bituminous pavements.

Taken together, the results obtained in this study suggest that, at least when using tensoactive additives, it is possible to manufacture cleaner asphalt mixtures at lower temperatures (WMA), with the material showing comparable mechanical behaviours to that recorded for conventional hot mixtures. In addition, it was also shown that this type of mixture could be manufactured in a conventional asphalt plant without the need to modify the equipment.

Acknowledgments: The authors wish to offer a special acknowledgement to the company Construcciones Pérez Jiménez SA, whose involvement in the study has made possible the manufacture and evaluation of the materials in a real plant and trial section scale.

Author Contributions: Miguel Sol-Sánchez has collaborated in carrying out the experimental plan and writing the paper; Fernando Moreno-Navarro has taken part in the experimental plan and writing the paper; Mª Carmen Rubio Gámez has supervised the experimental plan while revising and correcting the paper.

Conflicts of Interest: The authors declare no conflict of interest.

References

1. Asphalt in Figures. Available online: www.eapa.org (accessed on 21 July 2017).
2. Kristjansdottir, O. Warm Mix Asphalt for Cold Weather Paving. Ph.D. Thesis, University of Washington, Seattle, WA, USA, 2006.
3. Almeida-Costa, A.; Benta, A. Economic and environmental impact study of warm mix asphalt compared to hot mix asphalt. *J. Clean. Prod.* **2016**, *112*, 2308–2317.
4. Moretti, L.; Mandrone, V.; D'Andrea, A.; Caro, S. Comparative from cradle to gate life cycle assessments of Hot Mix Asphalt (HMA) materials. *Sustainablility* **2017**, *9*, 400. [CrossRef]
5. Rubio, M.C.; Martínez, G.; Baena, L.; Moreno, F. Warm mix asphalt: An over-view. *J. Clean. Prod.* **2012**, *24*, 76–84. [CrossRef]
6. Capitão, S.D.; Picado-Santos, L.G.; Martinho, F. Pavement engineering materials: Review on the use of warm-mix asphalt. *Constr. Build. Mater.* **2012**, *36*, 1016–1024. [CrossRef]
7. Behl, A.; Kumar, G.; Sharma, G.; Jain, P.K. Evaluation of field performance of warm-mix asphalt pavements in India. *Procedia Soc. Behav. Sci.* **2013**, *104*, 158–167. [CrossRef]
8. Blankendaal, T.; Schuur, P.; Voordjik, H. Reducing the environmental impact of concrete and asphalt: A scenario approach. *J. Clean. Prod.* **2014**, *66*, 27–36. [CrossRef]
9. D'Angelo, J.; Harm, E.; Bartoszek, J.; Baumgardner, G.; Corrigan, M.; Cowsert, J.; Harman, T.; Jamshidi, M.; Jones, W.; Newcomb, D.; et al. *Warm-Mix Asphalt: European Practice*; Report No. FHWA-PL-08-007; International Technology Scanning Program: Washington, DC, USA, 2008.

10. Pérez-Martínez, M.; Moreno-Navarro, F.; Martín-Marín, J.; Ríos-Losada, C.; Ruubio-Gámez, M.C. Analysis of cleaner technologies based on waxes and surfactant additives in road construction. *J. Clean. Prod.* **2014**, *65*, 374–379. [CrossRef]
11. Dinis-Almeida, M.; Afonso, M.L. Warm mix recycled asphalt—A sustainable solution. *J. Clean. Prod.* **2015**, *107*, 310–316. [CrossRef]
12. Jamshidi, A.; Golchin, B.; Hamzah, M.O.; Turner, P. Selection of type of warm mix asphalt additive based on the rheological properties of asphalt binders. *J. Clean. Prod.* **2015**, *100*, 89–106. [CrossRef]
13. Banerjee, A.; Smit, A.F.; Prozzi, J.A. The effect of long-term aging on the reology of warm mix asphalt binders. *Fuel* **2012**, *97*, 603–611. [CrossRef]
14. Bower, N.; Wen, H.; Willoughby, K.; Weston, J.; DeVol, J. Evaluation of the performance of warm mix asphalt in Washington state. *WSDOT Res. Rep.* **2012**, *789*, 1. [CrossRef]
15. Wen, H.; Wu, S. Performance of WMA technologies: Stage II Long-term field performance. Research Project NCHRP 09-49. In Proceedings of the Asphalt Mixture ETG Meeting, Fall Rivers, MA, USA, 7–8 April 2015.
16. Gu, F.; Luo, X.; Zhang, Y.; Lytton, R.L. Using overlay test to evaluate fracture properties of field-aged asphalt concrete. *Constr. Build. Mater.* **2015**, *101*, 1059–1068. [CrossRef]
17. Sol-Sánchez, M.; Moreno-Navarro, F.; García-Travé, G.; Rubio-Gámez, M.C. Analysing the industrial manufacturing in-plant and in-service performance of asphalt mixtures cleaner technologies. *J. Clean. Prod.* **2016**, *121*, 56–63. [CrossRef]
18. Zhang, J.; Apeagyei, A.; Airey, G.D.; Grenfell, J.R.A. Influence of aggregate mineralogical composition on water resistance of aggregate-bitumen adhesion. *Int. J. Adhes. Adhes.* **2015**, *62*, 45–54. [CrossRef]
19. Hurley, G.; Prowell, B. *Evaluation of Evotherm for Use in Warm Mix*; National Center for Asphalt Technology: Auburn, AL, USA, 2006.
20. Kuang, Y. Evaluation of Evotherm as a WMA Technology Compaction and Anti-Strip Additive. Bachelor's Thesis, Iowa State University, Ames, IA, USA, 2012.
21. EN 12697-26. *Bituminous Mixtures—Test Methods for Hot Mix Asphalt—Parte 26: Stiffness*; AENOR (Asociación Española de Normalización y Certificación): Madrid, Spain, 2012.
22. EN 12697-12. *Bituminous Mixtures—Test Methods for Hot Mix Asphalt—Parte 12: Determination of the Water Sensitivity of Bituminous Specimens*; AENOR (Asociación Española de Normalización y Certificación): Madrid, Spain, 2009.
23. EN 12697-25. *Bituminous Mixtures—Test Methods for Hot Mix Asphalt—Parte 25: Cyclic Compression Test*; AENOR (Asociación Española de Normalización y Certificación): Madrid, Spain, 2006.
24. EN 12697-6. *Bituminous Mixtures—Test Methods for Hot Mix Asphalt—Parte 6: Determination of Bulk Density of Bituminois Specimens*; AENOR (Asociación Española de Normalización y Certificación): Madrid, Spain, 2012.
25. EN 12697-23. *Bituminous Mixtures—Test Methods for Hot Mix Asphalt—Parte 23: Determination of the Indirect Tensile Strength of Bituminous Specimens*; AENOR (Asociación Española de Normalización y Certificación): Madrid, Spain, 2004.
26. Brown, E.R. Density of asphalt concrete—How much is needed? In Proceedings of the 69th Annual Meeting of the Transportation Research Board, Washington, DC, USA, 8–9 January 1990.
27. Mogawer, W.S.; Austerman, A.J.; Engstrom, B.; Bonaquist, R. Incorporating High Percentages of Recycled Asphalt Pavement (RAP) and Warm Mix Asphalt (WMA) Technology into Thin Hot Mix Asphalt Overlays to be Utilized as a Pavement Preservation Strategy. In Proceedings of the Transportation Research Board, Washington, DC, USA, 11–15 January 2009.
28. Tao, M.; Mallick, R.B. Effects of Warm Mix Asphalt additives on workability and mechanical properties of reclaimed asphalt pavement material. *J. Transp. Res. Board* **2009**, *2126*, 151–160. [CrossRef]
29. Ahmed, T.A.H. Investigating the Rutting and Moisture Sensitivity of Warm Mix Asphalt with Varying Contents of Recycled Asphalt Pavement. Bachelor's Thesis, University of IOWA, Iowa City, IA, USA, 2014.
30. Buss, A.; Christopher, R.; Schram, S. Evaluation of moisture susceptibility tests for war mix asphalt. *Constr. Build. Mater.* **2016**, *102*, 358–366. [CrossRef]

Article

Laboratory Evaluation of Rejuvenating Agent on Reclaimed SBS Modified Asphalt Pavement

Jie Wang *, Wei Zeng *, Yongchun Qin, Songchang Huang and Jian Xu

Research Institute of Highway Ministry of Transport, Beijing 100088, China; yc.qin@rioh.cn (Y.Q.); sc.huang@rioh.cn (S.H.); j.xu@rioh.cn (J.X.)

* Correspondence: j.wang@rioh.cn (J.W.); w.zeng@rioh.cn (W.Z.); Tel.: +86-010-6207-9525 (J.W. & W.Z.)

Academic Editors: Zhanping You, Qingli (Barbara) Dai and Feipeng Xiao

Received: 13 June 2017; Accepted: 19 July 2017; Published: 21 July 2017

Abstract: To evaluate the effect of rejuvenating agent on reclaimed SBS (styrene-butadiene-styrene) modified asphalt pavement (RSMAP) material, different tests of asphalt and mixtures were conducted. Firstly, the improvement effect of rejuvenating agents on the aged modified asphalt was tested at macroscopic and microscope level. Then the properties of hot mix asphalt (HMA) with different RSMAP contents (0%, 30%, 50% and 70%) were evaluated by conducting freeze-thaw split, semi-circular bending (SCB) and dynamic modulus (DM). The results indicate that rejuvenating agent can recycle the properties of aged modified asphalt effectively. The initial phase structure of the aged modified asphalt is not changed by adding rejuvenating agents. Moreover, the SBS particles area ratio of modified asphalt blends is significantly correlated with tenacity as the proportion of rejuvenating agent increases. For mixtures, RSMAP is harmful to moisture susceptibility and low-temperature cracking resistance of recycled mixture, especially with high RSMAP content. Moreover, the two properties can be improved by adding rejuvenating agents, but the recycled mixtures with high RSMAP content are not easy to recover to HMA mixture level. In general, the rejuvenating agent has an effect on the stiffness of the recycled mixture, but this is effect not obvious. When loading frequency reaches a higher value, the DM of recycled mixtures and HMA mixture tend to be consistent including high RSMAP content.

Keywords: road engineering; reclaimed SBS modified asphalt pavement; rejuvenating agent; phase structure; fracture energy; fracture toughness; stiffness

---

1. Introduction

Polymer modified asphalt has been widely applied in the construction and maintenance of asphalt pavement in China since the 1990s in order to adapt to the heavy load of modern traffic, and effectively resist the adverse effect of environmental conditions [1,2]. (styrene-butadiene-styrene) SBS modified asphalt is most widely used as it has the advantages of adequate stability under high temperature and low temperature, and satisfactory elastic resilience [3–5]. Especially in the 21st century, SBS modified asphalt has been extensively used in China for the surface layer construction of asphalt pavement, and it is also used for the construction of many asphalt middle layers [6,7].

A large number of SBS modified asphalt pavements have entered into or are approaching the overhaul period considering their designed life goal of 15 years, given the high traffic volume, and overloaded vehicles in China. The maintenance of modified asphalt pavements is facing the problem of disposal of reclaimed SBS modified asphalt pavement (RSMAP). If RSMAP cannot be used rationally, it will lead to serious environmental pollution and economic losses. Therefore, the recycling of RSMAP has become a focus of highway maintenance in China [8,9]. Currently, there are many in-depth studies on the asphalt recycling [10–12] with many successful applications in China. However, the studies on SBS modified asphalt recycling need to be further expanded. Some questions that need to be answered include whether the aged modified asphalt can be effectively recycled; the role of rejuvenating agent in aged modified asphalt; the effect of different types and proportion of rejuvenating agents on the properties of aged modified asphalt and HMA with different RSMAP contents, especially for high content [13–15]. In addition, the studies of the low-temperature cracking resistance, which is one of the most properties of recycled mixtures, are relatively few in number [16]. These questions make the recycling of aged modified asphalt uncertain.

SBS modified asphalt is a multiphase blend composed of polymer SBS and asphalt [17,18]. The two kinds of materials exist with differences in molecular weight and chemical structure as a thermodynamically incompatible system. Because of the interface interaction between different components, it is difficult for the blends to reach a homogeneous level and just achieve physical structure compatibility. The differences of compatibility has lead to a great differences in the properties of modified asphalt [19]. Therefore, the phase structure of SBS modified asphalt can generally be divided into three types: (1) single-phase continuous structure with SBS polymer as dispersed phase and asphalt phase as continuous phase. This type is more common in highway construction; (2) single-phase continuous structure with SBS polymer as continuous phase and asphalt phase as dispersed phase; (3) two-phase continuous structure with SBS polymer and asphalt phase both as continuous phase. The different phase structure has a significant impact on the properties of modified asphalt [20,21]. Ignoring the phase structure to study modified asphalt recycling is not appropriate. This study takes into consideration the phase characteristics of modified asphalt. It focuses on the role of rejuvenating agent in aged modified asphalt, and the effect of rejuvenating agent on properties of recycled mixtures, including high RSMAP content. This study is intended to further clarify the recycling mechanism of rejuvenating agent on RSMAP.

2. Materials and Methods

2.1. Materials

The RSMAP material used in this paper was obtained from Beijing, and the aged modified asphalt extraction from the mixture was performed by Abson method [22]. The aged asphalt content of RSMAP is 3.6%, and the main properties of the aged modified asphalt and virgin SBS modified asphalt used in the study are shown in Table 1.

Table 1. Main properties of modified asphalt.

Properties		Aged Modified Asphalt	Virgin Modified Asphalt
Penetration (25 °C)/0.1 mm		36.3	48.1
Softening point/°C		68.8	69.3
Ductility (5 °C)/cm		0.1	30.7
Viscosity (135 °C)/Pa·s		2.24	1.05
Toughness test	Maximum tension/N	599.9	332.7
	Toughness/N·m	17.74	13.52
	Tenacity/N·m	6.65	9.87

Three rejuvenating agents with different viscosity were selected in this study in order to evaluate the recycling effect on the aged modified asphalt, and represented by H, S, G respectively. The main properties of the rejuvenating agents were tested according to ASTM D4552 and the results are shown in Table 2.

Table 2. Main properties of rejuvenating agent.

Type	Viscosity (60 °C)/10^{-3} Pa·s	Flash Point/°C	Saturates Content/%	Aromatic Content/%	Density/g·cm^{-3}	Tests on Residue from TFOT	
						Viscosity Ratio *	Weight Change/%
H	123	256	17.07	62.07	1.002	2.3	−3.34
S	1202	267	23.38	44.35	1.013	1.8	−2.62
G	2607	278	25.98	49.81	1.028	1.4	−2.73

* Viscosity ratio = Viscosity (60 °C) of residue from TFOT/Virgin viscosity (60 °C).

In the process of aggregate gradation design of recycled mixtures with different RSMAP contents, the gradation should be as similar as possible. AC-13 is selected as the type of recycled mixtures. The virgin coarse aggregate is basalt, the crushing value is 14.3%, and the Los Angeles abrasion value is 13.5% [23]. The virgin fine aggregate is limestone. Table 3 shows the aggregate gradation and asphalt content of recycled mixtures with different RSMAP contents (0%, 30%, 50% and 70%). The type of rejuvenating agent for recycled mixtures used in this study is S.

Table 3. Aggregate gradation and asphalt content of recycled mixtures with different reclaimed SBS (styrene-butadiene-styrene) modified asphalt pavement (RSMAP) contents.

Sieve Size/mm	Type of Recycled Mixtures			
	0%	30%	50%	70%
16	100.0	100.0	100.0	100.0
13.2	97.3	97.2	97.0	96.9
9.5	75.1	75.0	73.4	72.6
4.75	41.9	41.9	40.4	40.7
2.36	28.1	29.1	28.2	28.4
1.18	18.2	18.7	18.4	18.6
0.6	13.6	13.8	13.7	14.0
0.3	10.2	10.3	10.3	10.6
0.15	8.0	7.9	8.0	8.4
0.075	6.3	6.2	6.4	6.7
Virgin asphalt content/%	0.0	3.7	3.1	2.4
Total asphalt content/%	4.8	4.8	4.9	5.0

2.2. Methods

The aged modified asphalt was blended with three types of rejuvenating agents with the proportion of 3%, 6% and 9% by mass of aged modified asphalt, respectively. The tests include the macroscopic properties of 25 °C penetration, softening point, 5 °C ductility, 135 °C Brookfield viscosity, toughness and tenacity [22]. The phase characteristics of the modified asphalt were evaluated using fluorescence microscope.

The properties of recycled mixtures in this study mainly focus on moisture susceptibility, low-temperature cracking resistance and stiffness. Conducted tests included freeze-thaw split [22], semi-circular bending (SCB) and dynamic modulus (DM).

2.3. Semi-Circular Bending Test

The SCB test is used as a method for evaluating asphalt mixture cracking resistance [24]. The test configuration and sample size are illustrated in Figure 1. Test samples were fabricated using SGC at 7 ± 0.5% air void. SGC-compacted samples were 10 cm in height and 15 cm in diameter; each sample was trimmed to half-disc-shaped samples 15 cm in diameter and 5 cm thick. The notch is cut in the middle point of the lower surface of the sample to ensure that the crack will initiate at the desired point. In this study, the notch depth was 1.5 cm. The test was conducted at −20 °C and −10 °C at constant loading rate of 1 mm/min.

In order to evaluate the crack resistance of recycled mixtures, fracture energy (G_f) is used, which can be obtained by dividing fracture work with ligament area [25], as provided in Equation (1). The total fracture work can be calculated by the area under the P-u curve. Figure 1d shows a typical SCB load versus displacement curve.

$$G_f = \frac{W_f}{A_{lig}} \tag{1}$$

where G_f = fracture energy (J/m^2); A_{lig} = ligament area (m^2) = $(r - a) \times t$; r = specimen radius (m); a = notch depth (m); t = specimen thickness (m); and W_f can be calculated as follows in Equation(2).

$$W_f = \int P du \tag{2}$$

where P = applied load (kN); and u = load line displacement (mm).

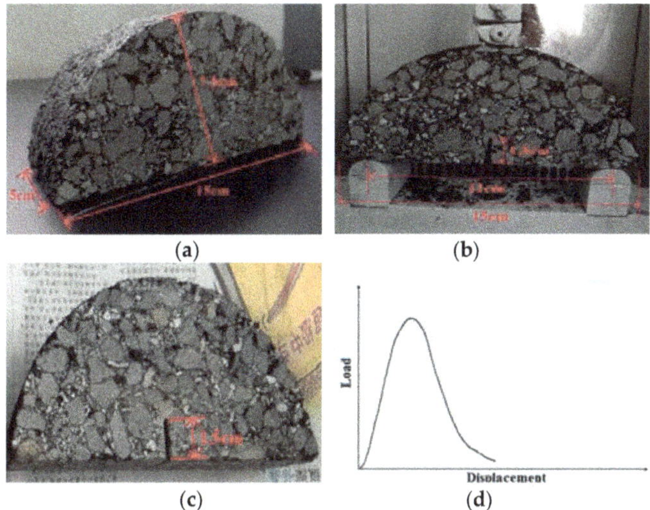

Figure 1. (**a**) Sample dimensions; (**b**) semi-circular bending (SCB) test configuration; (**c**) samples with a notch depth; (**d**) typical SCB load versus displacement (P-u) curve.

Fracture toughness (K_c) is considered as the stress intensity factor at the critical load (P_c), and provides a single parameter characterization that includes the effect of specimen configuration, boundary conditions, and load. In addition, K_c is qualified as an intrinsic material property, describing the ability to resist cracking [26]. In this paper, K_c is calculated using the following equation.

$$K_c = Y_I\, \sigma_c \sqrt{\pi a} \tag{3}$$

where $\sigma_c = P_c/2rt$; $Y_I = 4.782 + 1.219(a/r) + 0.063\exp[7.045(a/r)]$; P_c = peak load; Y_I = normalized stress intensity factor.

2.4. Dynamic Modulus Test

Dynamic modulus (E^*) is a fundamental property of asphalt mixture that defines the viscoelastic stiffness as a function of loading rate and temperature. In this study, test samples were fabricated using SGC at 7 ± 0.5% air void. SGC-compacted samples were 17 cm in height and 15 cm in diameter, and the samples were cored and trimmed to 15 cm in height and 10 cm in diameter. DM tests were conducted at three temperatures (4 °C, 20 °C, 40 °C) and six frequencies (25 Hz, 10 Hz, 5 Hz, 1 Hz,

0.5 Hz, 0.1 Hz) using asphalt mixture performance test (AMPT, IPC Global, Victoria, Australia) and following AASHTO T342-11, as shown in Figure 2.

The stress level and the number of loading cycles in the test are shown in Tables 4 and 5 respectively.

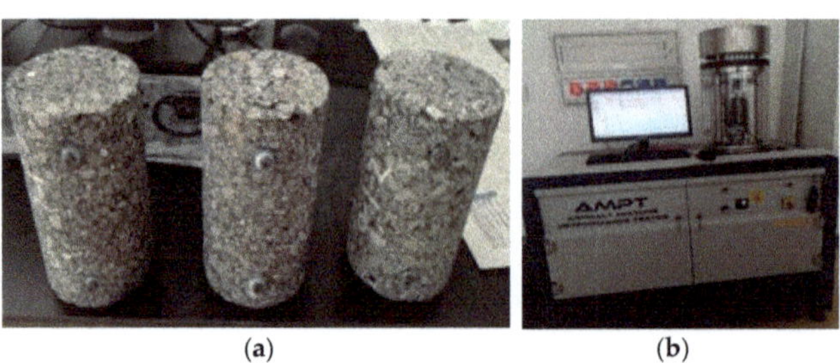

(a) **(b)**

Figure 2. (**a**) Three parallel sample; (**b**) asphalt mixture performance test (AMPT) equipment.

Table 4. The stress level (kPa) of dynamic modulus (DM) test.

Loading Frequency/Hz	Test Temperature/°C		
	4	20	40
25	1400	700	250
10	1260	630	228
5	1120	560	206
1	980	490	184
0.5	840	420	162
0.1	700	350	140

Table 5. The number of loading cycles of DM test.

Loading Frequency/Hz	Number of Loading Cycles
25	200
10	200
5	100
1	20
0.5	15
0.1	15

Using the dynamic modulus test results measured at three different temperatures and six different loading frequencies, the master curves of recycled mixtures based on the time-temperature superposition principle were constructed for a reference temperature of 20 °C [27]. The sigmoidal function is used to fit and develop the master curve, as provided in Equation (4).

$$\log|E^*| = \delta + \frac{a}{1+\exp(\beta+\gamma(\log f_r))} \tag{4}$$

where $|E^*|$ = dynamic modulus (MPa); f_r = reduced frequency at reference temperature; δ = minimum value of $|E^*|$; $\delta + \alpha$ = maximum value of $|E^*|$; β, γ = fitting parameters.

Equation (5) shows the temperature shift factor.

$$\log(f_r) = \log(a(T)) + \log(f) \tag{5}$$

where $a(T)$ = temperature shift factor; f = frequency at a particular temperature.

The temperature shift factor may be written in the form of Equation (6) by using the Arrhenius time-temperature superposition model [28].

$$\log(a(T)) = c\left(\frac{1}{T_k} - \frac{1}{T_{ref}}\right) \quad (6)$$

where T_k = the test temperature in Kelvin temperature; T_{ref} = the reference temperature in Kelvin temperature; c = nonlinear fitting curve coefficient.

A nonlinear optimization program in 1stOpt software was used to determine the master curve coefficients.

3. Results and Analyses

3.1. Macroscopic Properties of Rejuvenating Agents on Aged Modified Asphalt

It is well known that the stiffness of modified asphalt will increase after aging, resulting in attenuation of the low-temperature property and improvement for the high-temperature property. Consequently, modified asphalt recycling should focus on the recovery of low-temperature properties. Table 6 and Figures 3 and 4 show the results of different types of rejuvenating agents on aged modified asphalt.

Table 6. The effect of rejuvenating agent on aged modified asphalt.

Indexes	Aged Modified Asphalt	H			S			G		
		3%	6%	9%	3%	6%	9%	3%	6%	9%
Penetration/0.1 mm	36.3	50	85	108	46	50	62	39	49	57
Softening point/°C	68.8	64.9	59.4	59.2	64.5	64.3	62	67.2	64.2	62.3
Ductility/cm	0.1	19.3	25.9	39.3	17.8	21.2	29.6	13.7	20.7	35.7
Viscosity/Pa·s	2.24	1.07	0.83	0.64	1.07	1.01	0.88	1.26	1.01	0.81

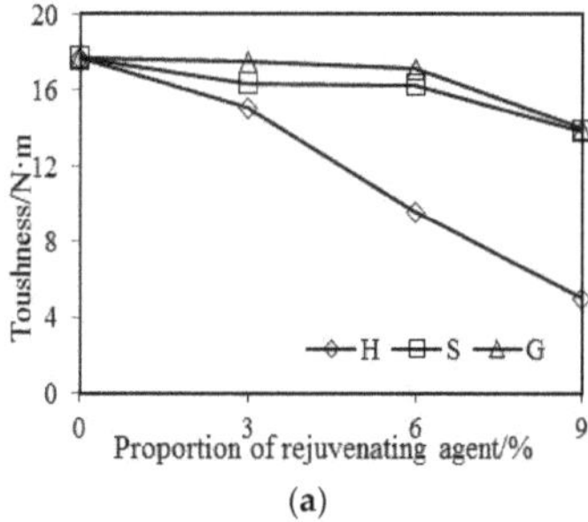

(a)

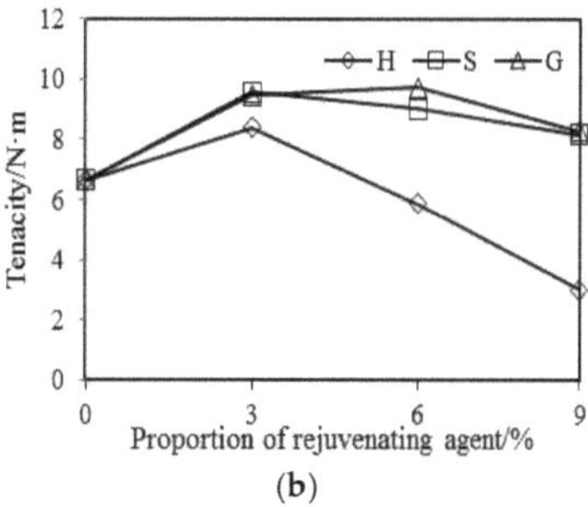

(b)

Figure 3. Relationship between (**a**) toughness, (**b**) tenacity and proportion of Rejuvenating Agent.

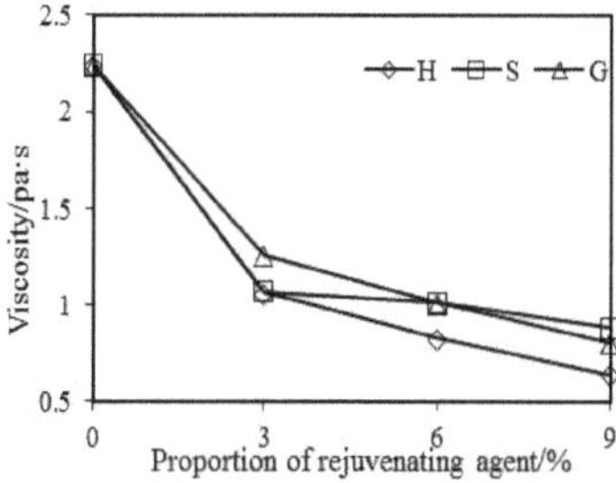

Figure 4. Relationship between penetration and viscosity.

Table 6 shows that: with the increasing of rejuvenating agent proportion, softening point and viscosity decrease, while penetration and ductility increase. When the proportion of H, S, G is 9%, the ductility of modified asphalt blends increases from brittle fracture to 39.3 cm, 29.6 cm and 35.7 cm respectively, which indicates that the rejuvenating agent can recover the low-temperature property of modified asphalt. Satisfactory results can be achieved which indirectly prove the sufficient feasibility of modified asphalt recycling.

From Figure 3, the toughness decreases continuously as the rejuvenating agent proportion increases, and the trend of tenacity is different. When the proportion of rejuvenating agent is small, the tenacity increases at first, and then decreases when the proportion increases continuously, probably because the rejuvenating agent is beneficial to the improvement of swelling effect between polymer SBS and asphalt. Moreover, it can partially improve the mechanical properties of modified asphalt. The tenacity increase can also indirectly prove the feasibility of modified asphalt recycling. When the proportion of rejuvenating agent increases further, the tenacity of modified asphalt blends begins to decrease due to reduced viscosity.

Figure 4 indicates that the penetration of modified asphalt blends increases with the decreasing of viscosity, but the trend is not linear. When the viscosity of modified asphalt blends is more than 1.1 Pa·s, the change of penetration along with the viscosity is not obvious. However, the penetration increases noticeably with the decreasing of viscosity, roughly showing linear relationship when the viscosity of modified asphalt blends is less than 1.1 Pa·s. This phenomenon indicates that the viscosity must be less than a certain value for the change of penetration to be obvious, and this result is similar to the conclusion of reference [29].

In conclusion, the rejuvenating agent can partially or completely recycle the properties of aged modified asphalt. Increasing the proportion of rejuvenating agent can better improve the low-temperature ductility. However, the correlation of penetration, ductility and softening point indicates that increasing the proportion of rejuvenating agent to improve ductility will lead to a sharp increase of penetration and a decrease of softening point. For example, when the proportion of H is 6%, the penetration of modified asphalt blend is 85 (0.1 mm). However, the ductility is still relatively small. The results indicate that the recycling of penetration and ductility is not synchronous and it is not appropriate to use the penetration and ductility to carry on the design of modified asphalt recycling at the same time.

3.2. Anti-Aging Property of Modified Asphalt Blends

To examine the anti-aging property of modified asphalt blends, the thin film oven test (TFOT) was conducted. According to the recycling effect and economy of the above-mentioned three rejuvenating agents H, S, G, the proportion of rejuvenating agent for TFOT were set to 3%, 6%, 6% respectively. The test results are shown in Table 7.

Table 7. Anti-aging property of modified asphalt blends.

Properties		Specification Requirement	Modified Asphalt Blends		
			3%H	6%S	6%G
Penetration/0.1 mm		40~60	50.0	50.3	49.3
Softening point/°C		≥60	64.9	64.3	64.2
Ductility/cm		≥20	19.3	21.2	20.7
Tests on residue from TFOT	Penetration ratio/%	≥65	78.0	78.3	79.3
	Ductility/cm	≥15	10.8	8.1	9.7
	Weight change/%	≤±1.0	0.302	0.438	0.369

Table 7 shows that the penetration ratio and weight change of modified asphalt blends after TFOT meet the requirements of China's specifications, except in terms of ductility. The results indicate that, although the properties of aged modified asphalt blended with rejuvenating agent can be recycled, the low-temperature property often cannot meet the requirements after aging, indicating that the modifier has lost part of the original modified effect.

Therefore, modified asphalt recycling should focus on its low-temperature property, requiring rejuvenating agent not only to effectively recover the colloidal structure of aged modified asphalt, but also to repair and supplement the aged modifier. Ultimately, the objective is to recover both the low-temperature cracking resistance and high-temperature rutting resistance of modified asphalt.

3.3. Phase Characteristics of Rejuvenating Agents on Aged Modified Asphalt

Polymer SBS in asphalt absorbs light oil to form the polymer phase. Under the fluorescence excitation, the polymer phase reflects longer wavelength light, appearing white. In contrast, the asphalt phase does not reflect any light, appearing black. So the asphalt phase and polymer phase are clearly distinguished by fluorescence microscope. As the reflected light field is applied, there is no damage to the polymer phase distribution in the asphalt. Hence, the fluorescence microscope gets highly reproducible asphalt micrographs. By analyzing the morphological structure and dispersion state of polymer SBS in asphalt, it is possible to effectively evaluate the role of rejuvenating agent on the aged modified asphalt, and establish the relationship between macroscopic mechanical properties and microscopic characteristics of modified asphalt blends.

To ensure the reliability of fluorescence images and avoid the particularity of individual samples, three slides were made in parallel for each sample. Four representative fluorescence images were taken for each slide.

Table 8 shows the fluorescent micrographs of aged modified asphalt and modified asphalt blends with different types and proportions of rejuvenating agents.

Table 8 shows that the polymer SBS in aged modified asphalt are spherical particles distributed in the asphalt, and this modified asphalt belongs to a single-phase continuous structure, with polymer SBS as dispersed phase and asphalt phase as continuous phase. Compared with modified asphalt blended with different types and proportions of rejuvenating agents, recycling did not change its phase structure, still belonging to single-phase continuous structure. The SBS particle's uniformity is satisfactory. With the naked eye only, one cannot see the difference of SBS particle in appearance and shape. The images have to be further processed to quantify potential differences.

Professional image processing software (Image-Pro Plus) is applied to analyze and process the fluorescence images of modified asphalt blends, which can segment, count and measure the images. The area ratio of polymer SBS particles is the quantitative evaluation indexes to describe the phase characteristics and recycling mechanism of modified asphalt. Here the area ratio refers to the percentage of polymer SBS particles area to total area, indirectly reflecting the SBS swelling degree in the asphalt.

Table 8. Fluorescence micrographs of modified asphalt.

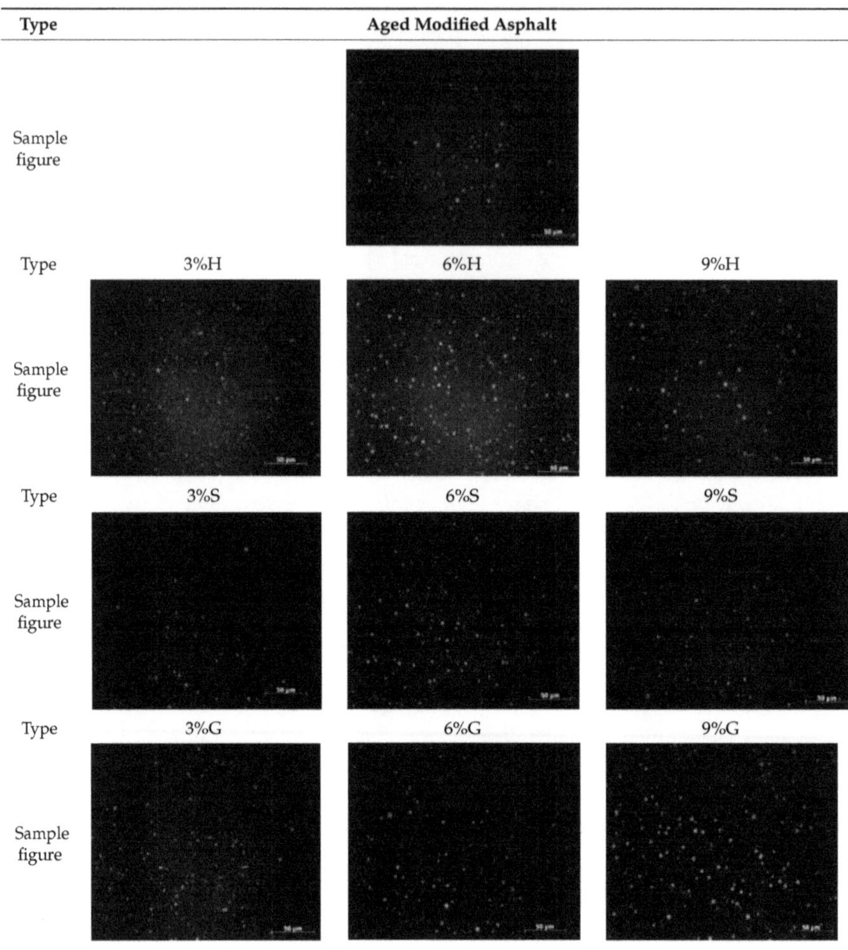

Table 9 shows the phase characteristics of modified asphalt blended with different proportions of rejuvenating agents.

Table 9. Phase characteristics of modified asphalt blends.

Indexes	Aged Modified Asphalt	Types of Modified Asphalt Blends								
		3%H	6%H	9%H	3%S	6%S	9%S	3%G	6%G	9%G
Area ratio of SBS particles/%	2.753	3.679	2.980	2.030	3.832	3.672	3.147	4.207	4.539	4.081
Standard deviation/%	0.38	0.51	0.29	0.31	0.54	0.57	0.48	0.46	0.59	0.32

From Table 9, the area ratio of polymer SBS particles increases at first and then decreases, presumably because the proportion of rejuvenating agent was small, and the swelling effect of SBS particles continued with the action of light oil. With the increasing of the proportion of rejuvenating agent, the swelling effect reaches a certain extent, and excess of rejuvenating agent decreases the relative content of SBS particles in asphalt, resulting in the area ratio of SBS particles beginning to decrease.

A comparison was made of the relationship of macroscopic properties shown in the Section 3.1 and the microscopic characteristics of modified bitumen blends shown in Table 9 based on the connection points of rejuvenating agent proportion. The comparison indicates that the variation of area ratio of SBS particles is consistent with the variation in tenacity.

It is concluded that with an increase of proportion of rejuvenating agent, the trend of area ratio is similar to tenacity, but this may be just an apparent phenomenon. Software statistical analysis was conducted at bivariate significance level of 0.05. The results of modified asphalt blends are shown in Table 10.

Table 10. Correlation of area ratio of SBS particles and tenacity with proportion of rejuvenating agents.

Rejuvenating Agent Type	Correlation Coefficient	Significance
H	0.953	0.047
S	0.983	0.017
G	0.957	0.043

Table 10 shows that the correlation coefficient between area ratio of SBS particles and tenacity are all over than 0.95, indicating that the positive correlation is strong. The significance parameters are all less than 0.05, indicating that the two indexes have significant correlation.

According to the above research into penetration, softening point, 5 °C ductility, viscosity, and tenacity of recycled asphalt and economy, the proportion of rejuvenating agent S for recycled mixtures is set to 6%.

3.4. Freeze-Thaw Split Test Results

The moisture susceptibility of recycled mixtures is evaluated by freeze-thaw split test, and the tensile strength ratio (TSR) is used as the evaluation index. The results are shown in Table 11, and * in the study represents the addition of rejuvenating agent.

Table 11. Freeze-thaw split test results of recycled mixtures.

RSMAP Content	No Freeze-Thaw Split Strength/MPa	Freeze-Thaw Split Strength/MPa	TSR/%
0%	0.95	0.87	91.1
30%	1.40	1.12	80.0
30% *	1.28	1.12	87.2
50%	1.28	1.02	79.7
50% *	1.31	1.19	90.8
70%	1.33	0.89	66.9
70% *	1.09	0.9	82.6

*: the addition of rejuvenating agent.

Table 11 shows that TSR of recycled mixtures decreases significantly with increasing of RSMAP contents. When the RSMAP content is 70%, the TSR is 66.9%, which could not meet China's specification. By adding rejuvenating agent, TSR can increase significantly. This indicates that RSMAP will decrease moisture susceptibility, especially with high RSMAP contents; adding rejuvenating agent can improve the moisture susceptibility of recycled modified mixtures including high RSMAP contents. However, the moisture susceptibility of recycled mixtures is not easy to recover to HMA level.

3.5. Semi-Circular Bending Test Results

Figure 5 graphically illustrates the fracture energy (G_f) and fracture toughness (K_c) of recycled mixtures in SCB test, and RA in the Figure 5 represents rejuvenating agent.

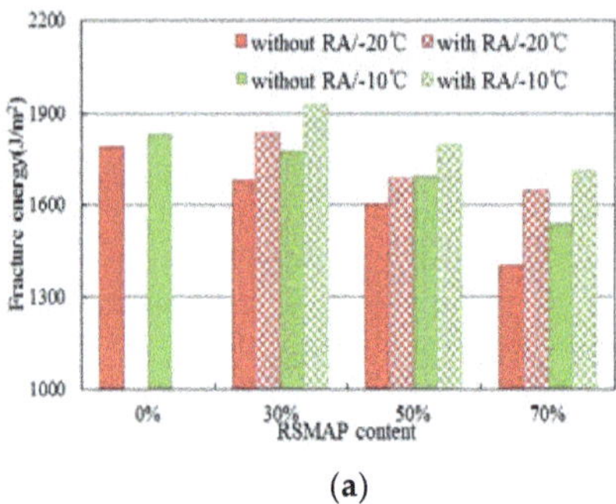

(a)

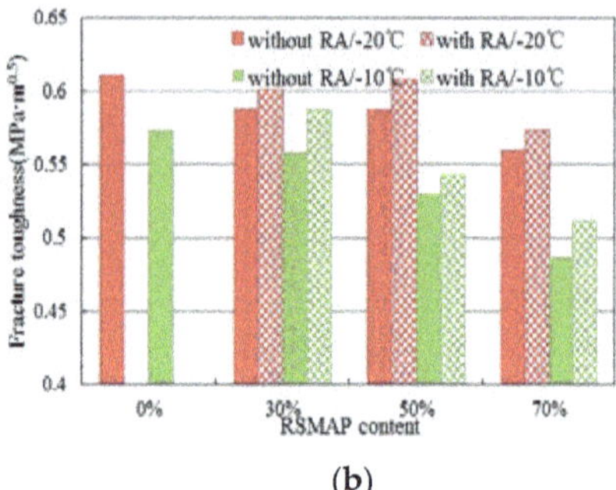

(b)

Figure 5. SCB test results of recycled mixtures. (**a**) Fracture energy; (**b**) Fracture toughness.

Figure 5 shows that the trends for SCB fracture energy and fracture toughness are decreasing with increasing RSMAP content in the mixture at the two temperatures. As the test temperature increases, the fracture energy increases and fracture toughness decreases. When the RSMAP content is 70%, the fracture energy and fracture toughness of recycled mixture without rejuvenating agent decrease by 22% and 9% compared with the control mixtures (RSMAP is 0%). By adding rejuvenating agent, both indexes are improved. However, the two indexes of recycled mixtures with high RSMAP contents are not easy to recover to HMA level.

This indicates that RSMAP has a negative impact on low-temperature properties of recycled mixtures, and the lower the temperature is, the greater the negative impact is. The rejuvenating agent can improve the fracture energy and fracture toughness of recycled mixtures including high RSMAP contents.

3.6. Dynamic Modulus Test Results

The dynamic modulus master curve model parameters (Equation (4)), developed for different recycled mixtures, are presented in Table 12.

Table 12. Dynamic modulus master curve model parameters of recycled mixtures.

RSMAP Content	$\|E^*\|$ Master Curve Parameters					Correlation Coefficient
	δ	α	β	γ	c	R^2
0%	2.074	2.311	−0.986	−0.689	8516.5	0.99
30%	1.573	2.935	−1.109	−0.487	8356.7	0.99
30% *	1.901	2.609	−0.809	−0.515	7249.7	0.99
50%	1.116	3.458	−1.360	−0.389	7943.0	0.99
50% *	1.717	2.809	−1.094	−0.452	8084.1	0.99
70%	0.338	4.223	−1.603	−0.339	8728.1	0.99
70% *	1.358	3.107	−1.229	−0.39	8218.9	0.99

*: the addition of rejuvenating agent.

From Table 12 and based on the nonlinear fitting statistics, it is evident that the dynamic modulus models used for developing the master curves are highly related. The temperature shift factors for a reference temperature 20 °C are shown in Table 13.

Table 13. Temperature shift factors (log$a(T)$) of recycled mixtures.

RSMAP Content	Temperature Shift Factors (log$a(T)$)		
	4 °C	20 °C	40 °C
0%	1.678	0	−1.857
30%	1.647	0	−1.822
30% *	1.429	0	−1.581
50%	1.566	0	−1.732
50% *	1.594	0	−1.763
70%	1.721	0	−1.903
70% *	1.620	0	−1.792

*: the addition of rejuvenating agent.

The dynamic modulus master curves of recycled mixtures with different RSMAP contents are provided in Figure 6. To facilitate comparison, the master curve of control mixture is also provided in Figure 6.

From Figure 6a–c, it's observed that dynamic modulus of all mixtures tested increases with the increase of the loading frequency. In general, the dynamic modulus values of recycled mixtures are significantly higher than control mixture, and the difference is greater with higher RSMAP content. By adding rejuvenating agent, the dynamic modulus decreases slightly. Figure 6d shows that at higher loading frequency, there is no notable difference between the dynamic modulus of the recycled mixtures and control mixture. The results indicate that the rejuvenating agent has an effect on the stiffness of the recycled mixtures generally, but the effect is not obvious. In addition, when the loading frequency reaches a greater value, the difference of recycled mixtures and control mixture tend to be consistent, and include high RSMAP content.

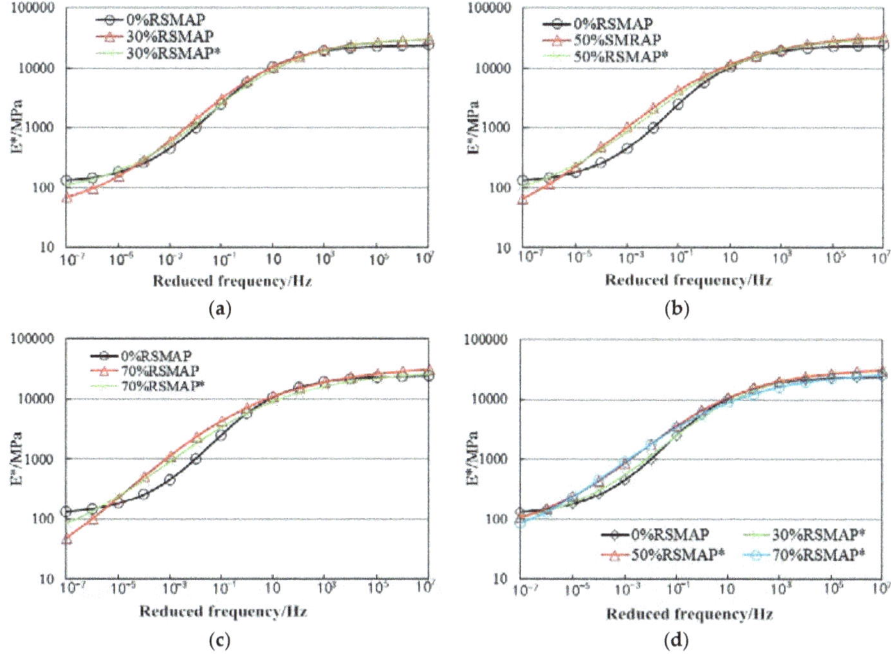

Figure 6. (**a**) Master curves of recycled mixtures with 30% RSMAP; (**b**) Master curves of recycled mixtures with 50% RSMAP; (**c**) Master curves of recycled mixtures with 70% RSMAP; (**d**) Master curves of recycled mixtures with all RSMAP contents.

4. Conclusions

Rejuvenating agent can effectively recycle the properties of aged modified asphalt, and improve the swelling effect between modifier and asphalt. However, the anti-aging property of modified asphalt blends is poor. The relationship between penetration and viscosity of modified asphalt blends shows that the change of penetration will be obvious when the viscosity is less than a certain value.

The recovery degree of penetration and ductility of modified asphalt blends is poorly synchronized when applying rejuvenating agent. It is not appropriate to apply the viscosity and ductility tests to carry out the recycling design at the same time.

Rejuvenating agent cannot change the phase structure of aged modified asphalt, and the SBS particle's uniformity is still satisfactory. On the basis of microscopic analysis, the SBS particles area ratio of modified asphalt blends is significantly correlated with tenacity as the proportion of rejuvenating agent increases.

RSMAP can decrease moisture susceptibility of HMA mixture, especially with high RSMAP content, and adding rejuvenating agent can improve the property. However, the property of recycled mixtures with high RSMAP content is not easy to recover to HMA mixture level.

RSMAP has a negative impact on low-temperature cracking resistance of recycled mixtures, and the lower the temperature is, the greater the negative impact is. Rejuvenating agent can improve the fracture energy and fracture toughness of recycled mixtures including high RSMAP contents. However, the recycled mixtures with high RSMAP content are not easy to recover to HMA mixture level.

In general, the rejuvenating agent has an effect on the stiffness of the recycled mixture, but not an obvious one. When the loading frequency reaches higher value, the difference of recycled mixtures and control mixture tend to be consistent, and include high RSMAP content.

Acknowledgments: This research was supported by the national Science & Technology Support Program "The technology, equipment and demonstration for the application of reclaimed asphalt pavement (No. 2014BAC07B01)" and "Low-carbon construction technology and project demonstration of asphalt pavement (No. 2014BAC07B05)". We state that we have received funds for covering the costs to publish in open access.

Author Contributions: Jie Wang, Yongchun Qin, Songchang Huang and Jian Xu conceived and designed the experiments; Jie Wang and Wei Zeng performed the experiments; Jie Wang and Wei Zeng analyzed the data; Wei Zeng contributed materials; Jie Wang wrote the paper.

Conflicts of Interest: The authors declare no conflict of interest was reported. The founding sponsors had no role in the design of the study; in the collection, analyses, or interpretation of data; in the writing of the manuscript, and in the decision to publish the results".

References

1. Jin, H.; Gao, G.; Zhang, Y.; Zhang, Y.X.; Sun, K.; Fan, Y.Z. Improved properties of polystyrene-modified asphalt through dynamic vulcanization. *Polym. Test.* **2002**, *21*, 633–640. [CrossRef]
2. Zhang, F.; Yu, J. The research for high-performance SBR compound modified asphalt. *Constr. Build. Mater.* **2010**, *24*, 410–418. [CrossRef]
3. Khodaii, A.; Mehrara, A. Evaluation of permanent deformation of unmodified and SBS modified asphalt mixtures using dynamic creep test. *Constr. Build. Mater.* **2009**, *23*, 2586–2592. [CrossRef]
4. Khodaii, A.; Moghadas Nejad, F.; Forough, S.A.; Ahari, A.S. Investigating the effects of loading frequency and temperature on moisture sensitivity of SBS-modified asphalt mixtures. *J. Mater. Civ. Eng.* **2014**, *26*, 897–903. [CrossRef]
5. Kök, B.V.; Çolak, H. Laboratory comparison of the crumb-rubber and SBS modified bitumen and hot mix asphalt. *Constr. Build. Mater.* **2011**, *25*, 3204–3212. [CrossRef]
6. Huang, W.D.; Zheng, M.; Tang, N.P.; Shan, Z. Comparson of evaluation parameters for high temperature performance of SBS modified asphalt. *J. Build. Mater.* **2017**, *20*, 139–144.
7. Fu, H.Y.; Xie, Z.D.; Yu, M.; Dou, D.Y.; Li, L.F.; Yao, S.D. Dynamic shear rheologic properties of SBS modified asphalt. *J. Highw. Transp. Res. Dev.* **2005**, *22*, 9–12.
8. Hu, Z.W.; Sun, H.W.; Ma, S.J. Hot in-place recycling performance of SBS modified SMA mixture. *J. Highw. Transp. Res. Dev.* **2012**, *29*, 29–47.

9. He, Z.Y.; Ran, L.F.; Cao, Q.X. Mechanism study of recycle of SBS modified asphalt based on spectrum analysis. *J. Build. Mater.* **2015**, *18*, 900–904.
10. Jamshidi, A.; Hamzah, M.O.; Shahadan, Z.; Yahaya, A.S. Evaluation of the rheological properties and activation energy of virgin and recovered asphalt binder blends. *J. Mater. Civ. Eng.* **2015**, *27*. [CrossRef]
11. Huang, S.C.; Turner, T.F. Aging Characteristics of RAP Blend Binders: Rheological Properties. *J. Mater. Civ. Eng.* **2014**, *26*, 966–973. [CrossRef]
12. Zhou, Z.G.; Yang, Y.P.; Zhang, Q.P.; Gao, J.Y. Recycling behavior of recycling agent on aged asphalt. *J. Traffic Transp. Eng.* **2011**, *11*, 10–16.
13. Cong, P.; Luo, W.; Xu, P.; Zhao, H. Investigation on recycling of SBS modified asphalt binders containing fresh asphalt and rejuvenating agents. *Constr. Build. Mater.* **2015**, *91*, 225–231. [CrossRef]
14. Sun, L.; Wang, Y.; Zhang, Y. Aging mechanism and effective recycling ratio of SBS modified asphalt. *Constr. Build. Mater.* **2014**, *70*, 26–35. [CrossRef]
15. Cong, P.; Zhang, Y.; Liu, N. Investigation of the properties of asphalt mixtures incorporating reclaimed SBS modified asphalt pavement. *Constr. Build. Mater.* **2016**, *113*, 334–340. [CrossRef]
16. Winkle, C.V.; Mokhtari, A.; Lee, H.; Williams, R.C.; Schram, S. Laboratory and field evaluation of HMA with high contents of recycled asphalt pavement. *J. Mater. Civ. Eng.* **2017**, *29*. [CrossRef]
17. Goli, A.; Ziari, H.; Amini, A. Influence of Carbon Nanotubes on Performance Properties and Storage Stability of SBS Modified Asphalt Binders. *J. Mater. Civ. Eng.* **2017**, *29*, 654–660. [CrossRef]
18. Kang, A.H.; Kou, C.J.; Liu, X.Y. Evaluation methods of dispersed phase uniformity in SBS modified asphalt based on digital image analysis technology. *J. Sichuan Univ.* **2014**, *46*, 172–176.
19. Huang, W.; Sun, L. Phase structure of modified asphalt. *J. Highw. Transp. Res. Dev.* **2001**, *18*, 1–3.
20. Sengoz, B.; Isikyakar, G. Evaluation of the properties and microstructure of SBS and EVA polymer modified bitumen. *Constr. Build. Mater.* **2008**, *22*, 1897–1905. [CrossRef]
21. Zhang, D.; Zhang, H.; Shi, C. Investigation of aging performance of SBS modified asphalt with various aging methods. *Constr. Build. Mater.* **2017**, *145*, 445–451. [CrossRef]
22. JTG E20-2011. *Standard Test Methods of Bitumen and Bituminous Mixtures for Highway Engineering*; China Communication Press: Beijing, China, 2011.
23. JTG E42-2005. *Test Methods of Aggregate for Highway Engineering*; China Communication Press: Beijing, China, 2005.
24. Zegeye, E.T.; Moon, K.H.; Turos, M.; Clyne, T.R.; Marasteanu, M.O. Low temperature fracture properties of polyphosphoric acid modified asphalt mixtures. *J. Mater. Civ. Eng.* **2012**, *24*, 1089–1096. [CrossRef]
25. Wu, Z.; Mohammad, L.N.; Wang, L.B.; Mull, M.A. Fracture resistance characterization of superpave mixtures using the semi-circular bending test. *J. ASTM Int.* **2006**, *2*, 1–15. [CrossRef]
26. Li, X.J.; Marasteanu, M.O.; Kvasnak, A.; Jason, B. Factors study in low-temperature fracture resistance of asphalt concrete. *J. Mater. Civ. Eng.* **2010**, *22*, 145–152. [CrossRef]
27. Khosravifar, S.; Haider, I.; Afsharikia, Z.; Schwartz, C.W. Application of time–temperature superposition to develop master curves of cumulative plastic strain in repeated load permanent deformation tests. *Int. J. Pavement Eng.* **2015**, *16*, 214–223. [CrossRef]
28. Ghabchi, R.; Singh, D.; Zaman, M.; Hossain, Z. Laboratory characterisation of asphalt mixes containing RAP and RAS. *Int. J. Pavement Eng.* **2016**, *17*, 829–846. [CrossRef]
29. Kupolati, W.K. Characterization of bitumen extracted from used asphalt pavement. *Eur. J. Sci. Res.* **2010**, *7*, 1663–1672. [CrossRef]

Article

Three Dimensional Digital Sieving of Asphalt Mixture Based on X-ray Computed Tomography

Chichun Hu [1,*], **Jiexian Ma** [1] **and M. Emin Kutay** [2]

[1] College of Civil and Transportation Engineering, South China University of Technology, Wushan Road, Guangzhou 510641, China; ma.jiexian@mail.scut.edu.cn

[2] Department of Civil and Environmental Engineering, Michigan State University, 3554 Engineering Blvd., East Lansing, MI 48824, USA; kutay@egr.msu.edu

* Correspondence: cthu@scut.edu.cn; Tel.: +86-20-8711-1030

Academic Editor: Feipeng Xiao
Received: 30 June 2017; Accepted: 12 July 2017; Published: 18 July 2017

Abstract: In order to perform three-dimensional digital sieving based on X-ray computed tomography images, the definition of digital sieve size (DSS) was proposed, which was defined as the minimum length of the minimum bounding squares of all possible orthographic projections of an aggregate. The corresponding program was developed to reconstruct aggregate structure and to obtain DSS. Laboratory experiments consisting of epoxy-filled aggregate specimens were conducted to investigate the difference between mechanical sieve analysis and the digital sieving technique. It was suggested that concave surface of aggregate was the possible reason for the disparity between DSS and mechanical sieve size. A comparison between DSS and equivalent diameter was also performed. Moreover, the digital sieving technique was adopted to evaluate the gradation of stone mastic asphalt mixtures. The results showed that the closest proximity of the laboratory gradation curve was achieved by calibrated DSS, among gradation curves based on calibrated DSS, un-calibrated DSS and equivalent diameter.

Keywords: asphalt mixture; aggregate gradation; sieve analysis; image processing; X-ray computed tomography

---

1. Introduction

Aggregate microstructure and gradation have great impacts on the performance of asphalt mixture, which also influences the pavement durability, stiffness, and fatigue behavior [1–3]. Traditionally, gradation is evaluated by passing the aggregates through a series of sieves. The sieve retains particles larger than the sieve opening, while smaller ones pass through. In practice, the non-uniform distribution of coarse and fine aggregate components within the asphalt mixture is called aggregate segregation [4,5]. The solvent extraction method or the ignition oven method is often used to measure segregation. However, these laborious and time-consuming tests involve a specially-equipped laboratory, as well as the use of solvents hazardous to workers' health [6].

With the development of the digital image processing (DIP) technique, which has advantage of high efficiency, convenience, automation, and reducing manpower, many researchers have attempted to evaluate the gradation of asphalt mixture through the DIP technique.

In early years, the research was toward two dimensional (2D) images of vertical or horizontal plane cross-sections that were cut from field cores or laboratory-prepared asphalt concrete (AC) specimens. Different parameters were used to describe the size of aggregate in order to obtain a size distribution. Yue et al. applied the DIP technique to quantify the distribution, orientation, and shape of coarse aggregates in AC mixtures [7]. Area gradations on different cross-sections were obtained based on the minor axis length, major axis length, and equivalent circular diameter.

As the gradation estimated from a cross-section is dependent on the position at which the slice was taken, the average gradations of multiple sections were used to evaluate gradation. Masad et al. characterized the internal structure of AC in terms of aggregate gradation. The volume gradation based on the equivalent circular diameter was obtained from the analysis of 2D vertical sections [8]. In volume calculations, the thickness of the aggregate was taken as the average diameter measured in two dimensions. It was found that two specimens (six sections) was sufficient to satisfactorily capture the actual gradation of coarse aggregate. Bruno et al. estimated mixture gradation using the length of the minor axis of the ellipse that had the same normalized second central moments as the aggregate [6]. The average gradation curve obtained from 22 planar slices was compared with the laboratory gradation.

One of the major problems with the two dimensional DIP technique is that only the area gradation is computed. Since volume gradation has a better correlation with traditional weight gradation, a stereological method was developed to acquire the volume gradation [9,10]. Guo et al. adopted the minor axis of the circumscribed ellipse of the aggregate to calculate the planar gradation of the mixture [10]. Then the planar area gradation was transformed to a three-dimensional volume gradation by the stereological method with the assumption of ellipsoidal particles. The results indicated that aspect ratio of the aggregate affected the stereological coefficients significantly.

Although the X-ray computed tomography (CT) technique has been applied to the study of the internal structure of asphalt mixtures since 1999, it was not until 10 years ago that CT images began to be used in calculation of the three dimensional (3D) size and volume gradation. The improvement in image resolution and particle segmentation algorithms were the main reasons. Kutay et al. presented the filtered watershed transform method to overcome the problem of touching aggregates [11]. Gradations by equivalent diameter, major principal axis, minor principal axis, and average axial length were estimated from the CT images. The results showed that the closest match was achieved by the gradation based on the equivalent diameter. In other areas, Wang et al. developed an image segmentation method for multi-size particles and calculated dimensions along three principal axes, which was determined from the analysis of moments [12]. The gradation curves based on traditional sieve analysis, equivalent diameter, maximum dimension, intermediate dimension, and minimum dimension were compared.

However, these techniques, based on indirect parameters, utilized only partial characteristics of aggregate shape to evaluate gradation, making them unable to exactly capture the real sieve opening that the aggregate passed through. In addition, there is a need to study the potential error brought by indirect parameter in describing particle size.

The aim of this study is to develop 3D digital sieving technique used to evaluate the gradation of the asphalt mixture in X-ray CT images, which considers both the shape of aggregate particles and the square opening of the mechanical sieve. For this purpose, the definition of digital sieve size was proposed and the corresponding program was developed using Matlab™ (Natick, MA, USA) to obtain the digital sieve size. After that, laboratory experiments consisting of epoxy-filled aggregate specimens were conducted to calibrate the program. Moreover, the digital sieving technique was used to estimate the gradation of stone mastic asphalt mixtures.

2. X-ray Computed Tomography

X-ray CT is a non-destructive technique that can be used to conduct 3D reconstruction of a sample and obtain information about its geometry. Compact-225 (YXLON, Hamburg, Germany) industrial CT equipment at the road material laboratory of South China University of Technology was used in this work (Figure 1). X-rays from different directions were detected through detectors, processed by electronic components, and then saved in a computer. The X-ray intensities are measured before they enter the specimen and after they penetrate through it. The attenuation intensity depends on the overall linear attenuation properties of the penetrated material. The attenuation or intensity variation was identified as the contour information, which was processed to obtain the CT image for each layer.

The CT image is the spatial distribution of the linear attenuation coefficients, and brighter regions indicate a higher value of the attenuation coefficient. Therefore, the differentiation of features within the specimen is possible because the linear attenuation coefficient at each point depends directly on the density of the specimen at that point [13,14].

Figure 1. Industrial CT.

The linear attenuation coefficient varies with the component and density of the material. CT is sensitive for characterizing materials with different densities. In a typical CT image of an asphalt mixture, the aggregate is the brightest, followed by the asphalt mastic, and then the air voids. The CT slices can be put together for 3D reconstruction. There are four parameters that are critical to CT image quality: spatial resolution, comparison resolution, noise, and ring artifact [15].

3. Three Dimensional Reconstruction of Aggregates

After CT slices are obtained from industrial CT equipment, the traditional image threshold method can be used to separate the aggregates from the mastic and air voids. Based on the binary image of aggregates, connected components will be detected and labeled. However, in an asphalt mixture where the aggregates are in close proximity to each other, multiple aggregates are often incorrectly considered and labeled as a single particle. In order to eliminate the problem of clustering of the aggregates, the digital image processing (DIP) method proposed by Kutay et al. [11] was adopted to separate aggregates in this paper. There were two parameters (standard deviation of the Gaussian filter and the height of the H-maxima transform) that needed careful selection in this segmentation method. Hence, a trial and error process was used to determine these parameters until the best separation was achieved based on visual inspection.

4. Three-Dimensional Digital Sieving

In the laboratory, a gradation test is performed by mechanical size analysis. A representative sample of aggregates is poured into a column of sieves, which is typically placed in a mechanical sieve shaker. During the shaking process, the mechanical device would move the sieves to cause the particles to bounce, tumble, or otherwise turn so as to present different orientations to the sieving surface [16]. As a result, aggregate larger than the sieve opening remains on the sieve, while the smaller ones pass through. Due to numerous particles in the mixture, high-frequency interaction between the sieve mesh and particles, and the very large computation consumption, it is impractical to simulate the laboratory sieve analysis based on the mechanical model. However, if the sieving process is possible to be simplified as a mathematical problem, the digital gradation test could be solved by numerical calculation.

In order to simplify the problem, an assumption related to passage judgment of aggregate is made first. When an aggregate is right above a sieve opening, whether it can pass that sieve opening is assumed to be equivalent to whether its orthographic projection on the sieve plate can be surrounded by four sides of the square opening. Apparently, this could be determined by the length of the minimum bounding square (MBS) of the projection area. If the length of the MBS of the projection is smaller than the sieve size, the aggregate passes through. It should be noted that the MBS is obtained when the length of circumscribed square achieves the minimum value. Though the problem of particle passage judgment is solved, the displacement of aggregate during shaking is not considered. In the actual mechanical sieving process, the aggregate will experience rotation. Thus, orthographic projection in every direction might be the critical projection that determines the smallest sieve opening that aggregate passes through. In this paper, digital sieve size (DSS) is defined as the minimum length of the minimum bounding squares of all possible orthographic projections. Ideally, each aggregate has only one DSS because it corresponds to the smallest MBS. It could be thought as a virtual sieve opening that aggregate is just able to fit in. Hereinafter, whether an aggregate is able to pass through certain sieve opening is totally depend on the value of DSS of the aggregate.

The numerical method used to obtain the DSS of an aggregate is described as follows:

1. Assume a local rectangular coordinate system $x'y'z'$, as shown in Figure 2, where xyz is the global rectangular coordinate system of the voxel data of aggregate particle, (δ, θ) is used to describe the direction of z' axis, δ is the angle between x axis and the projection of z' axis on xy plane, and θ is the angle between the z' axis and the xy plane.
2. Calculate Euler angles α, β, γ based on (δ, θ). Euler angles are used to describe the rotation of a coordinate system [17].
3. Calculate the rotation matrix R through Euler angles α, β, γ, as shown in Equation (1). Perform coordinate conversion according to Equation (2). Hence, voxel data of the particle can be expressed in the local coordinate system $x'y'z'$ (see Figure 3).

$$R = \begin{bmatrix} \cos\beta\cos\gamma & -\cos\beta\sin\gamma & \sin\beta \\ \sin\alpha\sin\beta\cos\gamma + \cos\alpha\sin\gamma & -\sin\alpha\sin\beta\sin\gamma + \cos\alpha\cos\gamma & -\sin\alpha\cos\beta \\ -\cos\alpha\sin\beta\cos\gamma + \sin\alpha\sin\gamma & \cos\alpha\sin\beta\sin\gamma + \sin\alpha\cos\gamma & \cos\alpha\cos\beta \end{bmatrix}, \quad (1)$$

$$\begin{bmatrix} x' \\ y' \\ z' \end{bmatrix} = R \begin{bmatrix} x \\ y \\ z \end{bmatrix}. \quad (2)$$

4. Project voxel data to $x'y'$ plane (see Figure 3). The three-dimensional problem is successfully converted into a two-dimensional problem.
5. Calculate the minimum bounding square for the aggregate projection (see Figure 3). The process of finding the minimum bounding square is the same as the linear time algorithm for the minimum bounding rectangle [18]. First, the convex polygon of a particle projection area is solved using the Matlab function convhull. This is based on the observation that a side of a minimum enclosing box must be collinear with a side of the convex polygon [18]. After corner points of the convex polygon are obtained, two vertical calipers are rotated [19] around the aggregate projection until the minimum square is found. The length of the minimum square is recorded.
6. Assume another local rectangular coordinate system. Repeat Steps 1–5 until the minimum length is obtained, which is the DSS.

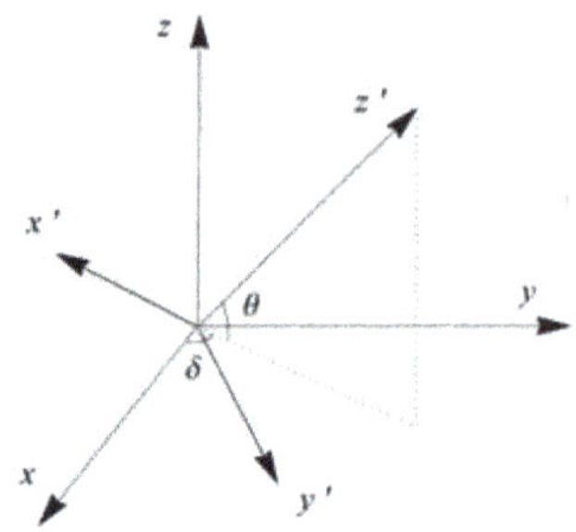

Figure 2. Rotation of the rectangular coordinate system.

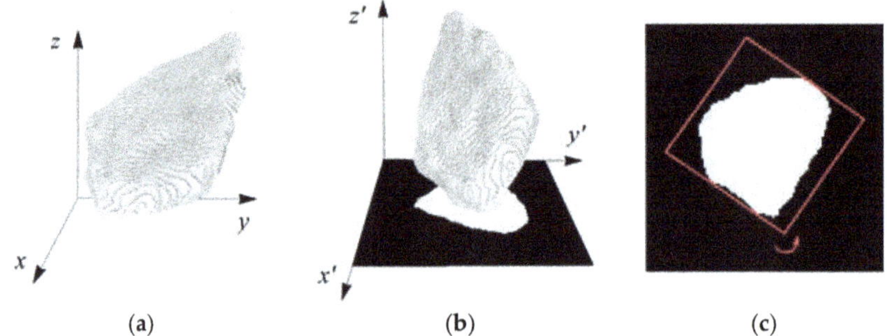

(a) (b) (c)

Figure 3. Illustration of 3D digital sieving of an aggregate: (**a**) voxel data in global coordinate; (**b**) perform coordinate conversion and obtain projection; (**c**) solve the minimum bounding square.

It is noted that, in order to improve the efficiency of numerical method used to calculate DSS, the voxel data of each aggregate is simplified to a point set of its convex hull using Matlab function convhull before the above mentioned steps. The possible error of this simplification will be examined in the following section. The average calculation time of DSS for an aggregate with 100,000 voxels is 1.4 s on a 3.2 GHz single-core processor. Since the numerical method is based on heuristics only, its run time can be further improved by an iterative approach [20].

Two examples of DSS are shown in Figure 4.

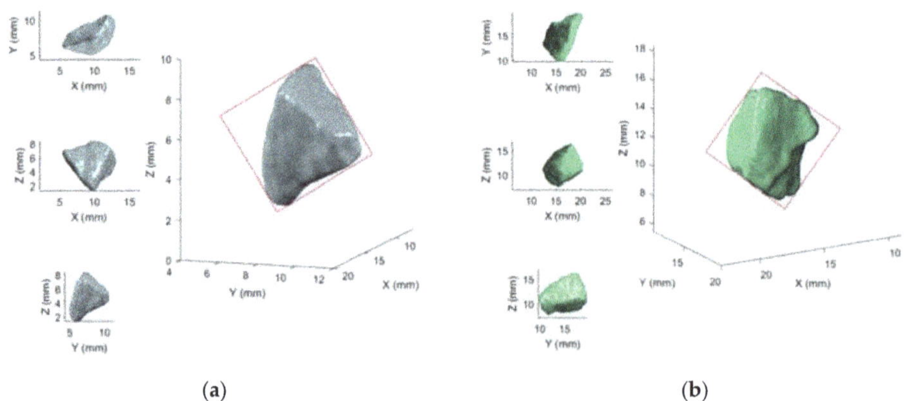

(a) (b)

Figure 4. Examples of digital sieve size: (**a**) aggregate 1; (**b**) aggregate 2.

5. Calibration of Digital Sieving Technique

As the shape of aggregate particles is quite complex and diverse, further investigation of the difference between mechanical sieve analysis and the digital sieving technique developed in this research is essential. In order to accomplish this, four groups of epoxy-filled aggregate specimens were prepared using aggregates retained on four different sieve sizes (4.75 mm, 9.5 mm, 13.2 mm and 16 mm, respectively). The number and type of aggregates for different specimens are shown in Table 1. The specimen dimension was 104 ± 1 mm in diameter, and the gaps between aggregates were filled using epoxy resin (see Figure 5) which would significantly enhance the contrast between the aggregates and background in the X-ray CT images.

Table 1. The number and type of aggregates in specimens.

Grouping	Specimen 1	Specimen 2	Type
Group 1 (16~19 mm)	35	36	Limestone
Group 2 (13.2~16 mm)	73	80	Granite
Group 3 (9.5~13.2 mm)	104	138	Granite
Group 4 (4.75~9.5 mm)	211	-	Granite

Figure 5. Epoxy-filled aggregate specimens.

After the voxel data of each aggregate was extracted from X-ray CT images of epoxy-filled aggregate specimens, the properties such as volume, equivalent diameter, and DSS were computed. The following equations were used to calculate volume and equivalent diameter:

$$V = N_{\text{voxels}} dxdydz, \tag{3}$$

$$ED = 2\left(\frac{3V}{4\pi}\right)^{\frac{1}{3}}, \tag{4}$$

where V is the volume, ED is the equivalent diameter, N_{voxels} is the number of voxels that the given aggregate is made of, and dx, dy and dz are the dimensions of one voxel (mm).

Figure 6 shows the results of the DSS for four different groups. It can be seen that, in each specimen, there was a certain percent of aggregates whose DSS was larger than the upper limit of their mechanical size. That percentage (counted by number) ranged from 5.69% to 31.43% in seven specimens, with the maximum ratio achieved by Group 1, Specimen 1. Additionally, those aggregates having the largest DSS in Group 2, Specimen 2 and Group 3, both specimens, should be noted because their DSS were over the mid-value of their superior sieve size range. For instance, the largest DSS in Group 3, Specimen 1 was 14.8985 mm, and its superior sieve size range was 13.2~16 mm. As the smallest DSS in Group 3, Specimen 1 was quite close to the lower limit (9.5 mm), the above phenomenon was not caused by the selection of the threshold value in the segmentation process that may increase the dimension of aggregate data. It was also observed that there were overlaps between the DSS distribution ranges of three different groups of granite.

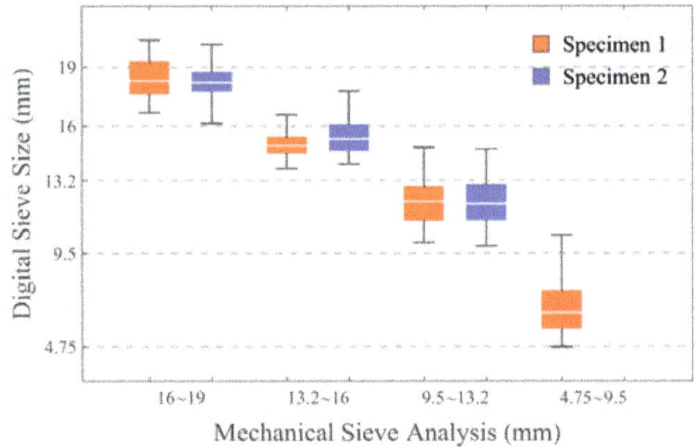

Figure 6. Size of seven specimens based on the digital sieving technique.

In order to investigate the simplification of voxel data of each aggregate to its convex hull, 70 aggregates (10 from each specimen) were chosen to calculate the DSS with or without the simplification. It was found that the relative error brought by the simplification was less than one billionth, indicating that treating the aggregate as convex hulls was entirely feasible for the DSS computation. Considering the definition of DSS involved orthographic projection and minimum bounding square, both hid the convex basis, the above approaching-zero relative error was not unexpected. It can be concluded that the concave surface of aggregate was the possible reason for the disparity between DSS and mechanical sieve size, as shown in Figure 7.

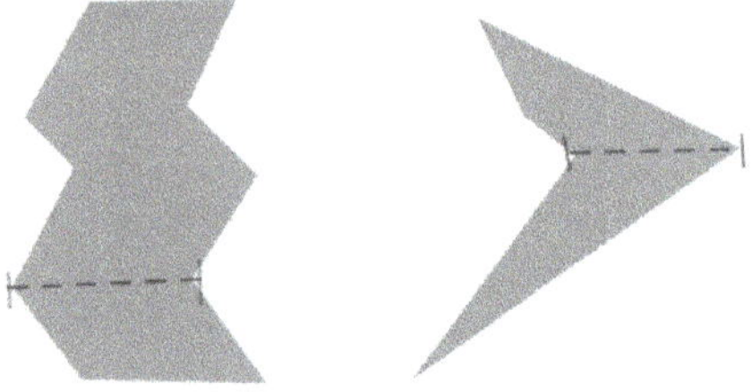

Figure 7. Effect of concave surface on sieving.

Due to difference between the digital and mechanical sieve size, the digital sieving technique was required to be calibrated before it can be employed to evaluate the gradation of the asphalt mixture. For granite, the maximum and minimum DSS in each specimen and their corresponding upper and lower bounds of mechanical sieve size were used for calibration. The result of linear fitting is shown in Figure 8.

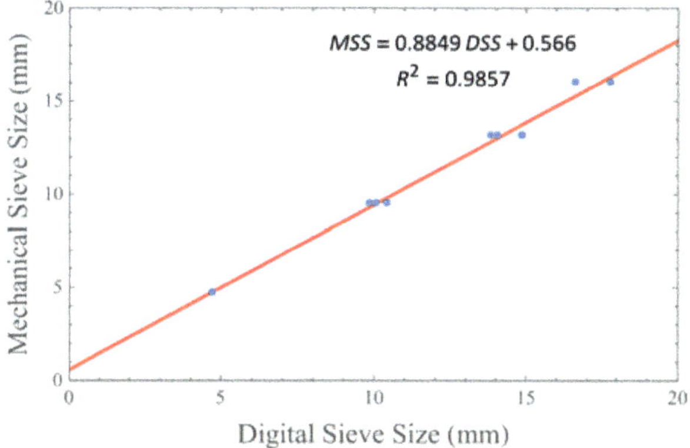

Figure 8. Linear calibration curve for granite.

A comparison between un-calibrated DSS and equivalent diameter was also performed, as demonstrated in Figure 9. The maximum value of absolute difference between DSS and equivalent diameter for Group 1 to Group 4 was 4.272 mm, 3.638 mm, 2.871 mm and 1.398 mm, respectively. As particle size increased, the absolute difference increased. Therefore, estimation of the gradation of the coarse-grained asphalt mixture based on equivalent diameter might lead to significant errors. However, it might be feasible to obtain gradation of the fine-grained mixture based on equivalent diameter with rigorous calibration.

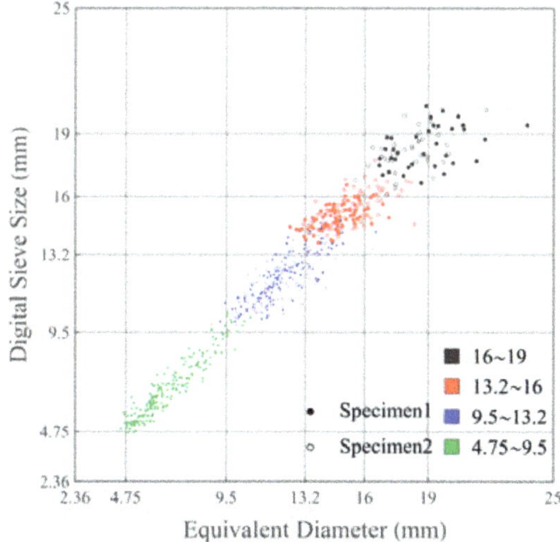

Figure 9. Comparison between digital sieve size and equivalent diameter.

6. Application

The 3D digital sieving technique developed in this paper was adopted to estimate the gradation of two specimens of stone mastic asphalt mixture (SMA-13). The gradation of the asphalt mixture used is listed in Table 2. The granite aggregates utilized here were from the same construction site (Zhanjiang City) as the granite mentioned in last section, thus the calibration curve can be used directly. The specimens were 101.6 mm (diameter) by 63.5 mm (height) in dimension and were compacted using the Marshall compactor. After demolding, specimens were sent for X-ray CT scanning. The resolutions of the resulting images were approximately 0.0878 mm/pixel in the horizontal and vertical direction. The results of 3D reconstruction are shown in Figure 10. It was observed that most of clustering coarse aggregates were separated successfully, whereas a certain amount of fine aggregates still adhered to the surface of coarse aggregate. The emphasis of future study should be placed upon the segmentation of aggregates in X-ray CT images. In addition, the aggregate structure (see Figure 10) can also be used to establish mechanical model for micromechanical behavior analysis. For example, after aggregate and mastic phases are identified from 3D images, triangular surface meshes for each phase can be created, and then finite element meshes can be generated [21].

Considering the breakage of aggregates during Marshall compaction has an effect on gradation [22], aggregates were extracted from mixtures by ignition method (T0735-2011 [23]) after scanning, and laboratory gradations were determined (T0725-2000 [24]) for further analysis.

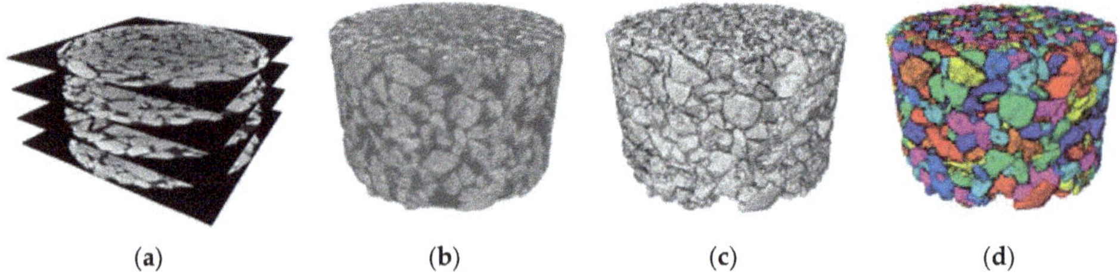

(a) (b) (c) (d)

Figure 10. 3D reconstruction of aggregate structure: (**a**) CT slices; (**b**) volume rendering; (**c**) binary image after thresholding; and (**d**) connected-component detection.

Table 2. Gradation of the mixture.

Sieve size (mm)	16	13.2	9.5	4.75	2.36	1.18	0.6	0.3	0.15	0.075
Passing ratio (%)	10	95	62.5	27	20	19	16	13	12	10

To convert the graphical volume-based gradation to a realistic weight-based gradation, it was assumed that all the aggregates in the specimen have the same specific gravity. Thus, the weight-based gradation can be directly estimated from X-ray CT images. After each aggregate was classified into a different standard size range based on their equivalent diameter or DSS, the total volume of the aggregates in each standard size range and the accumulative passing percentage were computed. It is noted that the aggregates smaller than 4.75 mm will not be considered in the calculation, which is acceptable because they are components of asphalt mortar.

The gradation curves based on different methods are presented in Figures 11 and 12.The results showed that the gradation based on calibrated DSS was closer to the laboratory gradation curve than the curve based on DSS or equivalent diameter, indicating the calibration of DSS was necessary. In the case of specimen 1 (see Figure 11), although a significant difference existed between the mix design and laboratory gradation, the calibrated DSS was able to capture the variability. Moreover, the gradation curves based on DSS and equivalent diameter were quite close. This result was owing to the finding in Figure 9 that there was a good statistical correlation between these two parameters.

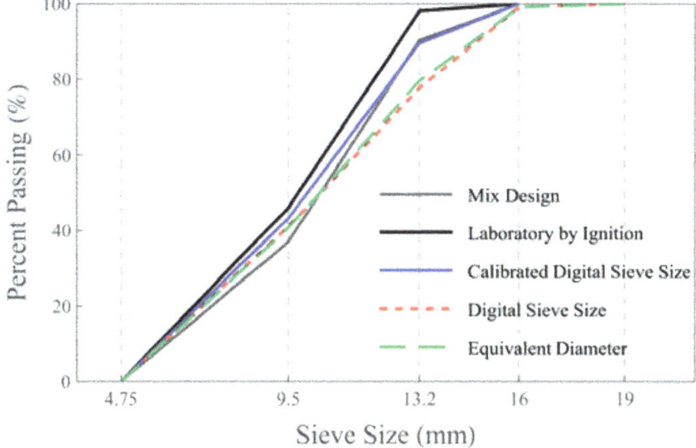

Figure 11. Gradation curves for Specimen 1.

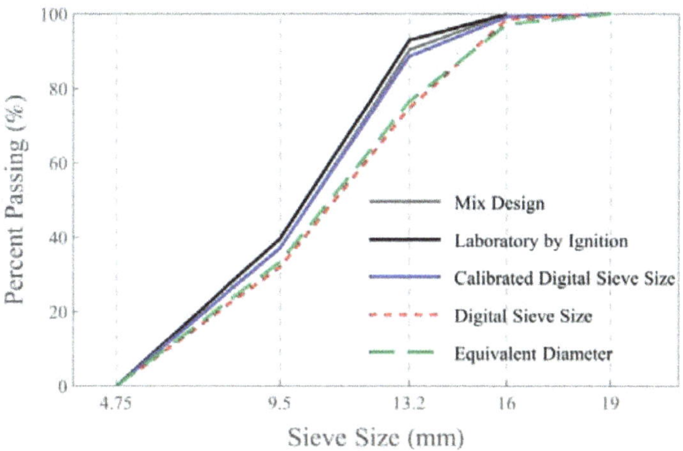

Figure 12. Gradation curves for Specimen 2.

7. Conclusions

The aim of this study was to develop 3D digital sieving technique used to estimate the gradation of asphalt mixture in X-ray CT images. For this purpose, a concept named digital sieve size (DSS) was proposed, which was defined as the minimum length of the minimum bounding squares of all possible orthographic projections of an aggregate. The corresponding numerical method used to obtain DSS was introduced. In order to investigate the difference between mechanical sieve analysis and the digital sieving technique, four groups of epoxy-filled aggregate specimens were scanned and analyzed. As a convex basis lurked in the definition of DSS, there were a certain percent of aggregates in each specimen with their DSS larger than the upper limit of their mechanical size in the result. It was suggested that the concave surface of aggregate was the possible reason for the disparity between DSS and mechanical sieve size. After a linear calibration of DSS was performed for granite, the digital sieving technique was adopted to evaluate the gradation of stone mastic asphalt mixtures. The results showed that the closest proximity of the laboratory gradation curve was achieved by calibrated DSS among the gradation curves based on calibrated DSS, un-calibrated DSS, and equivalent diameter.

Apart from good reliability after calibration and being easy to understand, the major advantage of DSS is that it considers the shape of the aggregate and the square opening of the mechanical sieve, as compared to the equivalent diameter. Although the benefit of DSS for a dense-grained asphalt mixture is weakened by the large quantity and volume summation of aggregates in each standard size range, DSS is promising in capturing size distribution of coarse/medium-grained mixture. Moreover, DSS is much more sensitive to the clustering of coarse aggregates than the equivalent diameter since clustering particles often have strange shapes and large aspect ratios. The gradation curve based on DSS can be used to select the optimum parameters for the segmentation algorithm.

Acknowledgments: Supports provided by National Natural Science Foundation (51578248) and Pearl River S&T Nova Program of Guangzhou were greatly appreciated.

Author Contributions: Chichun Hu proposed the idea, designed the experiments and simulation; Jiexian Ma performed the experiments and simulation; M. Emin Kutay contributed image analysis tools and provided simulation suggestions.

Conflicts of Interest: The authors declare no conflict of interest.

References

1. Masad, E.; Olcott, D.; White, T.; Tashman, L. Correlation of fine aggregate imaging shape indices with asphalt mixture performance. *Transp. Res. Rec.* **2001**, *1757*, 148–156. [CrossRef]
2. Sefidmazgi, N.R.; Tashman, L.; Bahia, H. Internal structure characterization of asphalt mixtures for rutting performance using imaging analysis. *Road Mater. Pavement* **2012**, *13*, 21–37. [CrossRef]
3. Haddock, J.; Pan, C.; Feng, A.; White, T. Effect of gradation on asphalt mixture performance. *Transp. Res. Rec.* **1999**, *1681*, 59–68. [CrossRef]
4. Williams, R.; Duncan, G.; White, T. Hot-mix asphalt segregation: Measurement and effects. *Transp. Res. Rec.* **1996**, *1543*, 97–105. [CrossRef]
5. American Association of State Highway and Transportation Officials (AASHTO). *Segregation: Causes and Cures for Hot Mix Asphalt*; AASHTO: Washington, DC, USA, 1997.
6. Bruno, L.; Parla, G.; Celauro, C. Image analysis for detecting aggregate gradation in asphalt mixture from planar images. *Constr. Build. Mater.* **2012**, *28*, 21–30. [CrossRef]
7. Yue, Z.Q.; Bekking, W.; Morin, I. Application of digital image processing to quantitative study of asphalt concrete microstructure. *Transp. Res. Rec.* **1995**, *1492*, 53–60.
8. Masad, E.; Muhunthan, B.; Shashidhar, N.; Harman, T. Internal structure characterization of asphalt concrete using image analysis. *J. Comput. Civ. Eng.* **1999**, *13*, 88–95. [CrossRef]
9. Wu, W.; Wang, D.; Zhang, X. Estimating the gradation of asphalt mixtures using X-ray computerized tomography and stereology method. *Road Mater. Pavement* **2011**, *12*, 699–710. [CrossRef]
10. Guo, Q.; Bian, Y.; Li, L.; Jiao, Y.; Tao, J.; Xiang, C. Stereological estimation of aggregate gradation using digital image of asphalt mixture. *Constr. Build. Mater.* **2015**, *94*, 458–466. [CrossRef]
11. Kutay, M.E.; Arambula, E.; Gibson, N.; Youtcheff, J. Three-dimensional image processing methods to identify and characterise aggregates in compacted asphalt mixtures. *Int. J. Pavement Eng.* **2010**, *11*, 511–528. [CrossRef]
12. Wang, Y.; Lin, C.L.; Miller, J.D. 3D image segmentation for analysis of multisize particles in a packed particle bed. *Powder Technol.* **2016**, *301*, 160–168. [CrossRef]
13. Masad, E.; Jandhyala, V.K.; Dasgupta, N.; Somadevan, N.; Shashidhar, N. Characterization of air void distribution in asphalt mixes using X-ray computed tomography. *J. Mater. Civ. Eng.* **2002**, *14*, 122–129. [CrossRef]
14. Hu, C.; Youtcheff, J.; Wang, D.; Zhang, X.; Kutay, E.; Thyagarajan, S. Characterization of asphalt mixture homogeneity based on X-ray computed tomography. *J. Test. Eval.* **2012**, *40*, 1–7. [CrossRef]
15. Krumm, M.; Kasperl, S.; Franz, M. Reducing non-linear artifacts of multi-material objects in industrial 3D computed tomography. *NDT & E Int.* **2008**, *41*, 242–251.
16. ASTM C136/C136M-14. *Standard Test Method for Sieve Analysis of Fine and Coarse Aggregates*; ASTM International: West Conshohocken, PA, USA, 2014.
17. Goldstein, H. *Classical Mechanics*, 3rd ed.; Addison-Wesley Publishing: Reading, PA, USA, 2001.

18. Freeman, H.; Shapira, R. Determining the minimum-area encasing rectangle for an arbitrary closed curve. *Commun. ACM* **1975**, *18*, 409–413. [CrossRef]
19. Toussaint, G.T. Solving geometric problems with the rotating calipers. In Proceedings of the IEEE Melecon'83, Athens, Greece, 24–26 May 1983.
20. Chan, C.K.; Tan, S.T. Determination of the minimum bounding box of an arbitrary solid: An iterative approach. *Comput. Struct.* **2001**, *79*, 1433–1449. [CrossRef]
21. Hu, J.; Qian, Z.; Wang, D.; Oeser, M. Influence of aggregate particles on mastic and air-voids in asphalt concrete. *Constr. Build. Mater.* **2015**, *93*, 1–9. [CrossRef]
22. Airey, G.D.; Hunter, A.E.; Collop, A.C. The effect of asphalt mixture gradation and compaction energy on aggregate degradation. *Constr. Build. Mater.* **2008**, *22*, 972–980. [CrossRef]
23. T 0735-2011. *Test Method for Asphalt Content of Asphalt Mixture (Ignition Method)*; China Communication Press: Beijing, China, 2011.
24. T 0725-2000. *Test. Method for Examining Mineral. Gradation of Asphalt Mixture*; China Communication Press: Beijing, China, 2000.

Article

Permeability and Stiffness Assessment of Paved and Unpaved Roads with Geocomposite Drainage Layers

Cheng Li [1,*], Jeramy Ashlock [1], David White [2] and Pavana Vennapusa [2]

1 Department of Civil Construction and Environmental Engineering, Iowa State University, Ames, IA 50011, USA; jashlock@iastate.edu
2 Ingios Geotechnics, Inc., Northfield, MN 55057, USA; david.white@ingios.com (D.W.); pavana.vennapusa@ingios.com (P.V.)
* Correspondence: cheng@iastate.edu; Tel.: +1-515-509-6050

Academic Editor: Zhanping You
Received: 10 June 2017; Accepted: 9 July 2017; Published: 13 July 2017

Abstract: Poor subsurface drainage is frequently identified as a factor leading to the accelerated damage of roadway systems. Geocomposite drainage layers offer an alternative to traditional methods but have not been widely evaluated, especially in terms of the impact of changes on both drainage capacity and stiffness. In this study, both paved and unpaved test sections with and without an embedded geocomposite drainage layer were constructed and tested. The geocomposite layers were installed directly beneath the roadway surface layers to help the rapid drainage of any infiltrated water and thus prevent water entering the underlying foundation materials. The laboratory, field, and numerical analysis results showed that the geocomposite layers increased the permeability of roadway systems by two to three orders of magnitude and that it can effectively prevent the surface and foundation materials from becoming saturated during heavy rainfall events. For the stiffness of the sections, the paved sections with and without a geocomposite layer showed that the composite modulus values measured at the surface were more reflective of the foundation layer support conditions beneath the geocomposite layer than the geocomposite layer itself. The unpaved road section with the geocomposite layer yielded lower composite modulus values than the control section but showed overall better road surface conditions after a rain event due to the improved subsurface drainage condition.

Keywords: geocomposite; permeability; drainage; stiffness; elastic modulus; concrete pavement; asphalt pavement; unpaved road; water infiltration model

1. Introduction

Poor subsurface drainage is frequently identified as a factor leading to the accelerated damage of both paved and unpaved roadway systems. Typically, subsurface drainage is controlled with ditches, edge drains, drainable aggregate layers, and/or roadway crowns. These methods typically require maintenance, and the drainage capacity degrades with time. Geocomposite drainage layers offer an alternative to traditional methods but have not been widely field tested, especially in terms of the impact of changes on both drainage capacity and pavement foundation stiffness.

Geocomposites usually consist of two geotextile outer layers and an internal drainage layer (i.e., geonet) and are typically designed to provide three-dimensional subsurface drainage, soil separation, and filtration [1,2]. Previous lab and field studies have shown the potential for using geocomposite drainage layers to reduce drainage-related damage for both paved and unpaved roads [3–7]. However, the influence of geocomposite layers on the composite stiffness of pavement systems and the drainage performance under transient water flow conditions are not well documented. A numerical analysis showed that geotextiles on either side of a geocomposite layer can decrease plastic deformation through combined mechanistic and hydraulic actions for both paved and unpaved roads, but increasing the surface course or base course thickness reduces the reinforcement benefits [3]. The effects of the location of a geocomposite drainage layer in an asphalt pavement system were studied by Christopher et al. [4], in which it was found that geocomposite drainage layers were quickest at removing water during spring thaws when placed on or within the subgrade. Falling weight deflectometer (FWD) test results from the same study showed that a section with geocomposite in the subgrade had a higher stiffness than sections with geocomposites embedded at higher positions in the pavement foundation.

In this study, paved (with either concrete or asphalt surfaces) and unpaved roadway test sections with and without an embedded geocomposite drainage layer were constructed and tested in Iowa, USA. The geocomposite drainage layers were placed directly beneath the roadway surface layers to help the rapid drainage of any infiltrated water through the pavement joints, cracks, or unpaved road surface materials and thus prevent water entering into the underlying foundation layers. However, since the locations of the relatively soft geocomposite layers are close to the roadway surfaces, the impact of changes on both the drainage capacity and stiffness of the systems need to be evaluated. A laboratory large-scale horizontal permeameter test (HPT) device was developed to measure the horizontal saturated hydraulic conductivities of the system with and without an embedded geocomposite drainage layer. Core-hole permeameter (CHP) and air permeameter test (APT) devices were also used to evaluate the in situ drainage conditions of the test sections. To assess the influence of the geocomposite layers on the composite stiffness of the roadway systems, falling weight deflectometer (FWD) tests were used to measure the composite elastic modulus of the test sections. A two-dimensional (2D) water infiltration model was also developed based on Richard's equation and Haverkamp's soil water retention characteristic (WRC) model to simulate a heavy rain event (i.e., transient flow condition) and to compare the effectiveness of the geocomposite drainage layer with the current practice of improving the drainage conditions of unpaved roads by building and maintaining a 4% crown to enable lateral water flow.

2. Descriptions of Materials and Test Sections

In this study, the two different geocomposite materials (GC-1 and GC-2) consisting of two layers of non-woven geotextile with synthetic polymer geonet cores were evaluated. The properties of the two materials provided by the manufactures are shown in Table 1. Geocomposite GC-1 (Figure 1a,b) possessed a thicker non-woven geotextile and a stiffer geonet core than GC-2 (Figure 1c,d). The stiffness values of the geocomposite materials were not provided by the manufacturers.

Table 1. Properties of the GC-1 and GC-2 geocomposite materials.

Property	GC-1	GC-2	Testing Method
Thickness (mm)	11	9	NA (used Calipers)
Strength (kN)	Exceeds Class 2	17	AASHTO [2] M 288 [8] for GC-1 ASTM [3] D5035 [9] for GC-2
Water Flow Rate (L/min/m^2)	4481	4100	ASTM D 4491 [10]
AOS [1] of the Geotextile (mm)	0.212	0.212	ASTM D 4751 [11]

[1] AOS stands for apparent opening size; [2] AASHTO is the abbreviation of American Association of State Highway and Transportation Officials; [3] ASTM is the abbreviation of American Society for Testing and Materials.

Figure 1. Photos of the (**a**,**b**) Roadrain RD-5 (GC-1) and (**c**,**d**) Macdrain W 1091 (GC-2) geocomposite materials placed beneath paved and unpaved surface courses, respectively.

A total of six test sections were constructed at two different sites, with the Portland cement concrete (PCC) and warm-mixed asphalt (WMA) sections located at the same site. The nominal cross-section profiles of the test sections are shown in Figure 2. For the paved test sections, GC-1 was installed at the top of the base layer (directly beneath the pavement layer) prior to paving. For the unpaved test sections, GC-2 was placed at the interface of the subgrade and surface aggregate layers. The soil index properties of the geomaterials used in this study are summarized in Table 2.

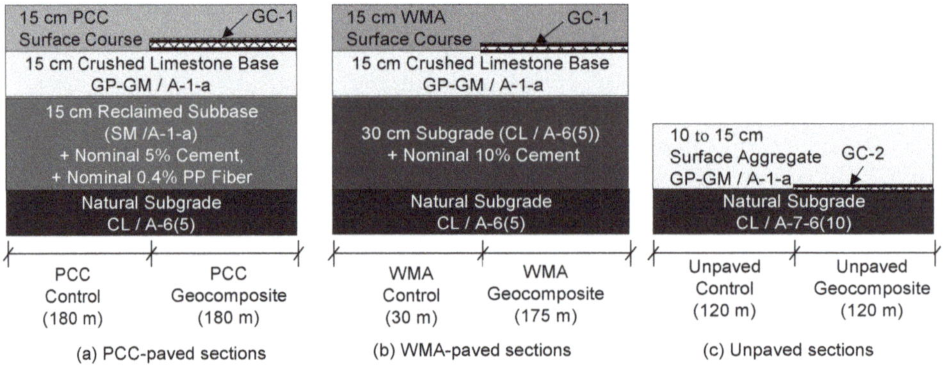

Figure 2. Nominal cross-section profiles (not to scale) of the (**a**) Portland cement concrete (PCC) paved; (**b**) warm-mixed asphalt (WMA) paved and (**c**) unpaved test sections.

Table 2. Summary of the gradation, plasticity, and classifications of the geomaterials used in this study.

Parameter	Crushed Limestone Base	Subgrade of Paved Sections	Unpaved Road Surface Aggregate	Subgrade of Unpaved Sections
Gravel content (%) (>4.75 mm)	65.2	5.3	57.9	0.9
Sand content (%) (4.75–0.075 mm)	27.7	39.7	30.3	39.8
Silt content (%) (4.75–0.005 mm)	3.9	29.3	9.2	30.6
Clay content (%) (<0.005 mm)	3.2	25.7	2.6	28.7
D_{10} (mm)	0.3	–	0.049	–
D_{30} (mm)	3.6	0.08	1.731	0.003
D_{60} (mm)	10.1	0.12	9.720	0.081
Coefficient of uniformity, c_u	33.7	–	198.96	–
Coefficient of curvature, c_c	4.36	–	6.31	–
Liquid limit (%)	Non-plastic	33	Non-plastic	43
Plastic limit (%)		15		22
AASHTO classification	A-1-a	A-6(5)	A-1-a	A-7-6(10)
Unified soil classification system (USCS) group symbol	GP-GM	CL	GP-GM	CL
USCS group name	Poorly graded gravel with silt and sand	Sandy lean clay	Poorly graded gravel with silt and sand	Sandy lean clay

For the PCC-paved sections, discrete fibrillated polypropylene (PP) fibers with 5% Portland cement (PC) were used to stabilize the reclaimed subbase layer. Two WMA sections were constructed at the same site as the PCC sections (Figure 2). The top 30 cm of natural subgrade of the WMA sections were stabilized with 10% Portland cement. The detailed test section designs, construction processes, material properties, and costs are beyond the scope of this paper but are reported in [12,13].

The unpaved road test sections were constructed at a different site. The final thickness of the surface aggregate layer after construction ranged between 10 and 15 cm, and the subgrade had a slightly higher fines content than the subgrade of the paved test sections. The GC-2 geocomposite was placed at the interface of the surface aggregate layer and subgrade. The construction processes and costs of the test sections are described in detail in [14].

3. Test Methods

To compare the drainage conditions of the roadway systems with and without a geocomposite drainage layer and determine inputs for the numerical simulations, a large-scale laboratory horizontal permeameter test (HPT) was conducted to evaluate the saturated horizontal hydraulic conductivity of the systems. For the field-test sections, two innovative devices, core-hole permeameter (CHP) and air permeameter test (APT) devices, designed at Iowa State University, were used to measure the in situ hydraulic conductivities of the different test sections. The influences of the geocomposite layers on the composite stiffness of the roadway systems were assessed using the falling weight deflectometer (FWD) test. The descriptions, theories, and testing procedures of the different testing methods are described in the following sections.

3.1. Laboratory Large-Scale Horizontal Permeameter Test (HPT)

The large-scale horizontal permeameter test (HPT) was developed to measure the saturated hydraulic conductivity of the aggregate materials under horizontal flow conditions because most permeability tests employ vertical flow, which does not accurately represent how water typically drains or flows through the pavement base layers horizontally in the field. The device can simulate direct horizontal flow situations under different pressure heads and is large enough (soil tank dimensions are 1 m × 0.46 m × 0.33 m) to effectively test multiple material layers, as shown in Figure 3. The inside wall of the HPT soil tank has several ribs (1 cm in height) installed perpendicular to the flow direction to prevent water flowing through the interface of the soil specimen and the inside wall of the soil tank.

Figure 3. Photo of the large-scale horizontal permeameter test (HPT) device used in this study.

To quantify the improvement in horizontal drainage offered by the geocomposite layer, representative base material (Unified soil classification system (USCS) group symbol: GP-GM) were collected from the field and compacted in the HPT soil tank with and without a layer of geocomposite. Three HPTs were conducted; one specimen consisted of the base material only, and the other two had one of the two types of geocomposite drainage materials embedded (i.e., GC-1 and GC-2). The testing specimens were compacted to the standard Proctor maximum dry unit weight of 22.0 kN/m^3, determined in accordance with ASTM D698 [15]. The saturated horizontal hydraulic conductivities of the three specimens were measured under four different constant water heads of 50, 100, 150 and 200 mm. In the field, the geocomposite materials are placed at the interface of material layers. However, for the HPT test, the geocomposite layer was embedded at the middle of the test specimen to prevent any boundary issues such as the water flowing through the gap between the geocomposite layer and the bottom of the soil tank.

3.2. Core-Hole Permeameter (CHP) Test

The CHP device (Figure 4a) was developed to assess the in situ drainage capacities of pavement base systems. The test method follows the procedure in ASTM D6391 [16]. In this study, the CHP test results were used to compare the relative drainage capacities of the foundation system with and without a geocomposite layer placed at the pavement/base layer interface. The condition this simulates is the ability to drain any infiltrated water through the pavement surface via joints/cracks. Vennapusa et al. [17] showed that by using the American Association of State Highway and Transportation Officials (AASHTO) 1993 design guide PCC pavement design procedure, an increase in the drainage coefficient (C_d) value of the base material from 1.0 to 1.2 can reduce the required PCC pavement thickness by up to 10%. The CHP test uses the falling head method to measure the in situ hydraulic conductivity of the drainable base and geocomposite drainage layers of the paved test sections after drilling cores, as shown in Figure 4b,c. The test involves drilling a 15 cm diameter core hole through the pavement surface to the underlying base or geocomposite layer, inserting the device and sealing it against the bottom foam ring and interior of the core hole using a rubber tube inflated to 135–175 kPa, and recording the rate of water head loss from the device over a period of 30 to 60 min. The hydraulic conductivity (K) of the tested layer was calculated using Equation (1). The testing procedure is described in detail by Zhang et al. [18].

$$K = R_t G_1 \frac{\ln(H_1/H_2)}{(t_2 - t_1)} \tag{1}$$

$$R_t = 2.2902\left(0.9842^T\right)/T^{0.1702} \tag{2}$$

$$G_1 = \left(\pi d^2/11D_1\right)[1 + a(D_1/4b_1)] \tag{3}$$

where H_1 and H_2 are the effective heads (cm) at time t_1 and t_2 (s), respectively; R_t is the ratio of the kinematic viscosity of the permeant at the temperature of the test during time increment t_1 to t_2 to that of water at 20 °C; T is the temperature of the test permeant (°C); d is the inside diameter of the standpipe (3.6 cm for the top standpipe and 33 cm for the middle standpipe); D_1 is the inside diameter of bottom casing (12.7 cm); b_1 is the thickness of tested layer (cm); and a is equal to +1 for an impermeable base at b_1, 0 for the infinite depth of the tested layer (i.e., $b_1 > 20D_1$), and −1 for a permeable base at b_1.

Figure 4. Photos of (**a**) core-hole permeameter (CHP) device; (**b**) CHP test conducted on the paved test sections and (**c**) GC-1 material after the CHP test.

3.3. Air Permeameter Test (APT)

The APT device was developed for rapid in situ saturated hydraulic conductivity determination for granular materials (Figure 5a). The APT device consists of a contact ring, differential pressure gauges, precision orifices, and a programmable digital display (Figure 5b).

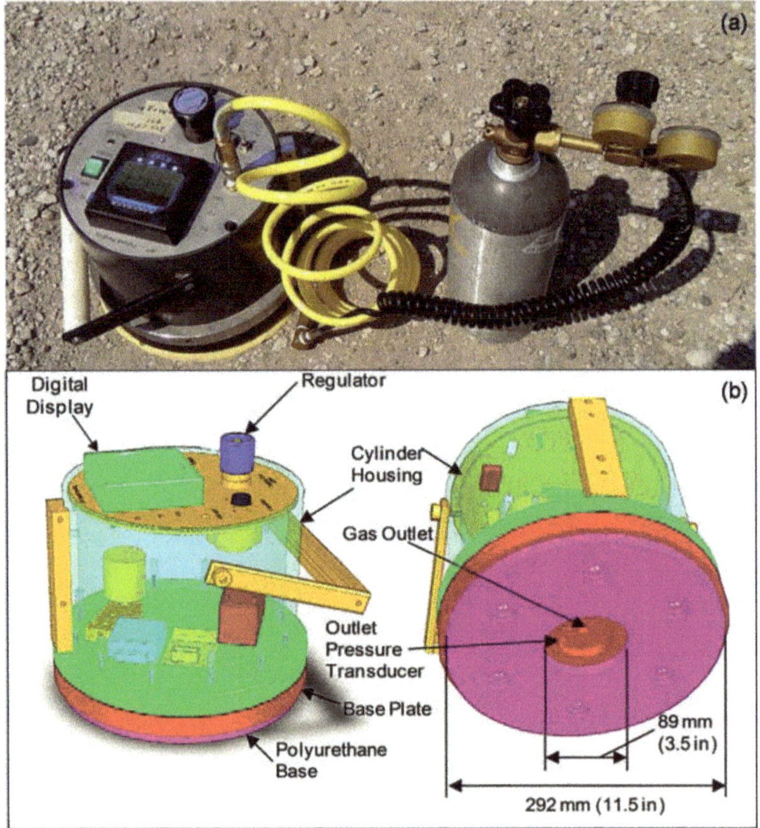

Figure 5. (**a**) Air permeameter test (APT) test conducted on unpaved test section and (**b**) schematic showing the major components of the APT device [13].

The APT device measures the gas pressure on the inlet and outlet sides of the precision orifice and calculates the gas flow rate. From these measurements and the material parameters, the gas permeability of the material being tested can be derived and converted to water saturated hydraulic conductivity (K_{sat}) by taking into account the effect of partial saturation (Equation (4)).

$$K_{sat} = \left[\frac{2\mu_{gas} Q P_{2a}}{r G_0 ({P_{2a}}^2 - {P_{atm}}^2)}\right] \times \frac{\rho g}{\mu_{water}(1 - S_e)^2 \left(1 - {S_e}^{((2+\lambda)/\lambda)}\right)} \tag{4}$$

where K_{sat} is the saturated hydraulic conductivity (cm/s), μ_{gas} is the kinematic viscosity of the gas (Pa-s), Q is the flow rate (cm^3/s), P_{2a} is the absolute gas pressure on the surface (Pa), r is the radius at the outlet (cm), G_0 is the geometric factor determined from test layer thickness, P_{atm} is the atmospheric pressure (Pa), ρ is the density of water (g/cm^3), g is the acceleration due to gravity (cm/s^2), μ_{water} is the kinematic viscosity of water, S_e is the effective saturation, and λ is the Brooks-Corey pore size distribution index.

The derivation of Equation (4) involves an expansion of Darcy's Law considering the compressibility and viscosity of the gas, the gas flow under partially saturated conditions, and the Brooks-Corey pore size distribution index. The development of the testing method, description of the device, and derivations of the theoretical relationship to calculate the hydraulic conductivity from the gas flow are detailed by White et al. [19]. The repeatability of the APT measurements, expressed as a coefficient of variation (COV), is ≤1% [20]. In this study, the APT test was conducted to measure the in situ saturated hydraulic conductivity as a function of depth for the unpaved road test sections.

3.4. Falling Weight Deflectometer (FWD) Test

FWD tests were conducted using a Kuab Model 150 2m device with a 300-mm diameter segmented loading plate to apply a relatively uniform stress distribution. A 40 kN impact load was applied on the roadway surface, which is the AASHTO standard axle load [21]. The induced roadway surface deflection at the center of the loading plate was recorded by a seismometer. For each test location, a single equivalent composite elastic modulus of the roadway system was calculated based on Boussinesq's solution using Equation (5);

$$E_{\text{Composite}} = \frac{(1 - \nu^2)\sigma_0 A}{d_0} \times f \quad (5)$$

where $E_{\text{Composite}}$ is the composite elastic modulus (MPa), d_0 is the measured deflection under the center of the loading plate (mm), σ is the Poisson's ratio (assumed to be 0.4), σ_0 is the normalized applied peak stress (MPa), A is the radius of the plate (mm), and f is the shape factor, assumed to be 2 because of the assumed uniform stress distribution.

4. Laboratory Horizontal Permeameter Test (HPT) Results

The saturated horizontal hydraulic conductivities of the specimens with and without an embedded geocomposite layer were measured under four different constant water heads (i.e., 50, 100, 150 and 200 mm). The relationships between the hydraulic conductivity (K) and hydraulic gradient (i) of the three specimens are shown in Figure 6. The horizontal saturated hydraulic conductivity of the specimen without a geocomposite layer at 20 °C ($K_{20\,°C}$) was approximately 0.8 m/day and remained relatively constant as the hydraulic gradient increased. Compared to the specimen without a geocomposite layer, the geocomposite drainage layer increased the saturated hydraulic conductivity by two to three orders of magnitude.

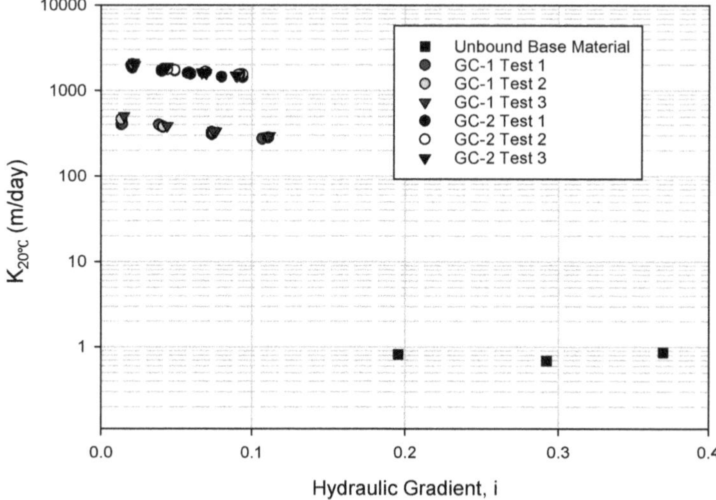

Figure 6. Large-scale horizontal permeameter test (HPT) results of the specimens with and without an embedded geocomposite drainage layer.

Figure 6 also shows that the GC-2 geocomposite had a higher horizontal hydraulic conductivity (~1500 m/day) than the GC-1 geocomposite (~350 m/day). The total thickness of GC-1 is greater than that of GC-2 due to the thicker geotextile on each side of the geonet core for GC-1 However, the geonet core of GC-2 is more porous and has the approximately the same thickness as GC-1.

5. In Situ Permeability and Stiffness of The Test Sections

5.1. In Situ Permeability Assessment

The CHP test was conducted on the PCC and WMA sections to compare the subsurface drainage conditions of the GC-1 geocomposite layer and conventional base materials. The CHP test results showed that the in situ hydraulic conductivity of the GC-1 layer varied between 637 and 695 m/day with a coefficient of variation (COV) of 5.8%, whereas that of the crushed limestone base layer was between 0.1 and 34.2 m/day with a COV of 110%. The hydraulic conductivity of the geocomposite varied within a much smaller range compared to the base material, which may indicate that the segregation of the traditional base material can cause significant variation in the drainage performance of roadway systems.

The APT device was used to quantify the improvement in drainage offered by the geocomposite layer (GC-2) for the unpaved road test sections. For each test location, APTs were performed at different depths within the overlying aggregate layer to determine the variation of saturated hydraulic conductivity (K_{sat}) with depth. The test results are summarized in Figure 7. For the control section, the K_{sat} values at the three test locations were similar (~19 m/day) and remained relatively constant with depth. However, for the geocomposite section, K_{sat} in the surface aggregate layer increased consistently with depth from the roadway surface. The K_{sat} of the GC-2 drainage layer is approximately 1920 m/day, which is more than two orders of magnitude greater than that of the control section. The field-measured K_{sat} of the GC-2 agreed very well with the HPT test results shown in Figure 6.

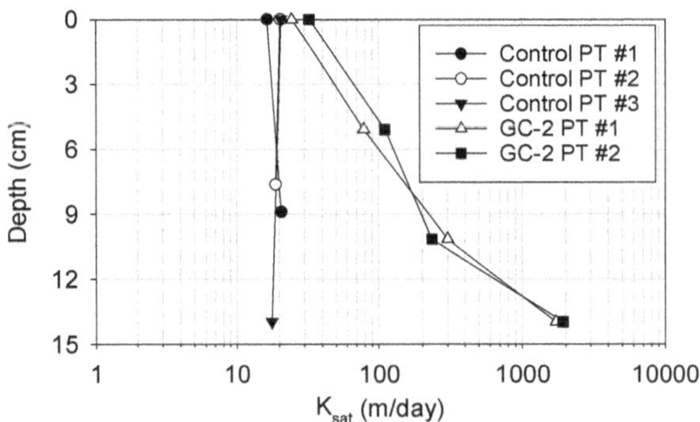

Figure 7. APT-measured saturated hydraulic conductivity versus depth for the surface aggregate layers of the unpaved geocomposite and control sections.

5.2. In Situ Stiffness Assessment

To assess the influence of the geocomposite layer on the composite stiffness of the PCC- and WMA-paved sections, two groups of FWD tests were conducted; (1) on the base layer prior to placing the geocomposite layers on the base layer and (2) on the pavement surfaces after paving, with the geocomposite layer installed at the pavement/base layer interface. The FWD tests were conducted at the same test locations on both the base and the pavement surface layers. A comparison of the two groups of test results shows the influence of the geocomposite on the composite stiffness (Figure 8).

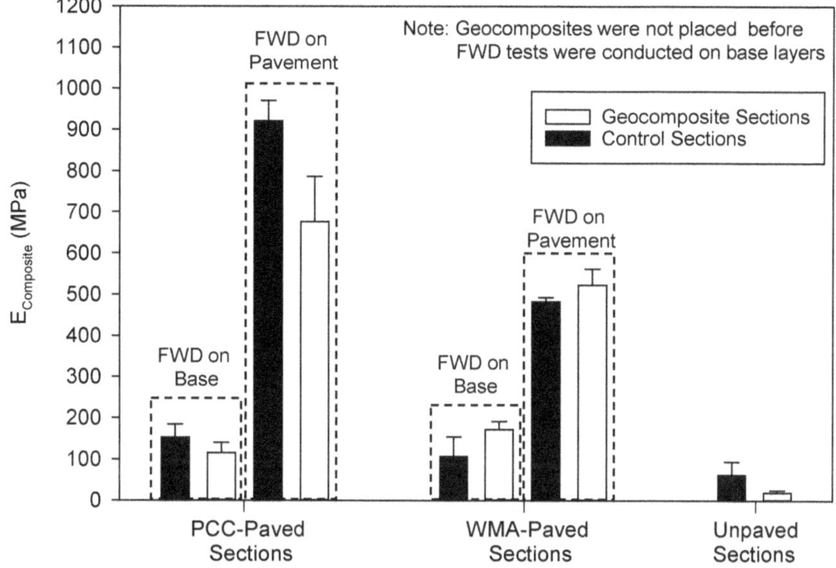

Figure 8. Summary of the falling weight deflectometer (FWD) test results of the PCC-paved, WMA-paved, and unpaved sections.

For the PCC sections prior to paving, the foundation layer in the geocomposite section had a lower average $E_{Composite}$ value than the control section. After placing the geocomposite layer and paving, the FWD tests conducted on the pavement surface showed similar trends, wherein the control section had a higher average $E_{Composite}$ value than the geocomposite section. For the WMA sections, the composite modulus values in the geocomposite section were higher than in the control section for tests conducted both on the foundation layer and the pavement layer. These FWD results from the paved road test sections, although limited, suggest that the composite stiffness measurements obtained at the pavement surface were more reflective of the foundation layer support conditions beneath the geocomposite layer than of the stiffness of the geocomposite layer itself.

FWD tests were also conducted on the unpaved sections after the spring thaw in 2015 because the spring thaw is when unpaved roads are most susceptible to moisture-related damage. Figure 8 shows that the average $E_{Composite}$ of the geocomposite section is less than half that of the control section. However, the roadway surface performance of the geocomposite section was much better than that of the control section based on observations after a heavy rainfall and subsequent traffic, as can be clearly seen in Figure 9. The control section suffered much greater rutting and moisture retention than the geocomposite section. These field observations after the rain event indicate that the improved subsurface drainage conditions offered by the geocomposite layer can effectively prevent the surface material from becoming saturated during a heavy rain event, reduce the amount of water infiltrating into the subgrade, and thus reduce the surface damage. To further examine this hypothesis, a numerical analysis is conducted in the following section to compare the drainage performance of the geocomposite and control sections after a simulated heavy rain event.

Figure 9. Surface conditions of the unpaved road sections after a heavy rainfall (photos taken on 2 September 2015).

6. Numerical Simulation of Drainage Performance under Heavy Rainfall

To improve the drainage conditions of unpaved roads, most agencies choose to build and maintain a 4% roadway crown to enable lateral water flow to drain excess water from the surface course into side drains or ditches. However, this practice requires frequent maintenance, and the effectiveness of the method needs to be quantified. In this study, a numerical simulation was performed to compare the drainage capacity offered by the geocomposite drainage layer with that of a 4% roadway surface crown for unpaved roads. A 2D water infiltration model was programmed in Matlab using Richard's equation and Haverkamp's soil water retention characteristic (WRC) model [22,23].

Richards' equation (Equation (6)) is used to predict volumetric water content (θ) and matric potential (h) changes of the unpaved road surface material during transient water flow;

$$\frac{\partial\theta}{\partial t} = \frac{\partial}{\partial z}\left[K(h)\left(\frac{\partial h}{\partial z} + 1\right)\right] \tag{6}$$

Liquid water flow in the vertical and horizontal directions is expressed using the well-known Darcy-Buckingham flux equations:

$$q = -K(h)\left(\frac{dh}{dz} - \cos\alpha\right) \text{ (for vertical direction)} \tag{7}$$

$$q = -K(h)\left(\frac{dh}{dx} - \sin\alpha\right) \text{ (for horizontal direction)} \tag{8}$$

where q is the water flux in cm/s, $K(h)$ is the unsaturated hydraulic conductivity in cm/s, h is the matric potential in cm, $\frac{dh}{dz}$ or $\frac{dh}{dx}$ is the matric potential gradient, α is the slope of the soil layer in degrees, and the material is assumed to be isotropic and homogenous.

Haverkamp's water retention characteristic (WRC) model was used to estimate relationships between unsaturated hydraulic conductivity, matric potential, and the water content of the unpaved surface material during transient flow, as shown in Equations (9) and (10):

$$\theta(h) = \theta_r + \frac{\alpha(\theta_s - \theta_r)}{\alpha + |h|^\beta} \tag{9}$$

where $\theta(h)$ is the volumetric water content at the corresponding matric potential (h), θ_r is the residual volumetric water content, θ_s is the saturated volumetric water content, α is equal to 1.611×10^6, and β is equal to 3.96.

$$K(h) = K_s\frac{A}{A + |h|^\beta} \tag{10}$$

where $K(h)$ is the unsaturated hydraulic conductivity at the corresponding matric potential (h) in cm/s, K_s is the saturated hydraulic conductivity in cm/s, A is equal to 1.175×10^6, and, β is equal to 4.74.

In the above two equations, the coefficient values (α, β, and A) were those suggested by Haverkamp et al. for granular materials based on infiltration experiments [22]. The residual and saturated volumetric water contents (0.034 and 0.27 cm^3/cm^3) and the saturated hydraulic conductivity (34.56 cm/day) of the base material were measured in the laboratory using the HPT tests. The saturated hydraulic conductivity of the subgrade of the unpaved road sections was measured using the laboratory rigid-wall compaction mold permeameter test in accordance with ASTM D5856 [24]. The measured saturated hydraulic conductivity of the subgrade is 1.8×10^{-8} cm/day, which can be considered impermeable relative to the surface aggregate material.

In the present numerical analysis, the rate of the simulated rainfall event was taken as 12.7 cm/h, which can be considered a violent rain event for the local weather conditions. This rainfall rate is less than the saturated hydraulic conductivity (K_{sat}) of the surface aggregate so surface runoff cannot occur. The rainfall duration was 30 min, and the total simulation time was 90 min. The rainfall direction is assumed to be perpendicular to the ground surface. The left side boundary condition is fixed at zero flux due to symmetry about the centerline of the road section; therefore water can only drain laterally to the bottom and the ditch side, which is open to the air. Two different bottom boundary conditions were prescribed to simulate the test sections with and without a geocomposite layer. For the section without the geocomposite, the natural subgrade (Table 2) was assumed to be impermeable due to the extremely low saturated hydraulic conductivity. For the test section with a geocomposite drainage layer, the hydraulic conductivity of the bottom boundary was updated after each time step of 0.25 s and set equal to the unsaturated hydraulic conductivity of the bottom layer of the surface material at the previous time step because the flow capacity of the geocomposite is much greater than that of the overlying surface aggregate. Therefore, the hydraulic conductivity of the bottom boundary (the geocomposite layer) is controlled by that of the surface course material.

The numerical analysis results in terms of the 2D volumetric water content distributions of half of the roadway cross-section with and without the geocomposite layer are shown in Figure 10. At the end of the simulated rainfall event, the volumetric water content of the control section surface aggregate is close to the lab-measured saturated volumetric water content of 0.27 cm^3/cm^3 and increases with depth (Figure 9a). One hour after the rain stopped, the water content of the bottom layer remained relatively unchanged, as shown in Figure 10a,b, and the shoulder material was not significantly drier than the centerline material.

Compared to the control section with the 4% crown, the system with a geocomposite layer shows a slightly lower water content at the end of the rain event as shown in Figure 10c, but the water content is greatly reduced one hour after the rain stops (Figure 10d). For the particular rainfall event simulated, these comparisons clearly demonstrate that the 4% road crown cannot remove water from the unpaved road surface material as effectively as the geocomposite drainage layer, which significantly improves the subsurface drainage conditions and effectively prevents the unpaved road surface material from becoming saturated.

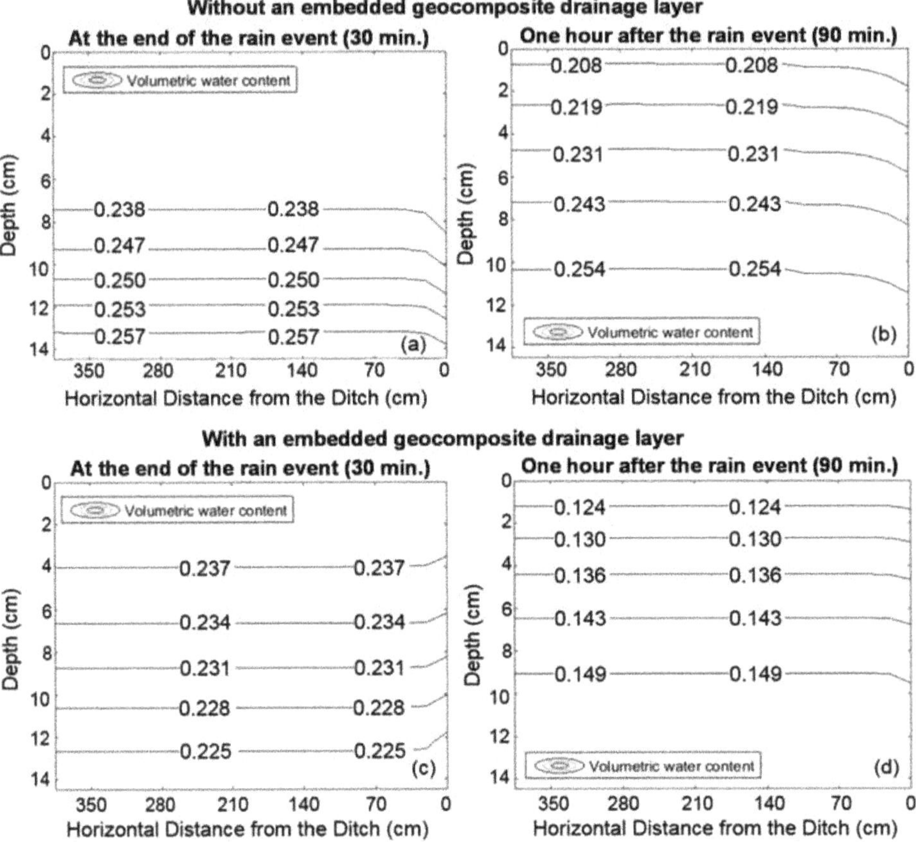

Figure 10. Comparisons of volumetric water content distributions of unpaved road system (**a**,**b**) without and(**c**,**d**) with an embedded geocomposite drainage layer, (**a**,**c**) immediately and (**b**,**d**) one hour after the simulated rainfall event.

7. Summary and Conclusions

This study evaluated the influence of an embedded geocomposite drainage layer on the permeability and stiffness of both paved and unpaved roadway systems using laboratory tests, field tests, and numerical simulations. Both the paved and unpaved test sections constructed using two geocomposite materials were tested in Iowa, USA. A two-dimensional water infiltration model was developed to compare the drainage performance of the geocomposite drainage layer with the 4% roadway surface crown after a simulated heavy rain event. Based on the experimental and numerical analysis data, although limited, some conclusions can be drawn from this study.

For the PCC and WMA sections, the GC-1 geocomposite layer increased the horizontal and vertical saturated hydraulic conductivities by two to three orders of magnitude without significant changes in composite stiffness of the pavement systems. For the unpaved test section, the GC-2 geocomposite layer yielded a lower composite stiffness than the control section due to the relatively softer geonet core of the GC-2. However, the visual observations, field permeability test, and numerical analysis results indicate that the unpaved road section with a geocomposite drainage layer performs better than the control section after a heavy rainfall event and traffic loading, which is attributed to the geocomposite drainage layer rapidly draining water out of the system and effectively preventing the overlying surface material and top layer of subgrade from becoming saturated.

The field permeameter test results indicate that the drainage performance can be significantly influenced by the segregation of the traditional base material, but the geocomposite drainage layer can provide a more uniform subsurface drainage condition. The numerical analysis also showed that the current practice of using the 4% surface crown cannot remove water from the unpaved road surface material as effectively as the geocomposite drainage layer.

The initial construction costs of using the geocomposite for pavement systems might be considered high, but a breakeven cost analysis conducted for the unpaved road project showed that the geocomposite will begin to provide cost savings after 11 years relative to continuation of the current maintenance practices [25]. In addition, the better performance of the roadway system with a geocomposite layer can provide benefits beyond economic ones such as significantly improved ride quality.

Acknowledgments: The research projects were sponsored by the Iowa Department of Transportation and the Federal Highway Administration. The authors would like to thank the Hamilton County Secondary Roads Departments for assistance with constructing the test sections. The assistance of Yang Zhang, Jinhui Hu, and Ji Lu in performing the field tests is also greatly appreciated.

Author Contributions: David White, Pavana Vennapusa, Jeramy Ashlock, and Cheng Li conceived and designed the experiments; David White, Jeramy Ashlock, Pavana Vennapusa, and Cheng Li designed and supervised the constructions of the test sections. Cheng Li, Pavana Vennapusa and Jeramy Ashlock performed the experiments; Cheng Li, Pavana Vennapusa, and Jeramy Ashlock analyzed the data; David White and Pavana Vennapusa contributed materials and analysis tools; Cheng Li, Jeramy Ashlock, David White, and Pavana Vennapusa wrote the paper.

Conflicts of Interest: The authors declare no conflict of interest.

References

1. Han, J. Design of Planar Geosynthetic-Improved Unpaved and Paved Roads. In *First International Symposium on Pavement and Geotechnical Engineering for Transportation Infrastructure*; Huang, B., Bowers, B.F., Mei, G.-X., Luo, S.-H., Zhang, Z., Eds.; American Society of Civil Engineers (ASCE): Nanchang, China, 2011; pp. 31–41.
2. Holtz, R.D.; Christopher, B.R.; Berg, R.R. Geosynthetic design and construction guidelines. In *Highway Capacity Manual: A Guide for Multimodal Mobility Analysis*, 6th ed.; Federal Highway Administration, U.S. Department of Transportation: Washington, DC, USA, 2008; p. 592.
3. Bahador, M.; Evans, T.M.; Gabr, M.A. Modeling effect of geocomposite drainage layers on moisture distribution and plastic deformation of road sections. *J. Geotech. Geoenviron. Eng.* **2013**, *139*, 1407–1418. [CrossRef]
4. Christopher, B.R.; Hayden, S.A.; Zhao, A. Roadway base and subgrade geocomposite drainage layers. In *The Symposium of 'Testing and Performance of Geosynthetics in Subsurface Drainage*; ASTM: Seattle, WA, USA, 2000; pp. 35–51.
5. Henry, K.S.; Holtz, R.D. Geocomposite Capillary Barriers to Reduce Frost Heave in Soils. *Can. Geotech. J.* **2001**, *38*, 678–694. [CrossRef]
6. Stormont, J.C.; Ramos, R.; Henry, K.S. Geocomposite Capillary Barrier Drain Systems with Fiberglass Transport Layer. *Transp. Res. Rec. J. Transp. Res. Board* **2001**, *1772*, 131–136. [CrossRef]
7. Sweet, J.G. Vertical Stiffness Characterization of a Geocomposite Drainage Layer for PCC Highway Pavements. Master's Thesis, West Virginia University, Morgantown, WV, USA, 2005.
8. AASHTO. Geotextile Specification for Highway Applications. In *Standard Specifications for Transportation Materials and Methods of Sampling and Testing, AASHTO M 288*; American Association of State Highway and Transportation Officials: Washington, DC, USA, 2014.
9. ASTM. Standard Test Method for Breaking Force and Elongation of Textile Fabrics (Strip Method). In *Annual Book of ASTM Standards ASTM D5035-11*; ASTM International: West Conshohocken, PA, USA, 2013.
10. ASTM. Standard Test Methods for Water Permeability of Geotextiles by Permittivity. In *Annual Book of ASTM Standards ASTM D4491-99*; ASTM International: West Conshohocken, PA, USA, 2013.

11. ASTM. Standard Test Methods for Determining Apparent Opening Size of a Geotextile. In *Annual Book of ASTM Standards ASTM D4751-12*; ASTM International: West Conshohocken, PA, USA, 2013.
12. White, D.J.; Becker, P.; Vennapusa, P.K.; Dunn, M.J.; White, C.I. Assessing Soil Stiffness of Stabilized Pavement Foundations. *Transp. Res. Rec. J. Transp. Res.* **2013**, *2335*, 99–109. [CrossRef]
13. White, D.J.; Vennapusa, P.K.; Becker, P.J.; Zhang, Y.; Dunn, M.J. Performance Assessment of Cement Stabilized, Polymer Fiber-Reinforced Pavement Foundation Layers. In Proceedings of the Geosynthetics Conference, Portland, OR, USA, 15–18 February 2015.
14. Li, C.; Ashlock, J.C.; White, D.J.; Vennapusa, P. Mechanistic-based comparisons of stabilised base and granular surface layers of low-volume roads. *Int. J. Pavement Eng.* **2017**. [CrossRef]
15. ASTM. Standard Test Methods for Laboratory Compaction Characteristics of Soil Using Standard Effort (12,400 ft-lbf/ft3 (600 kN-m/m3)). In *Annual Book of ASTM Standards ASTM D698-07*; ASTM International: West Conshohocken, PA, USA, 2013.
16. ASTM. Standard Test Method for Field Measurement of Hydraulic Conductivity Using Borehole Infiltration. In *Annual Book of ASTM Standards ASTM D6391-11*; ASTM International: West Conshohocken, PA, USA, 2013.
17. Vennapusa, P.; White, D.J.; Jahren, C.T. In-Situ Permeability of Unbound Granular Bases Using the Air Permeameter Test. In Proceedings of the 85th Annual Transportation Research Board Conference, Washington, DC, USA, 22–26 January 2006.
18. Zhang, J.; White, D.; Taylor, P.C.; Shi, C. A case study of evaluating joint performance in relation with subsurface permeability in cold weather region. *Cold Reg. Sci. Technol.* **2015**, *110*, 19–25. [CrossRef]
19. White, D.J.; Vennapusa, P.K.R.; Suleiman, M.T.; Dahren, C.T. An in-situ device for rapid determination of permeability for granular bases. *Geotech. Test. J.* **2007**, *30*, 282–291.
20. White, D.J.; Vennapusa, P.; Zhao, L. Verification and repeatability analysis for the in situ air permeameter test. *Geotech. Test. J.* **2014**, *37*, 365–376. [CrossRef]
21. AASHTO. *AASHTO Guide for Design of Pavement Structures*; American Association of State Highway and Transportation Officials: Washington, DC, USA, 1993.
22. Haverkamp, R.; Vauclin, M.; Touma, J.; Wierenga, P.J.; Vachaud, G. A Comparison of Numerical Simulation Models for One-Dimensional Infiltration. *Soil Sci. Soc. Am. J.* **1977**, *41*, 285–294. [CrossRef]
23. Jury, W.A.; Horton, R. *Soil Physics*, 6th ed.; John Wiley & Sons: Hoboken, NJ, USA, 2004.
24. ASTM. Standard Test Method for Measurement of Hydraulic Conductivity of Porous Material Using a Rigid-Wall, Compaction-Mold Permeameter. In *Annual Book of ASTM Standards ASTM D5856-07*; ASTM International: West Conshohocken, PA, USA, 2013.
25. Li, C.; Ashlock, J.C.; White, D.J.; Vennapusa, P. *Low-Cost Rural Surface Alternatives: Demonstration Project*; Iowa Highway Research Board (IHRB) Project TR-664; Iowa State University: Ames, IA, USA, 2015.

Article

Evaluation of Aging Resistance of Graphene Oxide Modified Asphalt

Shaopeng Wu, Zhijie Zhao, Yuanyuan Li *, Ling Pang *, Serji Amirkhanian and Martin Riara

State Key Laboratory of Silicate Materials for Architectures, Wuhan University of Technology, Wuhan 430070, China; wusp@whut.edu.cn (S.W.); zzjcbms@126.com (Z.Z.); samirkhanian@eng.ua.edu (S.A.); mriara@seku.ac.ke (M.R.)

* Correspondence: liyuanyuan_09@126.com (Y.L.); lingpang@whut.edu.cn (L.P.); Tel.: +86-27-8716-2595 (Y.L. & L.P.)

Academic Editor: Feipeng Xiao

Received: 19 June 2017; Accepted: 4 July 2017; Published: 7 July 2017

Abstract: Graphene oxide (GO) has a unique layered structure with excellent gas and liquid blocking properties. It is widely used in many areas, such as gas sensors, carbon-based electronics, impermeable membranes, and polymeric composite materials. In order to evaluate whether GO (1% and 3% by weight of asphalt) can improve the aging resistance performance of the asphalt, 80/100 penetration grade asphalt (90 A) and styrene–butadiene–styrene modified asphalt (SBS MA) were used to prepare the GO modified asphalt by the melt blending method. The surface morphology of the GO was analyzed by scanning electron microscope (SEM). The UV aging test was conducted to simulate the aging during the service period. After UV aging test, the physical performances of GO-modified asphalts were tested, and the $I_{C=O}$ and $I_{S=O}$ increments were tested by Fourier transform infrared spectroscopy (FTIR) to evaluate the aging resistance performance of the GO modified asphalt. In addition, the rheological properties of GO modified asphalts were studied using a dynamic shear rheometer (DSR). The SEM analysis indicated that the GO exhibits many shared edges, and no agglomeration phenomenon was found. With respect to the physical performance test, the FTIR and the DSR results show that GO can improve the UV aging resistance performance of 90 A and SBS MA. In addition, the analysis indicated that the improvement effect of 3% GO is better than the 1% GO. The testing on the rheological properties of the modified asphalt indicated that the GO can also improve the thermo-oxidative aging resistance performance of asphalt.

Keywords: GO modified asphalt; aging resistance; chemical structure; physical performance; rheological property

---

1. Introduction

Asphalt concrete is widely used as the pavement of choice in many countries. However, asphalt binder will be aged by many factors, such as heat, light, and oxygen during the high-temperature production and natural environment during service. Aging makes the asphalt binder become harder, leads to early damage, and reduces the service life span of the pavement [1,2]. Therefore, to prolong the service life span of an asphalt pavement, it is important to improve the anti-aging performance of the asphalt.

Asphalt aging is mainly caused by oxidation [3]. After aging, the asphalt molecular weight will be increased or associated irreversibly to form macromolecules by absorbing oxygen, the asphalt colloid structure will, therefore, be changed, decreasing the asphalt pavement performance [4,5]. Oxygen not only reacts with the surface asphalt, generates hydroxyl- and the oxygen-containing groups, but also continues aging the interior asphalt by gradually diffusing into the internal asphalt. The main difference of the thermo-oxidative aging and optical oxygen aging is the different excitation source to generate the free radicals in the initial stage. The excitation sources are the thermal and light excitations [6,7]. There are many possible ways in preventing the asphalt from being aged or in reducing the rate of asphalt aging. Some of these techniques include the following: (a) by reducing the excitation energy (thermal energy or light energy); (b) by reducing the formation rate or concentration of the initial free radicals; or (c) by preventing contact of oxygen with the asphalt, which can reduce the reactant concentration (O_2) of the oxidation reaction in the aging process.

Inorganic nano-materials, such as layered double hydroxides (LDHs) [8], montmorillonite [9,10], and carbon black [11,12], have a layered structure and are excellent barriers for oxygen, which can be used in improving the anti-aging performance of asphalt. However, the compatibility between inorganic materials and organic asphalt is poor, making it difficult for inorganic materials to disperse well in the asphalt; because of the van der Waals forces or the electrostatic forces, they tend to aggregation easily [13]. These drawbacks decrease the barrier performances and reduce the modification effects of the modifiers. Graphene oxide (GO) is a type of layered inorganic nano-material with a molecular structure roughly the same as that of grapheme. GO shows excellent performance in gas and liquid blocking, which is similar to that of graphene [14,15]. The difference of GO compared with graphene is that there are many active oxygen-containing functional groups on the layers and the lamella edges of the GO, which endows GO with good compatibility toward inorganic and organic materials, and makes the graphene oxide-modified material have a low permeability of gas and liquid [16]. In another word, GO not only has the unique layer nanostructures with excellent oxygen barrier performance, but also has good compatibility with the asphalt. Therefore, it is important to conduct more research in this area to determine the effect of GO on the properties of virgin and modified binders.

In order to evaluate to effects of GO on the anti-UV aging performance of the asphalt, two types of asphalt and two levels of the GO contents (1% and 3% by weight of the binder) were utilized in preparing the GO-modified asphalt by the melt-blending method. First, the asphalts were aged by a thin film oven test (TFOT), and then aged by UV aging. After aging, the physical performance tests were conducted to study the performance attenuation during the aging process and the characteristic functional groups were analyzed by Fourier transform infrared spectroscopy (FTIR). The rheological properties were evaluated with a dynamic shear rheometer (DSR) in the temperature range from −10 to 30 °C.

2. Materials and Methods

2.1. Materials

2.1.1. Asphalt

Two types of asphalt binders were used in this research project, namely base asphalt with 80/100 penetration grade (simply referred to as 90 A) and SBS-modified asphalt (simply referred to as SBS MA). They were obtained from Inner Mongolia Xindalu Asphalt Co., Ltd. (Chifeng, Inner Mongolia, China). Technical information of these two asphalt binders are shown in Table 1.

Table 1. Properties of 90 A and SBS MA.

Asphalt	Technical Parameters	Unit	Test Results	Method
90 A	25 °C Penetration	0.1 mm	84.6	ASTM D5 [17]
	Softening point	°C	47.8	ASTM D36 [18]
	10 °C ductility	cm	>100	ASTM D113 [19]
	60 °C viscosity	Pa·s	206	ASTM D4402 [20]
SBS MA	25 °C Penetration	0.1 mm	62.3	ASTM D5 [17]
	Softening point	°C	57.5	ASTM D36 [18]
	5 °C ductility	cm	55.8	ASTM D113 [19]
	135 °C viscosity	Pa·s	1.337	ASTM D4402 [20]

2.1.2. Graphene Oxide

GO was applied to be an anti-aging modifier for the asphalt binders. The GO with 5–10 layers was provided by the Suzhou Heng Ball Graphene Technology Co., Ltd. (Suzhou, China). The purity of GO was higher than 95%, the specific surface area of GO was about 100–300 m^2/g, and the lamella diameter of the GO was about 10–50 um.

2.2. Preparation of GO Modified Asphalt

First, 90 A and SBS MA were heated to 155 °C and 170 °C, respectively; then the design contents of the GO (1.0 wt % or 3.0 wt %) were added to the asphalt binder by using a high-speed shear mixing machine (4000 r/min) for 30 min. Since the heating and shearing of the preparation process inevitably causes the aging of asphalt, and it is rational to evaluate the aging resistance performance of GO-modified asphalt binders at the same condition, the 90 A (and SBS MA) was also exposed to the same heating and shearing process as the GO-modified 90 A (and GO-modified SBS MA).

2.3. Microstructure of GO

The surface morphology of the GO was analyzed by the JSM-5610LV SEM (Tokyo, Japan), the SEM resolution was 3 nm in a high vacuum. The SEM pictures were taken on the microscope at the voltage of 5 KV where the magnification of the pictures could be obtained from 18–300,000 times. The GO, in general, would be affected by thermal shock for its structural metastability [21], so the GO samples were dried at 50 °C.

2.4. Evaluation of Anti-Aging Performance

2.4.1. Aging Procedures

The samples used for the ultraviolet aging were first aged by TFOT (Cangzhou, Hebei, China). The mass of TFOT-aged samples were 50 ± 0.5 g, the diameter of the sample plate was 140 mm, the asphalt film thickness was approximately 3.2 mm, and the test temperature was stable at 163 ± 0.5 °C for 5 h. Then, the UV aging simulation test was performed with a straight-pipe high-pressure mercury lamp, the UV radiation intensity was 2000 uw/cm^2, the UV wavelength was 365 nm, and the aging times were set up to three, six, and nine days. The sample mass of UV aging was 20 ± 0.1 g and the test temperature was stable at 50 ± 0.5 °C.

2.4.2. Physical Performance Tests

Physical performances tests, such as penetration, softening point, ductility, and viscosity, were conducted according to the standards ASTM D5 [17], ASTM D36 [18], ASTM D113 [19], and ASTM D4402 [20], respectively. The temperature of the penetration test was 25 °C, and the ductility tests of 90 A and SBS MA were set to be 10 °C and 5 °C, respectively. Three replicate samples were tested for each of the penetration, ductility, and softening point tests, and the average of the three results was taken as the final result.

2.4.3. Characteristic Functional Group Test

A Fourier transform infrared spectroscopy (FTIR) instrument (Columbus, OH, USA) was used to test the characteristic functional group changes of the asphalts after UV aging, the preparation procedures of asphalt samples for FTIR are described briefly: First, the asphalt CS_2 solution with 5 wt % concentration of asphalt was prepared, then two drops of asphalt CS_2 solution were placed on the KBr chip by the glue dropper, so the thin film asphalt sample could be obtained after the CS_2 was fully volatilized. The scan wave number range was from 4000 to 400 cm^{-1}, and scan times were 64 times.

2.4.4. Rheological Property Test

A SmartPave102 dynamic shear rheometer (DSR) (Graz, Austria) was utilized to test the rheological properties of asphalt binders. The test temperature was in the range of −10 to 30 °C, the rotor diameter was 8.0 mm, the sample thickness was 2.0 mm, frequency was set to 10 rad/s, and the strain was 0.05%.

3. Results and Discussion

3.1. Characterization of the GO

GO has multiple layers of structural material with oxygen functional groups on the basal plane and at the edges [22,23]. Most researchers believe that epoxy and hydroxyl groups are on the basal plane, whereas carbonyl and carboxylic acid groups are at the edges [24–27]. These oxygen functional groups allow the GO to be easily and evenly dispersed because they decrease the van der Waals forces of the GO molecules [21]. The microstructure of the GO was studied by SEM, Figure 1a,b are SEM images of the GO magnified 500 and 1000 times, respectively. From Figure 1, the GO particles are mostly independent states and exhibit lots of shared edges, and the high face-to-face interaction stability or the face-to-edge interactions are not observed in the SEM pictures. Therefore, the GO may show a good compatibility with asphalt. On the other hand, these oxygen functional groups cause some defects on its basal plane and decrease its barrier performance [28,29]. Thus, to improve the barrier property, in this paper, multilayered (5–10) GO was selected as the anti-aging modifier for asphalt.

Figure 1. SEM pictures of GO ((**a**) 500×; and (**b**) 1000×).

For its excellent gas-barrier property, GO can also improve the blocking performance of the modified materials to block the gas and the vapor diffusion. The barrier property can extend the travelling pathway of the O_2 when it diffuses to the internal asphalt, and also prolongs the travelling pathway of the volatile organic compounds (VOCs) when released from the asphalt [30–32]. Asphalt aging is mainly caused by oxidation, so the GO may be able to retard the aging of asphalt by decreasing the O_2 diffusion rate.

3.2. UV Aging Resistance Performance

3.2.1. Chemical Structure

The asphalt chemical structure will be changed during the aging process, the characteristic functional groups index, namely carbonyl groups (1700 cm^{-1}) and sulfoxide groups (1032 cm^{-1}), will increase with the increase of aging degree [16,33]. In order to analyze the anti-aging performance of asphalt, the FTIR was used to test the characteristic functional groups of asphalt binders before and after aging. The $I_{C=O}$ and $I_{S=O}$ were calculated according to the Equations (1) and (2), respectively:

$$I_{C=0} = \frac{S_{1700\ cm^{-1}}}{S_{2000\sim600\ cm^{-1}}} \quad (1)$$

$$I_{S=0} = \frac{S_{1030\ cm^{-1}}}{S_{2000\sim600\ cm^{-1}}} \quad (2)$$

where, $S_{1700\ cm^{-1}}$ expresses the areas of the 1700 cm^{-1} centered carbonyl group absorption band; $S_{1032\ cm^{-1}}$ expresses the areas of the 1032 cm^{-1} centered sulfoxide group absorption band; $S_{2000\sim600\ cm^{-1}}$ expresses the areas of all absorption bands between 2000 cm^{-1} and 600 cm^{-1}.

The $I_{C=O}$ and $I_{S=O}$ of the aged binders (i.e., TFOT and UV) studied in this research are shown in Figures 2 and 3, respectively. As shown in Figure 2, the $I_{C=O}$ of 90 A and SBS MA were gradually increasing with an increase in the UV aging time, where, 90 A + 1% GO, 90 A + 3% GO, SBS MA + 1% GO, and SBS MA + 3% GO correspond to 90 A with 1% GO, 90 A with 3% GO, SBS MA with 1% GO, and SBS MA with 3% GO, respectively. The results for the $I_{S=O}$ followed a similar trend (Figure 3). After UV aging, the $I_{C=O}$ and $I_{S=O}$ increments of 90 A and SBS MA with GO are smaller than the virgin binder (asphalt without GO), indicating that the GO can improve the UV aging resistance of 90 A and SBS MA, and inhibit the increase in $I_{C=O}$ and $I_{S=O}$ of 90 A and SBS MA during UV aging.

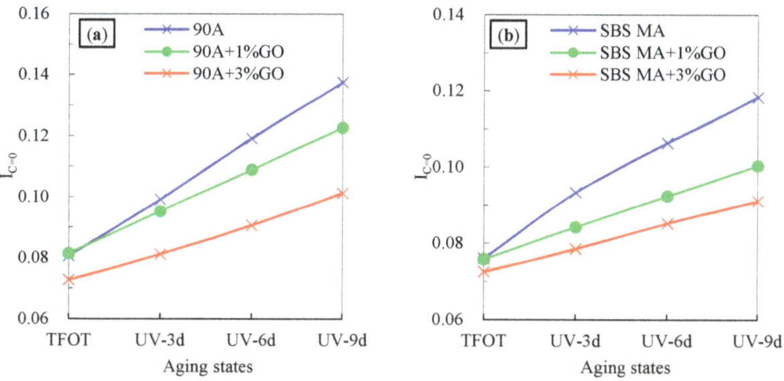

Figure 2. The $I_{C=O}$ of 90 A (**a**) and SBS MA (**b**) after UV aging.

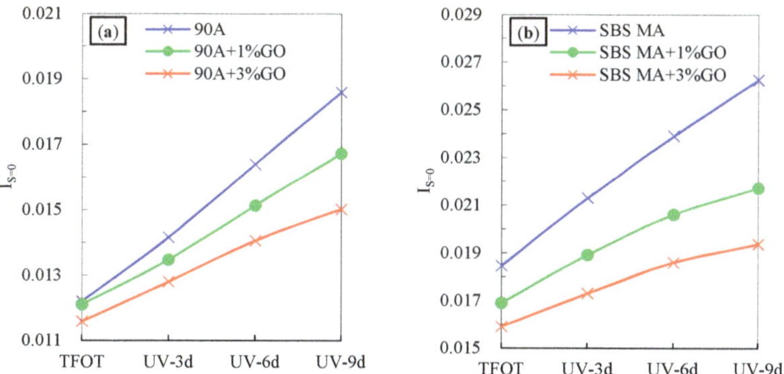

Figure 3. The $I_{S=O}$ of 90 A (**a**) and SBS MA (**b**) after UV aging.

3.2.2. Physical Performance

Three indicators, such as residual penetration ratio (PRR), softening point incremental (SPI), and residual ductility ratio (DRR) can characterize the physical properties attenuation amplitude during the UV aging process. Therefore, these three indicators were used in evaluating the UV aging degree of the asphalts used in this research project. The PRR, the SPI, and the DRR were calculated according to the Equations (3)–(5), respectively:

$$\text{PRR} = \frac{\text{P}_{\text{UV}}}{\text{P}_{\text{TFOT}}} \times 100\% \tag{3}$$

$$\text{SPI} = \text{SP}_{\text{UV}} - \text{SP}_{\text{TFOT}} \tag{4}$$

$$\text{DRR} = \frac{\text{D}_{\text{UV}}}{\text{D}_{\text{TFOT}}} \times 100\% \tag{5}$$

where, P_{TFOT}, D_{TFOT}, and SP_{TFOT} are the penetration (0.1 mm), the ductility (cm), and the softening (°C) after TFOT, respectively; P_{UV}, D_{UV}, and SP_{UV} are the penetration (0.1 mm), the ductility (cm), and the softening (°C) after UV aging, respectively. In general, it is expected that the asphalt with a good UV aging resistance should have higher PRR, lower SPI, and higher DRR values indicating that the physical performance of the asphalt will not change significantly after UV aging.

Figure 4 shows the PRR results after UV aging. It can be seen that the PRR value of all asphalts decrease gradually with the extension of UV aging time, when the UV ageing time increases from three days to nine days, the PRR value of 90 A decreases from 59.2% to 36.4%, and the PRR value of SBS MA decreases from 67.3% to 38.6%. The higher PRR value means a lower aging degree, and vice versa. After adding the GO, the PRR value of 90 A and SBS MA after UV aging increased, when the GO content was 3%, the PRR of 90 A after nine days of UV aging increased from 36.4% to 57.3%. In addition, the PRR value of SBS MA after nine days UV aging increased from 38.6% to 64.3%, which shows that the GO can alleviate asphalt penetration attenuation in the process of UV aging.

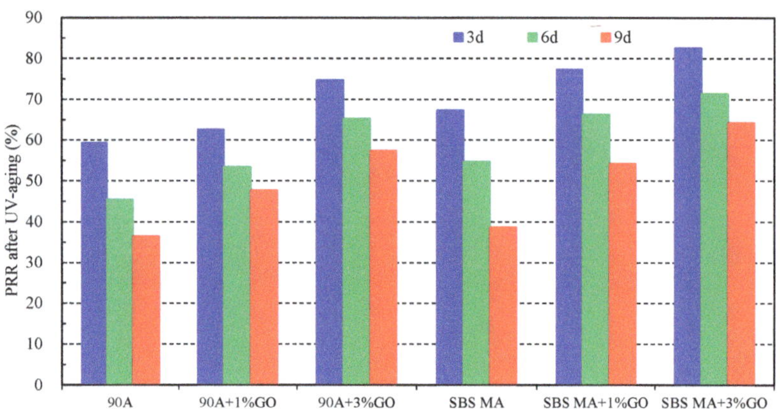

Figure 4. The PRR of asphalt after UV aging.

Figure 5 shows the SPI results after UV aging. The higher the SPI, the higher is the aging degree and vice versa. The SPI values of 90 A, SBS MA and the GO modified asphalts increased gradually with the extension of UV aging time. Compared to the asphalt without modifiers, after adding the GO, the SPI values of all asphalts decreased at the same UV aging condition, and the SPI value was lower when the GO content was 3% compared to 1%. This shows that the GO can retard the softening point increase in the UV irradiation process, and the retarding effect is more pronounced when the GO content is increased.

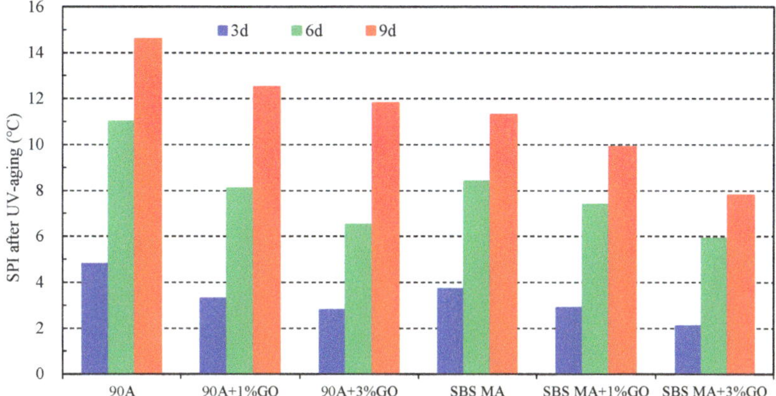

Figure 5. The SPI of asphalt after UV aging.

The DRR results after UV aging are shown in Figure 6. The lower value of DRR indicates an increase in the aging degree of the binder. With the extension of the UV times, the DRR values decrease obviously. At the same UV aging condition, the ductility attenuation rate decreases with the GO, and the decrease effect is more obvious when the GO content is increased from 1% to 3%. For instance, the DRR values of the GO modified 90 A will increase from 21.2% (without GO) to 38.1% (with 3% GO), and the DRR values of the GO modified SBS MA will increase from 25.0% (without GO) to 45.4% (with 3% GO).

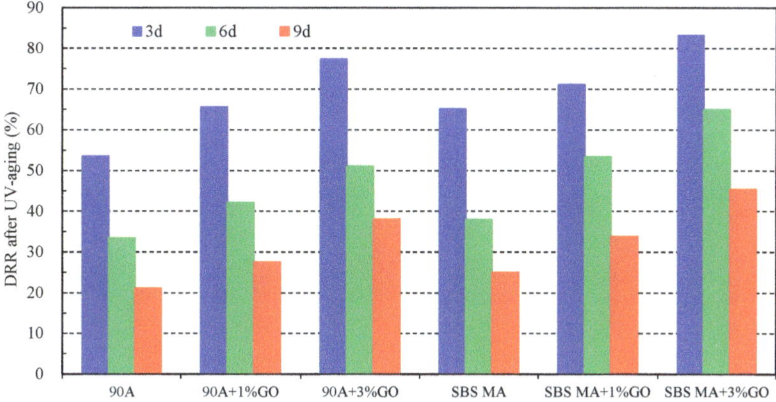

Figure 6. DRR of asphalt after UV aging.

The asphalt viscosity after TFOT and UV aging was tested at different temperatures, namely 30 °C, 45 °C, 60 °C, 75 °C, 90 °C and 105 °C. The viscosity aging indices (VAI) of all temperatures were calculated based on the Equation (6):

$$\mathrm{VAI} = \frac{\mathrm{V_{UV}} - \mathrm{V_{TFOT}}}{\mathrm{V_{TFOT}}} \times 100\% \tag{6}$$

where, $\mathrm{V_{UV}}$ is the viscosity after UV aging in Pa·s; and V_{TFOT} is the viscosity after TFOT in Pa·s. In general, a higher VAI value is an indication of a higher UV-aging degree. The VAI results of 90 A and GO-modified 90 A are shown in Figure 7, and the VAI results of SBS MA and the GO-modified SBS MA are shown in Figure 8.

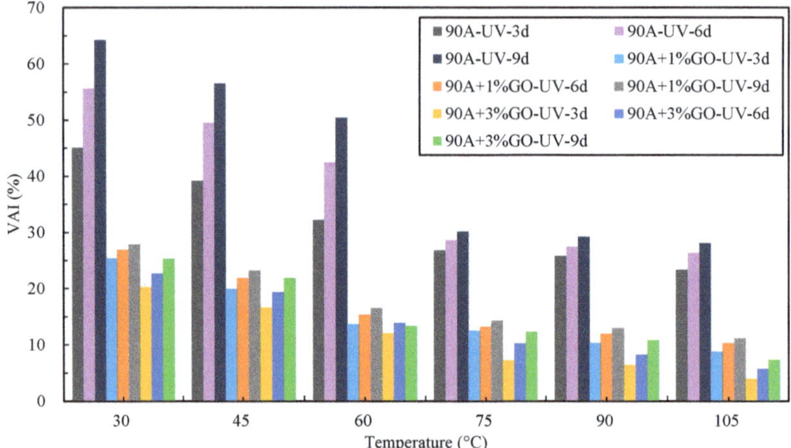

Figure 7. VAI of 90 A and GO modified 90 A after UV aging.

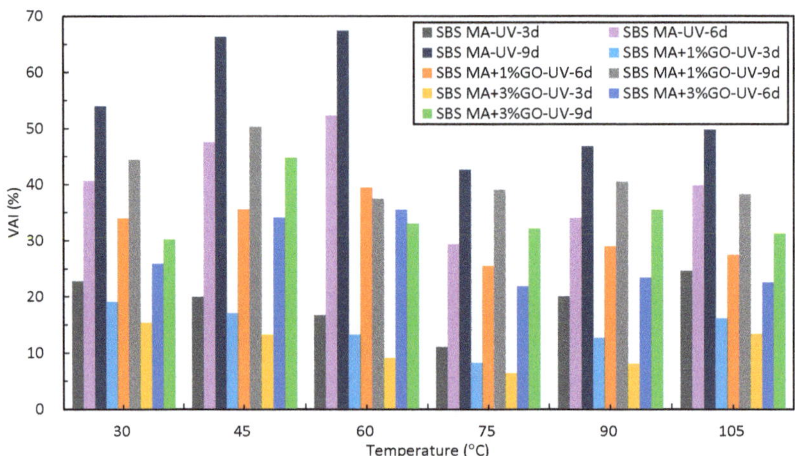

Figure 8. VAI of SBS MA and GO modified SBS MA after UV aging.

The results shown in Figure 7 indicate that the VAI values of all temperatures increase with the increase of the UV-aging time, which shows that the UV-aging degree is increased. For example, the VAI of the GO-modified 90 A decreased as the content of the GO increase. Therefore, for the 90 A binder tested, one can conclude that the GO can improve the UV aging resistance performance and, at the same UV-aging condition, the VAI of the 1% GO-modified 90 A has no significant difference with 3% GO-modified 90 A. However, the VAI value after adding 3% GO was slightly lower than that of 1% GO.

The VAI results of SBS MA in Figure 8 show the same trend as of 90 A, the VAI is increasing with an increase in UV-aging time. However, the VAI of the GO-modified SBS MA is much smaller than that of SBS MA. It is clear that the GO can improve the UV aging resistance performance of 90 A and

SBS MA, and the UV aging resistance performance will be enhanced by increasing the GO content from 1% to 3%.

3.2.3. Evaluating the Rheological Properties

The complex modulus (G*) and the phase angle (δ) of the 90 A and the GO-modified 90 A before aging and after TFOT aging are shown in Figure 9. In general, a nano-material powder always brings a hardening effect to the asphalt, and G* always increases.

From Figure 9, before aging, the G* increment of 90 A with 1% GO content is minimal, and the G* value of the 3% GO-modified 90 A is slightly lower than 90 A. This is possible due to the softening effect of the CO_2 sealed in the asphalt [33,34]. The δ value of the GO-modified 90 A is lower than the δ of 90 A, the difference is more notable from −10 °C to 0 °C. After TFOT, the ranking order of the G* of 90 A is 1% GO-modified 90 A < 3% GO-modified 90 A < 90 A, and there is little difference of the δ of 90 A and the GO-modified 90 A; the G* increment of the 3% GO-modified 90 A is almost the same as 90 A, but the δ reduction of 3% GO-modified 90 A is less than 90 A. This indicates that 3% GO can improve the thermo-oxidative aging resistance performance of 90 A; when the GO content is 1%, the G* is much smaller than that of 90 A. In addition, the δ reduction is also less than 90 A, so that 1% GO can obviously improve the thermo-oxidative aging resistance performance of 90 A. On the improvement in the thermo-oxidative aging resistance performance of 90 A, the 3% GO-modified effect is not as good as the 1% GO. This could be attributed to the residual CO_2 in the asphalt; the CO_2 will be released at the temperature of 163 °C during TFOT. The release process is equivalent to a stirring action and the asphalt film near the CO_2 bubble is very thin, which can, more or less, speed up the TFOT aging rate and weakens the improvement effect of the GO. In this paper, we simply name the process of releasing CO_2 as stirring and film action. When the GO content is 1%, this stirring and film action is less obvious than with the 3% GO content; hence, a better anti-aging performance.

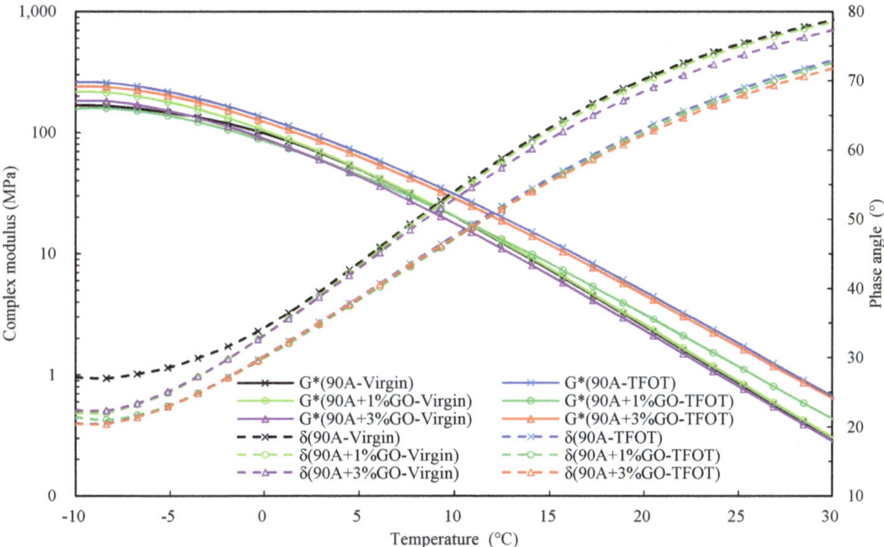

Figure 9. Complex modulus of 90 A and GO modified 90 A before and after TFOT.

The G* and the δ of 90 A and the GO-modified 90 A after TFOT and after nine days of UV aging are shown in Figure 10. After nine days of UV aging, the order of the G* of 90 A was found to be 3% GO-modified 90 A < 1% GO-modified 90 A < 90 A. The G* value is decreased by the utilization of the GO, especially the G* of the 3% GO-modified asphalt after nine days of UV aging is almost the same as before aging, which shows the GO has an excellent UV aging resistance performance. The GO can significantly improve the anti-UV aging performance of 90 A, and the improvement effect of the 3% GO is much better than 1% GO. The UV test temperature (50 °C) is only slightly higher than its softening point; on the other hand, after 5 h of TFOT aging, CO_2 is released completely out of the GO-modified asphalt, so the stirring and film action will not affect the improvement effect of the GO.

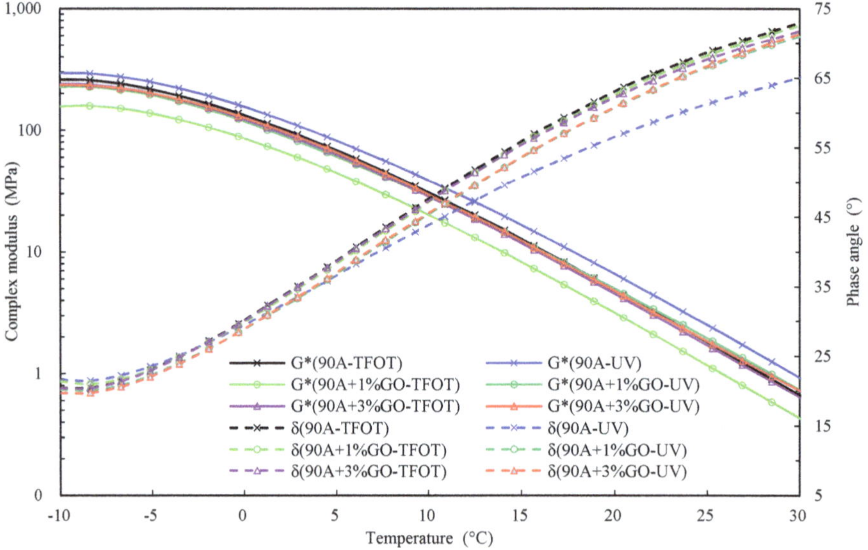

Figure 10. Complex modulus of 90 A and GO modified 90 A after TFOT and nine days of UV aging.

The fatigue cracking factor (G*sinδ) is used to evaluate the anti-fatigue performance of an asphalt binder where the lower value of the G*sinδ is an indication of better anti-fatigue performance. Superpave specifications specify that the G*sinδ should not be higher than 5.0 MPa [35]. In this paper, in order to quantitatively evaluate the effects of the GO on the anti-fatigue performance of asphalt at different aging states, the corresponding temperature fatigue cracking factor of 5.0 MPa is also calculated; this temperature is simply referred to as FFT.

Figure 11 shows the G*sinδ results of 90 A and the GO-modified 90 A. It can be found from this figure that the order of G*sinδ after UV aging is 3% GO-modified 90 A < 1% GO-modified 90 A < 90 A, indicating that the GO can improve the anti-fatigue performance of 90 A after UV aging, and the improvement effect of 3% GO is much better than 1% GO.

Table 2 shows the FFT of 90 A and SBS MA. It can be observed from Table 2 that, before aging, the FFT of the 1% GO-modified 90 A is the same as the 90 A, and the FFT of the 3% GO-modified 90 A is 0.7 °C lower than 90 A, so the GO can slightly improve the anti-fatigue performance of 90 A before aging. After TFOT, the FFT of 90 A and the GO-modified 90 A all increased, but the FFT of the 1% and 3% GO-modified 90 A were 2.6 °C and 0.5 °C lower than 90 A. This trend was the same for the G and δ values. After UV aging, the order of the TTF was found to be 3% GO-modified 90 A < 1% GO-modified 90 A < 90 A. The change above proves that GO can improve the anti-aging performance of an asphalt binder, the fatigue cracking resistance performance of the GO-modified 90 A are better than 90 A at any aging states, and the improvement effect will be more significant after aging.

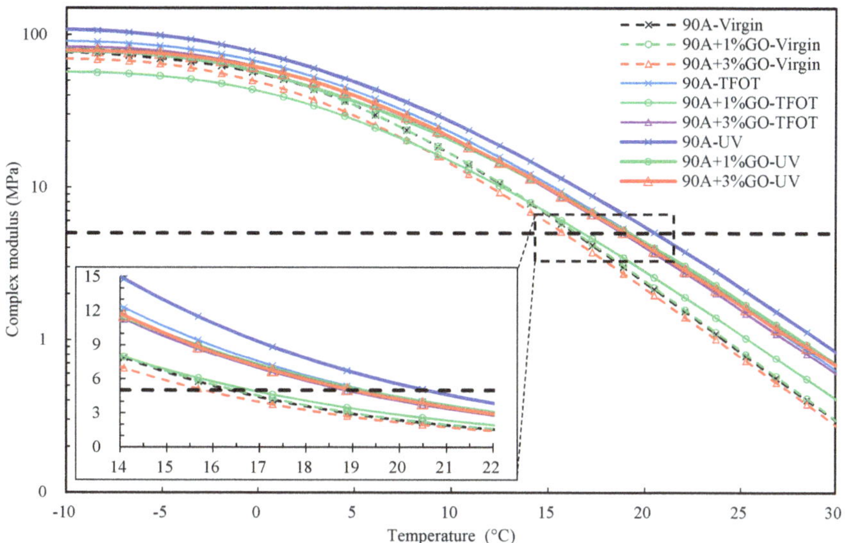

Figure 11. Fatigue cracking factor of 90 A and GO-modified 90 A.

Table 2. The FFT of 90 A and GO-modified 90 A.

Aging Degree	Virgin			TFOT Aging			UV Aging		
GO content (%)	0	1	3	0	1	3	0	1	3
FFT (°C)	16.5	16.5	15.8	19.3	16.7	18.8	20.6	19.2	19.0

The G* and the δ of SBS MA and the GO-modified SBS MA before aging and after TFOT are shown in Figure 12. Before aging, the GO did not increase the G* of the SBS MA significantly, the G* curve of GO-modified SBS MA almost overlap with the SBS MA without the GO. With the increase of GO content, the δ of the SBS MA only slightly decreases. The slight changes of the G* and the δ show the GO will not obviously decrease the low-temperature performance of the SBS MA. After TFOT aging, the G* increases while the δ decreases, which shows the increment of the asphalt aging degree. Compared to SBS MA, the G* increment and the δ decrement of the GO-modified SBS MA is almost the same, so the GO is not able to improve the thermo-oxidative aging resistance performance of SBS MA significantly. The improvement effect of the GO during TFOT may be weakened by the stirring and film action of the CO_2, and the viscosity of the SBS MA increase more sharply than the 90 A with the decrease of the temperature, which may seal more CO_2 in the asphalt. The CO_2 sealed in the asphalt will be released when the viscosity decreases at high temperature during TFOT. Thus, there is no obvious improvement of the SBS MA anti-TFOT aging performance.

The G* and the δ of SBS MA and the GO-modified SBS MA after TFOT and after nine days of UV aging are shown in Figure 13. After UV aging, the order of the G* is 3% GO-modified SBS MA < 1% GO modified SBS MA < SBS MA, and the G* increment of 3% GO-modified SBS MA is the lowest, 1% GO modified SBS MA is the second and SBS MA is the largest, there is no obvious difference of the δ of these three kinds of SBS MA binders. Thus, the order of the anti-UV aging performance is 3% GO-modified SBS MA > 1% GO-modified SBS MA > SBS MA. Therefore, one can conclude that, for these binders, GO can improve the anti-UV aging performance of SBS MA and the improvement of the 3% GO is better than that of the 1% GO.

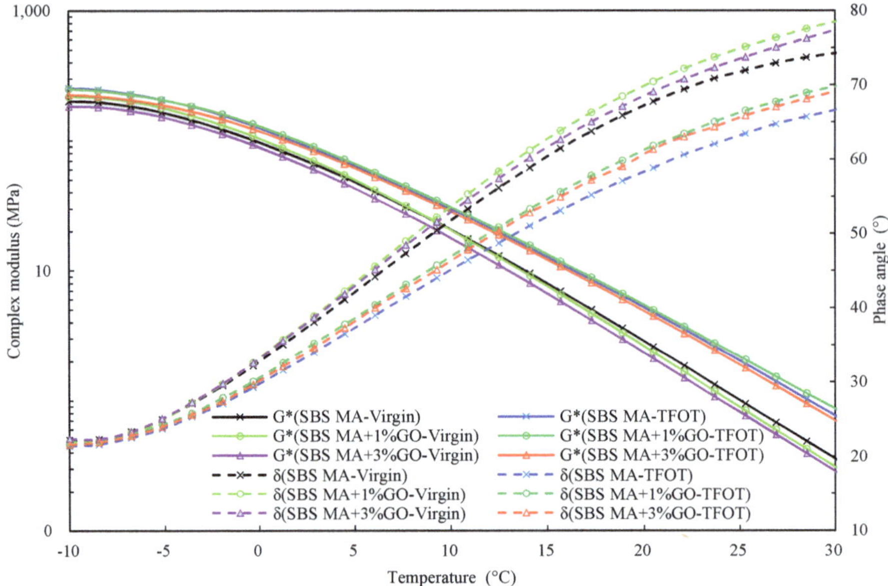

Figure 12. Complex modulus of SBS MA and GO modified SBS MA before and after TFOT.

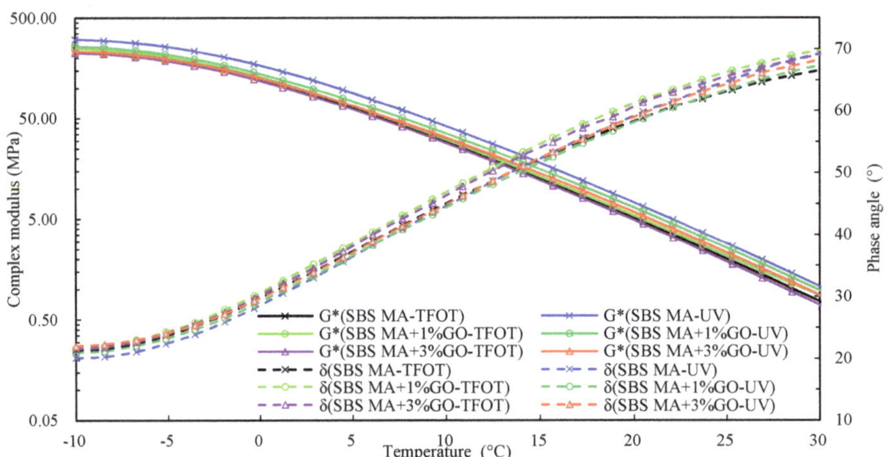

Figure 13. Complex modulus of SBS MA and GO-modified SBS MA after TFOT and nine days UV aging.

The G*sinδ of SBS MA and GO-modified SBS MA after UV aging are shown in Figure 14. With the increment of the GO content, the G*sinδ decreases, which shows that the fatigue property is improved after adding the GO, and the results of the 3% GO content is better than 1% GO content.

Table 3 gives the FFT results of SBS MA and the GO-modified SBS MA. Before aging, the FFT order of SBS MA is almost the same as 90 A, the1% GO and the 3% GO modified SBS MA are 0.2 °C and 1.0 °C smaller than SBS MA. Therefore, the GO can improve the fatigue cracking resistance performance of the SBS MA before aging. After TFOT, the FFT of the 3% GO-modified SBS MA are lower than the SBS MA; the FFT of the 1% GO-modified SBS MA are higher than SBS MA, but the increment is only 0.7 °C, which will not greatly reduce the low-temperature performance of a binder. Thus, 3% GO can slightly improve the anti-TFOT aging of SBS MA, while 1% GO slightly decreases the anti-TFOT aging of SBS MA. After UV aging, the order of the TTF is 3% GO-modified SBS MA < 1% GO SBS MA < SBS MA, the1% GO and the 3% GO-modified SBS MA are 0.8 °C and 1.3 °C lower than SBS MA. Fatigue cracking resistance performances of the modified SBS MA are better than 90 A after UV aging. Therefore, one can conclude that GO can improve the UV aging resistance performance of SBS MA, and the improvement effect of the 3% GO is better than the 1% GO.

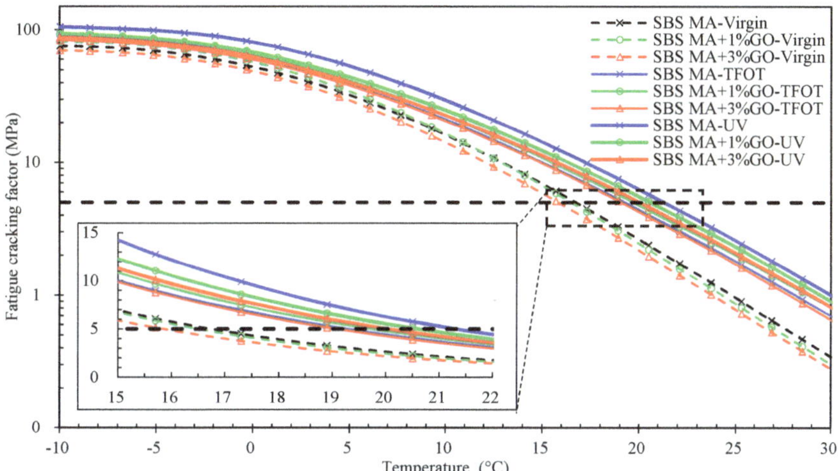

Figure 14. Fatigue cracking factor of SBS MA and GO-modified SBS MA.

Table 3. The FFT of the SBS MA and GO-modified SBS MA.

Aging Degree	Virgin			TFOT Aging			UV Aging		
GO content (%)	0	1	3	0	1	3	0	1	3
FFT (°C)	16.8	16.6	15.8	19.3	19.8	19.1	21.4	20.6	20.1

4. Conclusions

The $I_{C=O}$ and $I_{S=O}$ before and after UV aging were tested by the FTIR to study the chemical structure changes of the GO-modified asphalt during UV aging. The PRR, SPI, DRR, and the VAI were calculated and analyzed to estimate the physical properties changes of the GO-modified asphalt during UV aging. Finally, the DSR was used to study the rheological properties, G* and δ, before and after aging. After analyzing the results of these tests, the following conclusions can be made:

1. Before aging, the GO could decrease the G* and slightly change the δ of 90 A and SBS MA; in total, a smaller G*sinδ value was obtained after being modified by the GO, which shows that the GO could improve the fatigue cracking resistance performance of the 90 A and SBS MA binders.
2. According to the rheological property testing, before and after TFOT, the GO could improve the thermo-oxidative aging resistance performance of 90 A and SBS MA, and the improvement effect of the GO on 90 A was better than SBS MA.
3. After UV aging, the $I_{C=O}$ and $I_{S=O}$ increment of the GO modified asphalt were smaller than that of the asphalt without GO. The GO could retard the formations of the carbonyl and sulfoxide groups during UV aging, and decrease the aging degree of 90 A and SBS MA.
4. After UV aging, the GO could increase the PRR and DRR of 90 A and SBS MA, meanwhile decreasing the SPI and VAI; the order of the G* was 3% GO-modified asphalt < 1% GO-modified

asphalt < 90 A (or SBS MA). The increment of G* was obviously decreased after adding the GO, and the fatigue cracking resistance performance of GO-modified asphalt were better than that of the asphalt without the GO. The results showed that the GO could improve the stability of the asphalt physical performance. The GO could improve the UV aging resistance performance of 90 A and SBS MA obviously, and the improvement effect of the 3% GO was better than that of 1% GO.

Acknowledgments: The authors acknowledge the financial supported by the National Basic Research Program of China (973 program no. 2014CB932104), and the National Key Scientific Apparatus Development Program from the Ministry of Science and Technology of China (no. 2013YQ160501).

Author Contributions: Shaopeng Wu, Zhijie Zhao, and Ling Pang conceived and designed the experiments. Zhijie Zhao, Yuanyuan Li, and Martin Riara performed the experiments. Shaopeng Wu, Ling Pang, Serji Amirkhanian, and Zhijie Zhao analyzed the data. Yuanyuan Li and Martin Riara contributed reagents/materials/analysis tools. Shaopeng Wu, Zhijie Zhao, Ling Pang, and Yuanyuan Li wrote the paper. Martin Riara and Ling Pang designed the software used in analysis. Serji Amirkhanian reviewed the paper.

Conflicts of Interest: The authors declare no conflict of interest.

References

1. Rasool, R.T.; Wang, S.; Zhang, Y.; Li, Y.; Zhang, G. Improving the aging resistance of SBS modified asphalt with the addition of highly reclaimed rubber. *Constr. Build. Mater.* **2017**, *145*, 126–134. [CrossRef]
2. Behnood, A.; Olek, J. Rheological Properties of asphalt binders modified with styrene-butadiene-styrene (SBS), ground tire rubber (GTR), or polyphosphoric acid (PPA). *Constr. Build. Mater.* **2017**, *151*, 464–478. [CrossRef]
3. Islam, M.R.; Tarefder, R.A. Study of asphalt aging through beam fatigue test. *Transp. Res. Rec. J. Transp. Res. Board* **2015**, *2505*, 115–120. [CrossRef]
4. Pang, L.; Liu, K.; Wu, S.; Lei, M.; Chen, Z. Effect of LDHs on the aging resistance of crumb rubber modified asphalt. *Constr. Build. Mater.* **2014**, *67*, 239–243. [CrossRef]
5. Apeagyei, A.K. Laboratory evaluation of antioxidants for asphalt binders. *Constr. Build. Mater.* **2011**, *25*, 47–53. [CrossRef]
6. Petersen, J.C. A review of the fundamentals of asphalt oxidation: Chemical, physicochemical, physical property and durability relationships. Transportation research circular e-c140. *Transp. Res. E-Circ.* **2009**. [CrossRef]
7. Feng, Z.G.; Wang, S.J.; Bian, H.J.; Guo, Q.L.; Li, X.J. FTIR and rheology analysis of aging on different ultraviolet absorber modified bitumens. *Constr. Build. Mater.* **2016**, *115*, 48–53. [CrossRef]
8. Zhang, C.; Yu, J.; Xue, L.; Sun, Y. Investigation of γ-(2,3-Epoxypropoxy) propyltrimethoxy Silane Surface Modified Layered Double Hydroxides Improving UV Ageing Resistance of Asphalt. *Materials* **2017**, *10*, 78. [CrossRef]
9. Yu, J.Y.; Feng, P.C.; Zhang, H.L.; Wu, S.P. Effect of organo-montmorillonite on aging properties of asphalt. *Constr. Build. Mater.* **2009**, *23*, 2636–2640. [CrossRef]
10. Zhang, H.; Yu, J.; Wu, S. Effect of montmorillonite organic modification on ultraviolet aging properties of SBS modified bitumen. *Constr. Build. Mater.* **2012**, *27*, 553–559. [CrossRef]
11. Cong, P.; Xu, P.; Chen, S. Effects of carbon black on the anti aging, rheological and conductive properties of SBS/asphalt/carbon black composites. *Constr. Build. Mater.* **2014**, *52*, 306–313. [CrossRef]
12. Feng, Z.; Yu, J.; Wu, S. Rheological evaluation of bitumen containing different ultraviolet absorbers. *Constr. Build. Mater.* **2012**, *29*, 591–596. [CrossRef]
13. Nazari, M.H.; Shi, X. Polymer-Based Nanocomposite Coatings for Anticorrosion Applications. In *Industrial Applications for Intelligent Polymers and Coatings*; Springer International Publishing: New York, NY, USA, 2016.
14. Park, S.; Lee, K.S.; Bozoklu, G.; Cai, W.; Nguyen, S.T.; Ruoff, R.S. Graphene oxide papers modified by divalent ions—Enhancing mechanical properties via chemical cross-linking. *ACS Nano* **2008**, *2*, 572–578. [CrossRef] [PubMed]

15. Wang, L.; Wang, D.; Dong, X.Y.; Zhang, Z.J.; Pei, X.F.; Chen, X.J.; Chen, B.; Jin, J. Layered assembly of graphene oxide and Co–Al layered double hydroxide nanosheets as electrode materials for supercapacitors. *Chem. Commun.* **2011**, *47*, 3556–3558. [CrossRef] [PubMed]
16. Tang, Z.; Wu, X.; Guo, B.; Zhang, L.; Jia, D. Preparation of butadiene–styrene–vinyl pyridine rubber–graphene oxide hybrids through co-coagulation process and in situ interface tailoring. *J. Mater. Chem.* **2012**, *22*, 7492–7501. [CrossRef]
17. ASTM D5/D5M. *Standard Test Method for Penetration of Bituminous Materials*; American Society for Testing and Materials: West Conshohocken, PA, USA, 2013.
18. ASTM D36/D36M. *Standard Test Method for Softening Point of Bitumen (Ring and Ball Apparatus)*; American Society for Testing and Materials: West Conshohocken, PA, USA, 2012.
19. ASTM D113. *Standard Test Method for Ductility of Bituminous Materials*; American Society for Testing and Materials: West Conshohocken, PA, USA, 2007.
20. ASTM D4402. *Standard Test Method for Viscosity Determination of Asphalt at Elevated Temperatures Using a Rotational Viscometer*; American Society for Testing and Materials: West Conshohocken, PA, USA, 2013.
21. Yoo, B.M.; Shin, H.J.; Yoon, H.W.; Park, H.B. Graphene and graphene oxide and their uses in barrier polymers. *J. Appl. Polym. Sci.* **2014**, *131*. [CrossRef]
22. Lerf, A.; He, H.; Forster, M.; Klinowski, J. Structure of graphite oxide revisited. *J. Phys. Chem. B* **1998**, *102*, 4477–4482. [CrossRef]
23. He, H.; Klinowski, J.; Forster, M.; Lerf, A. A new structural model for graphite oxide. *Chem. Phys. Lett.* **1998**, *287*, 53–56. [CrossRef]
24. Eda, G.; Chhowalla, M. Chemically derived graphene oxide: Towards large-area thin-film electronics and optoelectronics. *Adv. Mater.* **2010**, *22*, 2392–2415. [CrossRef] [PubMed]
25. Kim, J.; Cote, L.J.; Kim, F.; Yuan, W.; Shull, K.R.; Huang, J. Graphene oxide sheets at interfaces. *J. Am. Chem. Soc.* **2010**, *132*, 8180–8186. [CrossRef] [PubMed]
26. Li, X.; Zhang, G.; Bai, X.; Sun, X.; Wang, X.; Wang, E.; Dai, H. Highly conducting graphene sheets and Langmuir–Blodgett films. *Nat. Nanotechnol.* **2008**, *3*, 538–542. [CrossRef] [PubMed]
27. Kudin, K.N.; Ozbas, B.; Schniepp, H.C.; Prud'Homme, R.K.; Aksay, I.A.; Car, R. Raman spectra of graphite oxide and functionalized graphene sheets. *Nano Lett.* **2008**, *8*, 36–41. [CrossRef] [PubMed]
28. Sun, P.; Zhu, M.; Wang, K.; Zhong, M.; Wei, J.; Wu, D.; Xu, Z.; Zhu, H. Selective ion penetration of graphene oxide membranes. *ACS Nano* **2012**, *7*, 428–437. [CrossRef] [PubMed]
29. Yu, L.; Lim, Y.S.; Han, J.H.; Kim, K.; Kim, J.Y.; Choi, S.Y.; Shin, K. A graphene oxide oxygen barrier film deposited via a self-assembly coating method. *Synth. Met.* **2012**, *162*, 710–714. [CrossRef]
30. Usuki, A.; Kojima, Y.; Kawasumi, M.; Okada, A.; Fukushima, Y.; Kurauchi, T.; Kamigaito, O. Synthesis of nylon 6-clay hybrid. *J. Mater. Res.* **1993**, *8*, 1179–1184. [CrossRef]
31. Priolo, M.A.; Gamboa, D.; Holder, K.M.; Grunlan, J.C. Super gas barrier of transparent polymer−clay multilayer ultrathin films. *Nano Lett.* **2010**, *10*, 4970–4974. [CrossRef] [PubMed]
32. LeBaron, P.C.; Wang, Z.; Pinnavaia, T.J. Polymer-layered silicate nanocomposites: An overview. *Appl. Clay Sci.* **1999**, *15*, 11–29. [CrossRef]
33. Zeng, W.; Wu, S.; Pang, L.; Sun, Y.; Chen, Z. The utilization of graphene oxide in traditional construction materials: Asphalt. *Materials* **2017**, *10*, 48. [CrossRef]
34. Schniepp, H.C.; Li, J.L.; McAllister, M.J.; Sai, H.; Herrera-Alonso, M.; Adamson, D.H.; Prud'homme, R.K.; Car, R.; Saville, D.A.; Aksay, I.A. Functionalized single graphene sheets derived from splitting graphite oxide. *J. Phys. Chem. B* **2006**, *110*, 8535–8539. [CrossRef] [PubMed]
35. AASHTO T315. *Determining the Rheological Properties of Asphalt Binder Using a Dynamic Shear Rheometer*; American Association of State Highway and Transportation Officials: Washington, DC, USA, 2010.

Article

Application of Finite Layer Method in Pavement Structural Analysis

Pengfei Liu [1], Qinyan Xing [2], Yiyi Dong [2], Dawei Wang [1,3,*], Markus Oeser [1] and Si Yuan [2]

1 Institute of Highway Engineering, RWTH Aachen University, Mies-van-der-Rohe-Street 1, D52074 Aachen, Germany; liu@isac.rwth-aachen.de (P.L.); oeser@isac.rwth-aachen.de (M.O.)
2 Department of Civil Engineering, Tsinghua University, 100084 Beijing, China; xingqy@tsinghua.edu.cn (Q.X.); dyy14@mails.tsinghua.edu.cn (Y.D.); yuans@tsinghua.edu.cn (S.Y.)
3 Institute of Highway Engineering, Paul-Bonatz-Street 9–11, University of Siegen, D57076 Siegen, Germany
* Correspondence: wang@isac.rwth-aachen.de; Tel.: +49-241-80-26742

Academic Editor: Zhanping You
Received: 25 April 2017; Accepted: 7 June 2017; Published: 13 June 2017

Abstract: The finite element (FE) method has been widely used in predicting the structural responses of asphalt pavements. However, the three-dimensional (3D) modeling in general-purpose FE software systems such as ABAQUS requires extensive computations and is relatively time-consuming. To address this issue, a specific computational code EasyFEM was developed based on the finite layer method (FLM) for analyzing structural responses of asphalt pavements under a static load. Basically, it is a 3D FE code that requires only a one-dimensional (1D) mesh by incorporating analytical methods and using Fourier series in the other two dimensions, which can significantly reduce the computational time and required resources due to the easy implementation of parallel computing technology. Moreover, a newly-developed Element Energy Projection (EEP) method for super-convergent calculations was implemented in EasyFEM to improve the accuracy of solutions for strains and stresses over the whole pavement model. The accuracy of the program is verified by comparing it with results from BISAR and ABAQUS for a typical asphalt pavement structure. The results show that the predicted responses from ABAQUS and EasyFEM are in good agreement with each other. The EasyFEM with the EEP post-processing technique converges faster compared with the results derived from ordinary EasyFEM applications, which proves that the EEP technique can improve the accuracy of strains and stresses from EasyFEM. In summary, the EasyFEM has a potential to provide a flexible and robust platform for the numerical simulation of asphalt pavements and can easily be post-processed with the EEP technique to enhance its advantages.

Keywords: finite layer method; asphalt pavement structural analysis; EasyFEM; element energy projection; super-convergence

---

1. Introduction

The analysis of stress states is of considerable importance for the design, construction, maintenance, and rehabilitation of asphalt pavements in practice. In the past several decades, a lot of computer software has been developed and is increasingly used routinely in pavement design and assessment processes.

In Germany, the guidelines for the analytical design of asphalt pavement superstructures RDO Asphalt 09 [1] propose the use of layer elastic theory (LET), e.g., the program BISAR, (Shell global, The Hague, The Netherlands) as the kernel of its pavement response model to calculate stresses, strains and displacements for different temperatures at critical locations within the pavement structure. Due to its simplicity, the LET has been utilized by pavement engineers for several decades. Currently, LET has been expanded to handle some problems with the material properties of viscoelasticity and nonlinearity as well as non-uniform loads [2]. However, the features of the software application based on the LET are still limited. For example, in BISAR, each layer is assumed homogeneous, isotropic, and linear elastic; all materials are weightless (no inertial effects are considered); all layers are assumed to be infinite in lateral extent and have a finite thickness except for the sub-grade, which is assumed to be infinite; the pavement systems are loaded statically over a uniform circular area; the compatibility of strains and stresses is assumed to be satisfied at all layer interfaces [3,4]. However, the reality of asphalt pavements may be very different to the assumptions made in the BISAR, e.g., finite geometrical scale in lateral extent, non-uniform loading conditions, and inelastic material properties. These differences may result in significant deviations between the calculated and the real responses of asphalt pavements.

With the finite element method (FEM), which is a numerical analysis technique proposed for obtaining approximate solutions for a wide variety of engineering problems, these specific requirements can be met and the asphalt pavement responses can be predicted more realistically. Some representative software applications have been widely used in pavement engineering, such as CAPA-3D (Delft University of Technology, Delft, The Netherlands) and APADS 2D (Austroads, Sydney, Australia). The development history of CAPA-3D goes back to the late 1980s. Currently, CAPA-3D is a linear/nonlinear, static/dynamic finite element (FE) system for the solution of very large scale three-dimensional (3D) solid models such as those typically encountered in pavement and soil engineering. It includes several constitutive material models, e.g., linear elasticity, hyperelasticity, elastoplasticity and viscoelasticity. As such, it can simulate a very broad range of engineering materials under various loading conditions [5]. Unfortunately, due to its 3D characteristics, the high hardware demands and long execution times render it suitable, primarily for research purposes [6]. APADS 2D was developed from 2008, in the scope of the Austroads project Developments of Pavement Design Models. It applies a two-dimensional (2D) axisymmetric concept to reduce the computational time and considers the effect of multiple loads through superposition. Due to the inherent limitations of axisymmetric models, it is difficult to simulate a pavement model with a determined geometry and complex loading conditions [7]. Other general-purpose FE software, such as ABAQUS (SIMULIA, Johnston, RI, USA) and ANSYS (ANSYS, Canonsburg, PA, USA), may provide more powerful capabilities to simulate the response of asphalt pavements to a certain extent. However, the expensive costs to get valid licenses and the time-consuming training process often render it impractical to be used by a road engineer.

One proposed method to overcome the aforementioned difficulties assumes that the displacements in one geometrical direction can be represented using Fourier series. Exploiting its orthogonal properties, a problem of such a class can be simplified into a series of 2D solutions. This method is so called semi-analytical FEM and was first developed in linear analysis by Wilson [8]. The analysis was extended to an elasto-plastic body by Meissner [9] based on Wilson′s work. A visco-plastic formulation was used by Winnicki and Zienkiewicz [10] to tackle material nonlinearity. To analyze the consolidation of elastic bodies subjected to non-symmetric loading, Carter and Booker [11] provided an effective solution by using the continuous Fourier series. Lai and Booker [12] successfully applied a discrete Fourier technique to analyze the stress state of solids with nonlinear behavior under 3D loading conditions. Further developments were made by Fritz [13] and Hu et al. [14], who programed simple FE codes to apply the semi-analytical FEM for the analysis of asphalt pavements. Recently, an FE code named SAFEM with more features was developed by Liu et al. [15–19]. Viscoelastic material property and dynamic analyses were integrated in this code. Here, the partial bonds between pavement layers can be considered and the infinite element was applied to reduce the influence of the boundary on the computational results.

The finite layer method (FLM) is proposed as another alternative way for analyzing the structural response of pavements, whilst reducing the computational requirements without decreasing the computational accuracy. This method combines a one-dimensional (1D) FEM discretization in the pavement depth direction with Fourier series in the two horizontal directions, which can be considered as a further development of the semi-analytical FEM. The idea is particularly suitable for multilayered structures. Furthermore, due to the use of Fourier series in two directions, the computational cost of FLM is significantly lower than that of conventional FE analysis methods. The FLM started to be applied in the analysis of elastic, horizontally layered foundations from 1979 [20], when the basic theory and application were proposed. Thereafter, the approach was further developed for the analysis of nonhomogeneous soils whose modulus increases linearly with depth [21]. An exact finite layer flexibility matrix was introduced to overcome the difficulties when the conventional finite layer stiffness approach was applied to incompressible materials [22,23]. Meanwhile, soil consolidation and surface deformation entailing the extraction of water was also investigated by using this approach [24–26]. In the next several years, researchers applied the FLM in groundwater flow models [27] and analyses of 3D Biot consolidation of layered transversely isotropic soils [28]. Recently, the FLM was applied for modeling the noise transmission through double walls [29]. analytical model, called 3D-Move uses a continuum-based FLM to compute pavement responses [30,31]. The 3D-Move model can account for important pavement response factors such as moving loads, 3D contact stress distributions (normal and shear) of any shape, and viscoelastic material characterization for the pavement layers.

Although the 3D-Move is freely available, it is not open source, and it still lacks some specific features such as the interlayer behavior for the pavement modeling. Therefore, a re-implementation of the FLM in code is necessary. As mentioned in the manual of 3D-Move, the accuracy of the calculated results is not guaranteed. Therefore, an algorithm or a technique that can provide more accurate results is of significant importance. In recent years, the research on super-convergence has increasingly drawn attention since it can provide higher order accuracy. A number of research studies have been conducted on this subject [32,33]. In recent years, a super-convergence technique named Element Energy Projection (EEP) method was proposed by Yuan et al. [34–36]. Its core assumption is to assume the energy projection theorem in the FEM mathematical theory [37] to be almost true for a single element. For various 1D problems, this assumption proves to be true and the convergent displacements and derivatives of the simplified form (i.e., with linear test/weight functions) converge at least one order higher than the conventional FEM results [38]. This method has been proven to be simple, convenient and effective. Moreover, it can be easily applied to other specific FE methods including FLM.

In the following sections, the basic theory of FLM is introduced, based on which the computer code named EasyFEM is self-developed. The EEP super-convergence strategy for EasyFEM is then described in detail to improve the accuracy of the results derived from EasyFEM. The EasyFEM is verified by comparing the results from BISAR and ABAQUS; the convergence rates between the results from ordinary EasyFEM and the EasyFEM with the post-processing with EEP are compared. The application of EasyFEM is to predict the mechanical responses of asphalt pavements. Lastly, a brief summary and outlook are provided.

2. Numerical Solution of EasyFEM

2.1. Finite Elements

A typical 3D EasyFEM model for the pavement problem was set up as shown in Figure 1.

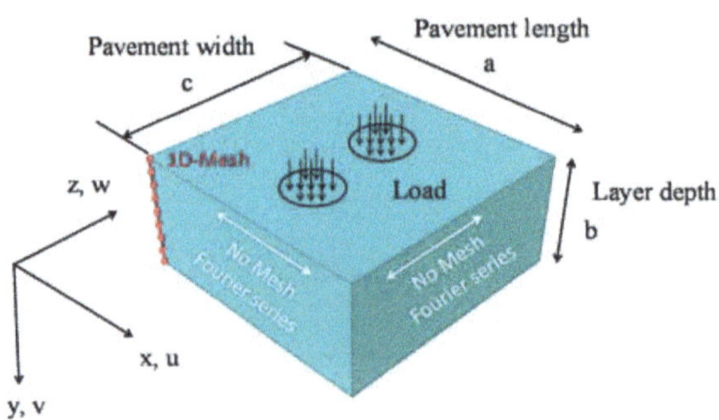

Figure 1. A typical EasyFEM model.

The system is discretized in the pavement depth direction; thus, the conventional 1D isoparametric elements of degree k can be adopted, e.g., the 3-node quadratic elements as shown in Figure 2:

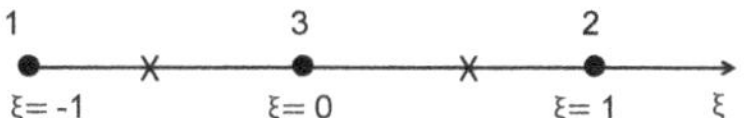

Figure 2. The natural coordinates for 1D quadratic elements.

where the shape function used was as follows:

$$\begin{aligned} N_1 &= \tfrac{-\xi(1-\xi)}{2} \\ N_2 &= \tfrac{\xi(1+\xi)}{2} \\ N_3 &= 1-\xi^2 \end{aligned} \tag{1}$$

with ξ being local coordinate.

2.2. FE Solution of Displacements

The pavement is assumed to be supported at side edges (x = 0, x = a, z = 0 and z = c) in a manner preventing all displacements in the vertical planes but permitting "unrestricted" movement in the x- and z-directions. The shape functions are re-written as the conventional 1D shape functions multiplied by the Fourier series, in which x ranges between zero and a; z ranges between zero and c. If the displacement functions with three components u, v and w are formulated in the following form, the boundary condition will meet this above-mentioned requirement:

$$\begin{gathered} d = \begin{Bmatrix} u \\ v \\ w \end{Bmatrix} = \sum_{m=1}^{M}\sum_{n=1}^{N} \begin{bmatrix} cos\frac{m\pi x}{a}sin\frac{n\pi z}{c} & 0 & 0 \\ 0 & sin\frac{m\pi x}{a}sin\frac{n\pi z}{c} & 0 \\ 0 & 0 & sin\frac{m\pi x}{a}cos\frac{n\pi z}{c} \end{bmatrix} \begin{Bmatrix} u^{mn} \\ v^{mn} \\ w^{mn} \end{Bmatrix} \\ = \textstyle\sum_{m=1}^{M}\sum_{n=1}^{N} N^{mn}.d^{mn}, \\ \text{with } N^{mn} = \begin{bmatrix} cos\frac{m\pi x}{a}sin\frac{n\pi z}{c} & 0 & 0 \\ 0 & sin\frac{m\pi x}{a}sin\frac{n\pi z}{c} & 0 \\ 0 & 0 & sin\frac{m\pi x}{a}cos\frac{n\pi z}{c} \end{bmatrix}, \\ d^{mn} = \begin{Bmatrix} u^{mn} \\ v^{mn} \\ w^{mn} \end{Bmatrix} = [N_k(y)]\{d_k^{mn}\}. \end{gathered} \tag{2}$$

N^{mn} is the matrix for the Fourier series expansion in the two horizontal directions, d^{mn} is the displacement vector without the Fourier series expansion, $[N_k(y)]$ is the conventional 1D shape function applied in the FEM model, and $\{d_k^{mn}\}$ represents the corresponding nodal displacements.

Similarly, the loading function for the pavement analysis can be formulated as:

$$\begin{gathered} f = \sum_{m=1}^{M}\sum_{n=1}^{N} p(y)sin\frac{m\pi x}{a}sin\frac{n\pi z}{c} = \sum_{m=1}^{M}\sum_{n=1}^{N}\{p\}^{mn}, \\ p(y) = \textstyle\sum_{s=1}^{S}\left(\frac{2P_s}{m\pi}\right)\left[cos\frac{m\pi}{a}X_{s1} - cos\frac{m\pi}{a}X_{s2}\right]\sum_{t=1}^{T}\left(\frac{2P_t}{n\pi}\right)\left[cos\frac{n\pi}{c}Z_{t1} - cos\frac{n\pi}{c}Z_{t2}\right], \end{gathered} \tag{3}$$

where m and n are the terms of Fourier series in the x and z directions, respectively. In this study, both M and N are adopted as 100, which has been previously proved to be sufficiently accurate to represent the load variation [16,17]. The product of P_s and P_t represents the tire load pressure, when m = n; X_{s1} and Z_{t1} are the x and z coordinates, where the tire load area starts, respectively; X_{s2} and Z_{t2} are the x and z coordinates where the tire load area ends, respectively.

At the nodes the strains are determined through displacements:

$$\varepsilon = \begin{Bmatrix} \varepsilon_x \\ \varepsilon_y \\ \varepsilon_z \\ \gamma_{xy} \\ \gamma_{yz} \\ \gamma_{zx} \end{Bmatrix} = \begin{bmatrix} \frac{\partial}{\partial x} & 0 & 0 \\ 0 & \frac{\partial}{\partial y} & 0 \\ 0 & 0 & \frac{\partial}{\partial z} \\ \frac{\partial}{\partial y} & \frac{\partial}{\partial x} & 0 \\ 0 & \frac{\partial}{\partial z} & \frac{\partial}{\partial y} \\ \frac{\partial}{\partial z} & 0 & \frac{\partial}{\partial x} \end{bmatrix} d = \left(L_1\frac{\partial}{\partial x} + L_2\frac{\partial}{\partial y} + L_3\frac{\partial}{\partial z}\right) \sum_{m=1}^{M} \sum_{n=1}^{N} N^{mn} \cdot d^{mn}$$
$$= \sum_{m=1}^{M} \sum_{n=1}^{N} E^{mn} d^{mn} = \sum_{m=1}^{M} \sum_{n=1}^{N} B^{mn} \cdot \{d_k^{mn}\}, \quad (4)$$

where L_1, L_2, L_3 are three different matrices of size 6 × 3 whose elements are either 0 or 1,

$$E^{mn} = L_1\frac{\partial N^{mn}}{\partial x} + L_2 N^{mn}\frac{\partial}{\partial y} + L_3\frac{\partial N^{mn}}{\partial z}, \quad (5)$$

and matrix

$$B^{mn} = \left[B_1^{mn}\ B_2^{mn} \cdots B_{k+1}^{mn}\right] = \left[\ E^{mn}N_1(y) \quad E^{mn}N_2(y) \quad \cdots E^{mn}N_{k+1}(y)\ \right] \quad (6)$$

is called the strain-displacement matrix.

The relation between stresses and strains is used to express the total potential energy with nodal displacements $\{d\}$:

$$\Pi(\{d\}) = \int_V \frac{1}{2}([B]\{d\})^T[D]([B]\{d\})dV - \int_V([N]\{d\})^T\{b\}dV - \int_S([N]\{d\})^T\{f\}dS\,. \quad (7)$$

The nodal displacements $\{d\}$ which correspond to the minimum of the functional Π are determined by the condition that the first variation of Π should be zero, i.e.,

$$\delta\Pi(\{d\}) = 0,$$
$$\delta\Pi(\{d\}) = \int_V\{\delta d\}^T[B]^T[D][B]\{d\}dV - \int_V\{\delta d\}^T[N]^T\{b\}dV - \int_S\{\delta d\}^T[N]^T\{f\}dS\,. \quad (8)$$

Then, the arbitrariness of $\{\delta d\}$ leads the following equilibrium equation for FE:

$$\int_V [B]^T[D][B]dV\{d\} - \int_V [N]^T\{b\}dV - \int_S [N]^T\{f\}dS = 0, \quad (9)$$

which is commonly given in the following form:

$$[k]\{d\} = \{f\},$$
$$[k] = \int_V[B]^T[D][B]dV \text{ and } \{f\} = \int_V[N]^T\{b\}dV + \int_S[N]^T\{f\}dS, begin \quad (10)$$

where $[k]$ is the element stiffness matrix; and $\{f\}$ is the load vector.

From Equations (4) and (10), the stiffness matrix of one element includes:

$$I_1 = \int_0^a \int_0^c \sin\frac{m\pi x}{a} \sin\frac{n\pi z}{c} \cdot \sin\frac{p\pi x}{a} \sin\frac{q\pi z}{c} \cdot dxdz,$$
$$I_2 = \int_0^a \int_0^c \cos\frac{m\pi x}{a} \sin\frac{n\pi z}{c} \cdot \cos\frac{p\pi x}{a} \sin\frac{q\pi z}{c} \cdot dxdz,$$
$$I_3 = \int_0^a \int_0^c \cos\frac{m\pi x}{a} \cos\frac{n\pi z}{c} \cdot \cos\frac{p\pi x}{a} \cos\frac{q\pi z}{c} \cdot dxdz. \quad (11)$$

The integrals exhibit orthogonal properties, which ensure that:

$$I_1 = I_2 = I_3 = \begin{Bmatrix} \frac{ac}{4}, & for\ m = p\ and\ n = q \\ 0, & for\ m \neq p\ or\ n \neq q \end{Bmatrix}. \quad (12)$$

This means that the matrix $(k^{mnpq})^e$ is diagonal. In other words, the non-zero values are only located on the diagonal, where $m = p$ and $n = q$. Thus, the stiffness matrix can be reduced to:

$$\left(k_{gk}^{mnmn}\right)^e = \frac{ac}{4}\int_{length}\left(B_g^{mnT}DB_k^{mn}\right)dy \qquad m = 1,\ 2\ldots M;\ n = 1, 2\ldots N, \tag{13}$$

where g and k represent the nodes of the element, respectively. The length corresponds to the length of the element.

A typical term for the force vector becomes:

$$(f^{mn})^e = \int\int\int_{vol}(N^{mn})^T\{p\}^{mn}dxdydz. \tag{14}$$

By assembling the stiffness matrix of each element to the global domain, the global linear system is achieved as follows:

$$\begin{bmatrix} K^{1111} & & & \\ & K^{1212} & & \\ & & \ddots & \\ & & & K^{MNMN} \end{bmatrix}\begin{Bmatrix} U^{11} \\ U^{12} \\ \vdots \\ U^{MN} \end{Bmatrix} + \begin{Bmatrix} F^{11} \\ F^{12} \\ \vdots \\ F^{MN} \end{Bmatrix} = 0. \tag{15}$$

Equation (15) shows that the large system of equations splits up into $M \times N$ separate problems:

$$K^{mnmn}U^{mn} + F^{mn} = 0, \tag{16}$$

where K is the global stiffness matrix; U is the global displacement vector; and F is the global loading vector.

3. EEP Super-Convergence Strategy for EasyFEM

When Equation (16) is solved for U^{mn}, both d^{mn} and d can be obtained from Equation (2). These two approximate solutions are denoted as d_h^{mn} and d_h to distinguish them from the exact solutions. It is clear, that if M and N are fixed, the accuracy of d_h mainly depends on the accuracy of d_h^{mn}. For conventional FEM models, a rather extensive mesh should be used to attain more accurate d_h^{mn}, which leads to a large number of degrees of freedom in the calculation. In this case, the newly-developed super-convergence technique EEP was introduced into EasyFEM as to improve the accuracy of the solutions for strains and stresses over the whole pavement model.

The original idea of the EEP technique came from a well-known mathematical theorem for FEM called the projection theorem [37]. Actually, the principle of minimum potential energy with respect to nodal displacements $\{d\}$ in Equation (8) can be rewritten in the same manner with respect to the continuous 1D displacement $d(y)$ as follows:

$$\delta\Pi(d(y)) = \int_V \delta d(y)^T[E]^T[D][E]d(y)dV - \int_V \delta d(y)^T[N(x,z)]^T\{b\}dV - \int_S \delta d(y)^T[N(x,z)]^T\{f\}dS = 0. \tag{17}$$

By replacing $\delta d(y)$ in Equation (17) with the test displacement $\widetilde{d}(y)$, an equivalent expression for the above principle of minimum potential energy is the virtual work equation as follows:

$$\begin{gathered} a(d(y),\widetilde{d}(y)) = (f,\widetilde{d}(y)), \forall\widetilde{d}(y) \in S^h, \\ \text{with } a\left(d(y),\widetilde{d}(y)\right) = \int_V \widetilde{d}(y)^T[E]^T[D][E]d(y)dV, \\ (f,\ \widetilde{d}(y)) = \int_V \widetilde{d}(y)^T[N(x,z)]^T\{b\}dV + \int_S \widetilde{d}(y)^T[N(x,z)]^T\{f\}dS, \end{gathered} \tag{18}$$

where $a(\cdot,\cdot)$ and $(\cdot,\cdot)$ are defined as the energy product and the linear form in mathematics, respectively, and S^h is the test space of FLM.

It can be easily understood that Equation (18) holds true both for the exact solution $d(y)$ and the approximate solution $d_h(y)$. Always taking the same test displacement for these two cases, the following projection theorem is derived for FLM:

$$a(d(y) - d_h(y), \tilde{d}(y)) = 0, \forall \tilde{d}(y) \in S^h. \tag{19}$$

This theorem holds true over the entire 3D pavement model. By assuming that it is approximately true over one element of FLM, the EEP equation was obtained:

$$a^e(d(y) - d_h(y), \tilde{d}(y)) \approx 0, \quad \forall \tilde{d}(y) \in S^h. \tag{20}$$

A typical element e of the FLM is shown in Figure 3, which is actually a small layer of the pavement with the y-coordinate ranging from y_1 to y_2.

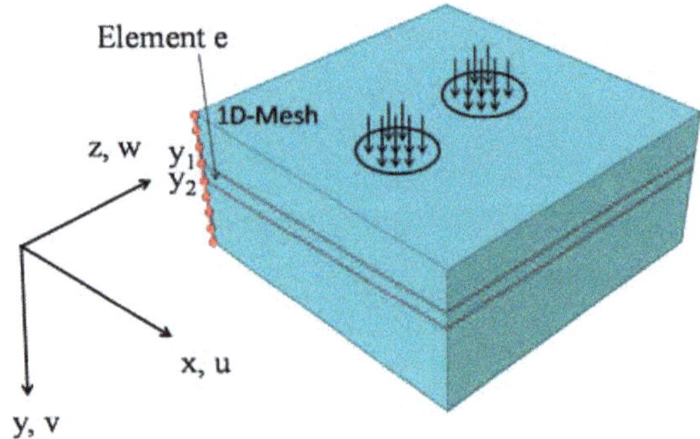

Figure 3. A typical EasyFEM element e in the pavement model.

Denote $e_h = d(y) - d_h(y)$. Then:

$$\begin{array}{c} a^e\left(d(y) - d_h(y), \tilde{d}(y)\right) = \int_{V_e} \tilde{d}(y)^T [E]^T [D][E] e_h dV \\ = \sum_{m=1}^{M} \sum_{n=1}^{N} \int_{V_e} \tilde{d}^{mn^T} [E^{mn}]^T [D][E^{mn}] e_h^{mn} dV = \sum_{m=1}^{M} \sum_{n=1}^{N} a^e(e_h^{mn}, \tilde{d}^{mn}) \approx 0 \end{array} \tag{21}$$

Denoting $E_{13}^{mn} = L_1 \frac{\partial N^{mn}}{\partial x} + L_3 \frac{\partial N^{mn}}{\partial z}$, $E_2^{mn} = L_2 N^{mn}$, and considering the zeros from integration of the Fourier series as shown in Equation (12), Equation (21) can be further rewritten as follows:

$$\begin{array}{c} \sum_{m=1}^{M} \sum_{n=1}^{N} \int_{V_e} \left(E_{13}^{mn} \tilde{d}^{mn} + E_2^{mn} \frac{\partial \tilde{d}^{mn}}{\partial y}\right)^T [D] \left(E_{13}^{mn} e_h^{mn} + E_2^{mn} \frac{\partial e_h^{mn}}{\partial y}\right) dV = \\ \sum_{m=1}^{M} \sum_{n=1}^{N} \int_{y_1}^{y_2} \left[\left(\frac{\partial \tilde{d}^{mn}}{\partial y}\right)^T A^{mn} \frac{\partial e_h^{mn}}{\partial y} + \left(\frac{\partial \tilde{d}^{mn}}{\partial y}\right)^T G^{mn} e_h^{mn} + \left(\tilde{d}^{mn}\right)^T (G^{mn})^T \frac{\partial e_h^{mn}}{\partial y} + \right. \\ \left. \left(\tilde{d}^{mn}\right)^T H^{mn} e_h^{mn} \right] dy \approx 0, \end{array} \tag{22}$$

where

$$\begin{array}{l} A^{mn} = \int_0^a \int_0^c (E_2^{mn})^T [D] E_2^{mn} dx dz, \\ G^{mn} = \int_0^a \int_0^c (E_2^{mn})^T [D] E_{13}^{mn} dx dz, \\ H^{mn} = \int_0^a \int_0^c (E_{13}^{mn})^T [D] E_{13}^{mn} dx dz, \end{array}$$

with integration by parts along the y-direction, the EEP can be equivalently converted into the following expression:

$$\begin{aligned}\sum_{m=1}^{M}\sum_{n=1}^{N}\int_{y_1}^{y_2}\left(-\widetilde{d}^{mn^T}A^{mn}\frac{\partial^2 e_h^{mn}}{\partial y^2}+\widetilde{d}^{mn^T}((G^{mn})^T-G^{mn})\frac{\partial e_h^{mn}}{\partial y}+\left(\widetilde{d}^{mn}\right)^T H^{mn}e_h^{mn}\right)dy \\ +\left(\widetilde{d}^{mn^T}(A^{mn}\frac{\partial e_h^{mn}}{\partial y}+G^{mn}e_h^{mn})\right)_{y_1}^{y_2}\approx 0.\end{aligned} \tag{23}$$

Since the test function $\widetilde{d}^{mn}$ can be arbitrarily selected for any m and n, the left side of Equation (23) is decoupled into

$$\int_{y_1}^{y_2}\widetilde{d}^{mn^T}\left(-A^{mn}\frac{\partial^2 e_h^{mn}}{\partial y^2}+\left(G^{mn^T}-G^{mn}\right)\frac{\partial e_h^{mn}}{\partial y}+H^{mn}e_h^{mn}\right)dy+\left(\widetilde{d}^{mn^T}(A^{mn}\frac{\partial e_h^{mn}}{\partial y}+G^{mn}e_h^{mn})\right)_{y_1}^{y_2}\approx 0. \tag{24}$$

Taking two linear polynomials:

$$\overline{N}_1=\frac{y_2-y}{y_2-y_1};\ \overline{N}_2=\frac{y-y_1}{y_2-y_1} \tag{25}$$

as two test functions $\widetilde{d}^{mn}$, respectively, and considering that:

$$d^{mn}(y_1)=d_h^{mn}(y_1),d^{mn}(y_2)=d_h^{mn}(y_2), \tag{26}$$

since the nodal displacements are super-convergent based on mathematical theory, one can obtain the following equation to calculate a recovered derivative solution at either of the two end-nodes of element e [39]:

$$\begin{aligned}\frac{\partial d_E^{mn}(y_1)}{\partial y} &= (A^{mn})^{-1}\left[(b,\overline{N}_1)-a^e(d_h^{mn},\overline{N}_1)-G_1 d_h^{mn}(y_1)\right],\\ \frac{\partial d_E^{mn}(y_2)}{\partial y} &= -(A^{mn})^{-1}\left[(b,\overline{N}_2)-a^e(d_h^{mn},\overline{N}_2)+G_2 d_h^{mn}(y_2)\right],\end{aligned} \tag{27}$$

where:

$$i=1,\ 2,\quad m=1,\cdots,M,\quad n=1,\cdots,N. \tag{28}$$

Substituting the derivative results calculated from Equation (28) into Equation (4), recovered strains and the corresponding stresses are obtained for the pavement model. This is the EEP super-convergence technique for FLM. When it is implemented in EasyFEM, stresses are obtained with much fewer elements and significantly less computational effort.

4. Numerical Analysis Using EasyFEM and EEP Techniques

4.1. Analytical Verification of EasyFEM

The analytical verification of the EasyFEM without EEP post-processing was carried out. The models of the asphalt pavement were created in BISAR, ABAQUS and EasyFEM, which are representative software applications developed based on LET, FEM and FLM, respectively. The pavement type is widely used in Germany according to the guidelines RStO-12 [40] and RDO-Asphalt-09 [1], as shown in Figure 4. The thicknesses of all layers except for the sub-grade were derived from RstO-12 [40]. The thickness of the sub-grade was defined to be 2000 mm in ABAQUS and EasyFEM after a previous mesh study in order to minimize the influence of the boundary on the computational results. Furthermore, the length and width of the pavement in the full-scale model were set to be 6000 mm based on the same reason. A full-scale model was created in the EasyFEM while the model in ABAQUS was one-fourth symmetrical model. The full-scale model exhibits advantages for the simulation of nonsymmetrical models with complex loading conditions in the further development of the EasyFEM. In BISAR, the thickness of the sub-grade as well as the length and width of the pavement were set to be infinite according to the LET. The pavement surface temperatures of −12.5 (winter) and 27.5 °C (summer) were assumed, and then the associated material properties were determined according to RDO-Asphalt-09 [1], as listed in Figure 4.

Surface course — 4 cm
Binder course — 8 cm
Asphalt base course — 14 cm
Road base course — 15 cm
Sub-base — 34 cm
Sub-grade

Layer	μ	Winter	Summer
		E [MPa]	E [MPa]
Surface course	0.35	22690	2902
Binder course	0.35	27283	6817
Asphalt base course	0.35	17853	4903
Road base course	0.25	10000	10000
Sub-base	0.5	100	100
Sub-grade	0.5	45	45

Figure 4. Geometrical data and material properties of the pavement.

The square load with the side length of 264 mm and uniformly distributed contact stress of 0.7 MPa was applied at the center of the full-scale pavement surface in the EasyFEM and a corresponding set-up was applied in ABAQUS. In BISAR, only circular loads can be defined and thus a uniform contact load of 0.7 MPa with a radius of 150 mm was applied. In all models, the three asphalt layers were totally bound; the two contact layers among the asphalt base course, road base course, sub-base and sub-grade were defined as being partially bound.

In order to attain a high accuracy, a very fine mesh was adopted in ABAQUS and EasyFEM after a mesh study. The mesh size in EasyFEM was uniformly set as 10 mm and thus 275 3-node quadratic elements with 551 nodes were generated. In order to reduce the computational consumption, the mesh size increases gradually from the top to the bottom and from the loading center to the boundary in the one-fourth symmetrical model in ABAQUS, as shown in Figure 5. The number of elements and nodes in ABAQUS are 162,728 and 193,601, respectively. There was no mesh generated from the BISAR.

Figure 5. Mesh generated from ABAQUS.

The computational results shown in Figure 6 are derived from five series of response points offset from the loading center to the boundary.

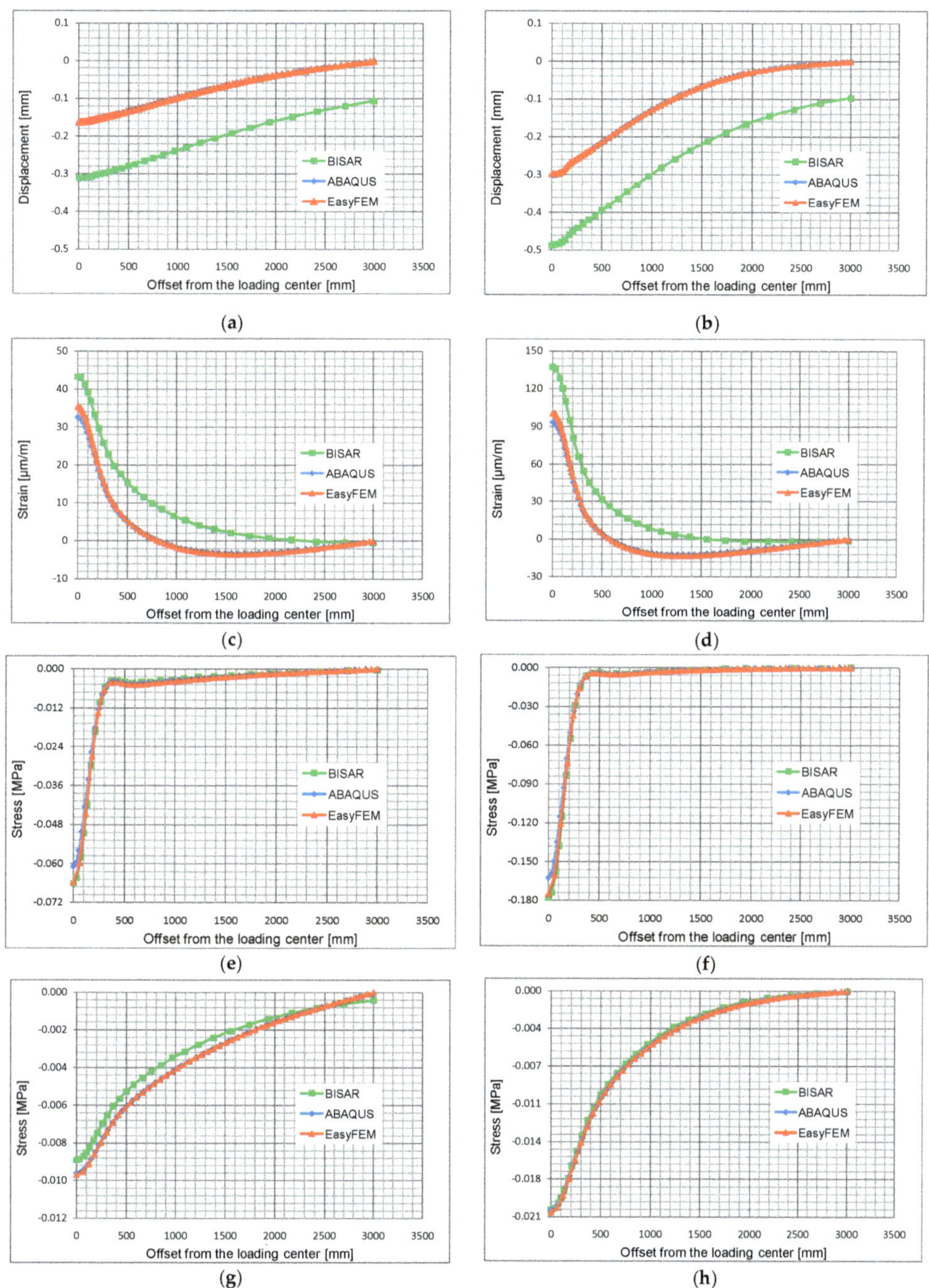

(a) (b) (c) (d) (e) (f) (g) (h)

Figure 6. *Cont.*

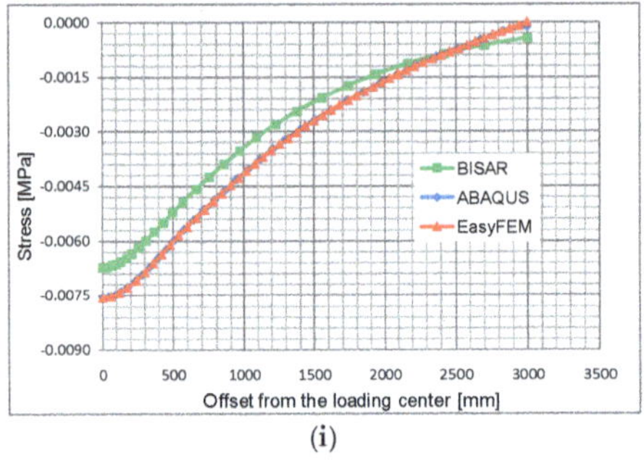

(i)

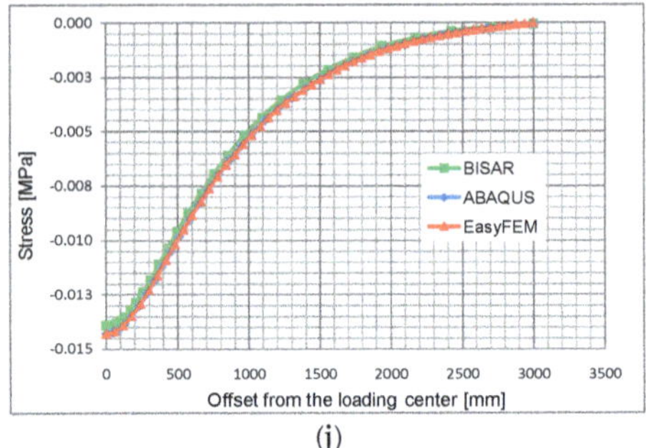

(j)

Figure 6. Comparison of vertical displacement at the top of the asphalt surface in (**a**) winter, (**b**) summer; comparison of horizontal strain along transverse direction at the bottom of the asphalt base course in (**c**) winter, (**d**) summer; comparison of vertical stress at the top of the road base layer in (**e**) winter, (**f**) summer; comparison of vertical stress at the top of the sub-base layer in (**g**) winter, (**h**) summer; comparison of vertical stress at the top of the sub-grade layer in (**i**) winter, (**j**) summer.

Given these figures, it can be concluded that the results derived from ABAQUS and EasyFEM are consistent with each other, yet the results from BISAR exhibit great deviations from those from the other two software applications, especially along the first two series of response points. The underlying reasons for the differences can be attributed to the different geometrical definitions of the models in BISAR due to its intrinsic limitation (circular load, infinite length and width of the layers and infinite thickness of the sub-grade); the deviations are the largest at the response points that are closest to the load.

In Table 1, the computational results from ABAQUS and EasyFEM are considered at five critical locations directly below the loading center shown in Figure 6, where maximum compressive or tensile values may occur.

Table 1. Comparison between ABAQUS and EasyFEM based on the results at critical points below the loading center.

Points	Results	Winter		
		EasyFEM	ABAQUS	Difference
1	Vertical displacement [mm]	-1.63×10^{-1}	-1.64×10^{-1}	7.72×10^{-4} (−0.47%)
2	Horizontal strain [-]	3.54×10^{-5}	3.25×10^{-5}	2.82×10^{-6} (8.68%)
3	Vertical stress [MPa]	-6.57×10^{-2}	-6.09×10^{-2}	-4.84×10^{-3} (7.95%)
4	Vertical stress [MPa]	-9.70×10^{-3}	-9.62×10^{-3}	-8.61×10^{-5} (0.90%)
5	Vertical stress [MPa]	-7.57×10^{-3}	-7.55×10^{-3}	-1.51×10^{-5} (0.20%)
Points	**Results**	**Summer**		
		EasyFEM	**ABAQUS**	**Difference**
1	Vertical displacement [mm]	-2.99×10^{-1}	-3.02×10^{-1}	3.24×10^{-3} (−1.07%)
2	Horizontal strain [-]	1.01×10^{-4}	9.33×10^{-5}	7.39×10^{-6} (7.92%)
3	Vertical stress [MPa]	-1.75×10^{-1}	-1.62×10^{-1}	-1.27×10^{-2} (7.85%)
4	Vertical stress [MPa]	-2.06×10^{-2}	-2.04×10^{-2}	-1.75×10^{-4} (0.86%)
5	Vertical stress [MPa]	-1.43×10^{-2}	-1.43×10^{-2}	-1.01×10^{-5} (0.07%)

It can be stated that the results from both programs have a high correlation except for a slightly larger difference in the horizontal strain at the bottom of the asphalt base course and vertical stresses at the top of the sub-base. Considering the different element types and mesh algorithms in the two programs, the differences are judged to be acceptable. Moreover, the computational time of the EasyFEM is much shorter than that of the ABAQUS. Both analyses were run on a computer with an Intel Core Duo 3.4 GHz, 32 GB RAM (Santa Clara, CA, USA). On average, the computational time required by the ABAQUS is about 420 s, whereas the EasyFEM model requires 120 s. It is worth mentioning that the model in ABAQUS is one-fourth symmetrical, but the one in EasyFEM is full-scale. With code optimization, the computation time of the EasyFEM can be reduced further. In summary, the accuracy and efficiency of the EasyFEM are proved by these comparisons.

4.2. Comparison of Convergence Rate between Ordinary EasyFEM and EasyFEM with EEP Post-Processing Techniques

As discussed above, by using 275 3-node quadratic finite elements along the y-direction, EasyFEM has performed fairly well compared to ABAQUS. If the number of elements can be further reduced without decreasing the computational accuracy, the EasyFEM will be even more suitable for application in pavement design and assessment from an engineering point of view. The EEP technique for recovering derivatives makes this possible. The convergence rates between ordinary EasyFEM and EasyFEM with the EEP post-processing technique are compared in this section to show the impact of the EEP technique on the solution accuracy.

To emphasize the significant improvement brought about by the application of EEP, a series of rough meshes with 2-node linear elements was adopted in the EasyFEM. In particular, three models with different meshes were created, i.e., each layer of the asphalt pavement was divided into one, two and four elements (N_e = 1, 2 and 4), respectively; the asphalt pavement had six layers resulting in a total of $6 \times N_e$ = 6, 12 and 24 linear elements in the three pavement models. EasyFEM with the EEP post-processing technique can produce strains and stresses with a higher accuracy, which is demonstrated by a number of numerical results. In Figure 7, vertical strains at the points located below the loading center and at the bottom of asphalt base course were compared between ordinary EasyFEM and EasyFEM with the EEP technique, where one, two and four linear finite elements were used in each layer along the y-direction, respectively. The results from the model with the fine mesh in ABAQUS created in Section 4.1 were taken as the reference. Since the stresses are calculated by multiplying strains with elastic constants, similar results can be obtained for vertical stresses as well. It can be seen that the EEP technique for FLM, which converges faster, can significantly improve the accuracy of vertical strains and stresses of a plain EasyFEM. With the EEP technique, the EasyFEM can offer reliable results much more quickly and efficiently and is thus suitable to be applied in pavement engineering practice.

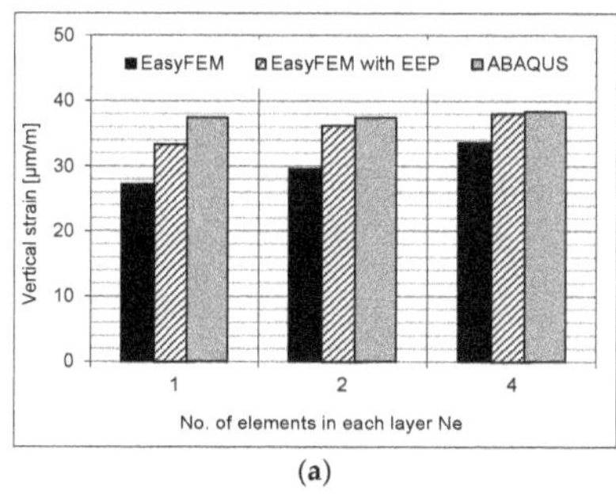

(**a**)

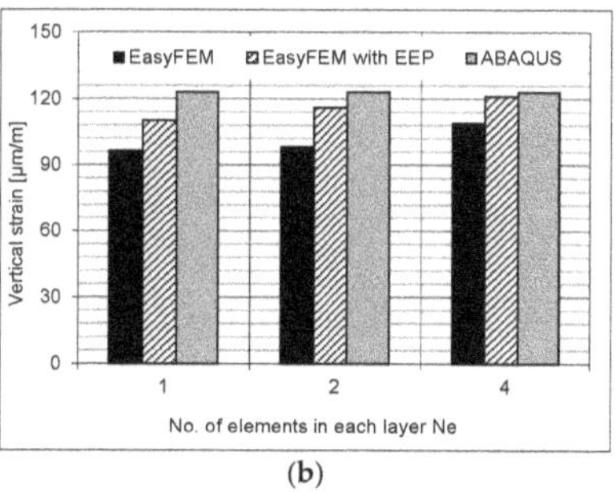

(**b**)

Figure 7. Comparison of the vertical strains at the points located below the loading center and at the bottom of asphalt base course obtained with ABAQUS, ordinary EasyFEM and EasyFEM with EEP (Element Energy Projection), (**a**) winter; and (**b**) summer.

4.3. Application of EasyFEM

The EasyFEM was used to predict the mechanical responses and the predictions were compared with the data derived from field measurements on the test track of the German Federal Highway Research Institute (BASt), as shown in Figure 8a. A truck driving at a speed of approximately 30 km/h was selected for the measurement to apply the loads, as shown in Figure 8b.

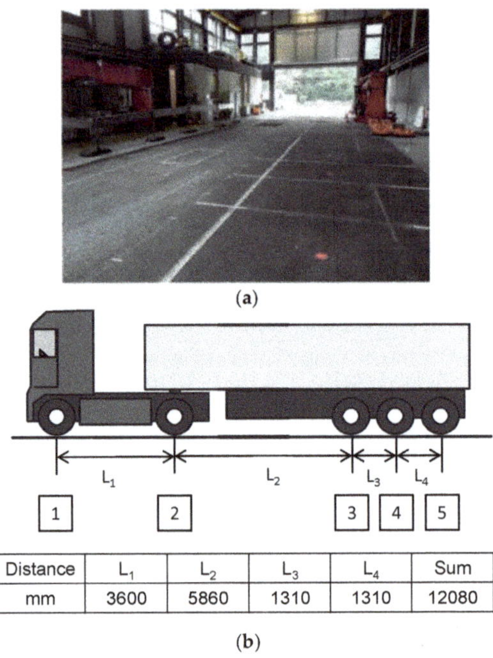

Figure 8. (**a**) The test track in BASt (the German Federal Highway Research Institute); (**b**) geometrical data and tires of the truck S23.

During the test track construction, strain gauges and pressure load cells were embedded at different depths of the test track along the center of the wheel path, which can measure strains and stresses along the corresponding directions when the truck passes. Due to the low speed of loads and the suggestion from [17], a static analysis with stationary loads is precise enough to compute the pavement responses. The material parameters of the pavement layers were derived from the laboratory tests on the specimens drilled from the test track. The thicknesses and the material properties of the test track are listed in Table 2 [17]. The length longitudinal to the direction of traffic (both directions) was defined as 20 times the loading radius to limit the time required for the computational calculation. The width of the pavement was defined to be 3750 mm.

Table 2. Layer thickness and material properties of the test track.

Layer	Thickness [mm]	μ	E [MPa]
Surface course	40	0.35	11150
Binder course	50	0.35	10435
Asphalt base course	110	0.35	6893
Gravel base layer	150	0.49	157.8
Frost protection layer	570	0.49	125.7
Sub-grade	2000	0.49	98.9

The loading parameters are listed in Table 3 [17]. All of the contact areas are assumed as squares with a side length of 264 mm. The distance between the center of the first left tire and the left edge of the test track along the traffic direction is 1100 mm.

Table 3. Loading parameters of the tire.

Axle	Wheel Number	Pressure [MPa]	Axle Load [kg]
1	2	0.522	7425
2	4	0.377	10725
3	2	0.513	7300
4	2	0.511	7275
5	2	0.519	7375

The upper three asphalt layers were totally bound. The two contact layers between the asphalt base course, gravel base layer, frost protection layer and sub-grade were defined as being partially bound. The 3-node quadratic elements with the mesh size of 10 mm were applied in the EasyFEM with the EEP technique.

The following computational values were compared with the measured data, which is used to verify the EasyFEM. The strain along the traffic direction at the bottom of the asphalt base course and the vertical tensile stress at the top of the gravel base layer from the location below the center of the left tire of each axle were used in the validation process. The values from the measurement and simulation can be seen in Figure 9. All computational strains are higher than the measured values and all computational stresses are lower than those obtained from the measurements. Due to the uncertainties and fluctuations, the error range is considered to be ±20% [17]. The computational values are all within this range. Thus, the EasyFEM is suited to simulate the response of the asphalt pavement under the traffic load with sufficient accuracy.

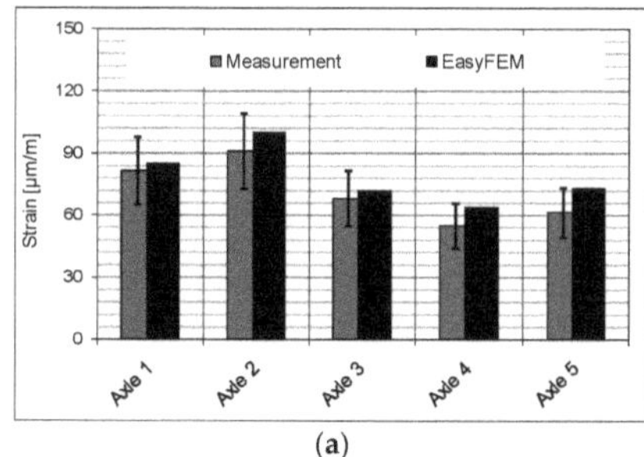

(a)

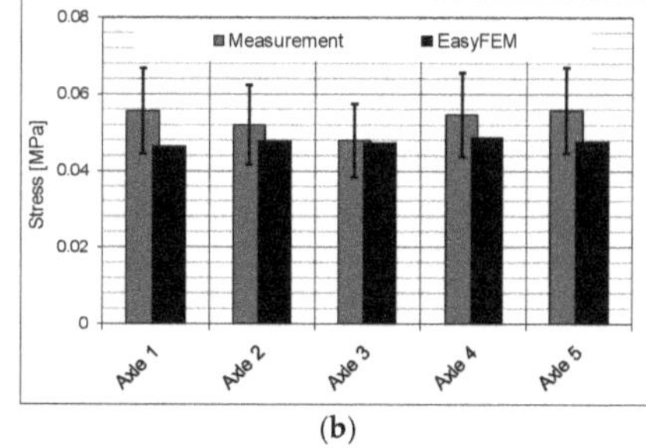

(b)

Figure 9. Comparison of the results derived from measurement and simulation. (**a**) horizontal strain along the traffic direction on the bottom of the asphalt base course; (**b**) vertical stress on the top of the gravel base layer.

5. Conclusions

Initially, the EasyFEM code is developed to predict the mechanical responses of asphalt pavements under stationary load. The accuracy of the program is verified by comparing the results to those obtained from BISAR and ABAQUS. The analytical verification showed that the pavement responses derived from EasyFEM and ABAQUS are in accordance with one another. Due to the different definitions in the models of the BISAR, some results appear to exhibit larger differences than those observed between EasyFEM and ABAQUS, which indicates that BISAR is not as flexible as the other two software applications when simulating the responses of the asphalt pavements. It should be emphasized that the computational time of the EasyFEM is much shorter than that of ABAQUS.

The EEP super-convergence strategy for FLM is introduced and applied in the EasyFEM to improve the accuracy of derivatives at two end-nodes on any element. The convergence rate of the EasyFEM with the EEP post-processing technique is much faster than without the EEP. The application of the EasyFEM proves that the current version of EasyFEM can provide a reliable prediction of the mechanical responses of asphalt pavements and thus represents a flexible and robust base for further development.

For the next step of development, various material properties, e.g., viscoelasticity and nonlinear elasticity, may be applied for asphalt layers and sub-base of the pavement, respectively. The EEP technique will be further developed for the new EasyFEM with more functions. Based on the EEP technique, adaptivity is to be included into the EasyFEM, i.e., an error tolerance for the solution is specified by the user before the calculation, and, finally, a numerical solution is obtained that satisfies the error tolerance, while the meshes used in the simulation are adaptively generated and adjusted automatically to produce a satisfactory solution with a high quality and accuracy. With these improvements, the EasyFEM should be more suited to predict the mechanical performances of the asphalt pavement.

Acknowledgments: This paper is based on a part of the research project carried out at the request of the German Research Foundation, under research project No. FOR 2089. It is also a part of the project No. 51508305 supported by the National Natural Science Foundation of China. The authors are solely responsible for the content.

Author Contributions: Pengfei Liu developed the FLM program EasyFEM; Qinyan Xing applied the EEP technique to EasyFEM; Yiyi Dong analyzed the data; Dawei Wang conceived the paper; Markus Oeser contributed programming algorithm; and Si Yuan directed the development of the EEP technique.

Conflicts of Interest: The authors declare no conflict of interest.

References

1. *Richtlinien für die Rechnerische Dimensionierung von Verkehrsflächen mit Asphaltdeckschicht: RDO Asphalt 09*; FGSV Publishing: Cologne, Germany, 2009.
2. Dong, Z.; Tan, Y. *Dynamic Response of Asphalt Pavement*; Science Press: Beijing, China, 2015.
3. *BISAR: Main Frame Computer Program, User Manual*; Shell International Oil Products: The Hague, The Netherlands, 1998.
4. Huang, H. *Pavement Analysis and Design*; Prentice Hall: New Jersey, NJ, USA, 1993.
5. Scarpas, A. CAPA-3D: A Mechanics Based Computational Platform for Pavement Engineering. Ph.D. Thesis, Delft University of Technology, Delft, The Netherlands, 2005.
6. Lytton, R.L.; Tsai, F.L.; Lee, S.I.; Luo, R.; Hu, S.; Zhou, F. Appendix Q: Finite Element Program to Calculate Stress Intensity Factor. In *NCHRP Report 669: Models for Predicting Reflection Cracking of Hot-Mix Asphalt Overlays*; The National Academies of Sciences, Engineering, and Medicine: Washington, DC, USA, 2010. [CrossRef]
7. Gonzalez, A.; Bodin, D.; Jameson, G.; Oeser, M.; Vuong, B. *Development of a Nonlinear Finite Element Pavement Response to Load Model*; Austroads Publication No. AP-T199-12; Austroads: Sydney, Australia, 2012.
8. Wilson, E.L. Structural analysis of axisymmetric solids. *AIAA J.* **1965**, *3*, 2269–2274. [CrossRef]
9. Meissner, H.E. Laterally loaded pipe pile in cohesionless soil. In Proceedings of the 2nd International Conference on Numeric Methods in Geomechanics, Virginia Polytechnical Institute and State University, Blacksburg, VA, USA, 20–25 June 1976.
10. Winnicki, L.A.; Zienkiewicz, O.C. Plastic (or visco-plastic) behaviour of axisymmetric bodies subjected to non-symmetric loading-semi-analytical finite element solution. *Int. J. Num. Meth. Eng.* **1979**, *14*, 1399–1412. [CrossRef]
11. Carter, J.P.; Booker, J.R. Consolidation of axi-symmetric bodies subjected to non axi-symmetric loading. *Int. J. Numer. Anal. Meth. Geomech.* **1983**, *7*, 273–281. [CrossRef]
12. Lai, J.Y.; Booker, J.R. Application of discrete Fourier series to the finite element stress analysis of axi-symmetric solids. *Int. J. Num. Meth. Eng.* **1991**, *31*, 619–647. [CrossRef]
13. Fritz, J.J. Flexible Pavement Response Evaluation Using the Semi-Analytical Finite Element Method. *Int. J. Mater. Pavement Des.* **2002**, *3*, 211–225.

14. Hu, S.; Hu, X.; Zhou, F. Using Semi-Analytical Finite Element Method to Evaluate Stress Intensity Factors in Pavement Structure. In Proceedings of the RILEM International Conference on Cracking in Pavements, Chicago, IL, USA, 16–18 June 2008; pp. 637–646.
15. Liu, P.; Wang, D.; Oeser, M.; Chen, X. Einsatz der Semi-Analytischen Finite-Elemente-Methode zur Beanspruchungszustände von Asphaltbefestigungen. *Bauingenieur* **2014**, *89*, 333–339.
16. Liu, P.; Wang, D.; Otto, F.; Hu, J.; Oeser, M. Application of Semi-Analytical Finite Element Method to Evaluate Asphalt Pavement Bearing Capacity. *Int. J. Pavement Eng.* **2016**. [CrossRef]
17. Liu, P.; Wang, D.; Hu, J.; Oeser, M. SAFEM – Software with Graphical User Interface for Fast and Accurate Finite Element Analysis of Asphalt Pavements. *JTE* **2017**, *45*, 1–15. [CrossRef]
18. Liu, P.; Wang, D.; Oeser, M. The Application of Semi-Analytical Finite Element Method to Analyze Asphalt Pavement Response under Heavy Traffic Loads. *JTTE* **2017**. [CrossRef]
19. Liu, P.; Wang, D.; Oeser, M. Application of semi-analytical finite element method coupled with infinite element for analysis of asphalt pavement structural response. *JTTE* **2015**, *2*, 48–58. [CrossRef]
20. Cheung, Y.K.; Fan, S.C. Analysis of pavements and layered foundations by finite layer method. In *S.A.E. Preprints, Proceedings of the Third International Conference on Numerical Methods in Geomechanics, Aachen, Germany, 2–6 April 1979*; Wittke, W., Balkema, A.A., Eds.; Society of Automotive Engineers: Rotterdam, The Netherlands, 1979; pp. 1129–1135.
21. Rowe, R.K.; Booker, J.R. Finite layer analysis of nonhomogeneous soils. *JEMD* **1982**, *108*, 115–132.
22. Small, J.C.; Booker, J.R. Finite layer analysis of layered elastic materials using a flexibility approach, 1, Strip loadings. *Int. J. Numer. Meth. Eng.* **1984**, *20*, 1025–1037. [CrossRef]
23. Small, J.C.; Booker, J.R. Surface deformation of layered soil deposits due to extraction of water. In Proceedings of the Ninth Australasian Conference on the Mechanics of Structures and Materials, Sydney, Australia, 29–31 August 1984; pp. 33–38.
24. Booker, J.R.; Small, J.C. Finite layer analysis of consolidation I. *Int. J. Numer. Anal. Meth. Geomech.* **1982**, *6*, 151–171. [CrossRef]
25. Booker, J.R.; Small, J.C. Finite layer analysis of consolidation II. *Int. J. Numer. Anal. Meth. Geomech.* **1982**, *6*, 173–194. [CrossRef]
26. Booker, J.R.; Small, J.C. Finite layer analysis of viscoelastic layered materials. *Int. J. Numer. Anal. Meth. Geomech.* **1986**, *10*, 415–430. [CrossRef]
27. Smith, S.S.; Allen, M.B.; Puckett, J.; Edgar, T. *The Finite Layer Method for Groundwater Flow Models*; Wyoming Water Research Center: Laramie, WY, USA, 1992.
28. Zai, J.M.; Mei, G.X. Finite layer analysis of three dimentional Biot consolidation. *CJGE* **2002**, *24*, 31–33.
29. Díaz-Cereceda, C.; Poblet-Puig, J.; Rodríguez-Ferran, A. The finite layer method for modelling the sound transmission through double walls. *J. Sound Vib.* **2012**, *331*, 4884–4900. [CrossRef]
30. Siddharthan, R.V.; Yao, J.; Sebaaly, P.E. Pavement Strain from Moving Dynamic 3-D Load Distribution. *J. Transp. Eng.* **1998**, *124*, 557–566. [CrossRef]
31. Siddharthan, R.V.; Krishnamenon, N.; Sebaaly, P.E. Pavement Response Evaluation using Finite-Layer Approach. *Transp. Res. Rec.* **2000**, *1709*, 43–49. [CrossRef]
32. Zienkiewicz, O.C.; Zhu, J. The super-convergence patch recovery (SPR) and a posteriori error estimates, Part 1: The recovery technique, Part 2: Error estimates and adaptivity. *Int. J. Num. Method. Eng.* **1992**, *33*, 1331–1382. [CrossRef]
33. Babuska, I.; Strouboulis, T.; Upadhyay, C.S.; Gangaraj, S.K. A posteriori estimation and adaptive control of the pollution error in the h-version of the finite element method. *Int. J. Num. Method. Eng.* **1995**, *38*, 4207–4235. [CrossRef]
34. Yuan, S.; Du, Y.; Xing, Q.; Ye, K. Self-adaptive one-dimensional nonlinear finite element method based on element energy projection method. *Appl. Math. Mech.* **2014**, *35*, 1223–1232. [CrossRef]
35. Yuan, S.; He, X. A self-adaptive strategy for one-dimensional FEM based on EEP method. *Appl. Math. Mech.* **2006**, *27*, 1461–1474. [CrossRef]
36. Yuan, S.; Xing, Q.; Wang, X.; Ye, K. Self-adaptive strategy for one-dimensional FEM based on EEP method with super-convergence order. *Appl. Math. Mech.* **2008**, *29*, 591–602. [CrossRef]
37. Strang, G.; Fix, G. *An Analysis of the Finite Element Method*; Prentice-Hall: London, UK, 1973.
38. Zhao, Q.H. The Mathematical Analysis on Element Energy Projection Method. Ph.D. Dissertation, Hunan University, Changsha, China, 2007. (In Chinese)

39. Yuan, S.; Xiao, J.; Ye, K. EEP Super-Convergent Computation in FEM Analysis of FEMOL Second Order ODEs. *Eng. Mech.* **2009**, *26*, 1–9.
40. *Guidelines for the Standardization of the Upper Structure of Traffic Areas: RStO-12*; FGSV: Cologne, Germany, 2012.

Article

A New Life for Cross-Linked Plastic Waste as Aggregates and Binder Modifier for Asphalt Mixtures

Liliana M. B. Costa [1], Joana Peralta [2], Joel R. M. Oliveira [1] and Hugo M. R. D. Silva [1,*]

[1] CTAC, Centre for Territory, Environment and Construction, University of Minho, Guimarães 4800 058, Portugal; b6100@civil.uminho.pt (L.M.B.C.); joliveira@civil.uminho.pt (J.R.M.O.)
[2] Wacker Chemical Corporation, 6870 Tilghman Street, Allentown, PA 18106-9346, USA; Joana.Peralta@wacker.com
* Correspondence: hugo@civil.uminho.pt; Tel.: +351-253-510200

Academic Editors: Zhanping You, Qingli Dai and Feipeng Xiao
Received: 26 May 2017; Accepted: 6 June 2017; Published: 10 June 2017

Featured Application: This work deals with recycling materials for pavement application purposes. It shows a new potential use for a specific plastic waste material with a limited demand (cross-linked polyethylene waste), which can be applied as an aggregate substitute and/or binder modifier for asphalt mixtures, thus adding value to the whole paving solution.

Abstract: Every year, millions of tons of plastic waste, with potential to be reused, are wasted in landfills. Based on a literature review and in a local market analysis, cross-linked polyethylene (PEX) waste arose as the material with the greatest potential to be tested for incorporation in asphalt mixtures due to the difficulty in its recycling and the lack of solutions for its reuse. Thus, in the present work, mixtures produced with and without PEX were tested in order to compare their performance, aiming at understanding if this waste could successfully be used as an alternative material for this type of application. Thus, water sensitivity, rutting resistance, stiffness modulus and fatigue cracking resistance tests were carried out on asphalt mixtures with up to 5% PEX. Based on the results obtained, it can be concluded that the incorporation of PEX in asphalt mixtures is a viable solution for paving works, especially when high service temperatures are expected. It also decreases the density of the mixture, which can be attractive to lighten structures. Thus, this technology contributes to give new life to cross-linked polyethylene plastic waste.

Keywords: plastic waste; cross-linked polyethylene; binder modification; aggregate substitution; mix design; asphalt mixture performance

---

1. Introduction

The contemporary Society is stimulated by consumption, resulting in increasing demands for the use of scarce natural resources and in the production of large amounts of waste. This unsustainable scenario is also observed in the road infrastructure sector, where significant amounts of construction and demolition waste are sent to landfills every year. Therefore, the development of solutions that promote road pavement durability and reduce the amount of waste shall be promoted. Furthermore, several other wastes are difficult to recycle and may even need some treatment before being disposed. Taking that into account, it would be interesting to use wastes to improve the performance of road pavements.

Among the wastes used in road pavements, reclaimed asphalt pavement (RAP) materials are those that have been most successfully incorporated [1,2], even though some sort of rejuvenation additive may be needed [3,4]. Nevertheless, there are other wastes with high potential to be used in this type of application. Plastic wastes are one of the most promising solutions [5], while millions of tons of are sent to landfills every year [6]. Most of these plastics have a simple process of recycling, but the thermosetting plastics cannot be reprocessed, so finding the proper use for the disposal of this plastic waste is an emerging need [7].

Polymers can be incorporated into asphalt mixtures by a wet process, as bitumen modifier, or by a dry process, as partial substitute of the aggregates [8–10]. The wet process is the most widely used [11], since this method requires a smaller amount of polymer, which in its virgin state significantly increases the price of the binder. There are several examples of the use of virgin polymers in bitumen modification. Among the most commonly used are styrene-butadiene-styrene (SBS), styrene-butadiene rubber (SBR), ethylene-vinyl acetate (EVA), polyethylene (PE) and polypropylene (PP) [12–14].

According to some authors, it is possible to use plastic wastes in substitution of virgin polymers to achieve that goal, which leads to ecological and possibly economic benefits [15,16]. There are some plastic wastes with potential to be used in bitumen modification, but polyethylene has the advantage of being the most used polymer and thus resulting in the plastic waste available in higher quantities [6,17].

Some authors concluded that the use of recycled polyethylene by the wet method promoted improvements in the resistance to permanent deformation, decreasing the thermal susceptibility [18], increasing the stiffness [19] and promoting the best fatigue performance, when compared to unmodified asphalt mixtures [20].

The dry method was also studied by some researchers, which added recycled and virgin polymers directly to the aggregates [11]. This technique may reduce pavement deformation, increasing fatigue resistance and providing a better adhesion between the bitumen and the aggregates [10,17,21].

As mentioned above, some other plastic wastes are more difficult to recycle due to their nature, namely the thermosets, which are not possible to melt again after being produced. Cross-linked polyethylene (PEX) is a material used in the production of pipes and cables. Every year, in Europe, 400,000 tons of plastics from cable recycling are sent to landfills, and 20% of those are PEX [22]. Thus, based on a literature review and a market survey, it was concluded that it would be interesting to include this waste material in the production of asphalt mixtures with superior performance, while giving a new life to this polyethylene-based polymer that is difficult to recycle.

Therefore, the possibility of using this thermoset plastic waste in asphalt mixtures as a partial substitute of the aggregates would promote the use of higher amounts of this waste material, increasing its value, while saving landfill space and reducing the extraction of mineral aggregates. Thus, this work intended to replace part of the mineral aggregates with recycled PEX and improve the performance of the asphalt mixtures, which was obtained by using 5% of PEX, by volume of mixture.

2. Materials and Methods

2.1. Materials Used in This Study

Three different constituent materials were used in this work in order to study new asphalt mixtures incorporating a plastic waste. As usual, mineral aggregates and asphalt binders were used as indispensable components of asphalt mixtures, and a specific plastic waste was the third material used, applied both as an aggregate substitute and/or asphalt binder modifier.

Crushed granite aggregates were supplied by a local quarry in the Northwest of Portugal, being divided into three fractions with the following dimensions: 6/14 gravel, 4/6 gravel and 0/4 dust. A small quantity of limestone filler was also used, as it is usual practice for the production of asphalt mixtures with those aggregates. These aggregates were selected to produce AC surf 14 mixtures for pavement surface layers.

Two commercially available binders typically applied in Portugal were used in this study, both supplied by Cepsa Corporation, Madrid, Spain. A 50/70 pen grade bitumen (according to EN 12591) was used as the base binder for comparison purposes, before and after incorporation of plastic waste in the mixtures. A polymer modified binder PMB 45/80-60 (according to EN 14023) was also used for an additional comparison, taking into account that the use of plastic waste may change the base 50/70 bitumen into a PMB.

The plastic waste applied in this study was a cross-linked polyethylene (PEX), mainly recovered from pipe, tubing and electrical cables, and it was supplied by Gintegral S.A., Póvoa de Varzim, Portugal. This material is composed of particles with a lamellar flake shape and an equivalent diameter ($Ø_{eq}$) ranging between 0.5 and 10.0 mm. In order to reduce their size and lamellar shape, PEX flaxes were ground to obtain smaller particles with a nominal diameter between $0.5 < Ø_{eq}$ (mm) < 4.0.

2.2. Characterization of the Plastic Waste Material Used

First, the PEX waste was characterized in order to assess its main properties, which may directly influence the design of asphalt mixtures or their future performance in the pavement.

Taking into account that PEX will be used as aggregate substitute, it is essential to evaluate its grading curve (according to EN 933-1 standard) and particle density (EN 1097-6 standard). In fact, the results of these tests will be taken into consideration when designing the asphalt mixture, namely to adjust its grading curve.

Almost all PEX material is made from high-density polyethylene, containing cross-linked bonds in the polymer structure that change the thermoplastic to a thermoset nature. The cross-linking process can be more or less effective, but it is never complete. In fact, a higher degree of cross-linking could result in brittleness of the material, while a lower degree of cross-linking could result in products with poorer physical properties, especially at elevated temperatures. In this study, the PEX waste will be mixed with bitumen and aggregates at very high temperatures. It is not expected that the cross-linked part of PEX will be changed during the mixing phase, thus working as an aggregate substitute, while the remaining part will certainly interact with the base bitumen, modifying its properties. Thus, it is very important to assess the GEL content, or degree of cross-linking, of PEX waste, which can be carried out according to the ASTM D2765 standard. In order to quantify the amount of PEX waste that cannot melt, a sample with a known mass is immersed in xylene and boiled for 24 h and then the remaining sample is dried and weighed.

Finally, DSC tests were performed in order to evaluate the melting temperature of the PEX portion that is not cross-linked, which should be lower than the asphalt mixing temperature to ease the bitumen modification. The DSC tests were carried out in a DSC Diamond Pyris, Waltham, MA, U.S.A. About 10 mg of sample was sealed in an aluminium capsule, and then the following protocol was applied: two heating and one cooling cycle ranging from −40 °C to 160 °C at a rate of 10 °C/min. The two heating cycles should provide the same thermal history for the entire sample in order to assure that this range of temperatures used for asphalt production does not change the polymer properties.

2.3. Definition of the Asphalt Mixtures to Be Studied and Their Aggregate Grading Curves

The mineral aggregates and the filler were also characterized regarding their grading curves, according to the EN 933-1 standard. By knowing the gradation of all aggregates, including the PEX waste, it is possible to calculate the amount of each material to be used in order to fit the standard grading limits of the asphalt mixture to be produced in this work, i.e., an AC 14 surf mixture (Portuguese annex of standard EN 13108-1).

The main objective of this work is to incorporate a significant amount of PEX in an asphalt mixture for a road pavement surface layer and compare its performance with a conventional mixture without PEX. Thus, two comparable asphalt mixtures were defined in this work to carry out the study: (i) a conventional mixture with only mineral aggregates; (ii) a new mixture incorporating PEX, with 5% of PEX by volume of the aggregates used in the mixture. Taking into account that polymer modifiers are typically included only as a small percentage of the bitumen, the amount of PEX used in this work is remarkably high, in order to highlight the outcomes of this solution.

The calculated amounts of each aggregate fraction and PEX waste used in both asphalt mixtures, with or without PEX, in order to fit their gradation within the limits of an AC 14 surf mixture are presented in the Table 1.

Table 1. Percentage of each fraction of aggregates used in this work for asphalt mixture production.

Asphalt Mixture	Aggregate Fractions Used in the Mixtures (% in Volume)				
	6/14 Gravel	4/6 Gravel	0/4 Dust	Filler	PEX Waste
Conventional mixture (only with mineral aggregates)	46.0	13.0	39.3	1.7	-
Mixture incorporating PEX (with mineral aggregates and PEX)	45.0	10.9	37.4	1.7	5.0

The aggregate grading curves of the different mineral aggregates, filler and PEX waste are presented in Figure 1a. Then, by using the amount of each aggregate fraction and PEX waste previously presented, the aggregate grading curves of the conventional mixture (without PEX) and the mixture incorporating PEX are presented in Figure 1b, perfectly adjusted to the standard limits of the AC 14 surf mixture.

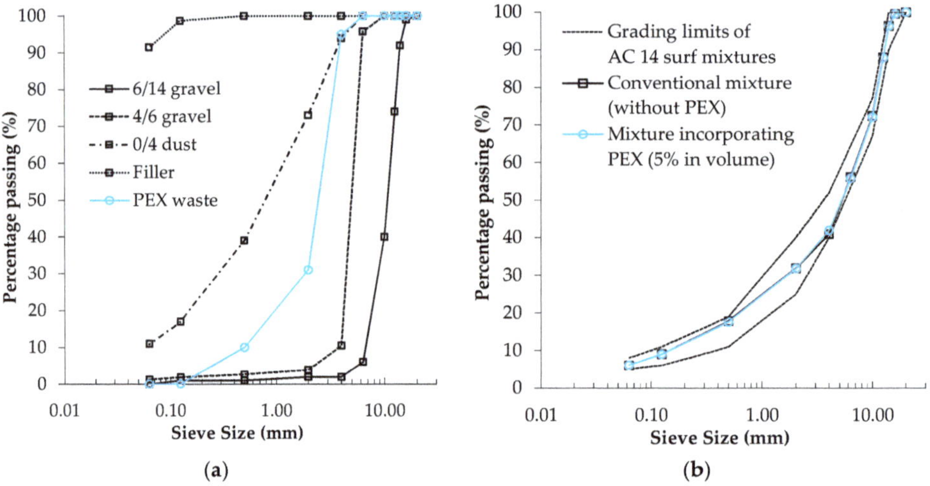

Figure 1. Aggregate grading curves of: (**a**) Mineral aggregates, filler and cross-linked polyethylene (PEX) waste and; (**b**) Asphalt mixtures (with and without PEX) fitted within the limits of an asphalt concrete mixture AC 14 surf.

PEX waste presents a 0.5/4 dimension, which will partially substitute the coarser part of 0/4 mineral dust. It should be noted that these dimensions could be changed in future studies, namely by further reducing the PEX size in order to increase the interaction with bitumen, but the industry costs associated with this operation shall be taken into consideration.

It can also be observed that the particle size distribution obtained for both asphalt mixtures, with or without PEX, is almost equivalent. Thus, the performance of both mixtures can be compared without direct influence of their gradation, which should only be influenced by PEX waste.

2.4. Evaluation of Asphalt Binder Characteristics after Interaction with the Plastic Waste

The use of PEX waste in an asphalt mixture, incorporated by the dry process as an aggregate substitute, will also cause the modification of the base bitumen used in the mixture because PEX is not totally cross-linked. Thus, the direct influence of the melting part of PEX waste in the properties of the base bitumen 50/70 was evaluated in this work because the polymer may be able to modify the binder during the short period of asphalt mixture production [13] when PEX and bitumen can interact at very high temperatures.

In order to simulate the interaction between the bitumen and the PEX waste during the asphalt mixture production, a PEX modified binder was produced by mixing the PEX with bitumen for 2 min at 180 °C (reproducing the time and temperature used for asphalt mixture production), in an IKA blender at 250 rpm (Figure 2a). The amounts of PEX waste and 50/70 bitumen used to produce the PEX modified binder were proportional to those used for asphalt mixture production. The amount of PEX used in the asphalt mixture is 5% by volume of aggregates, which corresponds to approximately 1.8% by weight due to the low density of PEX in comparison with that of the mineral aggregates. For an asphalt mixture with a bitumen content of 5% (typical value for AC 14 surf mixtures), the amount of PEX waste will be 1.71% by weight of the entire mixture, corresponding to 34.2% by weight of 50/70 bitumen. Thus, in order to produce 1000 g of PEX modified binder, 745 g of bitumen and 255 g of PEX were used. At the end of the mixing process, the binder was filtered with a metal mesh (sieve size of 0.5 mm) in order to separate the cross-linked part of PEX waste (Figure 2b) from the binder that was modified with the melted part of PEX, which will be further characterized to evaluate the actual modification process.

(a)

(b)

Figure 2. Evaluation of PEX interaction with bitumen: (**a**) Binder preparation for 2 min at 180 °C and; (**b**) The binder was filtered to separate the cross-linked part of PEX undissolved in bitumen.

At the end of this process, the filtered binder was characterized according to the EN 12591 standard. The basic characterization included penetration at 25 °C (EN 1426) and softening point or ring and ball (R&B) temperature (EN 1427) tests. Dynamic viscosity tests were also carried out in a Brookfield viscometer (EN 13302 standard) in order to evaluate the behaviour at mixing and compaction temperatures. The base bitumen (50/70) and a commercially available PMB (PMB 45/80-60) were also characterized using the same test protocol for comparison purposes and also because they will be used for the production of asphalt mixtures.

2.5. Design of the Asphalt Mixtures Incorporating PEX

The design of the conventional AC 14 surf mixtures without PEX has been previously presented [23,24], and the optimum binder content was found to be 5.0%. The same value was used in this work to produce the asphalt mixtures with mineral aggregates. However, the design of the asphalt mixtures incorporating PEX was performed for the first time in this work.

Taking into account the higher absorption of bitumen by the PEX waste in comparison with mineral aggregates, and the partial interaction of those two materials, it is expected that a higher binder content would be required to produce the new asphalt mixtures with PEX. Thus, the design of these mixtures was carried out according to the Portuguese annex of the NP EN 13108-1 standard, but only using three binder contents equal or higher than those of the conventional mixture (i.e., 5.0, 5.5 and 6.0%). The optimum binder content to be determined is the average value of the binder contents that results in the maximum Marshall stability (EN 12697-34 standard), an air void content of 4.0% for AC 14 surf mixtures (EN 12697-8) and the maximum bulk density (EN 12697-6, procedure B).

A complementary validation study was also performed to confirm the Marshall Mix design results by evaluating the water sensitivity (EN 12697-12 standard) of the mixtures with PEX for those three binder contents. The main result of this test is the ITSR, which is the ratio between the average indirect tensile strength test (EN 12697-23) results from a group of three specimens immersed in water (ITSw) and another group of three similar specimens tested under dry conditions (ITSd). The design and validation study of the asphalt mixtures with PEX allowed determination of the binder contents to use in the production of these mixtures for additional performance evaluation.

After defining the composition of all mixtures, two mixtures without PEX (with 50/70 and PMB 45/80-60 binders) and three mixtures with PEX (Figure 3a) were produced in order to obtain all test samples (Figure 3b) necessary to carry out several performance-related tests.

(a) (b)

Figure 3. Asphalt mixtures with PEX developed in this work: (**a**) Production stage applying the dry method of incorporation; (**b**) Tests samples with PEX particles clearly visible.

2.6. Performance Evaluation of Studied Asphalt Mixtures

The performance related tests included the determination of the water sensitivity, rutting resistance, stiffness modulus and fatigue resistance of the studied asphalt mixtures.

The design of a bituminous road pavement involves the definition of a structure that ensures the capacity of withstanding the loads applied by vehicles and by weather conditions. The mechanical properties of the bituminous mixtures to be used in the road structure should be evaluated, such as the stiffness modulus, phase angle, and pavement resistance to failure mechanisms such as cracking due to fatigue and permanent deformation.

Among the resistance to the weather conditions, the evaluation of asphalt mixture water sensitivity is very important, since this property is directly related to the performance of the road pavement. This property was determined by the EN 12697-12 standard. The indirect tensile strength test was carried out according to the EN 12697-23 standard, after a volumetric characterization of the specimens to determine the air void content, which significantly influences the results.

The rutting resistance of the mixtures was assessed by means of the wheel tracking test (WTT), according to the EN 12697-22 standard, using the procedure B (in air), with a standard wheel load of approximately 700 N. The test was carried at a temperature of 50 °C, being representative of our country's hotter summer days.

The stiffness modulus test was carried out on prismatic specimens, using the four-point bending configuration (4PB-PR), according to the EN 12697-26 standard. The test was carried out at 10, 20,

30 and 40 °C for a range of frequencies (0.1, 0.2, 0.5, 1, 2, 5, 8 and 10 Hz). Based on the principle of superposition time temperature, used to relate the equivalence between frequency and temperature, it was possible to construct master curves, in this case using 20 °C as temperature of reference. The results are represented by the stiffness modulus and the phase angle, which are the most relevant properties of viscoelastic materials such as asphalt mixtures.

The fatigue resistance of the studied mixtures was also determined using the four-point bending beam test procedure, according to the EN 12697-24 standard. The tests were carried out at 20 °C, in strain control mode, and using a frequency of 10 Hz. The fatigue laws of the studied mixtures can be represented by Equation (1):

$$N = \mathrm{a} \times (1/\varepsilon_t)^{\mathrm{b}}, \quad (1)$$

which relates the level of extension applied in the test (ε_t) with the number load cycles (N) that cause test specimen failure (reduction of stiffness to half of its initial value), and where a, b are coefficients determined experimentally.

3. Results and Discussion

3.1. Properties of the Plastic Waste Material

In this work, the PEX waste is the new material applied, originally, in asphalt mixtures. The results of its experimental characterization are presented below, according to the methodology presented in Section 2.2.

The particle density of PEX waste at 25 °C was evaluated as being 938.6 kg/m^3. This value is slightly lower than that of bitumen (nearly 1030 kg/m^3), but is clearly lower than that of mineral aggregates (nearly 2650 kg/m^3) that PEX is partially substituting. The knowledge of this value was crucial to calculate important volume and weight proportions between the different materials used in the asphalt mixtures. Furthermore, the grading curve of the ground PEX waste particles, previously presented in Figure 1a, was also essential to adjust the grading curve of the new asphalt mixtures with PEX within the standard limits of an AC 14 surf mixture.

The ability of PEX particles (Figure 4a) to partially melt under application of heat (180 °C) and pressure, specifically because they are not totally cross-linked, can be observed in Figure 4b. However, the cross-linked part of PEX is not changed by the presence of solvents, heat or pressure.

(a)

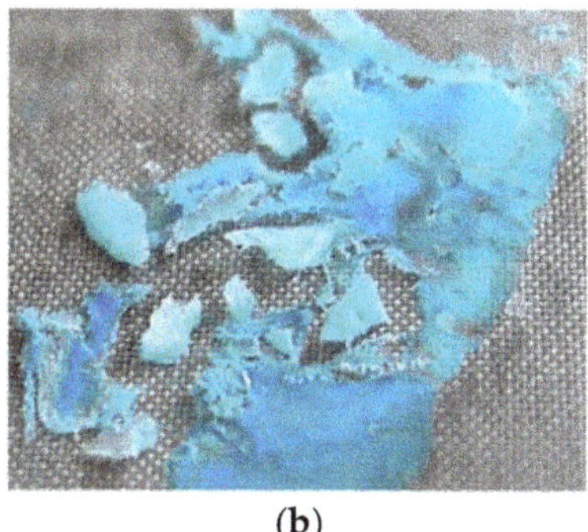

(b)

Figure 4. PEX waste particles: (**a**) Before being heated; (**b**) After applying heat and pressure at 180 °C.

The GEL content (or degree of cross-linking) of the PEX waste used in this work was determined to be 54%. This percentage of PEX will certainly not melt, working only as aggregate substitute, but the other part of PEX waste (nearly half of it) can melt and modify the asphalt binder during the production of asphalt mixtures, thus improving the mixture performance.

The calorimetry results (DSC) of PEX waste, presented in Figure 5, can be used to evaluate the melting temperature of this material. The melting temperature of PEX particles obtained in the DSC was nearly 130 °C, for both heating cycles, which is within the typical melting temperature range of polyethylene materials. This value is clearly lower than the temperature used for asphalt mixture production (180 °C). Thus, the PEX waste, introduced in the asphalt mixture by the dry process, as aggregate substitute, could easily interact with the asphalt binder at those temperatures.

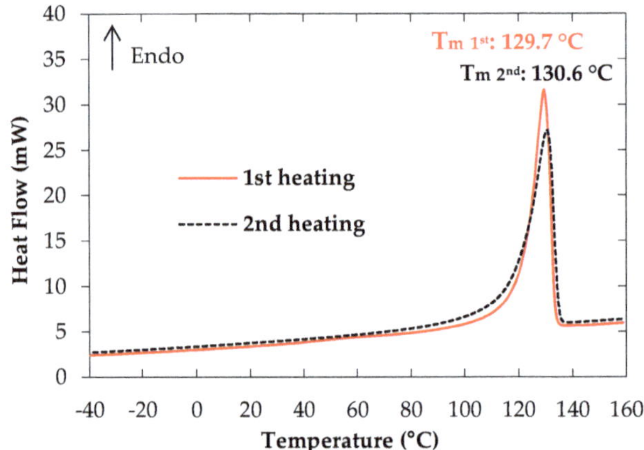

Figure 5. Differential scanning calorimetry (DSC) results of PEX waste.

Therefore, the binder properties should be assessed, before and after interaction with PEX, in order to quantify the binder modification attributed to this waste material.

3.2. Properties of the Asphalt Binders Used in This Study

Three binders were characterized in this part of the work: two commercially available binders selected for this study (base bitumen 50/70 and modified binder PMB 45/80-60) and the PEX modified binder obtained according to the process presented in Section 2.4. The basic properties (penetration and softening point or R&B temperature) of those binders are presented in Figure 6.

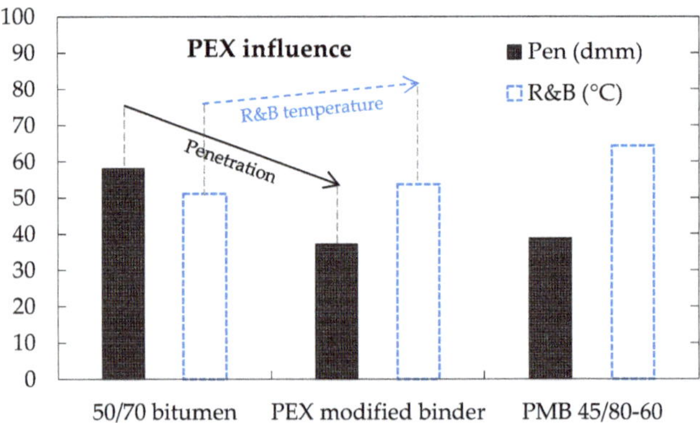

Figure 6. Basic properties of the asphalt binders used in this study, including the PEX modified binder.

Although PEX particles are only able to melt partially and they have been interacting with bitumen for only 2 min, it becomes clear that PEX waste certainly modified the 50/70 bitumen properties. In fact, the use of PEX reduced the penetration value and increased the softening point result in comparison with the base bitumen 50/70, which should also increase the stiffness modulus and permanent deformation resistance of the new mixtures incorporating PEX.

The effect of PEX is more evident in the penetration test than in the softening point test, which is usual when modifying binders with polyethylene-based polymers [18]. Actually, the penetration of PEX modified binder is lower than that of PMB 45/80-60, while its softening temperature is far from being similar to the commercially available PMB.

The viscosity of the studied binders could give valuable information about the performance at very high temperatures, being helpful to define the mixing temperatures to be used for production of asphalt mixtures. Figure 7 shows the evolution of the dynamic viscosity of the studied binders at different temperatures.

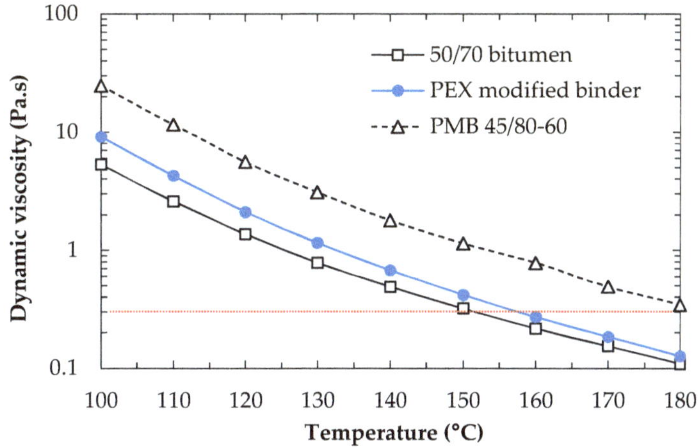

Figure 7. Dynamic viscosity results of the asphalt binders used in this study, including the PEX modified binder.

It can be observed that the interaction between PEX waste and 50/70 bitumen resulted in a modified binder with increased viscosity, even though it is clearly lower than that of PMB 45/80-60. Thus, the modification of 50/70 bitumen by PEX waste changes its behavior in the whole range of temperatures of asphalt mixtures in service or during paving works.

Moreover, by evaluating the mixing equiviscosity (fine and dotted line at 0.3 Pa·s) temperatures of all binders used in this study, it was concluded that the asphalt mixtures with 50/70 binder, PEX modified binder and PMB 45/80-60 should be produced, respectively, at the temperatures of 150 °C, 160 °C and 180 °C. However, in order to increase the interaction between the PEX and the asphalt binders, and also by aiming at using the same production conditions for all mixtures, a single mixing temperature of 180 °C was selected to produce all mixtures, independently of the used binder.

3.3. Marshall Mix Design of Asphalt Mixtures Incorporating PEX Waste

As explained previously, the mixtures without PEX have already been designed and presented an optimum binder content of 5.0%. Therefore, two mixtures with 5% PEX waste were used in this part of the work, namely a mixture produced with 50/70 bitumen and another mixture produced with PMB 45/80-60. The Marshal Mix design results obtained are presented in Table 2.

Table 2. Marshall Mix design results obtained for the mixtures incorporating PEX waste, either using 50/70 bitumen or polymer modified binder PMB 45/80-60.

Properties	Specifications	Design of Mixtures with PEX and 50/70 Bitumen			Design of Mixtures with PEX and PMB 45/80-60		
Binder content (%)	≥4.0	5.0	5.5	6.0	5.0	5.5	6.0
Bulk density (kg/m^3)	–	2284	2283	2308	2219	2257	2268
Air void content (%)	[3–5]	4.3	2.8	1.4	5.8	4.4	3.7
VMA (%)	≥14	15.4	15.0	14.8	16.6	16.2	16.9
Marshall stability (kN)	≥7.5	14.3	14.3	14.1	19.4	19.8	18.4
Deformation (mm)	[2–4]	2.9	3.2	3.5	3.7	3.9	4.3

These Marshall Test results showed that the optimum binder content should be between 5.0 and 5.5% for both mixtures with PEX, with the binder 50/70 and PMB 45/80-60. In fact, the results indicate that the optimum binder content of the mixture produced with the 50/70 pen bitumen may be closer to 5.0%, while the optimum binder content of the PMB 45/80-60 mixture should be 5.5%, taking into account the values of the air void content and permanent deformation obtained.

In order to better understand the effect of the binder content on the performance of these mixtures, water sensitivity tests were carried out on specimens produced with the same binder contents used in the Marshall Tests, i.e., 5.0%, 5.5% and 6.0%. The results of this additional water sensitivity study are presented in Figure 8.

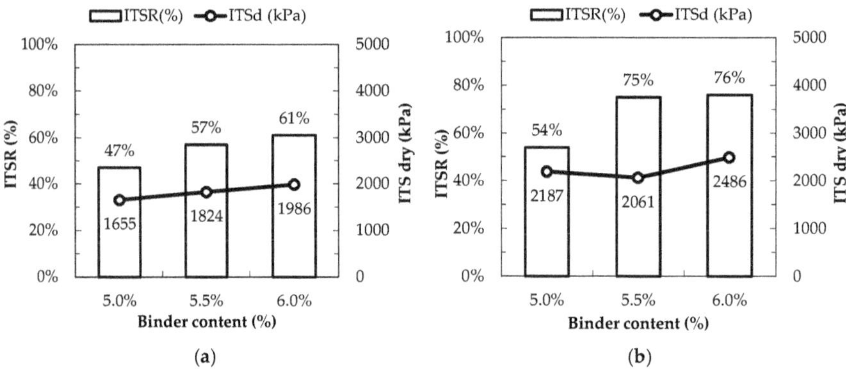

Figure 8. Influence of binder content in the water sensitivity test results for: (**a**) Asphalt mixture with PEX and 50/70 bitumen; (**b**) Asphalt mixture with PEX and PMB 45/80-60 binder.

The results presented in Figure 8 show that 5.0% binder content may not be enough to assure an adequate performance of the mixture. The ITSR test results have increased more significantly between the binder contents of 5.0 and 5.5% than between 5.5 and 6.0%. Therefore, based on both Marshall and water sensitivity tests results, the optimum binder content should be 5.5% for both mixtures with PEX. However, since the 50/70 binder is softer than the PMB 45/80-60, the former may result in a mixture more susceptible to permanent deformation. Thus, an additional mixture with the lower binder content (5.0%) was also produced and its performance evaluated, as presented in the ollowing section.

3.4. Performance of Studied Asphalt Mixtures

The performance of the five mixtures was evaluated as previously specified in Section 2.6, and the results will be presented and discussed next, taking into account the mixture composition presented in Table 3.

Table 3. Composition of the studied mixtures.

Mixture	Aggregate Volumetric Composition		Asphalt Binder	
	Mineral Aggregate	PEX Waste	Type	Content
M1	100%	0%	50/70 bitumen	5.0%
M2	100%	0%	PMB 45/80-60	5.0%
M3	95%	5%	50/70 bitumen	5.0%
M4	95%	5%	50/70 bitumen	5.5%
M5	95%	5%	PMB 45/80-60	5.5%

3.4.1. Volumetric Characteristics and Water Sensitivity of Studied Asphalt Mixtures

Before assessing the water sensitivity of the mixtures, the test specimens were volumetrically characterized. Thus, the maximum density of the mixtures was determined according to the procedure described in EN 12697-5, and the bulk density of all specimens was determined according to EN 12697-6, Procedure B. After that, the air void content of each specimen was calculated (EN 12697-8), and the average values of the five asphalt mixtures were assessed (Table 4).

Table 4. Volumetric characteristics of the studied mixtures.

Properties	Specifications	Mixture				
		M1	M2	M3	M4	M5
Binder content (%)	≥4.0	5.0	5.0	5.0	5.5	5.5
Air voids content (%)	[3–5]	2.8	4.8	5.0	2.9	5.3
Bulk density (kg/m^3)	–	2374	2369	2254	2285	2242

From the results obtained it was found that the volumetric characteristics of asphalt mixtures with PEX are different from those of mixtures with conventional aggregates. This is essentially related to the maximum density of the mixtures, which is much lower in the former due to the low density of PEX. This can be considered as one of the advantages of these mixtures, reducing its transportation costs and offering other advantages when applied on bridges or other structures intended to be lighter.

The results of the indirect tensile strength and water sensitivity tests obtained for the five mixtures are shown in Figure 9.

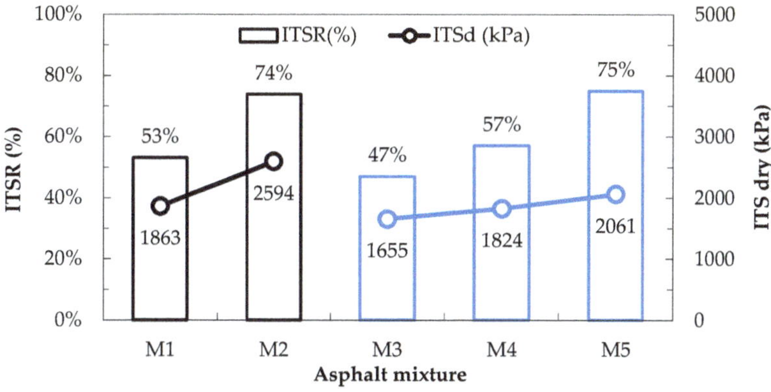

Figure 9. Water sensitivity test results of the studied mixtures.

Comparing the performance of the mixtures with the same binder type (50/70) and binder content (5.0%), but with and without PEX as aggregate, i.e., M1 and M3, the introduction of PEX slightly reduces the water sensitivity performance. However, when the binder content is increased by 0.5% (M4 or M5) in the mixtures with PEX, the situation is reversed, being the mixtures with the best performance in terms of water sensitivity.

3.4.2. Permanent Deformation Resistance of Studied Asphalt Mixtures

The rut resistance test evaluates the susceptibility of asphalt materials to permanent deformation. It consists of measuring the depth of a rut formed after a number of load applications of a wheel over an asphalt slab at high temperature conditions. Figure 10 represents the rut depth measured, for each mixture, up to the 10,000th cycle, and the wheel-tracking slope (WTS_{AIR}) calculated as the average rate at which the rut depth increases with repeated applications of a wheel load (mm/10^3 cycles), between the 5000th and the 10,000th cycles.

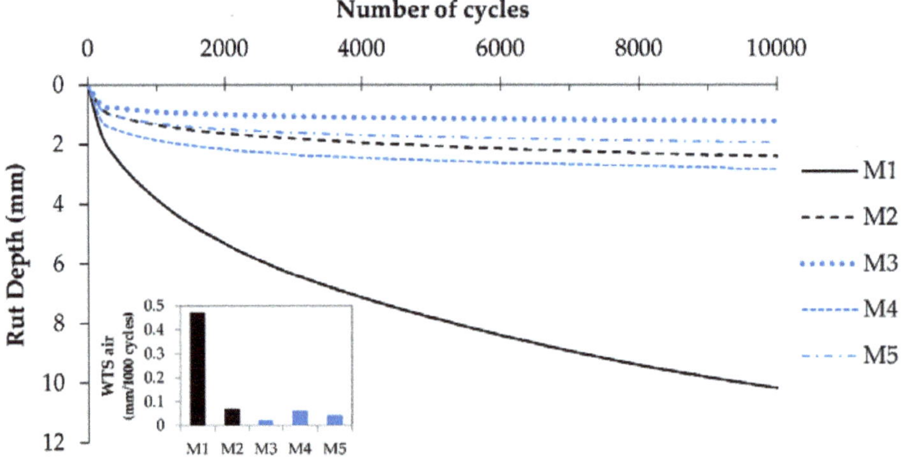

Figure 10. Evolution of the rut depth of the studied mixtures in the wheel tracking test (WTT) (the inset presents the resulting wheel tracking slope in air (WTS_{AIR}) values).

By analysing the results obtained for the rut resistance, it can be concluded that the mixtures with PEX deformed much less than the mixtures with mineral aggregates. Looking at the performance results of M1 and M3 (same binder and binder content but with and without PEX as aggregate) the reduction in terms of deformation is very significant, which is also visible in the reduction of the WTS_{AIR} value. When increasing its binder content by 0.5% (M4), the deformation increases, as expected when compared with M3, but the performance is still very good, being close to the performance to the PMB mixture without PEX (M2). Between the mixtures with PMB binder, the use of PEX also decreased the deformation, but in this case, its effect is not that significant.

Regarding the WTS_{AIR}, the conclusions are similar. Thus, the asphalt mixtures with PEX have much better permanent deformation performance than the conventional mixture (M1).

3.4.3. Stiffness Modulus of Studied Asphalt Mixtures

Based on the test results of stiffness modulus and phase angle, carried out for a range of frequencies (0.1 to 10 Hz) at different temperatures, it is possible to predict the same parameters over a larger range of frequencies, from a reference temperature (20 °C), using the time-temperature superposition principle. The resulting curves (master curves) are presented in Figure 11 and cover a frequency range far beyond the experimentally accessible values. They were obtained by translation of each isotherm curve (0, 10, 30, 40 °C) based on the reference curve of 20 °C. In order to evaluate the applicability of this method, the horizontal translation factor should be overlapped by viscoelastic functions. The Arrhenius function was used for the construction of these master curves, and this process was performed using IRIS 9.0 software.

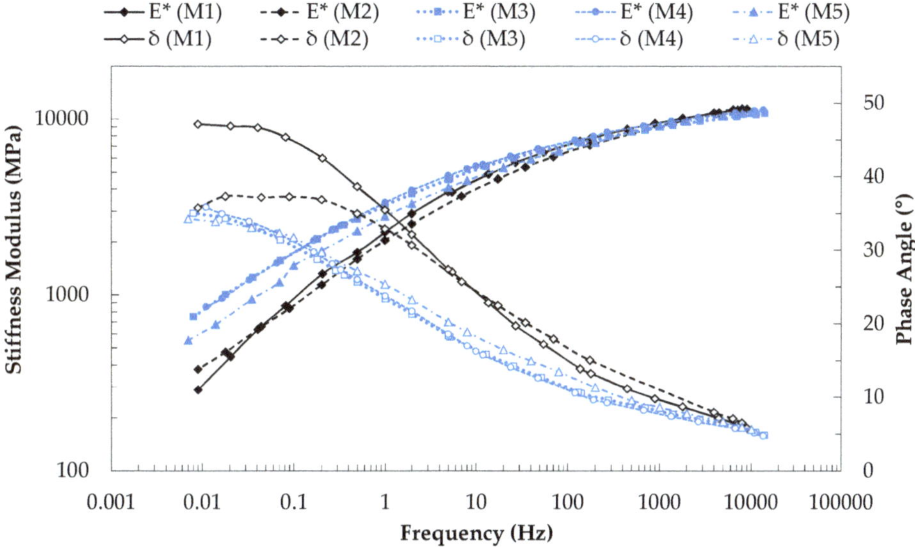

Figure 11. Master curves of the stiffness modulus and phase angle of the studied mixtures for a reference temperature of 20 °C.

The conventional mixture M1 has lower values for the stiffness modulus at lower frequencies (or high temperatures), which is in agreement with the permanent deformation results, followed by M2 (PMB mixture without PEX). The influence of the addition of PEX on stiffness modulus can be observed by comparing the results of M1 and M3 (50/70 pen bitumen, with and without PEX, respectively), these results showed that the addition of PEX increases the stiffness modulus, in a more noticeable way at low frequencies (or high operating temperatures). The same is observed for the mixtures with the PMB 45/80-60 binder (M2 and M5), and there is even an increase of 0.5% in the binder content. Looking at the results obtained for mixture M4, one could expect to obtain stiffness modulus values lower than those of mixture M3, due to its higher binder content. However, that behaviour was not observed, possibly due to the existence of a higher amount of bitumen available to interact with the PEX portion that was able to melt during the mixing operation, increasing its stiffness.

In terms of phase angle, only the conventional mixture (M1) showed values above 45° (tan δ higher than 1), i.e., M1 is the mixture that provides a more viscous behaviour, followed by mixture M2, while mixtures with PEX have a more elastic behaviour.

Generally, it becomes evident that PEX increased the stiffness moduli and reduced the phase angle values in comparison with the mixtures without this waste material.

3.4.4. Fatigue Cracking Resistance of Studied Asphalt Mixtures

In order to assess the fatigue cracking performance of the studied mixtures, 4 point bending beam tests were carried out at 20 °C. The results of those tests are presented in Figure 12.

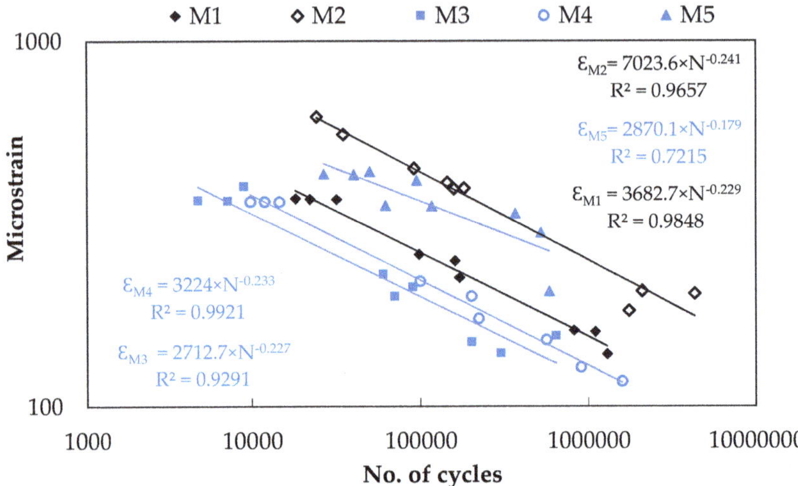

Figure 12. Fatigue cracking laws of the studied mixtures.

In Figure 12, it can be observed that the fatigue performance of the mixtures with PEX is not higher than that of the mixtures without PEX and a similar binder content. However, an increase of 0.5% in the binder content of the 50/70 pen bitumen is enough to improve the fatigue performance of the mixture (M3). This behaviour can be explained by the nature of the polymer that is contained in the PEX (plastomer), which does not perform in the same way as the elastomer used in the PMB 45/80-60 mixtures (M2 and M5). In fact, the mixture with the best fatigue performance was M2, which contains only the natural aggregates and the elastomer modified binder.

From the fatigue life equations (fatigue cracking laws) of the studied mixtures, two main fatigue performance parameters, ε_6 and N100, were calculated (as specified in EN 12697-24 standard) and are presented in Table 5.

Table 5. Fatigue parameters, N100 and ε_6, of the studied mixtures.

Fatigue Parameter	Mixture				
	M1	M2	M3	M4	M5
ε_6	156	252	117	129	242
N100 ($\times 10^6$)	6.9	45.9	2.1	3.0	140.0

The ε_6 results, representing the strain level at which the mixture would fail after 1 million cycles, showed that for the same binder type and content (comparing M1 and M3), the addition of PEX slightly reduced the fatigue resistance. The increased binder content used in mixture M4 was not enough to assure a fatigue performance equivalent to that of mixture M1, although the ε_6 value was not significantly lower.

As previously observed in Figure 12, the mixture with the best ε_6 value is M2, due to the increased flexibility given by the elastomeric additive used in the binder modification. However, if the binder content is increased by 0.5% a considerable amount of PEX could be introduced in the mixture, as aggregates, without compromising the performance of the mixture (M5), demonstrating that the use of PEX could be an interesting solution in ecological and technical terms.

Regarding the N100 values, which are the number of cycles corresponding to a test carried out at 100 microstrain, the mixtures produced with the PMB 45/80-60 binder are those that present the highest performance, which confirms the results presented in Figure 12. For those mixtures, the addition of

PEX, associated with a small increase in the binder content, resulted in the best performing mixture (M5), while in the mixtures produced with the conventional 50/70 pen bitumen, the addition of PEX did not have the same effect, even when a higher binder content was used.

4. Conclusions

According to the study presented in this manuscript, regarding the use of PEX as a partial substitute of aggregates and a binder modifier in asphalt mixtures, a series of conclusions can be drawn, as follows:

Cross-linked polyethylene (PEX) waste can be partially melted into hot bitumen, modifying the final properties of the resulting binder.

When incorporated into asphalt mixtures as an aggregate partial substitute, PEX reduces the density of the mixtures by about 5%, which may be an advantage in specific circumstances, namely, during transport and in the application over structural elements, like bridges.

Regarding the water sensitivity test results, the introduction of PEX was not able to improve the performance of the mixture, unless an increased binder content (0.5% higher) was used. Similar results were observed in the fatigue cracking resistance tests because polyethylene based polymers are not able to improve the flexibility of asphalt mixtures in the same way as the elastomeric polymers that were used in the PMB 45/80-60 mixtures.

Permanent deformation results of the mixtures with PEX were clearly better than those of the mixtures without PEX, even when higher binder contents were used. This is a result of the lower susceptibility to temperature variation of those mixtures, which can also be observed from their higher stiffness modulus and lower phase angle values. These properties are related to each other and can be seen as the main performance related improvements of incorporating PEX in asphalt mixtures.

The inclusion of PEX in asphalt mixtures is a viable solution that can assure adequate performance of a road pavement, especially when high service temperatures are expected. This technology also contributes to give new life to cross-linked polyethylene plastic waste. Nevertheless, further research should be carried out regarding low temperature and in situ performance, in order to validate this technology with a broader range of application, loading and climatic conditions.

Acknowledgments: The authors gratefully acknowledge the funding provided by the Portuguese Government (Portuguese Foundation for Science and Technology) and by the Competitiveness Factors Operational Programme (EU/ERDF funds), through the FCT project PLASTIROADS (PTDC/ECM/119179/2010).

Author Contributions: L.M.B.C. was in charge of the whole research, which is part or her PhD work. Nevertheless, all authors were involved in the conception and design of the experimental work. L.M.B.C. performed the experiments. Finally, all authors analyzed the data and contributed to the completion of the manuscript.

Conflicts of Interest: The authors declare no conflict of interest.

References

1. Abreu, L.P.F.; Oliveira, J.R.M.; Silva, H.M.R.D.; Fonseca, P.V. Recycled asphalt mixtures produced with high percentage of different waste materials. *Constr. Build. Mater.* **2015**, *84*, 230–238. [CrossRef]
2. Silva, H.M.R.D.; Oliveira, J.R.M.; Jesus, C.M.G. Are totally recycled hot mix asphalts a sustainable alternative for road paving? *Resour. Conserv. Recycl.* **2012**, *60*, 38–48. [CrossRef]
3. Kowalski, K.; Król, J.; Bańkowski, W.; Radziszewski, P.; Sarnowski, M. Thermal and fatigue evaluation of asphalt mixtures containing rap treated with a bio-agent. *Appl. Sci.* **2017**, *7*, 216. [CrossRef]
4. Xiao, Y.; Li, C.; Wan, M.; Zhou, X.; Wang, Y.; Wu, S. Study of the diffusion of rejuvenators and its effect on aged bitumen binder. *Appl. Sci.* **2017**, *7*, 397. [CrossRef]
5. Zhu, J.; Birgisson, B.; Kringos, N. Polymer modification of bitumen: Advances and challenges. *Eur. Polym. J.* **2014**, *54*, 18–38. [CrossRef]
6. PlasticsEurope; EPRO. *Plastics—The Facts 2015. An Analysis of European Plastics Production, Demand and Waste Data PlasticEurope*; European Association of Plastics Recycling & Recovery Organisations: Wemmel, Belgium, 2015.

7. Shang, L.; Wang, S.; Zhang, Y.; Zhang, Y. Pyrolyzed wax from recycled cross-linked polyethylene as warm mix asphalt (wma) additive for sbs modified asphalt. *Constr. Build. Mater.* **2011**, *25*, 886–891. [CrossRef]
8. Robinson, H.L. Polymers in asphalt. *Smith. Rapra Technol.* **2004**, *15*, 73.
9. Bocci, M.; Colagrande, S.; Montepara, A. Pvc and pet plastics taken from solid urban waste in bituminous concrete. In *Waste Management Series*; Woolley, G.R., Goumans, J.M., Wainwright, P.J., Eds.; Elsevier: Amsterdam, The Netherlands, 2000; Volume 1, pp. 186–195.
10. Zoorob, S.E.; Suparma, L.B. Laboratory design and investigation of the properties of continuously graded asphaltic concrete containing recycled plastics aggregate replacement (plastiphalt). *Cem. Concr. Compos.* **2000**, *22*, 233–242. [CrossRef]
11. Kalantar, Z.N.; Karim, M.R.; Mahrez, A. A review of using waste and virgin polymer in pavement. *Constr. Build. Mater.* **2012**, *33*, 55–62. [CrossRef]
12. Becker, Y.; Mendez, M.P.; Rodriguez, Y. Polymer modified asphalt. *Vis. Tecnol.* **2001**, *9*, 39–50.
13. Airey, G.D. Rheological evaluation of ethylene vinyl acetate polymer modified bitumens. *Constr. Build. Mater.* **2002**, *16*, 473–487. [CrossRef]
14. Zhang, F.; Yu, J. The research for high-performance sbr compound modified asphalt. *Constr. Build. Mater.* **2010**, *24*, 410–418. [CrossRef]
15. Murphy, M.; O'Mahony, M.; Lycett, C.; Jamieson, I. Recycled polymers for use as bitumen modifiers. *J. Mater. Civ. Eng.* **2001**, *13*, 306–314. [CrossRef]
16. García-Morales, M.; Partal, P.; Navarro, F.J.; Martínez-Boza, F.; Gallegos, C.; González, N.; González, O.; Muñoz, M.E. Viscous properties and microstructure of recycled eva modified bitumen. *Fuel* **2004**, *83*, 31–38. [CrossRef]
17. Awwad, M.T.; Shbeeb, L. The use of polyethylene in hot asphalt mixtures. *Am. J. Appl. Sci.* **2007**, *4*, 390–396. [CrossRef]
18. Fuentes-Audén, C.; Sandoval, J.A.; Jerez, A.; Navarro, F.J.; Martínez-Boza, F.J.; Partal, P.; Gallegos, C. Evaluation of thermal and mechanical properties of recycled polyethylene modified bitumen. *Polym. Test.* **2008**, *27*, 1005–1012. [CrossRef]
19. Ahmadinia, E.; Zargar, M.; Karim, M.R.; Abdelaziz, M.; Shafigh, P. Using waste plastic bottles as additive for stone mastic asphalt. *Mater. Des.* **2011**, *32*, 4844–4849. [CrossRef]
20. Casey, D.; McNally, C.; Gibney, A.; Gilchrist, M.D. Development of a recycled polymer modified binder for use in stone mastic asphalt. *Resour. Conserv. Recycl.* **2008**, *52*, 1167–1174. [CrossRef]
21. Vasudevan, R.; Sekar, A.R.; Sundarakannan, B.; Velkennedy, R. A technique to dispose waste plastics in an ecofriendly way—Application in construction of flexible pavements. *Constr. Build. Mater.* **2012**, *28*, 311–320. [CrossRef]
22. Rasmussen, E.; Boss, A. Plastsep—A new technology for sorting and recycling of cable polymers. In *Proceedings of the 21st Nordic Insulation Symposium, NORD-IS 09*; Institutionen för Material–och Tillverkningsteknik, Högspänningsteknik: Göteborg, Sweden, 2009.
23. Silva, H.M.R.D.; Oliveira, J.R.M.; Peralta, J.; Zoorob, S.E. Optimization of warm mix asphalts using different blends of binders and synthetic paraffin wax contents. *Constr. Build. Mater.* **2010**, *24*, 1621–1631. [CrossRef]
24. Oliveira, J.R.M.; Silva, H.M.R.D.; Abreu, L.P.F.; Gonzalez-Leon, J.A. The role of a surfactant based additive on the production of recycled warm mix asphalts—Less is more. *Constr. Build. Mater.* **2012**, *35*, 693–700. [CrossRef]

Article

Study of the Diffusion of Rejuvenators and Its Effect on Aged Bitumen Binder

Yue Xiao [1,*], Chao Li [1], Miao Wan [1], Xinxing Zhou [2], Yefei Wang [3] and Shaopeng Wu [1,*]

1. State Key Laboratory of Silicate Materials for Architectures, Wuhan University of Technology, Wuhan 430070, China; lic@whut.edu.cn (C.L.); alexwanm@163.com (M.W.)
2. Key Laboratory of Highway Construction and Maintenance in Loess Region, Shanxi Transportation Research Institute, Taiyuan 030006, China; zxx09432338@whut.edu.cn
3. Jinan Urban Construction Group Co., Ltd., Jinan 250031, China; wangyefei0803@163.com

* Correspondence: xiaoy@whut.edu.cn (Y.X.); wusp@whut.edu.cn (S.W.); Tel.: +86-27-8716-2595 (Y.X. & S.W.)

Academic Editors: Zhanping You, Qingli (Barbara) Dai and Feipeng Xiao
Received: 1 March 2017; Accepted: 7 April 2017; Published: 14 April 2017

Abstract: Aged asphalt mixture is heavily involved in pavement maintenance and renewed construction because of the development of recycling techniques. The aged bitumen binder has partially lost its viscous behavior. Rejuvenators are therefore designed and used in this recycling procedure to enhance the behavior of such aged reclaimed bitumen. However, tests have not yet been clearly specified to understand the diffusion characteristics of rejuvenators in aged bitumen. In this research, molecular dynamic simulation is proposed and conducted with Materials Studio software to study the diffusion behavior of rejuvenators in aged bitumen at the molecular level. Two rejuvenators, named R-1 and R-2, were included. The difference between these two rejuvenators is their chemical composition of C=O. The diffusion coefficient is determined by studying the molecular movement. Results illustrate that the proposed models can be used to study the diffusion of rejuvenators in aged bitumen sufficiently. In the meantime, a dynamic shear rheometer (DSR) is used to evaluate the recovery influence on aged bitumen resulting from rejuvenators. The experimental results strengthen the model simulations and indicate that the aging index of bitumen has a significant influence on the rejuvenating effect. Research results indicate that rejuvenators have a sufficient rejuvenating effect on the long-term aged bitumen and a limited effect on short-term aged bitumen.

Keywords: bitumen binder; molecular simulation; diffusion of rejuvenator; aged bitumen; rheological properties

---

1. Introduction

Reclaimed asphalt [1], which is of great importance for conserving resources and protecting the environment, is one of the largest fractions of recycled raw materials used in road construction [2–4]. Rejuvenators are employed in many cases to improve the flexibility, which means better stress relaxation and softer properties of such aged reclaimed asphalt. The performance of these reclaimed asphalts depends on the mixing degree of rejuvenators and aged bitumen [5]. Two of the main factors that assist in the mixing efficiency are the blending condition and the diffusion process [6,7]. This study therefore focused on the diffusion process of rejuvenators in aged bitumen during hot recycling.

Molecular diffusion plays an important role in bitumen recovery process as well as in numerous engineering applications [8]. Mazzotta used FTIR (Fourier transform infrared spectroscopy) to carry out chemical analysis of molecular changes during the bitumen aging process [9]. Indexes of $I_{C=O}$ and $I_{S=O}$ were defined and successfully used to characterize the changes of functional groups. Many studies on diffusion in bituminous materials have been successfully carried out recently. A method based on FTIR-ATR (Fourier Transform Infrared Spectroscopy using Attenuated Total Reflectance) was developed and described by Karlsson and Isacsson [10,11]. Using this method, diffusion rates are influenced by the temperature, size, and shape of the diffusion molecules. A positive outcome of that study was that aging showed very little influence on diffusion [12]. Karlsson and Isacsson [13] also did some laboratory studies on diffusion in bitumen using markers. They selected diphenyl-silane and t-butyldiphenylsilyl (DPS) as markers. They monitored the diffusion through the binder, and the importance of diffusant size and polarity on diffusion rate was demonstrated. Research on the diffusion of rejuvenators in aged bitumen has been further investigated using molecular dynamic simulation to calculate the diffusion coefficient.

Molecular simulation technology can be used to calculate the molecular conformations and diffusion process between different materials [14]. The establishment of the model mainly relies on system parameters such as the sizes of scale, the degree of light and shade, the colors, and the different structures of the microscopic particles themselves. The molecular simulation technique is an experimental technology based on computer analysis. This technology can simulate the static characteristics and motor behavior of molecules by setting up a reasonable molecular model to define the chemical and physical properties of the micro system [15]. Dong and Wang successfully used molecular dynamic simulation method to study the asphalt–aggregate interface [16,17]. Wang also used molecular dynamics simulation to characterize the oxidative aging effect in asphalt binder. Mean square displacement and diffusion coefficient were analyzed, and aged binder was found to have a higher activation energy barrier that decreased its viscoelastic behavior [18,19]. Huang applied molecular dynamic simulation to investigate the diffusion between virgin binder and aged binder, which was then verified with gel permeation chromatography [20,21]. They reported that the diffusion ability is a function of the diffusion ability itself and the diffusion environment. All of this research leads to a powerful method of studying the diffusion of rejuvenators in aged bitumen binder.

In this research, the chemical bond morphology of the investigated rejuvenators was first characterized to define structures for molecular dynamic simulation. Then, the molecular dynamic method was used, and the COMPASS force field was adopted to discuss the diffusion properties of the modeled rejuvenators in aged bitumen. Einstein's model was employed to analyze the results obtained from a computer and to calculate the diffusion coefficient [22]. At the same time, rheological properties of rejuvenated bitumen treated by different rejuvenators were characterized. These laboratory tests were conducted to further understand the findings from the molecular simulation on the diffusion and rejuvenating effect.

2. Materials and Methods

2.1. Materials

The 80/100 penetration grade bitumen provided by the Panjin asphalt company (Sichuan, China) was used in this research. The short-term aged bitumen used in this research was aged using the standard rolling thin-film oven test (RTFOT) at a temperature of 163 °C for 85 min. Long-term aged bitumen was obtained by means of a pressure aging vessel (PAV) and an ultraviolet aging oven (UV). The PAV was conducted at 95 °C with pressure of 2.1 MPa for 20 h. The ultraviolet aging oven was set with an ultraviolet intensity of 79.5 W/m^2 and aged for 168 h. Both short-term aged and long-term aged binders were then treated with rejuvenators. The rejuvenating contribution to short-term aged binder was studied to understand the recovering ability during hot mixing.

The rejuvenator-treated aged bitumen was prepared in two steps. Firstly, the aged bitumen was blended with 8% of R-1 or R-2, separately, using the propeller mixer (GS-1, Tianjin, China) for 30 min at 120 °C with a constant speed of 200 rpm. Secondly, the rejuvenator-treated bitumen was then allowed to sit at room temperature for 3 days before being used for further rheological analysis to ensure completed diffusion.

The rejuvenators, named R-1 and R-2, used throughout this research were provided by CRAFCO Company (Wuxi, P.R. China). Table 1 displays the basic technical data of R-1 and R-2. They are liquid and made from emulsified bitumen, containing about 61.8% to 76.5% of residue after fully cured.

Table 1. Technical data of Rejuvenators R-1 and R-2.

Parameter	R-1	R-2	Test Methods
Appearance	Viscose liquid	Viscose liquid	–
Color	tawny	brown	–
Viscosity (25 °C), SFS	40	15	ASTM D-244
Residue, wt %	65	60	ASTM D-244
Weight ratio for rejuvenate, %	8	8	ASTM D-2006-70
Asphaltenes, wt %	0.4	0.75	–

2.2. Molecular Dynamic Simulation

Molecular dynamic simulation was performed using Materials Studio software (Materials Studio 8.0, BIOVIA, Beijing, China). The compass force field was used only for describing atomic level interactions in molecular models. According to the literature, the saturates, aromatics, and asphaltenes in aged bitumen were modeled as a straight chain alkane $C_{22}H_{46}$, 1,7-dimethyl-naphthalene [23,24], and Groenzin's and Mullin's model $C_{72}H_{98}S$ [25], respectively.

Components of rejuvenators were first determined, and R-1 and R-2 were then defined as simplified monomer that consists of functional groups as Table 2 indicates. For simplified calculations, $C_{13}H_{15}NO_3$ and $C_{11}H_{13}NO_2$ were defined for R-1 and R-2, respectively.

Table 2. Functional groups in the rejuvenators used.

Rejuvenator	Functional Groups					
R-1	HC (>CH–)	–CH2 (>CH2)	–CH3	–NH2	–C(=O)–	benzene ring
R-2	HC (>CH–)	–CH2 (>CH2)	–CH3	–NH2	benzene ring	-

The simulation process of diffusion coefficients mainly includes two parts: energy minimization and a dynamic simulation process. After energy optimization, a canonical ensemble (NVT) and an isobaric–isothermal ensemble (NPT) were simultaneously applied for the first 50 picoseconds (ps) to ensure a density stable system [26]. NVT is a statistical ensemble that represents the possible states of a mechanical system in thermal equilibrium with a heat bath at a fixed temperature [27]. The system can exchange energy with the heat bath, so the states of diffusion system will differ in total energy. NPT is a statistical mechanical ensemble that maintains a constant temperature and a constant applied pressure [28].

The molecular dynamic simulation was running at a simulated temperature of 25 °C. Then, NVT dynamic simulation was used in the following 50 ps. The movement locus of the molecule was recorded to study the diffusion coefficient. Figure 1 presents the diffusion models. R-1 and R-2 are located in the boundary of the RTFOT aged asphalt at the initial stage, and are diffused into the center at a stable stage after full diffusion.

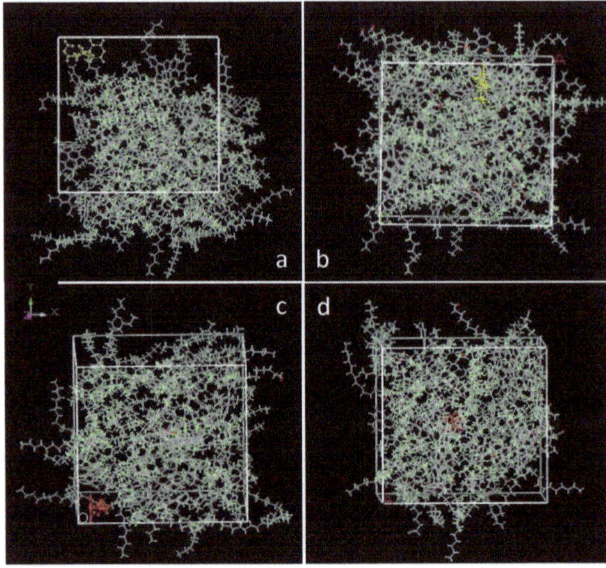

Figure 1. Diffusion models: the initial stage (**a**) and the stable stage (**b**) of R-1 in the rolling thin-film oven test (RTFOT) aged asphalt models; the initial stage (**c**) and the stable stage (**d**) of R-2 in the RTFOT aged asphalt models.

2.3. Rheological Properties Test

The dynamic shear rheometer (DSR) (Anton Paar, Graz, Austria) is usually used to characterize both viscous and elastic behaviors of bitumen by measuring the complex shear modulus and phase angle. A temperature sweep test was conducted. A sinusoidal signal of stress was applied. The complex shear modulus represents the sample's total resistance to the applied deformation during sinusoidal stress conditions. Phase angle is described as the lag between applied shear stress and the resulting shear strain. The higher the complex shear modulus value, the stiffer the bitumen binder is. The larger the measured phase angle, the more viscous the binder is [29]. The temperature sweep test was performed at a fixed frequency of 10 rad/s and in temperature ranges from −10 to 60 °C with temperature increments of 2 °C per minute.

In this research, DSR tests were conducted on virgin bitumen, aged bitumen (RTFOT, PAV, UV), and rejuvenated bitumen with 8% R-1 or 8% R-2. According to the test specification [30], different geometries of specimens are specified for different temperature conditions. The diameter of the plate that contains the bitumen binder is 8 mm, and the thickness is 2 mm when the temperature is lower than 20 °C. When the temperature is higher than 20 °C, the plate diameter should be 25 mm, and the thickness of bitumen binder should be 1 mm. Strain sweep tests with increasing applied strain were first conducted to study the viscoelastic range of virgin, aged, and rejuvenated bitumen binder. Based on strain sweep tests, strains were selected based on viscoelastic ranges of every type of binder for temperature sweep analysis, ensuring that all of the performed DSR moduli and phase angle evaluations are conducted in the viscoelastic range of the tested bitumen binder. Table 3 summarizes the applied strains.

Table 3. Applied shear stains during dynamic shear rheometer (DSR) analysis.

Temperature Ranges	Virgin Binder, Rejuvenated Binders (%)	RTFOT, PAV and UV Binders (%)
−10–30 °C	0.03	0.02
30–60 °C	1.5	1

3. Results and Discussion

3.1. Molecular Calculation Results

3.1.1. Mean Squared Displacement

The diffusion coefficients and diffusion mechanisms in R-1- and R-2-treated RTFOT, PAV, and UV aged asphalt were modeled and studied. The molecular dynamic simulation of the diffusion process can be investigated by recording the movement locus of the molecular particles. The mean squared displacement (MSD) of the analyzed particles can therefore be finalized by means of this movement.

Figure 2 compares MSD values along with diffusion time for rejuvenators in aged bitumen binder. In both R-1-modified and R-2-modified bitumen binder, the MSD values decrease in order of UV aged, PAV aged, and RTFOT aged binder. Rejuvenator R-1 performs higher MSD values than R-2 in bitumen binder with the same aging index, which means R-1 diffuses faster than R-2 in aged bitumen.

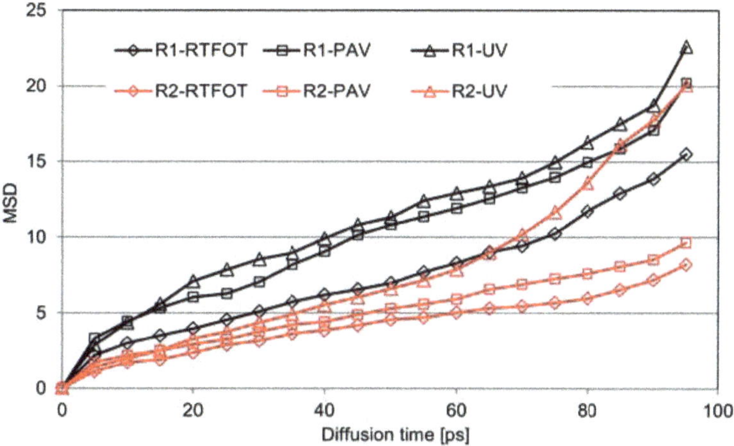

Figure 2. Mean squared displacement (MSD) curves of rejuvenators in aged bitumen at 150 °C.

3.1.2. Diffusion Coefficient

MSD values are then used to calculate the diffusion coefficient (D) via Einstein's formula as follows:

$$MSD = \lim_{t\to\infty}\frac{d}{dt}\sum_{i=1}^{N\alpha}\left\langle [r_i(t) - r_i(0)]^2 \right\rangle \quad (1)$$

where $r_i(t)$ is the displacement of particle i at time t, and $r_i(0)$ is the displacement at the starting time. In this formula, N represents the diffused particles of the system. When the MSD curve presents a linear trend, the equation can be simplified as follows:

$$D = \frac{1}{6T}MSD \quad (2)$$

where T is the total time of atoms or molecule movements. D is the $1/6$ slope of the simplified linear curve.

In the R-1-treated bitumen, the calculated D of the rejuvenator in RTFOT, PAV, and UV aged bitumen is 2.26×10^{-10}, 2.82×10^{-10}, and 3.09×10^{-10}, respectively. The calculated D of Rejuvenator R-2 in RTFOT, PAV, and UV aged bitumen is 1.18×10^{-10}, 1.35×10^{-10}, and 3.05×10^{-10}. Firstly, this illustrates that the severity level of aging has a significant influence on the diffusion of rejuvenators. Investigated rejuvenators diffuse easier and faster in the long-term aged binder, such as the UV aged binder, but diffuse slower in the short-term aged bitumen binder. Secondly, R-1 presents a higher diffusion coefficient in aged bitumen than does R-2, which indicates that R-1 has a better diffusion coefficient in the aged bitumen.

3.2. The Effect of Rejuvenators on the Rheological Properties

Figure 3 presents the complex shear modulus and phase angle of virgin bitumen, RTFOT aged bitumen, and rejuvenated bitumen with 8% R-1 or 8% R-2 in the higher temperature range. It is shown that the complex shear modulus of RTFOT aged bitumen is much higher, and the phase angle is much lower than that of the virgin bitumen. This indicates that aged bitumen is more elastic and stiffer. The introduction of 8% R-1 or 8% R-2 into aged bitumen decreases the complex shear modulus and obviously increases the phase angle. This means that the addition of R-1 or R-2 can reduce the aging degree of bitumen. Furthermore, the complex shear modulus of RTFOT aged bitumen treated with 8% R-1 is lower than that of the aged bitumen treated with 8% R-2, while the phase angle of R-1-treated bitumen is higher and the curve is closer to the virgin bitumen, compared to that of R-2-treated specimens. These results indicate that, although both R-1 and R-2 can recover the rheological properties of RTFOT aged bitumen, R-1 obviously has a stronger recovery capability of rheological properties than does R-2.

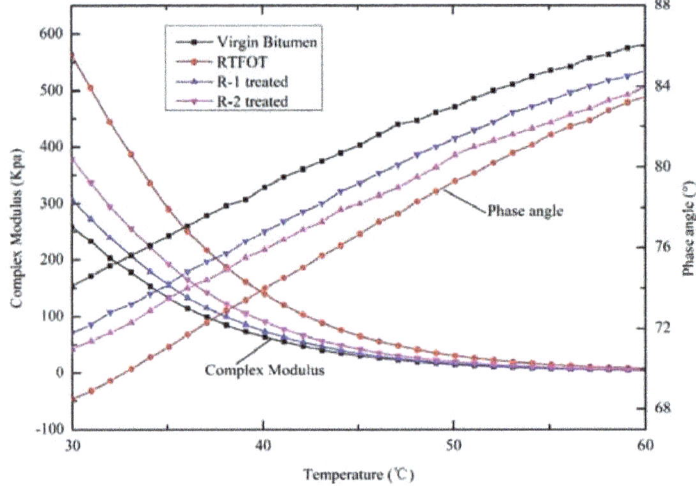

Figure 3. Effect of rejuvenators on RTFOT aged bitumen at higher temperatures.

Figures 4 and 5 reveal the complex shear modulus and phase angle of PAV aged bitumen and UV aged bitumen in the higher temperature range, respectively. The rejuvenating effect of the investigated rejuvenators on PAV aged bitumen is similar to the effect on RTFOT aged bitumen. While for the UV aged bitumen, the impact of R-2 is limited. Figure 5 shows that the phase angle curves of UV aged bitumen and R-2-treated bitumen overlap. The previous molecular calculation shows a higher diffusion coefficient of R-1 in aged binder than that of R-2. Figures 3–5 provide strong evidence that R-1 exhibits a much greater rejuvenating influence on the viscoelastic behavior at higher temperature conditions than does R-2.

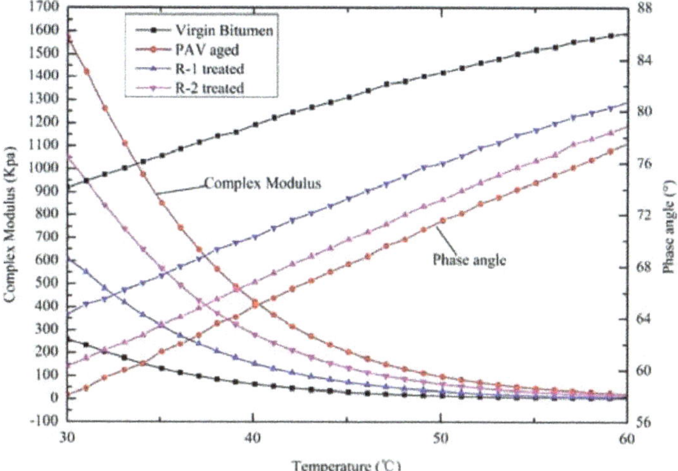

Figure 4. Effect of rejuvenators on pressure aging vessel (PAV) aged bitumen at higher temperatures.

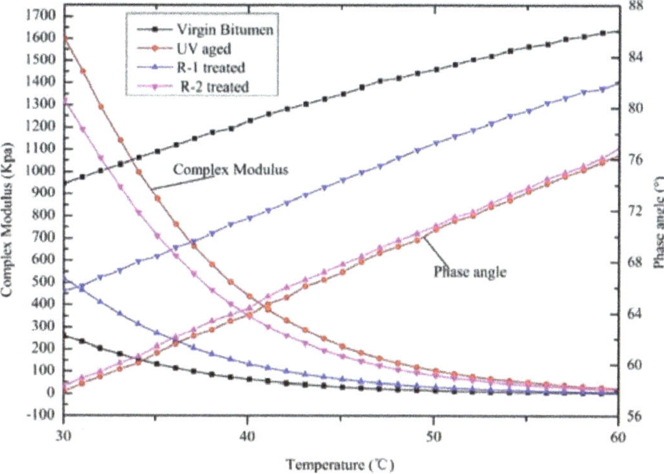

Figure 5. Effect of rejuvenators on UV aged bitumen at higher temperatures.

Figures 6–8 present the complex shear modulus and phase angle at lower temperatures of virgin bitumen, RTFOT (or PAV, or UV) aged bitumen, and rejuvenator-treated bitumen. The complex shear modulus of aged bitumen is higher than that of virgin bitumen. We all know that aged bitumen is very brittle and easily cracks under loading at low temperature. The addition of R-1 and R-2 can also decrease the complex shear modulus and increase the phase angle at lower temperatures. This does mean, however, that both R-1 and R-2 can significantly improve low-temperature properties of RTFOT aged bitumen and extend the service life of bitumen in a lower temperature condition. The modulus curve of aged bitumen treated with R-1 is closer to the curve of virgin bitumen, while the curve of aged bitumen treated with R-2 is closer to the curve of aged bitumen. This phenomenon indicates once again that R-1 has a stronger recovery capability of lower temperature behavior of aged bitumen than does R-2.

Furthermore, Figures 3–8 illustrate that the curve of aged bitumen treated with R-1 in the lower temperature range is closer to the curve of virgin bitumen than that in the higher temperature range. This indicates that the recovery capacity of R-1 on the lower temperature behavior is more significant than that on the higher temperature behavior. Figures 3–8 also show that the distinction between regenerative capacity of R-1 and R-2 is more obvious on the lower temperature properties than that on the higher temperature properties.

Practically speaking, RTFOT, which represents short-term aging, results in a minor property deterioration of bituminous materials. Therefore, in the long-term aged bitumen, such as UV aged binder, the rejuvenating effect resulting from R-1 and R-2 presents a significant difference from that in the case of short-term aged binder. In Figures 3 and 6, both R-1 and R-2 present an obvious rejuvenating contribution. Nevertheless, in Figures 5 and 8, R-1 shows a nice rejuvenating effect, while R-2 has a very limited influence on the properties of modulus and phase angle. This illustrates that the key to designing a good rejuvenator is the improvement of its effect on long-term aged binder, instead of focusing on short-term aged binder. An improved rejuvenator should have a sufficient effect on rejuvenating long-term aged bitumen.

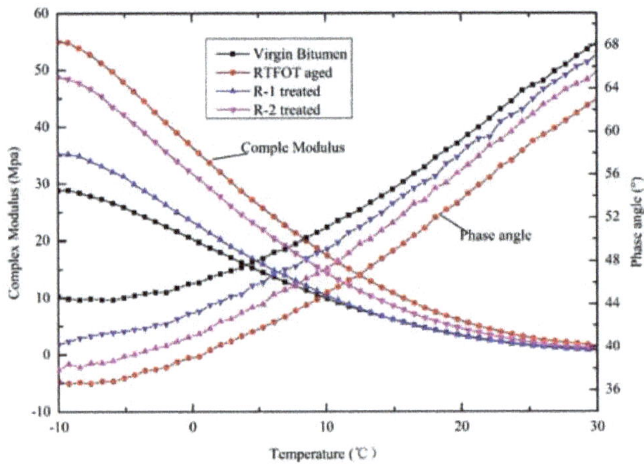

Figure 6. Effect of rejuvenators on RTFOT aged bitumen at lower temperatures.

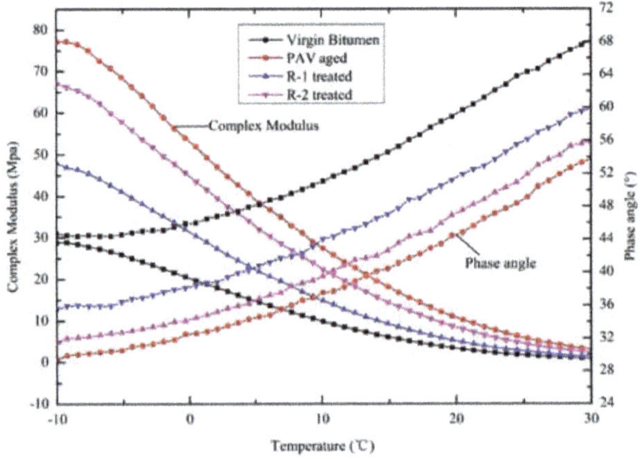

Figure 7. Effect of rejuvenators on PAV aged bitumen at lower temperatures.

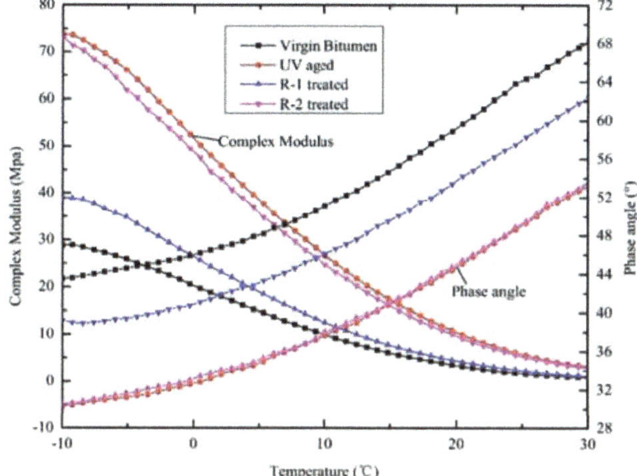

Figure 8. Effect of rejuvenators on UV aged bitumen at lower temperatures.

4. Conclusions

Molecular dynamic simulation was established in this research to simulate the actual movement of rejuvenators in aged bitumen and to characterize their diffusion coefficient. Then, DSR tests were carried out to further investigate the rejuvenating effect. Based on the discussed results, the following can be concluded:

1. Molecular dynamic simulation and rheological analysis results indicate that the severity level of aging has a significant influence on the diffusion coefficient and rejuvenating effect of rejuvenators. The investigated rejuvenators diffuse faster and more easily in the long-term aged binder, such as UV aged binder, but diffuse slower in the short-term aged bitumen binder. Consequently, there is a sufficient rejuvenating effect on the long-term aged bitumen and a limited effect on the short-term aged bitumen.
2. Both R-1 and R-2 can reduce the complex modulus and increase the phase angle of the long-term aged bitumen, but R-1 has a greater recovery capacity than R-2. The rejuvenators presented greater rejuvenate ability at lower temperatures than at higher temperatures.
3. Molecular simulation indicates that R-1 has a higher diffusion coefficient in aged bitumen binder than does R-2, while DSR proves that R-1 has greater recovery capacity. Molecular simulation and experiment results from the DSR test match quite well. This means that the molecular simulation proposed in this research can be successfully used to characterize the diffusion coefficient of rejuvenators in aged bitumen.

Acknowledgments: This study was supported by the National Natural Science Foundation of China (No. 51408447) and the National Key Scientific Apparatus Development Program from the Ministry of Science and Technology of China (No. 2013YQ160501).

Author Contributions: Yue Xiao, Chao Li and Yefei Wang conceived and designed the experiments; Chao Li, Miao Wan, and Yefei Wang performed the experiments; Yue Xiao, Chao Li, Miao Wan, and Yefei Wang analyzed the data; Xinxing Zhou and Shaopeng Wu contributed analysis tools; Yue Xiao and Chao Li wrote the paper.

Conflicts of Interest: The authors declare no conflict of interest.

References

1. Wu, S.; Qiu, J.; Mo, L.; Yu, J.; Zhang, Y.; Li, B. Investigation of temperature characteristics of recycled hot mix asphalt mixtures. *Resour. Conserv. Recycl.* **2007**, *51*, 610–620. [CrossRef]

2. Cui, P.; Wu, S.; Xiao, Y.; Wan, M.; Cui, P. Inhibiting effect of layered double hydroxides on the emissions of volatile organic compounds from bituminous materials. *J. Clean. Prod.* **2015**, *108*, 987–991. [CrossRef]
3. Li, X.-G.; Lv, Y.; Ma, B.-G.; Chen, Q.-B.; Yin, X.-B.; Jian, S.-W. Utilization of municipal solid waste incineration bottom ash in blended cement. *J. Clean. Prod.* **2012**, *32*, 96–100. [CrossRef]
4. Chen, Z.; Wu, S.; Xiao, Y.; Zeng, W.; Yi, M.; Wan, J. Effect of hydration and silicone resin on basic oxygen furnace slag and its asphalt mixture. *J. Clean. Prod.* **2016**, *112*, 392–400. [CrossRef]
5. Chen, M.; Xiao, F.; Putman, B.; Leng, B.; Wu, S. High temperature properties of rejuvenating recovered binder with rejuvenator, waste cooking and cotton seed oils. *Constr. Build. Mater.* **2014**, *59*, 10–16. [CrossRef]
6. Ma, T.; Wang, H.; Zhao, Y.; Huang, X.; Pi, Y. Strength mechanism and influence factors for cold recycled asphalt mixture. *Adv. Mater. Sci. Eng.* **2015**, *2015*, 181853. [CrossRef]
7. Ma, T.; Huang, X.; Zhao, Y.; Zhang, Y. Evaluation of the diffusion and distribution of the rejuvenator for hot asphalt recycling. *Constr. Build. Mater.* **2015**, *98*, 530–536. [CrossRef]
8. Sheikha, H.; Mehrotra, A.K.; Pooladi-Darvish, M. An inverse solution methodology for estimating the diffusion coefficient of gases in athabasca bitumen from pressure-decay data. *J. Pet. Sci. Eng.* **2006**, *53*, 189–202. [CrossRef]
9. Dondi, G.; Mazzotta, F.; Simone, A.; Vignali, V.; Sangiorgi, C.; Lantieri, C. Evaluation of different short term aging procedures with neat, warm and modified binders. *Constr. Build. Mater.* **2016**, *106*, 282–289. [CrossRef]
10. Karlsson, R.; Isacsson, U.; Ekblad, J. Rheological characterisation of bitumen diffusion. *J. Mater. Sci.* **2007**, *42*, 101–108. [CrossRef]
11. Karlsson, R.; Isacsson, U. Application of ftir-atr to characterization of bitumen rejuvenator diffusion. *J. Mater. Civ. Eng.* **2003**, *15*, 157–165. [CrossRef]
12. Karlsson, R.; Isacsson, U. Investigations on bitumen rejuvenator diffusion and structural stability (with discussion). *J. Assoc. Asph. Paving Technol.* **2003**, *72*, 463–501.
13. Karlsson, R.; Isacsson, U. Laboratory studies of diffusion in bitumen using markers. *J. Mater. Sci.* **2003**, *38*, 2835–2844. [CrossRef]
14. Mueller-Plathe, F.; Rogers, S.C.; Van Gunsteren, W.F. Diffusion coefficients of penetrant gases in polyisobutylene can be calculated correctly by molecular-dynamics simulations. *Macromolecules* **1992**, *25*, 6722–6724. [CrossRef]
15. Hofmann, D.; Fritz, L.; Ulbrich, J.; Paul, D. Molecular simulation of small molecule diffusion and solution in dense amorphous polysiloxanes and polyimides. *Comput. Theor. Polym. Sci.* **2000**, *10*, 419–436. [CrossRef]
16. Dong, Z.; Liu, Z.; Wang, P.; Gong, X. Nanostructure characterization of asphalt-aggregate interface through molecular dynamics simulation and atomic force microscopy. *Fuel* **2017**, *189*, 155–163. [CrossRef]
17. Xu, G.; Wang, H. Molecular dynamics study of interfacial mechanical behavior between asphalt binder and mineral aggregate. *Constr. Build. Mater.* **2016**, *121*, 246–254. [CrossRef]
18. Xu, G.; Wang, H. Molecular dynamics study of oxidative aging effect on asphalt binder properties. *Fuel* **2017**, *188*, 1–10. [CrossRef]
19. Pan, P.; Wu, S.; Xiao, Y.; Liu, G. A review on hydronic asphalt pavement for energy harvesting and snow melting. *Renew. Sustain. Energy Rev.* **2015**, *48*, 624–634. [CrossRef]
20. Ding, Y.; Huang, B.; Shu, X.; Zhang, Y.; Woods, M.E. Use of molecular dynamics to investigate diffusion between virgin and aged asphalt binders. *Fuel* **2016**, *174*, 267–273. [CrossRef]
21. Pan, P.; Wu, S.; Hu, X.; Liu, G.; Li, B. Effect of material composition and environmental condition on thermal characteristics of conductive asphalt concrete. *Materials* **2017**, *10*, 218. [CrossRef]
22. Sonnenburg, J.; Gao, J.; Weiner, J. Molecular dynamics simulations of gas diffusion through polymer networks. *Macromolecules* **1990**, *23*, 4653–4657. [CrossRef]
23. Zhang, L.; Greenfield, M.L. Analyzing properties of model asphalts using molecular simulation. *Energy Fuels* **2007**, *21*, 1712–1716. [CrossRef]
24. Bhasin, A.; Bommavaram, R.; Greenfield, M.L.; Little, D.N. Use of molecular dynamics to investigate self-healing mechanisms in asphalt binders. *J. Mater. Civ. Eng.* **2010**, *23*, 485–492. [CrossRef]
25. Groenzin, H.; Mullins, O.C. Molecular size and structure of asphaltenes from various sources. *Energy Fuels* **2000**, *14*, 677–684. [CrossRef]
26. Zhou, X.; Wu, S.; Liu, G.; Pan, P. Molecular simulations and experimental evaluation on the curing of epoxy bitumen. *Mater. Struct.* **2016**, *49*, 241–247. [CrossRef]
27. Sadus, R.J. *Molecular Simulation of Fluids*; Elsevier Science Ltd.: New York, NY, USA, 2002.

28. Bosko, J.T.; Todd, B.; Sadus, R.J. Molecular simulation of dendrimers and their mixtures under shear: Comparison of isothermal-isobaric (NPT) and isothermal-isochoric (NVT) ensemble systems. *J. Chem. Phys.* **2005**, *123*, 034905. [CrossRef] [PubMed]
29. Xiao, Y.; van de Ven, M.; Molenaar, A.; Su, Z.; Zandvoort, F. Characteristics of two-component epoxy modified bitumen. *Mater. Struct.* **2011**, *44*, 611–622. [CrossRef]
30. CEN-European Committee for Standardization. *Nen-en 14770 Bitumen and Bituminous Binders-Determination of Complex Shear Modulus and Phase Angle-Dynamic Shear Rheometer (DSR)*; BSI: London, UK, 2012.

Article

Simulation of Permanent Deformation in High-Modulus Asphalt Pavement with Sloped and Horizontally Curved Alignment

Mulian Zheng [1,*], Lili Han [1], Chongtao Wang [2], Zhanlei Xu [3], Hongyin Li [4] and Qinglei Ma [5]

1. Key Laboratory for Special Area Highway Engineering of Ministry of Education, Chang'an University, Xi'an 710064, China; 2014021034@chd.edu.cn
2. First Highway Consultants Co., Ltd., Xi'an 710075, China; chongtao0611@163.com
3. China Railway First Survey & Design Institute Group Co., Ltd., Xi'an 710054, China; xzl001001@163.com
4. Highway Administration Bureau of Transportation Department, Jinan 250002, China; chdmaqiang@163.com
5. Shandong College of Highway Technician, Jinan 250002, China; maqinglei@tom.com

* Correspondence: zhengml@chd.edu.cn; Tel.: +86-29-8233-4846

Academic Editors: Zhanping You, Qingli Dai and Feipeng Xiao
Received: 30 December 2016; Accepted: 24 March 2017; Published: 28 March 2017

Abstract: This study aims to evaluate the permanent deformation of high-modulus asphalt pavement in special road using viscoelastic theory. Based on the creep test, the Prony series representation of Burgers model parameters for different asphalt mixtures were obtained and used in the deformation simulation of a high-modulus asphalt pavement situated in a horizontally curved ramp. The orthogonal design method was used to show the effect of different factors on the deformation. Results reveal that rutting in curved ramp was greater than in straightaway. Further, evident upheaval was found on the downhill pavement surface and outer pavement parts of the curve due to longitudinal friction force and sideway force. In addition, the upper and middle asphalt courses in such road seemed more crucial to pavement anti-rutting performance, since inclusion of shear force changed pavement deformation characteristic and the potential rutting area tended to move up. Finally, a preliminary equation to predict rutting in sloped and curved road with widely accepted pavement structure in China was proposed.

Keywords: high-modulus asphalt concrete; permanent deformation; creep; sloped and horizontally curved road; prediction

JEL Classification: 580.1099

1. Introduction

Rutting, thermal cracking and fatigue cracking have long been regarded as serious issues confronting asphalt pavement due to increasing aggressiveness posed by huge traffic volume, heavy axle load and severe climate conditions [1]. It is widely accepted that asphalt mixture could be considered to be a viscoelastic material because its deformation relates closely to stress, time as well as temperature. Therefore, inclusion of viscoelastic theory in the permanent deformation investigation is a common practice [2–6].

With the development of computing technique and the publishing of a variety of commercial finite element software, modeling permanent deformation using experimentation-based viscoelastic mechanics has been preferred by many researchers [7–9]. However, their studies focused mainly on rutting in simple road, i.e., straightaways. Actually, it was found that rutting in special roads such as long steep uphill road, sharp horizontal curve, intersections, etc., were more serious, since pavement stress in these cases seems more intricate and vehicle speed is considerably reduced [10,11]. A case in point is that many asphalt pavements near signalized intersections in southern Nevada experienced severe rutting, according to historical documents [12]. Moreover, other investigations also verified the severity of rutting in special roads. For instance, Yang presented a full-thickness rutting test to evaluate rutting performance of asphalt pavement in continuous uphill section [13]. Yang et al. modeled the shear stress and vertical strain of asphalt pavement located in long steep longitudinal ramp and found that rutting in such condition was more susceptible to axle load [14]. Li et al. inspected the rutting depth of four signalized intersections in Nanjing, China and found that the deformations caused by tangential and vertical forces were more serious and their sensitivity to traffic and loading time was greatly enhanced [15]. Up to now, although some research was carried out on modeling rutting by considering both vertical and shearing load, models in these studies simply incorporated shear force caused by longitudinal slope, whereas the centrifugal action in curves was usually neglected. In addition, nowadays researchers have developed multiple materials to relieve rutting [16]. Among them, high-modulus asphalt concrete (HMAC), first developed in France, is a kind of hot asphalt mixture whose dynamic modulus (15 °C, 10 Hz) is greater than 14,000 MPa, and hence is more resistant to rutting than traditional asphalt mixture. To achieve huge mixture modulus several measures are frequently adopted, which involve introducing a special additive, using hard-grade bitumen and developing better aggregate blend, etc. Although some studies on the permanent deformation of this new material were reported [17,18], this new material in rutting modeling is rarely considered.

In response, this research simulated the permanent deformation of a HMAC pavement in a sloped and horizontally curved road by using the viscoelastic finite element method. The uniaxial static creep test was firstly performed to obtain the Burgers model parameters, which were subsequently used in the modeling. Meanwhile, adaptability of the Burgers model to the simulation was verified by simulating the creep test with ANSYS (ANSYS 2010, NASDAQ: ANSS, Canonsburg, PA, USA, 1970). Afterwards, deformation of the pavement was modeled and the effects of several factors on the deformation were analyzed based on the orthogonal design method. Particularly, a comparison between the deformations in straightaway and in sloped and curved road was made. Finally, a rutting prediction equation for HMAC pavement with sloped and curved alignment was proposed.

2. Materials and Methods

2.1. Binder and Aggregates

The widely-used A-70 bitumen [19] in China was employed as the asphalt binder, which has a 25 °C penetration from 60 to 80 (0.1 mm), and a ring and ball softening point greater than 45 °C.

Two optimized gradations were used to fabricate HMACs in this research. They were developed by the authors in preliminary studies [20], which are denoted by HMAC-16 and HMAC-20 (HMAC is short for high-modulus asphalt concrete, 16/20 is nominal maximum particle size). These gradations were experiment-based gradations in which the final gradation was selected by preparing asphalt mixture samples and conducting a series of performance tests such as the Marshall stability test, rutting test, low temperature mixture bending test and shearing test. Aside from these two gradations, the commonly adopted dense gradation AC-25 (AC is short for asphalt concrete) was also included for comparison. The three gradations are shown in Figure 1.

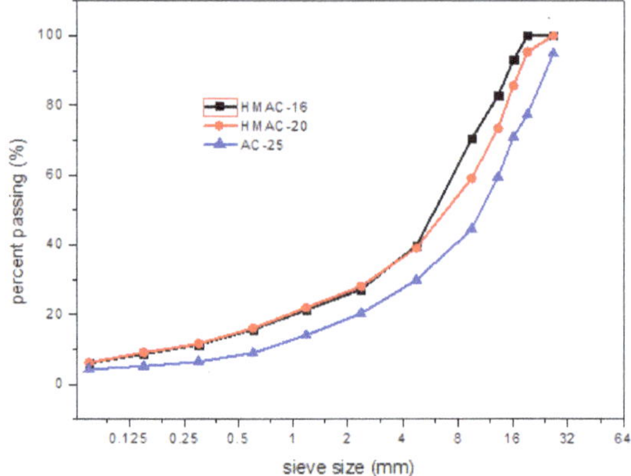

Figure 1. Gradation curves of the aggregates.

2.2. High-Modulus Additive

The high-modulus additive used is a blend of 4 mm dark blue or black cylindrical granules developed by PR INDUSTRIE, Paris, France (shown in Figure 2), whose main component is modified high-density polyethylene [20]. The high modulus additive content was determined as 0.4% by weight of whole asphalt mixture. It should be blended first with hot aggregates for 15 s and then mixed with hot asphalt and mineral filler to ensure uniform dispersion in the mixture.

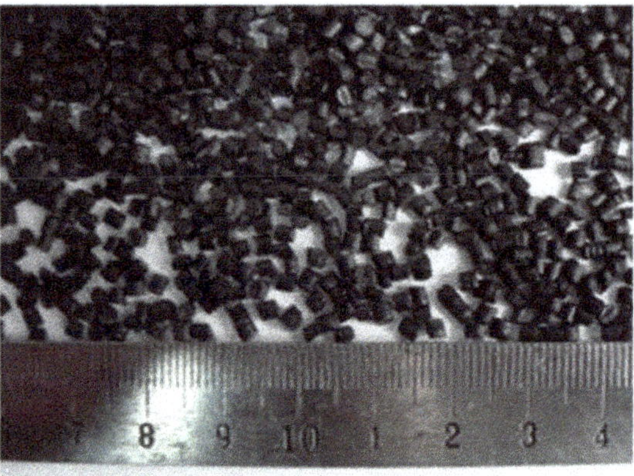

Figure 2. Appearance of high-modulus additive.

2.3. Creep Test

As it can be easily manipulated, the uniaxial static creep test is frequently preferred for asphalt mixture out of available test alternatives [21,22]. In this study, the creep test was performed on a group of Φ100 mm × 100 mm statically compacted cylindrical specimens using the MTS 810 system. The testing temperature was selected as 50 °C since rutting frequently occurs in the moderate to high temperature domain. For each mixture, three repetitions were undertaken to reduce experimental error. Before loading, all samples were kept in an oven for 4 h at 50 °C to reduce temperature deviation. In addition, before testing, a small amount of plaster was used on the sample top for leveling and the teflon sheets were placed between specimen end and the loading platen to reduce cyclo-hoop effect.

The loading protocol is: at the beginning of the loading, a 0.003 MPa stress lasting for 120 s was applied to maintain a positive contact between specimen and platen. Then, a 0.3 MPa stress lasting for 3600 s was applied to induce creep deformation. After that, the sample is unloaded for a rest period of 1800 s. During the whole process, the stress and strain were recorded automatically to develop viscoelastic parameter.

2.4. Viscoelastic Model and Its Adaptability

Since the linear viscoelastic theory has been widely applied in permanent deformation analysis of asphalt pavement [6,21], this research selected the Burgers model in the simulation, which was found effective in reflecting the instantaneous and lagged elastic strain as well as the viscous flow of asphalt mixture [22]. Its constitutive equation is expressed as follows.

$$\varepsilon(t) = \sigma_0\left[\frac{t}{\eta_1} + \frac{1}{E_1} + \frac{1}{E_2}\left(1 - e^{-\frac{tE_2}{\eta_2}}\right)\right] \quad (1)$$

where σ_0 is the original stress; $\varepsilon(t)$ is the creep strain at time t; and E_1, E_2, η_1 and η_2 are the model coefficients which could be obtained by regression on the creep data.

According to the creep test, the total deformation–time curve of the specimen was obtained automatically, as is shown in Figure 3. Based on this curve, the total strain–time curve could be obtained by dividing the deformation by specimen height. However, after subtracting the elastic strain from the total strain, the creep strain–time curve during the constant 3600 s loading period was developed as shown in Figure 4. Finally, the parameters of Burgers model (E_1, η_1, E_2 and η_2) for various mixtures could be acquired by conducting multivariate regression on creep data.

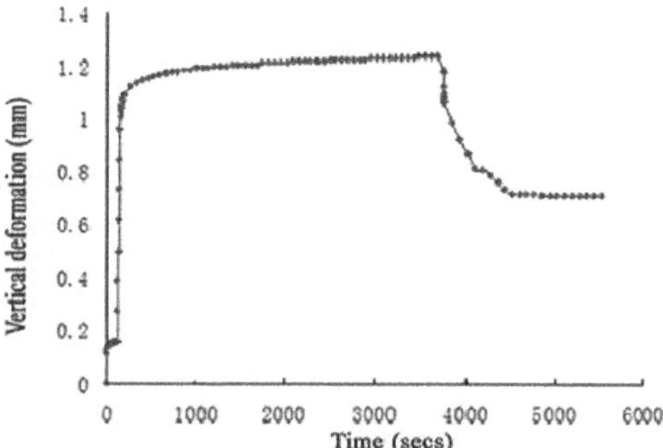

Figure 3. Deformation–time curve of the specimen.

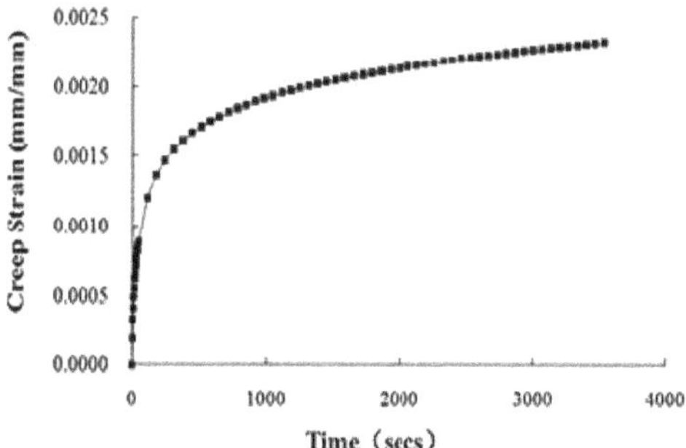

Figure 4. Creep strain–time curve of the specimen.

In ANSYS, the creep model for a linear viscoelastic solid are not represented in Equation (1) but instead in the Prony series representation shown in Equation (2), because the Prony series uses a series of decaying exponentials to reach remarkable computational efficiency and explicit physical basis [23].

$$G(t) = G_1[g_1 e^{-\frac{t}{\tau_1}} + g_2 e^{-\frac{t}{\tau_2}}] \quad (2)$$

where $G(t)$ is the shearing modulus at time t, G_1 is the initial shearing modulus; g_1, g_2, τ_1 and τ_2 are the Prony representation coefficients.

Since the Burgers model parameters have some connection with the Prony coefficients, these four parameters can be converted into the Prony representation coefficients (g_1, g_2, τ_1, τ_2), according to the method in the literature [24]. The fitted Burgers model parameters at 50 °C and the corresponding converted Prony series coefficients for different asphalt mixtures are shown in Table 1.

Table 1. Burgers model parameters and its Prony series coefficients of different mixtures. AC, asphalt concrete; HMAC, high-modulus asphalt concrete.

Temp.	Mixture	Model Parameter				Prony Series Coefficient			
		E_1 (MPa)	E_2 (MPa)	η_1 (GPa·s)	η_2 (GPa·s)	g_1	g_2	τ_1	τ_2
50 °C	AC-25	400	68	74	43	0.869	0.131	54.910	1706.277
	HMAC-16	700	166	190	57	0.932	0.068	63.290	1831.429
	HMAC-20	800	132	220	68	0.908	0.092	59.229	2399.987

To ensure the included model could characterize the deformation of HMAC adequately, verification of the model adaptability was conducted by modeling the creep test numerically and by comparing the simulation with the test results.

Since the creep test took the Φ100 mm cylinder, which was 100 mm high as the specimen, a three-dimensional model was constructed accordingly in the verification analysis. For the boundary condition the degree of freedom (DOFs) along the Z-axis on the model bottom was zero and the model was horizontally unconfined. The same loading program to the aforementioned creep test was employed in the simulating.

The creep compliance–time curves obtained from both the numerical simulation and the creep test are given in Figure 5. From this figure, a good fitness between them with a relative error no more than 3.2% was observed, indicating that the Burgers model is feasible in evaluating the permanent deformation of HMAC.

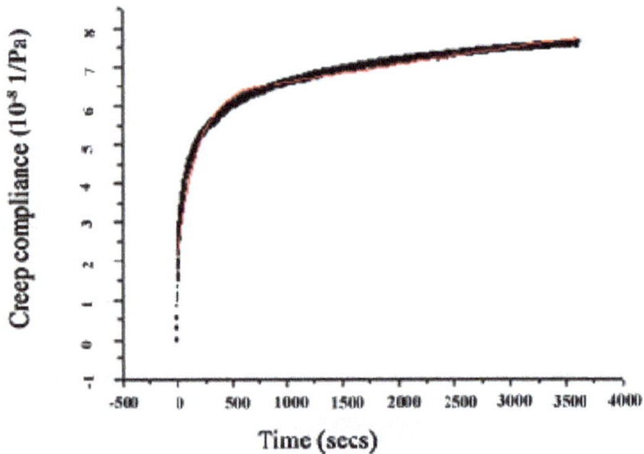

Figure 5. Creep compliance–time curve of the simulation and creep test.

3. Simulating HMAC Pavement Deformation in Curved Ramp

3.1. Road Alignment Feature

Field investigation revealed that rutting in special road was more serious than in straight and flat highways [25]. However, the terminology 'special road' includes profound implications. This section here presents four frequently encountered special roads and their geometric features as well as pavement stress characteristics. Moreover, the most unfavorable condition among them was selected for the following modeling.

Typical special roads involve: First, a flat straight way with dramatic speed change, which usually appears near a tollgate or an intersection. In this case, asphalt pavement is subjected to a shear force resulting from acceleration. The second is consecutive steep route in which the shear force applied to pavement mainly originates from the gravity component along the gradient. Speed change will also contribute to the shear force to which the pavement is subjected. The third is a horizontally curved alignment. Due to the centrifugal force, a force departing from the curvature centre is imposed on the pavement. This is called the sideway force, which increases with the decrease of radius. The last special road is the horizontally curved and sloped highway, which is the combination of the second and third condition. In this case, the forces imposed onto the pavement are rather complicated due to dual action of centrifugal force and longitudinal friction. The fourth condition had been proven to be the most unfavorable by the authors of [25], so it was employed to represent special road in the following model.

In curved ramp, the pavement unit bears two different shear forces, which are stemmed from the transverse and longitudinal friction between vehicle wheels and pavement surface as shown in Figure 6. One of them, the longitudinal friction force f_l should offset the component of wheel pressure p along the slope as well as the rolling friction force, according to Figure 6a. Based on the design speed and maximum gradient for roads at different levels in the Highway Alignment Design Specification [26], the longitudinal friction force f_l could be calculated as depicted in Table 2. The calculation shows that the longitudinal friction force was almost 0.6 of the standard 0.707 MPa vertical tire pressure at a profile of 5%.

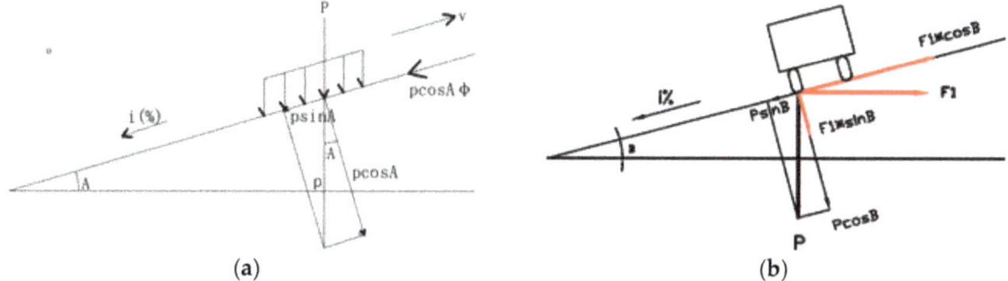

Figure 6. Pavement force analysis in curved and sloped road. (**a**) Longitudinal section; (**b**) transverse section.

Table 2. Pavement longitudinal friction force under different slopes (in pressure unit) [1].

Design Speed (km/h)	120	100	80	60	40	30	20
Max slope I (%)	3	4	5	6	7	8	9
Slope angle A (°)	1.718	2.291	2.862	3.434	4.004	4.574	5.143
$p\sin A$ (MPa)	0.021	0.028	0.035	0.042	0.049	0.056	0.063
$p\cos A$ (MPa)	0.707	0.706	0.706	0.706	0.705	0.705	0.704
$f_l = p\sin A + \phi p\cos A$ (MPa)	0.375	0.381	0.388	0.395	0.402	0.409	0.415

[1] ϕ is rolling friction coefficient. Wheel Contact Pressure p (MPa): Standard Axle Load BZZ100, p = 0.707 MPa, ϕ = 0.5.

Similarly, the transverse friction force f_t could also be determined by multiplying tire pressure by a factor that reflects the effect of radius. The factor is called sideway force coefficient (SFC). According to the Highway Alignment Design Specification, this analysis took the largest SFC of 0.2 for considering the worst condition [26].

3.2. ANSYS Model

The three-dimensional pavement model with dimensions in the longitudinal, transverse and depth direction of 2.1, 2.0 and 7.55 m respectively was constructed. This pavement unit is assumed to be located near the midpoint of the horizontal curve. The pavement structure and the thickness of each layer are shown in Table 3.

Table 3. Structural thickness and basic parameters of the pavement model.

Structural Layers		Thickness (cm)	Resilient Modulus (MPa)	Poisson Ratio
Asphaltic layers	Top layer	3–7	using viscoelastic parameters	using viscoelastic parameters
	Middle course	4–8		
	Binder course	5–9		
Semi rigid base		16–32	1400	0.25
Subbase		20	600	0.25
Subgrade		700	60	0.25

Based on the multi-layered pavement system, two different hypotheses—i.e., the elastic and linear viscoelastic hypotheses—were introduced for various layers. The former applied to semi-rigid base, subbase and subgrade which used the Solid 45 element for analyzing, while the viscoelastic hypothesis was applied to asphaltic layers which employed the 8-node Solid 185 element in the computation. Moreover, the Surf 154 element was adopted to simulate the shear force on pavement surface. For the asphaltic layers, the aforementioned viscoelastic parameters (shown in Table 1) were employed, while for the other layers, the elastic parameters are provided in Table 3.

Consideration of different constitutive relationships for different layers in this research can be justified as follows. In China today, asphalt pavement composed of asphaltic surface course as well as semi-rigid base and subbase is the overwhelmingly adopted structure. Numerous in-situ investigations indicate that during hot days, rutting deformation in such pavement mainly derives from viscous flow of asphaltic mixtures, whereas the contribution of semi-rigid base, subbase and subgrade is relatively small, provided that the pavement was well designed and constructed. Therefore, to highlight the deformation of asphaltic layers and to raise computation efficiency, this research applied viscoelastic theory to asphaltic layers while using elastic theory for the others.

The vertical load is the uni-axle-two-wheel load, which has a rectangular uniform pattern and a dual tire spacing of 31.95 cm. The wheel contact area is 20 cm long and 20 cm wide. Based on previous analysis, two shear forces (longitudinal and transverse force) were applied to the model top, which were determined through multiplying the vertical pressure by 0.6 and 0.2 respectively. In terms of loading time, since similar rutting predicting results under repeated loading–unloading cycles and single long time loading have been verified [27], in this study numerous individual loading was approximately accumulated into single long time load according to the Boltzmann Superposition principle. In this modeling, the equivalent single axle load (ESAL) number ranging from 200 to 2800 (10^4 times) and an operational velocity of 100 km/h were considered. The corresponding accumulated loading times were calculated as shown in Table 4. Additionally, the lateral wander factor of 0.4 was adopted in the calculation in accordance with current code [26]. Since the pavement deformation at the moment of unloading includes both elastic deformation and non-elastic, different recovering times were considered as shown in Table 4.

Table 4. Accumulated loading time and deformation recovering time. ELAL, equivalent single axle load.

ESALs (10^4 times)	Accumulated Loading Time (s)	Recovery Time (s)
200	5760	2400
700	20,160	8400
1300	37,440	15,600
2000	57,600	24,000
2800	80,640	33,600

4. Results and Discussion

4.1. Contrast of Permanent Deformation in Flat Straight Road and Curved Ramp

As illustrated in nephograms, the permanent deformations in straight and flat road as well as those in the horizontally curved and sloped section are shown in Figures 7 and 8 respectively.

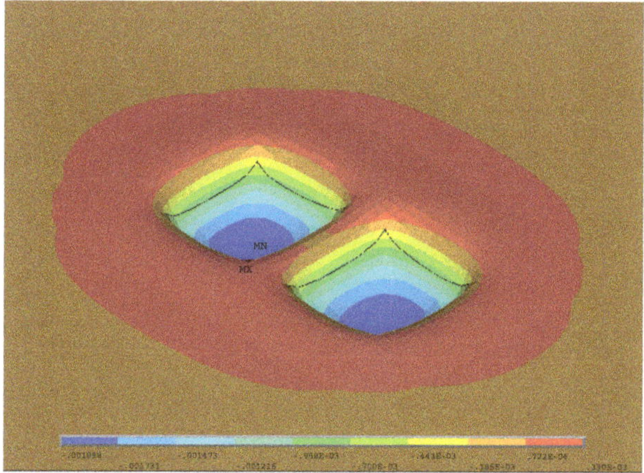

Figure 7. Nephogram of permanent deformation in simple road.

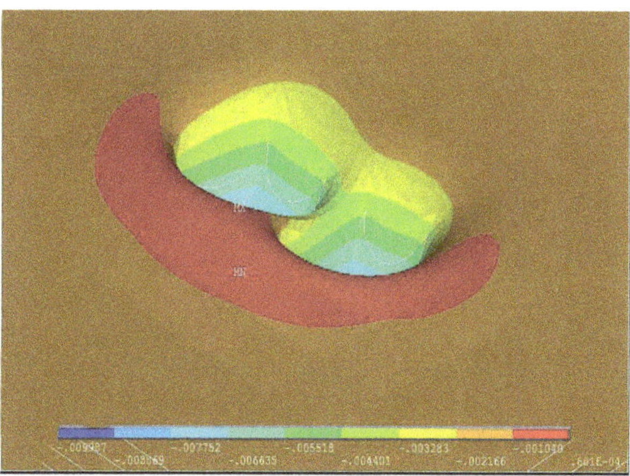

Figure 8. Nephogram of permanent deformation in curved ramp.

It was found that asphaltic mixtures in the pavement surface courses under two different alignments showed slow viscous flow after prolonged loading and unloading cycles, and different locations featured quite different deformations. Overall, the deformation decreased with the distance from axle center increasing. It was observed that the greatest vertical displacement occurred right below the axle load since the vertical compressive stress herein reaches the maximum value. In contrast, asphaltic mixture that is around the wheel contact area showed swelling due to considerable lateral extruding force.

However, the deformation in curved ramp showed some differences in terms of magnitude and deformation features. On the one hand, it is seen from the above two figures that the compressive deformation in curved and sloped roadway was greater than in the simple straight highway. Detailed calculation shows that the maximum vertical deformation, maximum bump height and total rutting depth in curved slope is 9.87, 0.68 and 10.55 mm respectively whereas that in simple road is 8.93, 0.46 and 9.39 mm respectively. The deformation in the first condition is about 1.1 times that in the second condition. This result is consistent with the in-situ rutting observation [11,28]. Another remarkable deformation characteristic is its asymmetric upheaval surrounding the rut groove, which may have resulted from the longitudinal and transverse friction force. As shown in Figure 8, the pavement area along the downward direction showed evident upheaval due to longitudinal friction force. While the asphaltic mixture upward showed a much smaller bump. Moreover, similar deformations could be found with respect to the transverse section. Figure 8 shows that the upheaval developed on the outside pavement of the horizontal curve seemed larger than the inner side. This is probably due to the sideway friction force.

According to current literature, Pei and Kong et al. had investigated the rutting deformation of long sloped asphalt pavements. Their field inspections revealed that most serious rutting occurred on the top area of profile, and the rutting depth in slopes was greater than in tangential flat road [11,28]. By comparing, it was found that the authors' simulation agreed with the previous research. However, the downward upheaval is not obvious in actual sloped road mainly because actual loading is not applied statically within a local spot but rather a stripped dynamic load. Therefore, the longitudinal upheaval tends to be smooth under the repeated rolling of vehicles.

4.2. *Influential Factor Analysis Using Orthogonal Design Method*

This section presents the discussion on the influence of different factors on the deformation developed in the HMAC pavement with sloped and horizontally curved road alignment. In view of the massive calculating runs resulting from full factorial design (5^6 = 15,625), the Taguchi array was employed to reduce experimental run. The Taguchi method or orthogonal design method (ODM) developed by Dr. Genichi Taguchi in Japan is a partial fraction experiment and has found wide use in many regions [29]. This method uses a special set of arrays called orthogonal array (OA) to choose the level combination of the variables for each experiment and to determine the minimal number of experiments to give full information of the factors that affect the objective index [30].

This research considered the permanent deformation as the objective index and chose six main factors each with five levels to reflect their influence. The factors and their levels are shown in Table 5.

Table 5. Six factors and the corresponding levels.

Factor	Level				
	1	2	3	4	5
P1-tire pressure (MPa)	0.4	0.7	1.0	1.3	1.6
P2-ESAL (10^4 times)	200	700	1300	2000	2800
P3-top layer thickness (cm)	3	4	5	6	7
P4-middle course thickness (cm)	4	5	6	7	8
P5-binder course thickness (cm)	5	6	7	8	9
P6-base thickness (cm)	16	20	24	28	32

An OA is usually denoted by $L_N(l^k)$, where N is the number of performed experiments, l is the number of levels per factor and k is the number of factors. This study selected the $L_{25}(5^6)$ OA (shown in Table 6) to design factor-level combination and to conduct trial calculations, which needs 25 experimental runs to give necessary data for further analysis. The calculated deformations under each of the 25 experiments are shown in the rightmost column of Table 6.

Table 6. Orthogonal array and results of intuitive analysis on 25 trials.

Experiment No.	Factor						Deformation (mm)
	P1	P2	P3	P4	P5	P6	
1	1	1	1	1	1	1	1.308
2	1	2	2	2	2	2	3.894
3	1	3	3	3	3	3	6.350
4	1	4	4	4	4	4	7.136
5	1	5	5	5	5	5	10.643
6	2	1	2	3	4	5	2.603
7	2	2	3	4	5	1	6.467
8	2	3	4	5	1	2	10.033
9	2	4	5	1	2	3	13.476
10	2	5	1	2	3	4	28.690
11	3	1	3	5	2	4	3.550
12	3	2	4	1	3	5	10.128
13	3	3	5	2	4	1	13.940
14	3	4	1	3	5	2	24.299
15	3	5	2	4	1	3	29.557
16	4	1	4	2	5	3	5.095
17	4	2	5	3	1	4	8.055
18	4	3	1	4	2	5	19.172
19	4	4	2	5	3	1	25.092
20	4	5	3	1	4	2	49.035
21	5	1	5	4	3	2	7.977
22	5	2	1	5	4	3	14.072
23	5	3	2	1	5	4	37.823
24	5	4	3	2	1	5	39.232
25	5	5	4	3	2	1	46.460
K_{j1}	29.331	20.533	87.541	111.770	88.185	93.267	-
K_{j2}	61.269	42.616	98.969	90.851	86.552	95.238	-
K_{j3}	81.474	87.318	104.634	87.767	78.237	68.550	-
K_{j4}	106.449	109.235	78.852	70.309	86.786	85.254	-
K_{j5}	145.564	164.385	54.091	63.390	84.327	81.778	-
$\overline{K}_{j1}$	5.866	4.107	17.508	22.354	17.637	18.653	-
$\overline{K}_{j2}$	12.254	8.523	19.794	18.170	17.310	19.048	-
$\overline{K}_{j3}$	16.295	17.464	20.927	17.553	15.647	13.710	-
$\overline{K}_{j4}$	21.290	21.847	15.770	14.062	17.357	17.051	-
$\overline{K}_{j5}$	29.113	32.877	10.818	12.678	16.865	16.356	-
Range	23.247	28.770	10.109	9.676	1.990	5.338	-

The Taguchi method usually uses intuitive analysis to analyze experimental data since it could give an explicit depict of the influence of different factors. The result of intuitive analysis in this study is also shown in Table 6. To guide the readers, several supplementary instructions to this table are presented as follows.

In Table 6, K_{ji} is the sum of the deformations with factor j at i level (j = P1 to P6, i = 1 to 5) out of all trials. In addition, the average value of K_{ji} and the range of each parameter is calculated as follows.

$$\overline{K}_{ji} = \frac{1}{5}K_{ji},\ Range_j = \max_i\{\overline{K}_{ji}\} - \min\{\overline{K}_{ji}\} \quad (3)$$

According to the Taguchi method, the range of a factor could reflect its influence on the objective index. The larger the range value of a factor, the greater its effect on the process [31]. Based on the intuitive analysis, the ranking of factors according to their effect on pavement deformation in descending order is P2 → P1 → P3 → P4 → P6 → P5. According to Table 5, this indicates that ESAL and contact pressure showed greater impact on the deformation developed in special road, and the thickness of upper and middle asphaltic layers were also crucial to the pavement deformation whereas the binder asphalt layer and semi rigid base demonstrated less influence.

The intuitive analysis diagram that indicates the change of pavement deformation with various factors is shown in Figure 9.

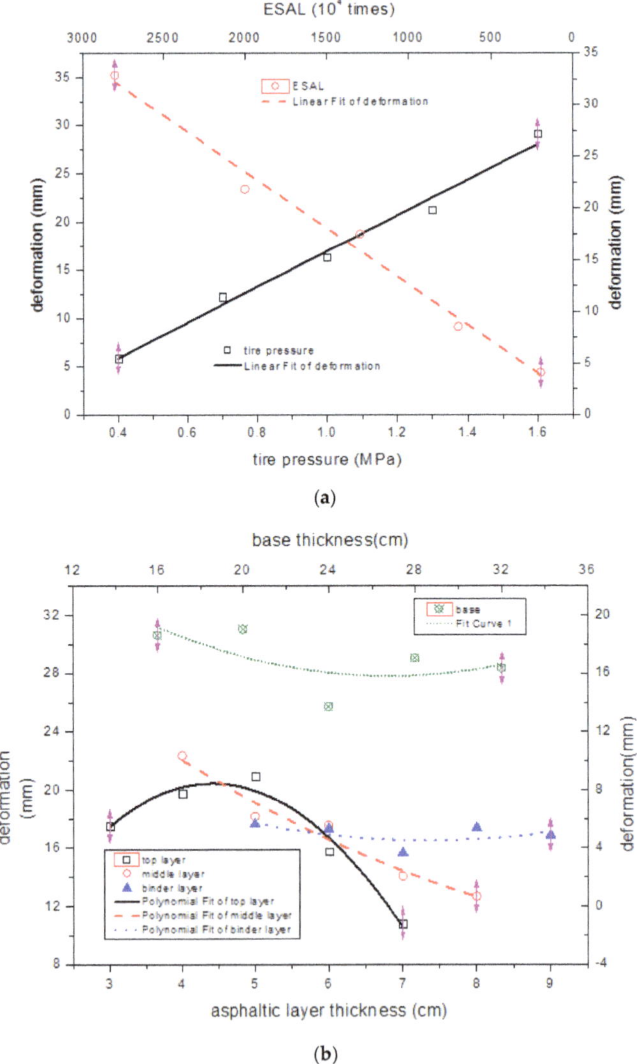

Figure 9. Change of permanent deformation with factors and levels, (**a**) with tire pressure and ESAL; (**b**) with different layer thickness.

From Figure 9a, it could be seen that the deformation (left Y-axis) increased linearly with the rise in tire pressure (bottom X-axis). The deformation under the pressure of 1.6 MPa was approximately five times that under 0.4 MPa. The pavement deformation (right Y-axis) also increased linearly with the increase in ESAL (top X-axis). The deformation under ESAL of 2800 (10^4 times) is eight times that under ESAL of 200 (10^4 times). According to the authors' previous research on HMAC pavement rutting with simple alignment, the deformation correlated well with ESAL in a convex power function whose exponent was 0.303 [32]. In contrast, Figure 9a shows that the pavement deformation in horizontally curved and sloped road increased linearly with ESAL, meaning the deformation susceptibility to traffic load in special road is increased. This result further confirms that heavy duty as well as overloading poses significant negative influence on pavement performance and should be therefore strictly controlled.

Also, Figure 9b indicates the deformation in HMAC pavement with sloped and curved alignment changed differently with the change in depth of each structural layer. In detail, the deformation (left Y-axis) increased by 19.5% when the upper layer depth changed from three to five centimeters (bottom X-axis), whereas it decreased by 48.3% with the depth increasing to seven centimeters further. This suggests that an upper asphaltic course no less than five centimeters in complicated road is necessary considering the anti-rutting performance of entire pavement. In contrast, the deformation (left Y-axis) decreased by 43.3% when the middle HMAC course depth (bottom X-axis) doubled. It could be thereof inferred that increasing thickness of the middle course contributed most to the reduction in pavement permanent deformation since the HMAC was used in this layer. Unlike the upper and middle asphalt courses, increase in the binder course depth (bottom X-axis) had little effect on the change of permanent deformation (left Y-axis), according to Figure 9b. It was concluded that the upper and middle asphalt courses were crucial to the rutting resistant performance of HMAC pavement under complicated route conditions. According to the literature, Xu suggested that modified asphalt binder be used in the middle and binder asphaltic layers to relieve pavement rutting in simple alignment [3]. However, this study shows that as far as special road is concerned, the upper and middle courses were more crucial to the pavement anti-rutting performance. This indicates that inclusion of the longitudinal and transverse shear force changed pavement deformation characteristic and the potential rutting area had a tendency to move up in curved ramp.

Similarly, Figure 9b shows the pavement deformation (right Y-axis) changed little with the increase in the depth of semi-rigid base (top X-axis). Calculation shows that the deformation decreased simply by 12.3% when the base thickness doubled, which suggests that an appropriate thickness that is not too great would be suitable for semi-rigid base.

4.3. Rutting Prediction Model for HMAC Pavement in Curved and Sloped Road

Since there is no rutting predicting equation applicable to HMAC pavement in curved and sloped road in current literature, this research conducted multivariate regression on the simulating data and presented the following model. The correlation coefficient shows this model fits well with the simulating results.

$$\begin{aligned} RD &= 0.0435P^{1.021}N^{0.831} - 3.089{H_u}^{0.5} - 7.650\ln H_m + 1.685\ln^2 H_b + \tfrac{0.133}{H_{ba}} + 13.677 \\ (R^2 &= 0.959) \end{aligned} \tag{4}$$

where, RD is the rutting depth (mm),

P is the tire pressure (MPa),

N is the ESAL number (10^4 times),

H_u is the thickness of top asphaltic course (cm),

H_m is the thickness of middle HMAC course (cm),

H_b is the thickness of binder asphaltic course (cm),

H_{ba} is the thickness of semi rigid base (cm).

5. Conclusions

This study modeled the permanent deformation of HMAC pavement in sloped and horizontally curved road using experimentation-based viscoelastic mechanics. Some primary conclusions are as follows.

(1) The maximum rutting depth and bump height of HMAC pavement in curved and sloped road were greater than those in straightaway. The downward pavement areas of curved and sloped road developed evident upheaval due to longitudinal friction force. In addition, the outside pavement of horizontal curve presented larger upheaval than the inner side due to sideway force.
(2) Analysis shows the upper and middle asphaltic courses were more crucial to pavement anti-rutting performance in curved and sloped road since inclusion of longitudinal and transverse shear force changed pavement deformation characteristic and the potential rutting area had a tendency to move up.
(3) A preliminary model to predict rutting in HMAC pavement with sloped and horizontally curved alignment was presented based on material and structural features of asphalt pavement in China today.

It should be noted that in this study there are still some issues that need to be further addressed. First, this research simply considered the deformation at 50 °C. Permanent deformations in a wider temperature range should be investigated for a better understanding of HMAC pavement behavior. Second, the dynamic loading effect as well as the intermittent loading was not considered in the simulation for simplicity, which could possibly lead to a larger deformation. The last point is that only six factors were chosen as the variable in the prediction model, and other factors such as modulus of the base, subbase and subgrade did not appear in the equation but were taken as fixed values. However, though limited to some extent, the aforementioned model could be used for engineering precision especially when confronted with common practice in China.

Finally, from the viewpoint of life cycle cost [33], technical measures such as putting strict traffic control on special road, using effective anti-rutting asphalt mixtures were strongly suggested to be taken into account in the construction of special road to cut the life cycle pavement cost.

Acknowledgments: This research was supported by the Fundamental Research Funds for the Central Universities in China(No. 310821163502), the 2016 Outstanding Doctoral Dissertation project of Fundamental Research Funds for the Central Universities in China (No. 310821165008), the Transportation Department of Shandong Province (No. 2008Y007) and the National Natural Science Foundation of China (No. 51008033).

Author Contributions: Mulian Zheng was in charge of the whole research plan and paper submission matters. The first author extends gratitude to Lili Han, Chongtao Wang and Zhanlei Xu for performing the experiment and analyzing test data, to Hongyin Li and Qinglei Ma for offering help in doing the experiment.

Conflicts of Interest: The authors declare no conflict of interest.

References

1. Drescher, A.; Kim, J.R.; Newcomb, D.E. Permanent deformation in asphalt concrete. *J. Mater. Civ. Eng.* **1993**, *5*, 112–128. [CrossRef]
2. Li, C.; Li, L. Criteria for controlling rutting of asphalt concrete materials in sloped pavement. *Constr. Build. Mater.* **2012**, *35*, 330–339. [CrossRef]
3. Xu, T.; Huang, X. Investigation into causes of in-place rutting in asphalt pavement. *Constr. Build. Mater.* **2012**, *28*, 525–530. [CrossRef]
4. Katicha, S.W.; Apeagyei, A.K.; Flintsch, G.W.; Loulizi, A. Universal linear viscoelastic approximation property of fractional viscoelastic models with application to asphalt concrete. *Mech. Time-Depend. Mater.* **2014**, *18*, 555–571. [CrossRef]
5. Hofko, B. Addressing the permanent deformation behavior of hot mix asphalt by triaxial cyclic compression testing with cyclic confining pressure. *J. Traffic Transp. Eng.* **2015**, *2*, 17–29. [CrossRef]

6. Zhao, Y.; Wang, L.; Chen, P.; Zeng, W. Determination of surface viscoelastic response of asphalt pavement. *J. Eng. Mech.* **2015**, *141*, 1–8. [CrossRef]
7. Blab, R.; Harvey, J. Modeling measured 3D tire contact stress in a viscoelstic FE pavement model. *Int. J. Geomech.* **2002**, *2*, 271–290. [CrossRef]
8. Fang, H.; Haddock, J.E.; White, D.E.; Hand, A.J. On the characterization of flexible pavement rutting using creep model-based finite element analysis. *Finite Elem. Anal. Des.* **2004**, *41*, 49–73. [CrossRef]
9. Abed, A. Evaluation of rutting depth in flexible pavements by using finite element analysis and local empirical model. *Am. J. Eng. Appl. Sci.* **2012**, *5*, 163–169.
10. Wu, S.; Wang, J.; Chen, T. Stress analysis of asphalt pavement on large longitudinal slope. *J. Wuhan Univ. Technol. Transp. Sci. Eng. Ed.* **2006**, *30*, 969–972. (In Chinese).
11. Pei, J.Z.; Chen, Y.; Chang, M.F. Mechanism of rutting formation in long and steep climbing sections of asphalt pavement. In Proceedings of the ICCTP 2009: Critical Issues in Transportation Systems Planning, Development, and Management, Harbin, China, 5–9 August 2009; ASCE: St. Louis, MO, USA, 2009.
12. Hajj, E.Y.; Tannoury, G.; Sebaaly, P.E. Evaluation of rut resistant asphalt mixtures for intersection. *Road Mater. Pavement Des.* **2011**, *12*, 263–292. [CrossRef]
13. Yang, X.; Guan, H.; Zhang, Q.; Zhou, L. The rutting resistant surface course combination for continuous uphill section of expressway. In Proceedings of the GeoShanghai 2010 International Conference: Paving Materials and Pavement Analysis, Shanghai, China, 3–5 June 2010; ASCE: St. Louis, MO, USA, 2010.
14. Yang, J.; Li, W. Finite element analysis of asphalt pavement on long-steep longitudinal slope. *J. Traffic Transp. Eng.* **2010**, *10*, 20–31. (In Chinese).
15. Li, L.; Huang, X.; Wang, L.; Li, C. Integrated experimental and numerical study on permanent deformation of asphalt pavement at intersections. *J. Mater. Civ. Eng.* **2013**, *25*, 907–912. [CrossRef]
16. Zheng, M.; Han, L.; Wang, F.; Mi, H.; Li, Y.; He, L. Comparison and analysis on heat reflective coating for asphalt pavement based on cooling effect and anti-skid performance. *Constr. Build. Mater.* **2015**, *93*, 1197–1205. [CrossRef]
17. Capitão, S.; Picado-Santos, L. Assessing permanent deformation resistance of high modulus asphalt mixtures. *J. Transp. Eng.* **2006**, *132*, 394–401. [CrossRef]
18. Petho, L. High modulus asphalt mix (EME) for heavy duty applications and preliminary laboratory test results in Australia. In Proceedings of the AAPA 15th International Flexible Pavements Conference, Brisbane, Australia, 22–25 September 2013.
19. Ministry of Transport (MOT). *Technical Specification for Construction of Highway Asphalt Pavement*; JTG F40-2004; Ministry of Transport: Beijing, China, 2004.
20. Espersson, M. Effect in the high modulus asphalt concrete with the temperature. *Constr. Build. Mater.* **2014**, *71*, 638–643. [CrossRef]
21. Ebrahimi, M.G.; Saleh, M.; Gonzalez, M. Interconversion between viscoelastic functions using the Tikhonov regularisation method and its comparison with approximate techniques. *Road Mater. Pavement Des.* **2014**, *15*, 820–840. [CrossRef]
22. Saboo, N.; Kumar, P. A study on creep and recovery behavior of asphalt binders. *Constr. Build. Mater.* **2015**, *96*, 632–640. [CrossRef]
23. Park, S.W.; Kim, Y.R. Fitting prony-series viscoelastic models with power-law presmoothing. *J. Mater. Civ. Eng.* **2001**, *13*, 26–32. [CrossRef]
24. Chen, J.; Zhou, C.; Wang, Z. Data processing and viscoelastic computation for creep test of asphalt mixture. *J. Southeast Univ. Nat. Sci. Ed.* **2007**, *37*, 1091–1095. (In Chinese).
25. Che, F. HMAC Performance Evaluation and Numerical Simulation for Special Section Pavement. Ph.D. Thesis, Chang'an University, Xi'an, China, 2011.
26. Ministry of Transport (MOT). *Design Specification for Highway Alignment*; JTG D20-2006; Ministry of Transport: Beijing, China, 2006.
27. Huang, F. Permanent Deformation Simulation and Rutting Depth Prediction of Asphalt Pavement. Master's Thesis, Southeast University, Nanjing, China, 2006.
28. Kong, H. Study on the Optimum Design of Asphalt Pavement Structure in Long Steep Longitudinal Slope Section Based on Rutting Resistance. Master's Thesis, Chang'an University, Xi'an, China, 2012.
29. Zhang, Q.; Zeng, S.; Wu, C. Orthogonal design method for optimizing roughly designed antenna. *Int. J. Antennas Propag.* **2014**, *2014*, 586360. [CrossRef]

30. Otto, K.N.; Antonsson, E.K. Extensions to the Taguchi method of product design. *J. Mech. Des.* **1993**, *115*, 5–13. [CrossRef]
31. Fraley, S.; Oom, M.; Terrien, B.; Zaleskwi, J. Design of Experiments via Taguchi Methods: Orthogonal Arrays. Available online: https://zh.scribd.com/document/133377816/Design-of-Experiments-via-Taguchi-Methods-Orthogonal-Arrays (accessed on 10 December 2016).
32. Zheng, M.; Han, L.; Qiu, Z.; Li, H.; Ma, Q.; Che, F. Simulation of permanent deformation in high modulus asphalt pavement using Bailey-Norton creep law. *J. Mater. Civ. Eng.* **2016**, *28*, 1–11. [CrossRef]
33. Praticò, F.G.; Casciano, A.; Tramontana, D. Pavement life cycle cost and asphalt binder quality: A theoretical and experimental investigation. *J. Constr. Eng. Manag.* **2011**, *137*, 99–107. [CrossRef]

Article

Fatigue Life Prediction of High Modulus Asphalt Concrete Based on the Local Stress-Strain Method

Mulian Zheng [1,*], Peng Li [1,*], Jiangang Yang [2], Hongyin Li [3], Yangyang Qiu [4] and Zhengliang Zhang [5]

1 Key Laboratory for Special Area Highway Engineering of Ministry of Education, Chang'an University, South Erhuan Middle Section, Xi'an 710064, Shaanxi, China
2 Civil Engineering Materials Laboratory, East China Jiaotong University, Shuang Gang East Street, Nanchang 330000, Jiangxi, China; mjgchd@163.com
3 Shandong Highway Administration Bureau, Shungeng Road of Jinan City, Jinan 250000, Shandong, China; chdmaqiang@163.com
4 Jinhua Traffic Planning and Design Institute, Songlian Road of Jin Dong District, Jinhua 321015, Zhejiang, China; qiu_yangyang@126.com
5 Anhui Highway Administration Center, Da Bie Shan Road of Hefei City, Hefei 230088, Anhui, China; xuhaileilie@163.com
* Correspondence: zhengml@chd.edu.cn (M.Z.); lp042820321@126.com (P.L.); Tel.: +86-29-8233-4846 (M.Z.)

Academic Editors: Zhanping You, Qingli (Barbara) Dai and Feipeng Xiao
Received: 23 January 2017; Accepted: 15 March 2017; Published: 20 March 2017

Abstract: Previously published studies have proposed fatigue life prediction models for dense graded asphalt pavement based on flexural fatigue test. This study focused on the fatigue life prediction of High Modulus Asphalt Concrete (HMAC) pavement using the local strain-stress method and direct tension fatigue test. First, the direct tension fatigue test at various strain levels was conducted on HMAC prism samples cut from plate specimens. Afterwards, their true stress-strain loop curves were obtained and modified to develop the strain-fatigue life equation. Then the nominal strain of HMAC course determined using finite element method was converted into local strain using the Neuber method. Finally, based on the established fatigue equation and converted local strain, a method to predict the pavement fatigue crack initiation life was proposed and the fatigue life of a typical HMAC overlay pavement which runs a risk of bottom-up cracking was predicted and validated. Results show that the proposed method was able to produce satisfactory crack initiation life.

Keywords: pavement engineering; high modulus asphalt concrete; local stress-strain; modified Neuber equation; fatigue life

1. Introduction

Fatigue damage refers to the accumulation of damage in asphalt concrete, causing cracks and fracture extensions, and ultimately resulting in loading failure. Fatigue failure includes the formation of the crack nucleation stage (fatigue), crack extension (stability) of crack growth, and fracture failure (unstable crack propagation). Because fracture failure happens rapidly, the fatigue life of pavement generally includes crack formation and propagation. The term "stage of fatigue crack initiation" usually refers to the fatigue crack nucleation and propagation, which require examination during engineering processes. In road engineering the period from crack nucleation to the crack reaching a length of 5 mm is treated as the life of the initiation of the fatigue crack.

There are different approaches to quantify the fatigue crack process in asphalt concrete mixtures. Among the different approaches, practitioners primarily use the strain approach, which was first introduced by Monismith, and the energy approaches because these approaches are simple and consist of experimentally oriented procedures [1–3]. The most common fatigue model adopted by the Mechanistic Empirical Pavement Design Guide (MEPDG) is a strain approach that considers the material's properties using the value of dynamic modulus forms [4]. Al-Qadi described the dissipated energy changes during loading and proposed four parameters to define fatigue performance for 13 mm dense-graded Hot Mix Asphalt (HMA) using bending beams at various strain amplitudes [5]. Molayem used continuum damage theory to characterize the fatigue properties of asphalt binders and their effect on fatigue resistance. This proposed model can be used as an alternative to time-consuming beam fatigue tests [6].

Metcalf evaluated the Strategic Highway Research Program (SHRP) fatigue crack initiation life prediction model with crack measurements in laboratory slab fatigue tests. The study showed that the SHRP crack initiation life model significantly underestimates the fatigue crack initiation life for slabs [7].

Kim developed a system for predicting the fatigue life of asphalt mixtures. This work indicates that the fatigue life of a mixture subjected to sinusoidal strain loading can be determined through the model's damage principles using IDT (indirect tensile test) test results [8]. Ali Khodaii employed a large scale experimental setup to examine the influence of the most important parameters on delaying reflection cracking in geogrid reinforced overlay in bending mode. Regression equations were developed to estimate fracture parameters which can be used to predict the crack growth rate in order to design asphalt overlay that is protected from reflective cracking [9].

One of the major achievements of the SHRP asphalt mix research projects is the development of a fatigue life prediction model based on the results of intensive bending beam fatigue tests. Several researchers have conducted fatigue testing on asphalt concrete in the laboratory to investigate fatigue life through bending beam fatigue tests [10–16]. Although this model covers a wide range of asphalt mix related fatigue variables, it has to apply a shift factor to deal with the stress concentration of cracks in High Modulus Asphalt Concrete (HMAC), including the state of stress. It would be helpful to determine how real stress state contributes to fatigue life, because local stress and direct tensile tests more realistically simulate the multiaxial state of stress and cracking in pavements than bending beam tests.

A basic principle of the local stress-strain method is that the nominal stress and strain of a structure under various loadings are firstly converted into local stress and strain at the most unfavorable points (i.e., the stress concentration points) based on the stress-strain loop curve obtained from the fatigue test. Combined with the established strain-fatigue life curve, these local stresses and strains are then corrected and used to estimate fatigue life of the structure.

Local stress-strain fatigue analysis focuses on determining local stress and strain. In the local stress concentration area, there is a complex nonlinear relationship between the stress, strain, and load. Methods for determining the local strain and stress include the test method, the elastic-plastic finite element method, and the approximate calculation method.

In this paper, we conduct direct tensile fatigue tests and cyclic stress-strain tests at different strain levels to determine the fatigue properties of HMAC, and propose a method to predict the fatigue crack initiation life of HMAC. This method is based on a modified Neuber equation and the local stress-strain method.

2. Process for Fatigue Life Estimation Based on the Local Stress-Strain Method

The fatigue life prediction method consists of three aspects: fatigue performance of the material, transformation from nominal stress and strain into local stress and strain, and fatigue cumulative damage theory. A modified Neuber equation was used to transform nominal stress-strain to local stress-strain. The Miner theory and Manson-Coffin equation have been widely used to characterize cumulative fatigue damage. Therefore, the cyclic stress-strain curve and strain-life curve can be combined as the local stress-strain method to estimate the fatigue crack life.

The raw data needed for calculation includes the elastic modulus, the steady cyclic stress–strain curve, the strain-life curve, the theoretical stress concentration factor, and load history.

Steps for fatigue life estimation, based on the local stress-strain method, are shown in Figure 1.

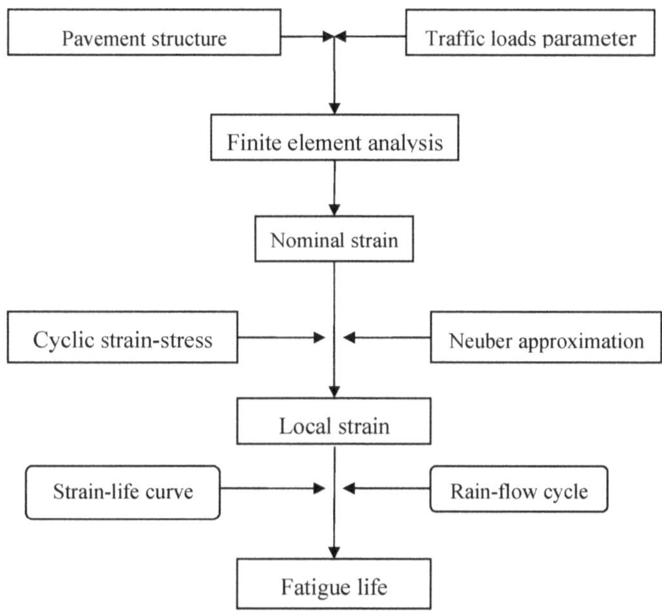

Figure 1. Process for fatigue life estimation based on local stress-strain method.

3. Direct Tensile Test of HMAC

3.1. Materials

90# asphalt was used to compound the high-modulus asphalt in the research. Details of the materials are listed in Table 1.

Table 1. Test results of 90# asphalt.

Items		Unit	Value
25 °C Penetration		0.1 mm	83
Penetration Index		...	−0.84
Ductility (10 °C)		cm	42
Ductility (15 °C)		cm	>150
Softening Point		°C	49.3
Flashing Point		°C	282
Paraffin Content (distillation)		%	0.91
Solubility (C_2HCl_3)		%	99.8
Density (15 °C)		g/cm^3	0.982
TFOT (163 °C, 5 h)	Quality Change	%	−0.005
	Penetration Ratio	%	68.5
	Ductility (15 °C)	cm	38.6
	Ductility (10 °C)	cm	9.6

PR PLASTS was chosen as the HMAC-20 additive with a dosage of 0.6%. The mineral aggregate gradation of the HMAC-20 is shown in Figure 2.

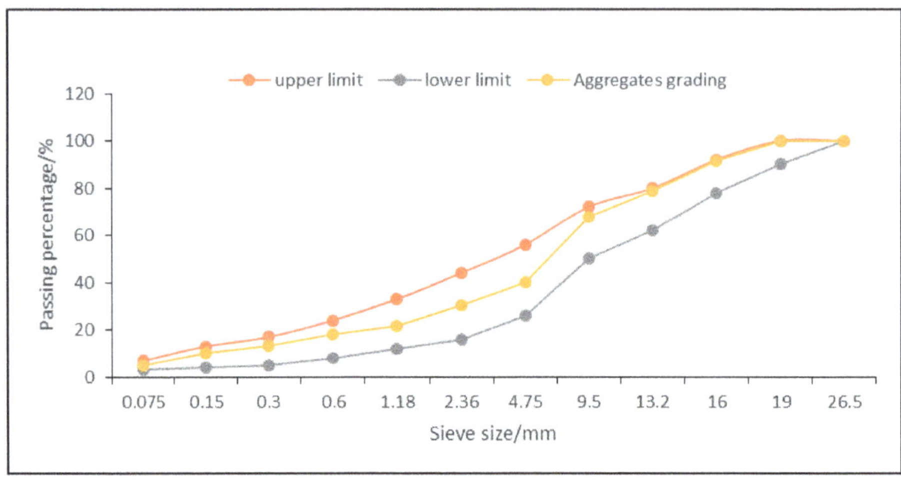

Figure 2. Gradation of the aggregates used in this study.

A plate was made according to the rutting test specifications and then cut into a suitable prism specimen.

3.2. Test Matrix

As shown in Figure 3, fatigue tests were performed by Material Test System (MTS) with a strain-controlled load on asphalt concrete prisms and were conducted in the laboratory.

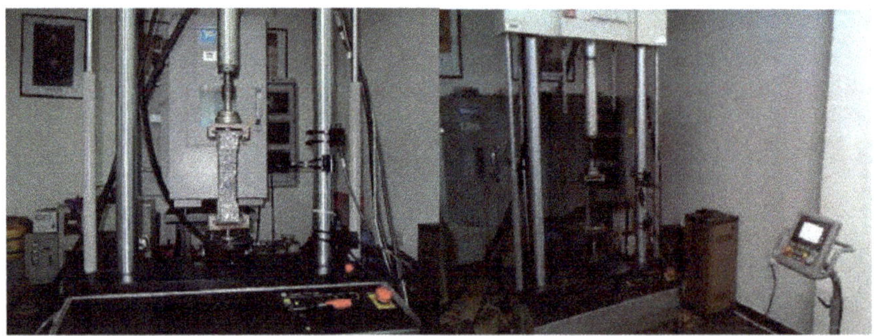

Figure 3. Material Test System (MTS).

Specimens must be a suitable shape and size for strain control and to ensure the reliability of test results. Several 300 mm × 300 mm × 50 mm plate specimens were firstly prepared and then they were cut into 250 mm × 50 mm × 50 mm prism samples in accordance with test procedures of direct tension testing in the literature [17].

As shown in Figure 4, both ends of the specimen were bonded with a steel plate, and a spherical hinge was set at the end of the active loading position, which ensured the central tension on the tip of the specimen. The strain-controlled cyclic loads were applied on a steel plate, and the spherical hinge could be automatically set back when tension was applied to the specimen. In order to avoid uncontrolled tension, the load was set on the middle of the prism.

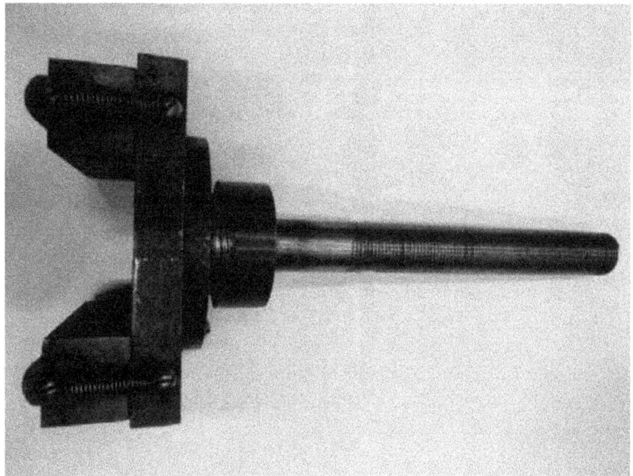

Figure 4. Specimen clamp for tests.

3.3. Loading Frequency

In order to make the experiment feasible and to comply with road surface conditions, the test was conducted at a variety of frequencies (1, 2, 5, 10 Hz). A load frequency of 2 Hz was determined for the test by taking into consideration influential factors including simulation results, the test time, and fixture performance.

3.4. Loading Waveform

In consideration of the fixed method of the specimen clamp, a gentle tug was exerted to avoid displacing the fixture gap. A non-intermittent triangular wave for the loading waveform was adopted (see Figure 5).

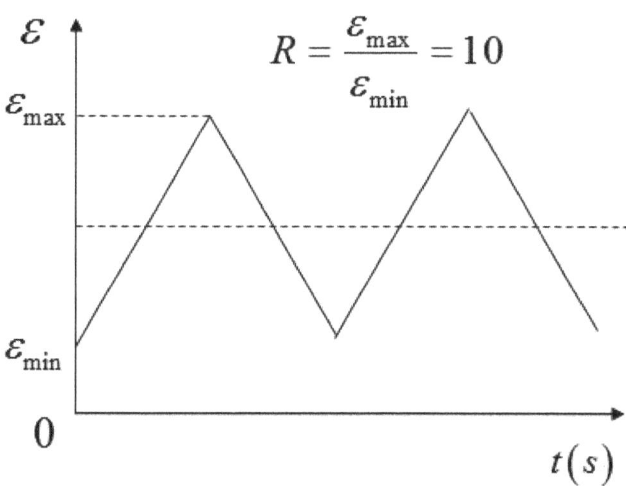

Figure 5. Loading waveform for fatigue test of High Modulus Asphalt Concretes (HMAC).

3.5. Test Temperature

In order to reflect the actual working state of high modulus asphalt pavement, a fatigue equivalent temperature was calculated and used as the test temperature.

SHRP presented the method to calculate the equivalent temperature of the fatigue test [18,19]:

$$T_{eff}(FC) = 0.8(MAPT) - 2.7 \quad (1)$$

where: $T_{eff}(FC)$—equivalent temperature of the fatigue test (°C); $MAPT$—annual average surface temperature in 1/3 Pavement depth (°C); 15 °C was determined to be the test temperature for the HMAC fatigue test.

3.6. Failure Criteria

Load descent failure criterion for the prism specimen was adopted after taking into consideration the asphalt mixture characteristics and test maneuverability. The specimen failed when the load decreased to 75% of the peak loading capacity [20,21].

The axial strain control mode was chosen and the high-low strain ratio of the specimens was set at 10. Four strain levels (0.3, 0.4, 0.5, 0.6) were adopted with three specimens tested in each level. Test conditions were determined, including a 2 Hz load frequency, a 1 mm/min stretching rate, a 15 °C test temperature, and a non-intermittent triangular wave for the loading waveform. These conditions were set in conjunction with the simulation effects, the test time, and clamp performance.

4. True Cyclic Stress-Strain Curve

The cyclic stress-strain curve, which is made by the vertices of the stable hysteresis loop at different strain levels, is a smooth curve. The following methods were used to determine the cyclic stress-strain curves:

(1) Multistage test method. Cyclic loading under several strain amplitude levels was conducted on the specimen, each cycle was loaded a number of times to achieve stability, and a smooth curve was drawn using the overlapping and stable hysteresis loop through the top. This loop was named the cyclic stress-strain curve. The method required few specimens and had a quick measurement speed; however, this method has low precision and it easily produces fatigue damage.

(2) Relegation-enhancement test method. When the specimen was loaded, the strain amplitude was gradually relegated and then enhanced. The other steps of this method are similar to the multistage test method

(3) Stretching after cycle stability method. The specimen underwent a series of strain decreases and increases, reducing the level of strain amplitude after cycle stability was reached. A specimen was then loaded to tensile failure to determine the stress-strain curve. Results of this method are in good agreement with those in the above two methods, however, it has low precision.

(4) Multistage multi-sample method. Multiple experiments controlled by a constant strain were conducted on specimens in multistage strain stages in order to obtain a stable hysteresis loop, then the top of the hysteresis loop was connected and a smooth cyclic stress-strain curve was obtained. This method can accurately reflect the cycle stress-strain properties of materials with high precision, but it requires more specimens and a longer test period than the other methods. The strength coefficient and cyclic strain hardening index are calculated using the cyclic stress-strain curve [22,23]:

$$\frac{\Delta\varepsilon}{2} = \frac{\Delta\sigma}{2E} + \left(\frac{\Delta\sigma}{2K'}\right)^{\frac{1}{n'}} \quad (2)$$

where: $\Delta\varepsilon$ = cyclic strain amplitude, $\Delta\sigma$ = cyclic stress amplitude.

The following equation can be obtained using the plastic component calculation:

$$\varepsilon = \frac{\sigma}{E} + \left(\frac{\sigma}{K'}\right)^{\frac{1}{n'}} \tag{3}$$

where: σ = stress, ε = strain, n' = Cyclic strain hardening exponent, K' = Cyclic stress hardening exponent.

The strain fatigue curve shows that the values of the hysteresis loop apex were calculated when fatigue life reached 50% of the specimen at strain levels ranging from 0.3 to 0.6 in order to determine the true strain and true stress amplitude at different strain levels and the plastic strain component (see Table 2).

Table 2. True stress amplitude and plastic strain component at different strain levels.

Strain Levels	$\Delta\varepsilon$	$\Delta\sigma$/MPa	$\Delta\varepsilon_p$
0.6	0.0014	0.213	0.00041
0.5	0.0013	0.184	0.00038
0.4	0.0011	0.176	0.00036
0.3	0.00081	0.143	0.00027

Table 2 shows the different fitting algorithms that were adopted to fit Equation (3). The fitting parameters K' = 26.847 and n' = 0.725 were then obtained and used in the following formula that predicts the strain of HMAC in the strain-controlled mode:

$$\varepsilon = \frac{\sigma}{E} + \left(\frac{\sigma}{26.847}\right)^{1.3793} \tag{4}$$

Stress can be obtained using Equation (4) with strains ranging from 0.0001~0.001; the results are presented in Figure 6.

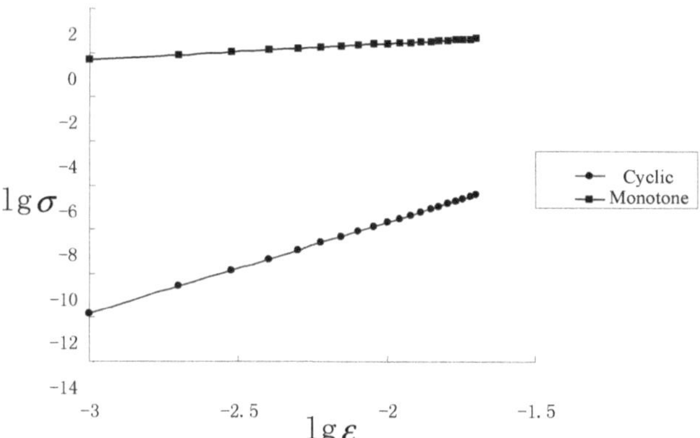

Figure 6. Comparison between cyclic curves and monotone curve.

The parameters K, n, and the HMAC monotone curve are compared, respectively, with the parameters K', n', and the cyclic curves. The cyclic curves are far higher than the monotone curve. Based on the position of the curve, HMAC was determined to be a cyclic hardening material [24], indicating that the stress required for producing the same strain during cyclic loading is higher than when under a static load. However, the hardening phenomenon gradually disappears when the cycling times increase.

Figure 7 shows the relationship between stress and time in the cyclic stress strain test with constant strain controlled. The results show that stress fluctuates as the length of time increases, whereas the general trend is that strain decreases along with the increasing length of time. It can be seen that the HMAC has cyclic relaxation properties [25,26], meaning strain gradually decreases when constant strain is controlled.

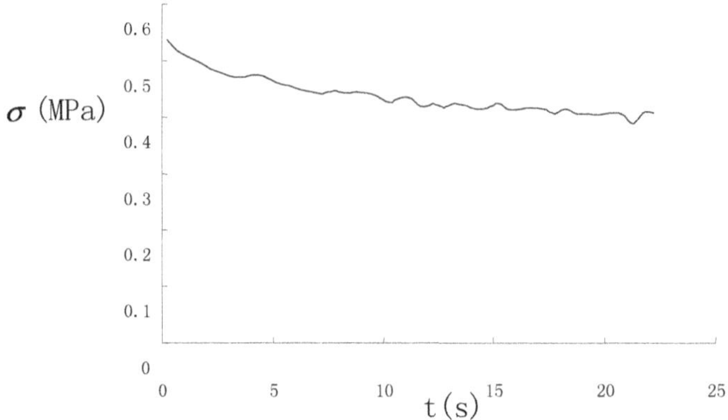

Figure 7. Stress-time history under control of strain.

5. Determination of Local Stress and Strain

5.1. Neuber Equation

The Neuber equation, a concise method for calculating local stress-strain of a notch [27,28], is as follows:

$$K_t = (K_\varepsilon K_\sigma)^{\frac{1}{2}} \tag{5}$$

$$K_\varepsilon = \frac{\varepsilon}{e} \tag{6}$$

$$K_\sigma = \frac{\sigma}{S} \tag{7}$$

where: K_t—theoretical stress concentration factor of notch, K_ε—strain concentration factors, K_σ—factor of stress concentration, e—nominal strain, S—nominal stress.

K_ε and K_σ were placed in Equation (5) and converted into an incremental form, and the Neuber hyperbolic equation was then obtained:

$$\Delta\sigma\Delta\varepsilon = K_t^2 \Delta S \Delta e \tag{8}$$

If the nominal stress is in the elastic range, then $\Delta e = \frac{\Delta S}{E}$ can be obtained, and the equation is as follows:

$$\Delta\sigma\Delta\varepsilon = (K_t \Delta S)^2 / E \tag{9}$$

The Neuber equation is derived from shear stress analysis of a prism. In order to be widely applied in the design and development of HMAC [29,30], the Neuber equation is modified by replacing the theoretical stress concentration factor K_t with the fatigue notch factor K_f. The general method for determining K_f is to fit the regression equation through fatigue test data of the specimen notch, and K_f is typically treated as a constant for most influencing factors. K_t is usually used when there is a lack of K_f data, which is a conservative method for estimating fatigue life. Values of K_t at different strain levels are shown in Table 3.

Table 3. Theoretical stress concentration factor of notch.

Strain Levels	0.6	0.5	0.4	0.3
K_t	1.973	1.947	2.046	1.938

5.2. Neuber Equation Correction

To conform to the plane strain problem, the cycle stress-strain curve of a small smooth specimen in the uniaxial stress state test must be modified. Moreover, parameters K_ε and K_σ in the biaxial stress state are different from those in the uniaxial stress state, meaning that the Neuber equations should be modified.

5.2.1. Cyclic $\sigma - \varepsilon$ Curve Correction

Hooke's law was applied to the elastic strain component, whereas the plastic deformation theory was used for the plastic component:

$$\sigma_a{}' = \sigma_a / \left(1 - \mu_1 + \mu_1^2\right)^{\frac{1}{2}} \tag{10}$$

$$\varepsilon_a{}' = \varepsilon_a \left(1 - \mu_1^2\right) / \left(1 - \mu_1 + \mu_1^2\right)^{\frac{1}{2}} \tag{11}$$

$$\mu_1 = \frac{\mu + E\varepsilon_{pa}/(2\sigma_a)}{1 + E\varepsilon_{pa}/\sigma_a} \tag{12}$$

where: σ'_a—the maximum main stress under plane strain condition, ε'_a—the maximum main strain under plane strain condition, μ_1—generalized Poisson's ratio, $\mu_1 = \dfrac{\mu + E\varepsilon_{pa}/(2\sigma_a)}{1 + E\varepsilon_{pa}/\sigma_a}$, μ—Poisson's ratio, ε_{pa}—plastic strain amplitude of cyclic $\sigma - \varepsilon$ curve under uniaxial stress condition, E_a—elastic modulus of cyclic stress-strain curve.

The equation is achieved by σ'_a—ε'_a fitting as follows:

$$\varepsilon'_a = \sigma'_a \left(1 - \mu^2\right) / E + \left(\sigma'_a / K'\right)^{\frac{1}{n'}} \tag{13}$$

where K'' is corrections of K', n'' is corrections of n'.

The parameter data found in Table 4 were put into Equation (13), and then the revised cyclic $\sigma - \varepsilon$ curve equation was obtained via Equation (14):

$$\varepsilon'_a = \frac{\sigma'_a(1-\mu^2)}{E_a} + \left(\frac{\sigma}{590.716}\right)^{1.014} \tag{14}$$

Table 4. Correction parameters of cyclic $\sigma - \varepsilon$ curve.

Strain Levels	ε'_a	σ'_a/MPa	μ_1	K''	n''
0.6	0.682	0.14	0.427		
0.5	0.594	0.097	0.419	590.716	0.986
0.4	0.473	0.083	0.417		
0.3	0.348	0.076	0.423		

5.2.2. Neuber Equation Correction

To show the plastic penetration level and equivalent stress and strain levels, the right end of the Neuber equation was multiplied by two coefficients, and the following corrections were made to the Neuber equation:

$$\Delta\sigma \cdot \Delta\varepsilon = m \cdot g \cdot \left(K_f \cdot \Delta S\right)^2 / E \quad (15)$$

$$m = \left(1 - \mu' + \mu'^2\right)^{\frac{1}{2}} \quad (16)$$

$$g = m / \left(1 - \mu'^2\right) \quad (17)$$

$$\mu' = 0.5 - (0.5 - \mu) E_s / E \quad (18)$$

where: m—plastic penetrable degree, g—coefficient of equivalent stress and strain, μ'—elastic-plastic Poisson's ratio, considering the compressibility of material, μ—Poisson's ratio, E—elasticity modulus, E_s—secant modulus, determined by the monotone tensile curve.

Parameters $\mu' = 0.4221$, $m = 0.87$, $g = 1.059$ can be obtained by putting $\mu = 0.25$, $E_s = 374$ MPa, and $E = 1200$ MPa into Equation (15), which becomes the following final form of the Neuber equation that predicts the fatigue life of HMAC in the strain-controlled mode:

$$\Delta\varepsilon \cdot \Delta\sigma = \frac{0.92\left(K_f \cdot \Delta S\right)^2}{E} \quad (19)$$

As nominal stress is in the elastic range, $\Delta e = \frac{\Delta S}{E}$ and $K_t = 1.976$ were substituted into Equations (9) and (19) to develop Equation (20):

$$\varepsilon\sigma = 0.92 \times 1.976^2 se \quad (20)$$

6. Strain-Life Fatigue Equation

The strain fatigue life equation is characterized by the fatigue curve taking the following form:

(1) $\varepsilon_a - N_f$ curve.

Strain fatigue tests were conducted at different strain levels and under equal strain control to obtain the fatigue data, of the control variables ε_a is the true strain amplitude.

(2) $\varepsilon_{eq} - N_f$ or $\sigma_{eq} - N_f$ curve.

Fatigue data are obtained from fatigue tests that combine stress and strain. The control variables are not equivalent strain amplitude ε_{eq} or equivalent stress amplitude σ_{eq}, which contain the mean stress.

(3) $S - N_f$ curve.

The fatigue data comes from existing data from the S-N curve. The influence of notch and strain fatigue should also be considered.

The second method needs a large amount of experimental data, and the third method was only applied to make a rough estimation, because of its low accuracy. The main $\varepsilon_a - N_f$ curves are shown in Table 5.

Table 5. Strain fatigue curve.

Forms	Author	Expression
	M. A. Manson L. F. Coffin	$\varepsilon_a = \frac{\sigma'_f}{E}\left(2N_f\right)^{2b} + \varepsilon'_f\left(2N_f\right)^{2c}$
	H. M. Fu	$\frac{C}{(\Delta\varepsilon_t - \Delta\varepsilon_o)^m} = N_f$
	R. W. Landgraf	$\frac{1}{N_f} = 2\left(\frac{\sigma'_f}{\varepsilon'_f E}\right)\left(\frac{\varepsilon_{pa}}{\varepsilon_{pa}}\right)\frac{\sigma'_f}{\sigma'_f - \sigma_m}$
	Ostergren	$\frac{1}{N_f} = 2\left(\frac{\sigma_{\max}\varepsilon_{pa}}{\sigma'_f\varepsilon'_f E}\right)^{1/(a+b)}$
$\Delta\varepsilon \sim N_f(R_e = -1)$	K. N. Smith	$\sigma_{\max}\varepsilon_a = \frac{\sigma'^2}{E}\left(2N_f\right)^{2b} + \sigma'_f\varepsilon'_f\left(2N_f\right)^{b+c}$
	Bergmann	$(\sigma_a + k\sigma_m)\varepsilon_a = \frac{\sigma'^2}{E}\left(2N_f\right)^{2b} + \sigma'_f\varepsilon'_f\left(2N_f\right)^{b+c}$
	F. Erdogan	$\sigma'_a\sigma_{\max}^{(1-r)} - \varepsilon_a = \frac{\sigma'^2}{E}\left(2N_f\right)^{2b} + \sigma'_f\varepsilon'_f\left(2N_f\right)^{b+c}$
	V. Adrov	$\frac{1}{N_f} = 2\left(4\sigma'_f\varepsilon'_f/A\right)^{b+c}$ $A = \frac{\beta}{a}I, a = f(\sigma_{\max}, \sigma_{\min})$ $I = \sum_{k=1}^{k_1}(j+1-2k)\left(\Delta\sigma_k - \Delta\sigma_{j+1} - k\right)\Delta\varepsilon$

The Manson-Coffin equation is widely used for all of the $\varepsilon_a - N_f$ curves by adopting a power function to characterize fatigue performance. The main parameter strain is calculated using the following equation:

$$\varepsilon_a = \frac{\sigma'_f}{E}\left(2N_f\right)^b + \varepsilon'_f\left(2N_f\right)^c \quad (21)$$

Based on the proposed Manson-Coffin equation, direct tensile fatigue tests for HMAC were conducted to obtain the fatigue parameters. Estimation methods for fatigue parameters included the general slope method, modified four-points correlation, and four-points correlation [31,32] (see Table 6).

Table 6. Estimation method for fatigue constant.

Methods		Fatigue Constant
General slope method	b	-0.12
	C	-0.16
	σ'_f	$1.75\sigma_b/E$
	ε'_f	$0.5\varepsilon_f^{0.006}$
Four-points correlation	b	$-[0.0792 + 0.179\lg(\sigma_f/\sigma_b)]$
	c	$-\lg(3.37\varepsilon_f^{0.25}\left[1 - 81.8\frac{\sigma_b}{E}\left(\frac{\sigma_f}{\sigma_0}\right)^{1.179}\right]^{-1/3})$
	σ'_f	$1.12\sigma_b(\sigma_f/\sigma_b)^{0.893}$
	ε'_f	$0.413\varepsilon_f\left[1 - 81.8\frac{\sigma_b}{E}\left(\frac{\sigma_f}{\sigma_0}\right)^{1.179}\right]^{-1/3}$
Modified four-points correlation	b	Same as four-points correlation
	c	Same as four-points correlation
	σ'_f	$1.18\sigma_b(\sigma_f/\sigma_b)^{0.0946}$
	ε'_f	$0.538\left[1 - 81.8\frac{\sigma_b}{E}\left(\frac{\sigma_f}{\sigma_0}\right)^{1.179}\right]^{-1/3}$

Fatigue test regulation for Manson-Coffin equation requires that $R_e = -1$, otherwise, the equation should be modified, (see Table 7).

Table 7. Modified Manson-Coffin equation when $R_e \neq -1$.

Methods	Modified Manson-Coffin Equation
Morrow elastic stress Linear correction	$\varepsilon_a = \frac{\sigma'_f - \sigma_m}{E}\left(2N_f\right)^b + \varepsilon'_f\left(2N_f\right)^c$
Gerber elastic stress curvature correction	$\varepsilon_a = \frac{\sigma'^2_f - \sigma^2_m}{E\sigma'_f}\left(2N_f\right)^b + \varepsilon'_f\left(2N_f\right)^c$
Marrow total strain correction	$\varepsilon_a = \frac{\sigma'_f - \sigma_m}{\sigma'_f}\frac{\sigma'_f}{E}\left(2N_f\right)^b + \varepsilon'_f\left(2N_f\right)^c$
Sachs plastic correction	$\Delta\varepsilon_a = \frac{\sigma'_f}{E}\left(1 - \frac{\sigma_m}{\sigma_b}\right)^b\left(2N_f\right)^c$

In accordance with the first method, an equal strain fatigue test with strain levels ranging from 0.3 to 0.6 and a monotone direct tensile test were conducted on HMAC using the dynamic tester MTS810 to calculate the $\varepsilon_a - N_f$ curve of asphalt concretes [33]. The $\varepsilon_a - N_f$ curve was characterized by the Mason-Coffin equation, and the power function was adopted to describe the strain life curve:

$$\varepsilon_a = \frac{\Delta\varepsilon}{2} = \frac{\Delta\varepsilon_e}{2} + \frac{\Delta\varepsilon_p}{2} = \left(\frac{\sigma'_f}{E}\right)\left(2N_f\right)^b + \varepsilon'_f\left(2N_f\right)^c \tag{22}$$

In the above equations, σ'_f and ε'_f are approximately equal to the fracture stress σ_f and fracture strain ε_f [34]. Fracture stress and fracture strain for HMAC were listed in the monotone direct tensile test (see Table 8).

Table 8. HMAC parameters of monotone direct tensile test.

Specimen	Fracture Strain	Fracture Stress/MPa	Secant Modulus/MPa
SJ-1	0.00308	0.1588	351.55
SJ-2	0.00384	0.13672	385.60
SJ-3	0.0023	0.13465	413.54
SJ-4	0.0023	0.10162	345.18
Average	0.00288	0.13295	373.97

As seen in Table 8, σ'_f of HMAC is 0.13295 MPa and ε'_f is 0.00288. Constant strain fatigue tests at different stress-strain levels were conducted on HMAC, and the cyclic $\sigma - \varepsilon$ curve at the strain level of 0.6 is shown in Figure 8.

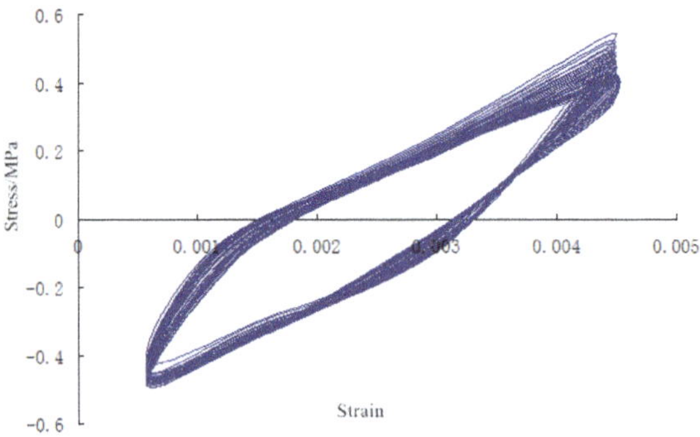

Figure 8. Cyclic $\sigma - \varepsilon$ curve at the strain level of 0.6.

The fatigue parameters strain and fatigue life of the specimen were then obtained, as shown in Table 9.

Table 9. Strain amplitude and fatigue life of HMAC in equal strain fatigue test.

Specimen	S1	S2	S3	S4	S5	S6	S7	S8
HMAC								
ε_a	0.32	0.53	0.74	0.91	1.15	1.392	1.56	1.66
N_f	18242	15124	7692	4543	3692	2420	1502	1031
Matrix asphalt concretes								
ε_a	1.28	2.17	3.04	3.75	4.18	5.37	6.17	6.89
N_f	4862	3941	1920	1137	1972	1143	108	84

As seen in Table 9, four common fitting algorithms were used to fit the data from Equation (1). The relationship between the fitting parameters and the equation was studied, as seen in Table 10.

Table 10. Parameters b and c using different fitting algorithms.

Fitting Algorithms	b	c	R^2
Marquardt method	−0.10234	−0.10263	0.981687
Quasi-newton method	−0.10233	−0.10264	0.981688
Evolution algorithm	−0.10231	−0.10267	0.981681
Max inherit optimization	−0.10237	−0.10261	0.981687

As shown in Table 10, the fitting parameters (b, c) of the four fitting methods are nearly the same—$b = -0.1023$ and $c = -0.1026$. When σ'_f, ε'_f, b, and c are put into Equation (22), then Equation (22) changes into the following form which predicts the fatigue life of HMAC in the strain-controlled mode:

$$\varepsilon_a = \frac{0.13295}{E} \times \left(2N_f\right)^{-0.1023} + 0.00288 \times \left(2N_f\right)^{-0.1026} \quad (23)$$

7. Fatigue Life Prediction Based on Local Stress-Strain Method

The fatigue life of HMAC was predicted based on the modified Neuber equation. First, the load-time process was transformed into a nominal-time history. Second, local stress and strain of the notched specimen were transformed to nominal stress and strain using the cyclic $\sigma - \varepsilon$ curve and the modified Neuber equation. Fatigue damage was calculated according to the strain fatigue life curve, and the fatigue life was then obtained based on cumulative damage theory. Specific steps for fatigue life prediction based on the modified Neuber equation and local stress-strain method are as follows:

(1) Local strain and local stress in the plane strain state were solved using the results of the load strain and temperature strain, cyclic $\sigma - \varepsilon$ curve and Neuber hyperbolic equations.

$$\varepsilon = \frac{\sigma}{E} + \left(\frac{\sigma}{26.847}\right)^{1.3793} \quad (24)$$

$$\varepsilon\sigma = 0.92 \times 1.976^2 se \quad (25)$$

(2) The fatigue life N_f was determined on the basis of the $\varepsilon_a - N_f$ curve equation.

$$\varepsilon_a = \frac{0.13295}{E} \times \left(2N_f\right)^{-0.1023} + 0.00288 \times \left(2N_f\right)^{-0.1026} \quad (26)$$

8. Case Study

8.1. Structure and Parameters of Pavement

Based on the proposed fatigue life prediction method, a typical HMAC overlay pavement situated in a long steep road in Shandong province of China was selected and its fatigue life was predicted below.

The pavement structure and material parameters are listed in Table 11. Five axle loads (80 kN, 100 kN, 120 kN, 150 kN, and 180 kN) and four varied temperature range (5, 10, 15, 20 °C) were considered in the analysis of nominal stress and strain. For each grade, fatigue life was calculated.

Table 11. Structure and parameters of pavement.

Structure Layer	Material Type	Thickness/cm	Modulus/MPa	Poisson's Ratio
Surface	High Modulus Asphalt Concrete	4	1200	0.25
	High Modulus Asphalt Concrete	6	1200	0.25
Base course	Lean concrete	16~28	10,000~40,000	0.15
Bed course	Cement Stabilized Crushed Stone	20	1700	0.2
Subbase	Graded broken stone	16	250	0.35
Subgrade	Soil	——	40	0.35

8.2. Results

8.2.1. Nominal Strain

According to the traffic axle load and varied temperature range, nominal strains at the bottom of HMAC were calculated using the finite element software, see Tables 12 and 13.

Table 12. Load strain at the bottom of HMAC.

Axle Load/kN	80	100	120	150	180
Load nominal strain	0.2319	0.2634	0.3048	0.3527	0.4156
Load local strain	0.3492	0.3751	0.4163	0.4806	0.5614

Table 13. Temperature strain at the bottom of HMAC.

Varied Temperature Range (°C)	5	10	15	20
Temperature nominal strain	0.1934	0.2486	0.3261	0.4835
Temperature local strain	0.2762	0.3294	0.4351	0.6214

8.2.2. Local Strain

Based on the nominal strain shown in Tables 12 and 13, local strain was calculated using the Equations (24) and (25) (see Tables 12 and 13).

8.2.3. The Analysis of Fatigue Damage

Based on linear cumulative damage hypothesis, damage caused by load and the maximal varied temperature range (20 °C) was calculated, as shown in Table 14. In this table, statistics of traffic loading times were collected from a traffic survey, loading times were calculated using Equation (26), and fatigue damage was the ratio of statistics of traffic loading times and calculated loading times.

Table 14. The analysis of fatigue damage.

Axle Load/kN	80	100	120	150	180
Local strain	1.2892	1.4009	1.4432	1.5057	1.5965
Calculated loading times	79,547	55,698	15,634	9846	5765
Statistics of traffic loading times	627	2568	1931	122	70
Fatigue damage	0.0079	0.0461	0.1235	0.0124	0.0121

8.2.4. Fatigue Life Prediction

As shown in Table 14, the cumulative fatigue damage could be calculated as 0.202, then the pavement fatigue life which is the reciprocal of cumulative fatigue damage was calculated as 4.95, meaning the HMAC overlay might generate reflective cracking in nearly 5 years. Moreover, our in situ survey also showed that a part of the pavement surface cracked almost 4 years after overlaying. This indicates that the proposed method to predict service life of pavement agreed with practical conditions. Furthermore, the current research suggested that a stress absorbing layer should be set before HMAC overlaid between HMAC and concrete pavement.

9. Conclusions

This study provides an experimental investigation into the fatigue performance of high modulus asphalt concretes, including the cyclic $\sigma - \varepsilon$ test and direct tensile fatigue test. A method to predict the fatigue crack initiation life of HMAC was proposed based on the modified Neuber equation and local $\sigma - \varepsilon$ method. The following conclusions can be drawn based on this work:

(1) The cyclic $\sigma - \varepsilon$ curve and modified $\sigma - \varepsilon$ curve of HMAC in plane strain state were established.
(2) Conversion from nominal stress and strain to local strain-stress is proposed based on a modified Neuber equation and the local stress-strain method.
(3) The strain fatigue equation for HMAC is presented based on fatigue tests.
(4) The fatigue life model is able to produce satisfactory crack initiation life predictions using the local stress-strain method based on fatigue test analysis of the specimens and the modified Neuber equation.

Acknowledgments: The research was sponsored by the National Natural Science Foundation (Grant No. 51008033), the Fundamental Research Funds for the Central Universities (Grant No. 310821163502), the Transportation Department of Shandong Province (Grant No. 2008Y007), the Transportation Department of Hainan Province (Grant No. 201000005), and the Transportation Department of Hubei Province of China (Grant No. Ejiaokejiao [2013] 731).

Author Contributions: Professor Mulian Zheng was in charge of the whole research plan and paper submission matters. The first author extends gratitude to Peng Li, Jiangang Yang and Zhengliang Zhang for performing the experiment and analyzing test data, to Hongyin Li and Yangyang Qiu for offering help in doing the experiment.

Conflicts of Interest: The authors declare no conflict of interest.

List of Symbols

ε_a = strain, $\Delta\varepsilon$ = cyclic strain amplitude, $\Delta\varepsilon_e$ = circular elastic, $\Delta\varepsilon_p$ = plastic strain amplitude, σ'_f = fatigue strength coefficient, ε'_f = ductility coefficient, N_f = fatigue life, b, c = regression constants, ε_{ea} = elastic strain, ε_{pa} = plastic strain, K_t = theoretical stress concentration factor of notch, K_ε = strain concentration factors, K_σ = factor of stress concentration, e = nominal strain, S = nominal stress, K_f = fatigue notch factor, M = plastic penetrable degree, g = coefficient of equivalent stress and strain, μ' = elastic-plastic Poisson's ration, considering the compressibility of material, μ = Poisson's ration, E = elasticity modulus, E_s = secant modulus, determined by the monotone tensile curve.

References

1. Forough, S.A.; Nejad, F.M.; Khodaii, A. Energy-based approach to predict thermal fatigue life of asphalt mixes using modified uniaxial test setup. *J. Mater. Civ. Eng.* **2016**, *28*. [CrossRef]

2. Zheng, M.L. Simulation of Permanent Deformation in High-Modulus Asphalt Pavement Using the Bailey-Norton Creep Law. *J. Mater. Civ. Eng.* **2016**. [CrossRef]
3. Monismith, C.L.; Epps, J.A.; Kasianchuk, D.A.; McLean, D.B. *Asphalt Mixture Behavior in Repeated Flexure*; Report TE 70-5; University of California: Berkeley, CA, USA, 1971.
4. Roque, R.; Ruth, B.E.; Dickison, S.W.; Reid, B. *Evaluation of SHRP Indirect Tension Tester to Mitigate Cracking in Asphalt Concrete Pavements and Overlays*; Final Report FDOT B-9885; University of Florida: Gainesville, FL, USA, 1997.
5. Yoo, P.J.; Al-Qadi, I.L. A strain-controlled hot-mix asphalt fatigue model considering low and high cycles. *Int. J. Pavement Eng.* **2010**, *11*, 565–574. [CrossRef]
6. Ameri, M.; Nowbakht, S.; Molayem, M. A study on fatigue modeling of hot mix asphalt mixtures based on the viscoelastic continuum damage properties of asphalt binder. *Constr. Build. Mater.* **2016**, *106*, 243–252. [CrossRef]
7. Li, Y.Q.; Metcalf, J.B. Crack Initiation model from asphalt slab tests. *J. Mater. Civ. Eng.* **2002**, *14*, 303–310. [CrossRef]
8. Seung, K.J.; Seung, K.C. Development of a predictive system for estimating fatigue life of asphalt mixtures using the indirect tensile test. *J. Transp. Eng.* **2012**, *138*, 1530–1540.
9. Fallah, S.; Khodaii, A. Developing a fatigue fracture model for asphalt overlay reinforced with geogrid. *Mater. Struct.* **2015**. [CrossRef]
10. Ker, H.W.; Lee, Y.H.; Wu, P.H. Development of fatigue cracking prediction models using long-term pavement performance database. *J. Transp. Eng.* **2008**, *134*, 477–482. [CrossRef]
11. Mohammad, S.; Kim, Y.R. *Development of a Failure Criterion for Asphalt Mixtures under Different Modes of Fatigue Loading*; Transportation Research Board of the National Academies: Washington, DC, USA, 2014; pp. 117–125.
12. Benedetto, H.D.; de la Roche, C.; Strom, R.L. Fatigue of bituminous mixtures. *Mater. Struct.* **2004**, *37*, 202–216. [CrossRef]
13. Karlaftis, A.G.; Badr, A. Predicting asphalt pavement crack initiation following rehabilitation treatments. *Transp. Res. C* **2015**, *55*, 510–517. [CrossRef]
14. Molenaar, A.A.A. Structure Performance and Design of Flexible Pavements and Asphalt Concrete Overlays. Ph.D. Thesis, Delft University of Technology, Delft, The Netherlands, 1983.
15. AASHTO T 321-07. *Determining the Fatigue Life of Compacted Hot Mix Asphalt (HMA) Subjected to Repeated Flexural Bending*; American Association of State and Highway Transportation Officials: Washington, DC, USA, 2007.
16. Shen, S.; Carpenter, S.H. Development of an asphalt fatigue model based on energy principles. *J. Assoc. Asphalt Paving Technol.* **2007**, *77*, 525–573.
17. Graham, J.A. *Fatigue Design Handbook*; Society of Automotive Engineers: New York, NY, USA, 1968.
18. Edward, T.H. *Performance Prediction Models in the Superpave Mix Design System*; Arizona State University: Tempe, AZ, USA, 2006.
19. Kim, T.W.; Baek, J.; Lee, H.J.; Choi, J.Y. Fatigue performance evaluation of SBS modified mastic asphalt mixtures. *Constr. Build. Mater.* **2013**, *48*, 908–916. [CrossRef]
20. De Jcorreia, A.M.P.; Correia, J.A.F.O. Critical Assessment of a Local Strain-Based Fatigue Crack Growth Model Using Experimental Data Available for the P355NL1 Steel. *J. Pressure Vessel Technol.* **2013**, *135*, 011404.
21. Hu, X.; Hu, S.; Walubita, L.F.; Sun, L. Investigation of fatigue cracking: Bottom-up or top-down. In *Pavement Cracking: Mechanisms, Modeling, Detection, Testing and Case Histories*; CRC Press: Leiden, The Netherlands, 2008; pp. 333–344.
22. Hofko, B. Combining performance based lab tests and finite element modeling to predict life-time of bituminous bound pavements. *Constr. Build. Mater.* **2015**, *89*, 60–66. [CrossRef]
23. Castro, M.; Sanchez, J. Estimation of asphalt concrete fatigue curves—A damage theory approach. *Constr. Build. Mater.* **2008**, *22*, 1232–1238. [CrossRef]
24. Khateeb, G.A.; Shenoy, A. A distinctive fatigue failure criterion. *J. Assoc. Asphalt Paving Technol.* **2004**, *73*, 585–622.
25. Topper, T.H.; Wetzel, R.M.; Morrow, J.D. Neubers rule applied to fatigue of notched specimens. *J. Met.* **1969**, *4*, 22–24.

26. Lee, H.J.; Lee, J.H.; Park, H.M. Performance evaluation of high modulus asphalt mixtures for long life asphalt pavements. *Constr. Build. Mater.* **2007**, *21*, 1079–1087. [CrossRef]
27. Shu, X.; Huang, B.S. Laboratory evaluation of fatigue characteristics of recycled asphalt mixture. *Constr. Build. Mater.* **2008**, *22*, 1323–1330. [CrossRef]
28. Schapery, R.A. A theory of crack initiation and growth in visco-elastic media; I: Theoretical development, II: Approximate methods of analysis, III: Analysis of continuous growth. *Int. J. Fract.* **1975**, *11*, 141–159, 369–388 and 549–562. [CrossRef]
29. Doh, Y.S.; Baek, S.H.; Kim, K.W. Estimation of relative performance of reinforced overlaid asphalt concretes against reflection cracking due to bending more fracture. *J. Constr. Build. Mater.* **2009**, *23*, 1803–1807. [CrossRef]
30. Zappalorto, M.; Lazzarin, P. Some remarks on the Neuber rule applied to a control volume surrounding sharp and blunt notch tips. *Fatigue Fract. Eng. Mater. Struct* **2014**, *37*, 349–358. [CrossRef]
31. Micaeloa, R.; Pereiraa, A.; Quaresman, L.; Cidade, M.T. Fatigue resistance of asphalt binders: Assessment of the analysis methods in strain-controlled tests. *Constr. Build. Mater.* **2015**, *15*, 703–712. [CrossRef]
32. Alireza, K.K.; Mahmoud, A. Laboratory evaluation of strain controlled fatigue criteria in hot mix asphalt. *Constr. Build. Mater.* **2013**, *47*, 1497–1502.
33. Salour, F.; Erlingsson, S. *Pavement Structural Behavior during Spring Thaw*; VTI Report 738 A; Swedish National Road and Transport Research Institute: Linköping, Sweden, 2012.
34. Jaeklin, E.P.; Scherer, J. Asphalt reinforcing using glass fiber grid 'Glasphalt'. In Proceedings of the 3rd International RILEM Conference on Reflective Cracking in Pavement: design and performance of overlay systems, Maastricht, The Netherlands, 2–4 October 1996.

Article

Low Temperature Performance Characteristics of Reclaimed Asphalt Pavement (RAP) Mortars with Virgin and Aged Soft Binders

Feipeng Xiao [1,*], Ruoyu Li [1], Henglong Zhang [2] and Serji Amirkhanian [3,*]

1 Key Laboratory of Road and Traffic Engineering of Ministry of Education, Tongji University, Shanghai 201804, China; 1632427@tongji.edu.cn
2 School of Civil Engineering, Hunan University, Changsha 410082, China; hlzhang@hnu.edu.cn
3 State Key Laboratory of Silicate Materials for Architectures, Wuhan University of Technology, Wuhan 430070, China
* Correspondence: fpxiao@tongji.edu.cn (F.X.); serji.amirkhanian@gmail.com (S.A.); Tel.: +86-135-0198-9548 (F.X.); +1-864-844-3145 (S.A.)

Academic Editor: Jorge de Brito
Received: 18 January 2017; Accepted: 16 March 2017; Published: 20 March 2017

Abstract: Reclaimed asphalt pavement (RAP) has many advantages and is utilized to improve the high temperature properties of asphalt mixtures. Low temperature cracking is a predominant distress in asphalt pavements containing RAP materials. Thus, the evaluation of fracture resistance for asphalt mixtures containing RAP is of interest. The objective of this research is to explore the low temperature performance characteristics of RAP mortars containing sieved RAP and soft binders at three aged states. The stiffness values and m-values from bending beam rheometer (BBR) tests at three test temperatures of −18 °C, −12 °C and −6 °C were obtained to conduct the minimum low temperature grades. RAP mortar with a higher aged binder content had a higher minimum low temperature regardless of RAP source. In addition, RAP mortars with virgin soft binder had the best low temperature resistance followed by the RAP mortars with rolling thin film oven (RTFO) and pressure-aged vessel (PAV) binders.

Keywords: reclaimed asphalt pavement; mortar; stiffness; m-value; low temperature determination

---

1. Introduction

Reclaimed asphalt pavement (RAP) is being broadly used as a component of asphalt mixture for new highway pavement. Many research studies indicate that there are plenty benefits of using RAP for the new asphalt mixture, including reduction of total cost in pavement construction, natural resource conservation, environment protection, and rutting resistance improvement [1–4].

In the U.S., the National Asphalt Pavement Association (NAPA) has tracked the use of RAP through annual industry surveys since 2009 and found that the utilization of RAP materials was clearly increasing [5]. In 2012, contractors in 12 states used less than 15% of their total tonnage to produce the new asphalt mixture for paving purposes [6]. This represents a total tonnage increase of 22% with respect to asphalt mixtures in terms of RAP applications from 2009 to 2012 in the U.S. (from 56 to 68.3 million tons) [5].

Hong et al. [7] indicated that the resistance to rutting of hot mix asphalt (HMA) with 35% RAP was better compared to only virgin asphalt due to the incorporation of aged binder. Daniel et al. [8] found that the high-temperature performance grade remains the same or increases only one grade for the various RAP percentages. Attia and Abdelrahman [9] found that the effect of moisture on RAP is similar to the effect of moisture on granular material. Dynamic Shear Rheometer (DSR) testing was used to evaluate recycled RAP and virgin asphalt and indicated that the shear modulus (G*) and the G*/sin δ increased with the increasing percentage of RAP at both the high and intermediate test temperatures [10]. In addition, it also could be found that RAP sources were vital factors to determine the shear moduli and other DSR parameters [11].

Low-temperature cracking is predominant result of distress in asphalt pavements because of the thermal stress that builds up in pavements in extreme climates [12]. These low temperature cracks result in transverse cracks and other distresses along the pavement and ultimately accelerate the deterioration of the asphalt pavement structure. Some research studies indicated that the involvement of RAP material in new asphalt pavement might result in noticeable damage of pavement surface [13–15]. Therefore, the evaluation of fracture resistance for asphalt mixtures containing RAP is of interest to owners and agencies seeking better performing pavements in cold climates [12,16].

Some articles reported that low temperature bending beam rheometer (BBR) stiffness increased with increasing RAP, the m-value decreased with the increasing percentage of RAP, and the magnitude of the changes were dependent on the RAP source [17]. In addition, it was reported that the critical low performance grade (PG) temperature increased with the increased RAP materials [10,11,18,19].

Mogawer et al. [3] found that the RAP mixtures performed similarly to their respective control mixture for all low-temperature cracking tests. These data suggest that plant-produced mixtures with up to 30% RAP may not be more susceptible to low temperature failures. Swiertz et al. [2] reported that when using RAP, the low-temperature PG grade depended on fresh binder grade and source. Testing also showed that RAP source was not a significant factor for dynamic modulus at low temperatures, although it significantly affected the dynamic modulus at high temperatures. The addition of 40% RAP also significantly decreased the low-temperature fracture resistance [16].

However, the conventional methods of classifying aged asphalt binders from RAP materials requires initial extraction of the asphalt binder from the RAP, which involves the use of harmful chemical solvents such as trichloroethylene. In recent years, a new testing procedure has been developed to estimate the low temperature properties of the RAP binder without extraction or chemical treatments [20]. This project provides a possibility to evaluate the properties of RAP binders by testing the RAP mortars (fresh binders blended with fine RAP materials) without extracting the RAP binders from them. With the respect to testing procedure, the modified bending beam rheometer test is employed with minor modifications to the equipment which do not alter the test method and general settings. The properties of the binder in RAP are then estimated from the mortar properties. Many initial trials of the materials and equipment involved were performed before conducting the testing procedures to determine the low temperature properties of the aged binders in RAP materials [17,21].

The objective of this study is to explore the low temperature performance characteristics of six RAP mortars blended with the soft binders at three aged states. The main properties of stiffness values and m-values from BBR tests at three test temperatures of −18 °C, −12 °C and −6 °C were obtained to conduct the minimum low temperature grade of these RAP mortars.

2. Materials, Test Methods and Analysis Methods

In this study, one soft binder PG 58-28 was used for blending with RAP mortars. The rheological properties are shown in Table 1. In addition, six RAP types including 2 high-stiffness RAPs, 2 medium-stiffness RAPs and 2 low-stiffness RAPs which were denoted as A through F were selected to yield the modified asphalt mortars. The extracted aged binders from six RAPs were tested in BBR first and categorized as high-, medium- and low-stiffness RAPs. These values were based on the stiffness values of these extracted aged binders.

These RAP materials were initially sieved to a size which passed through a #50 (0.3-mm) sieve and was retained on a #100 (0.15-mm) sieve, and then were mixed with base binders at various aged states accordingly. The aged binder contents of the total binder in these RAP materials were obtained by using an ignition oven to burn all the asphalt for each RAP source. These aged binder contents of the total binder in the RAP mortar are 6.07%, 7.10%, 6.50%, 5.86%, 5.25%, and 8.3% for RAP sources

A–F, respectively. These aged binder percentages were defined the ratio of burned aged binder to the total RAP before burning (in mass). It should be noted that the tested samples were produced with pure binder (PG 58-22 binder) and fine RAP mortar (filler and aged binder) in this study.

Table 1. Rheological properties of performance grade (PG) 58-28 binder.

Base Binder	Unaged			RTFO	PAV		
	Viscosity (135 °C)	Failure temp.	G*/sin δ (58 °C)	G*/sin δ (58 °C)	G*/sin δ (19 °C)	Stiffness (−18 °C)	m-Value (−18 °C)
	(cP)	(°C)	(kPa)	(kPa)	(kPa)	(MPa)	
PG 58-28	315	60.2	1.38	3.88	3595	249	0.281

Notes: G*: shear modulus; RTFO: rolling thin film oven; PAV: pressure-aged vessel.

In this research study, the base binders (virgin, rolling thin film oven (RTFO), and pressure-aged vessel (PAV)) blended with various sieved RAPs were used to produce RAP mortars, which were employed to fabricate the BBR beams. The trial and error procedures were performed to obtain the proper percentage of RAP (in terms of aged binder percentage). A percentage over 15% of aged binders was very stiff and could not be poured at a high temperature of over 165 °C. Therefore, in this study, a percentage of up to 15% aged binder (i.e., sieved RAP including 15% aged binder) was used to produce the modified mortar. The value of 15% was the ratio of the aged binder from RAP to the total binders (aged binder with pure soft binder). In addition, two more concentrations (5% and 10%) were utilized to help explore the performance characteristics of these mortars in this study.

The BBR test generally provides a low temperature measure of the stiffness and relaxation properties of an asphalt binder. The obtained results are typically used to provide an indication of an asphalt binder's ability to resist low temperature cracking. The creep stiffness of asphalt binder from the BBR test is usually as a function of time, which is a measure of the thermal stresses in the asphalt binder resulting from thermal contraction. A higher creep stiffness value indicates higher thermal stresses. Originally, the crucial values included creep stiffness values at 60 sand the slope of the master curve at 60 s, commonly defined the "m-value" in Superpave system.

In this study, three test temperatures of −18 °C, −12 °C, and −6 °C were utilized to test the stiffness/deflection values of various RAP mortars in terms of soft binder aging states. These stiffness values and m-values were used to determine the minimum low temperatures of these asphalt RAP mortars.

The virgin binder PG 58-28 was blended with RAP mortars concluding three aged binders (5%, 10% and 15%) from six RAP sources (A–F). The fabricated BBR samples were tested at three temperatures (−6 °C, −12 °C and −18 °C).

The Superpave criteria for characterizing low temperature cracking of an asphalt binder are based on the definition of a critical cracking temperature, which is the maximum temperature below which cracking occurs as a result of a single cooling cycle. Therefore, cracking would happen when an asphalt binder reached a critical stiffness value. This critical temperature is typically defined the limiting stiffness temperature.

The Superpave binder specification uses BBR to measure the stiffness of asphalt binder at specified temperatures. The temperature at which the stiffness value of an asphalt binder exceeds 300 MPa is called the limiting stiffness temperature. Meanwhile, to address various cooling rates, the slope of the creep curve (denoted as m) is also included in the binder specification. The temperature at which the m value drops below 0.30 is a factor in determining the limiting stiffness. For most asphalt binders the m-value is a controlling value for defining the limiting stiffness temperature.

3. Results and Discussions

3.1. RAP Mortars with Virgin Soft Binder

3.1.1. Stiffness Values and m-Values

Figure 1 showed the main stiffness values and m-values results at −6 °C. As shown in Figure 1, it can be found that, as expected, the stiffness values of RAP mortars with virgin binder generally decrease while m-values increase when the loading duration increases with logarithmic trends regardless of RAP source. In addition, the stiffness values and m-values of the RAP mortars, which were blended from virgin binder and RAP mortar, cannot be achieved when using 5% aged binder because these RAP mortars are too soft at the testing temperature of −6 °C.

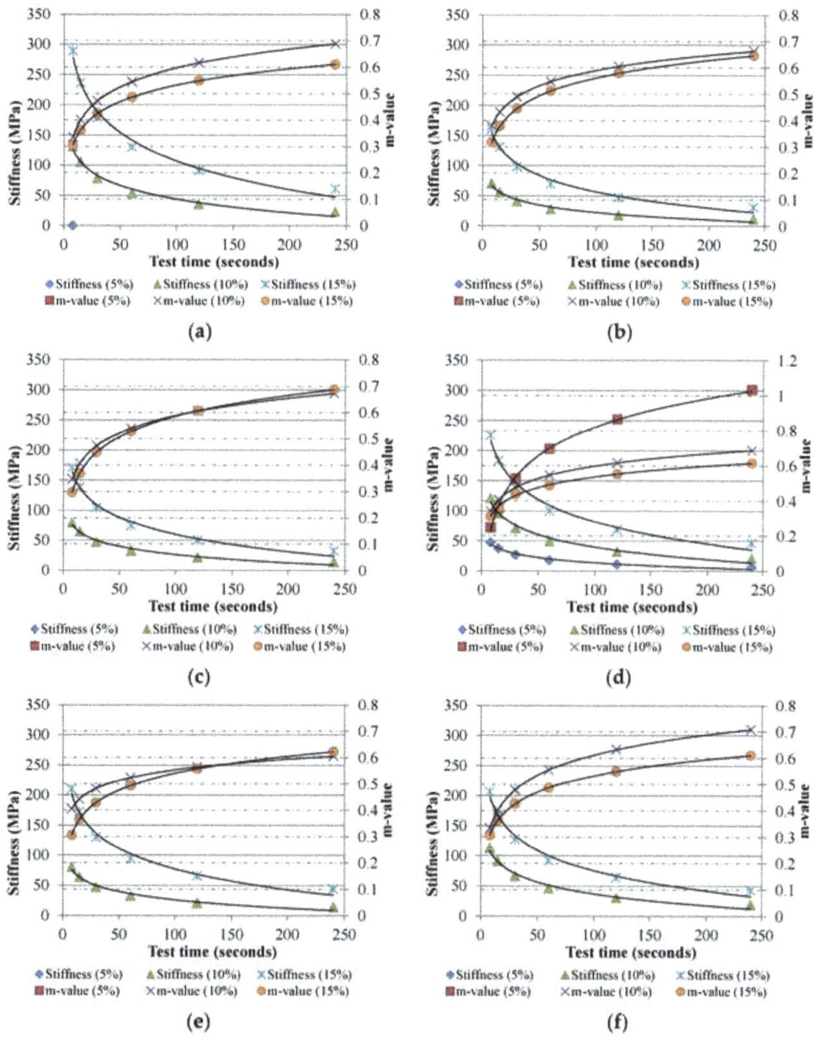

Figure 1. Stiffness values and m-values of reclaimed asphalt pavement (RAP) sources A–F modified with virgin binder PG 58-28 at −6 °C, (**a**–**f**) RAP sources A–F. A–F the names of RAP source in this study.

A higher percentage of aged binder results in a higher stiffness value and a lower m-value regardless of test duration and RAP source in Figure 1. Meanwhile, in terms of the stiffness values and m-values of RAP mortars A–F, it is noted that these values are significantly different at a same test time. The reason is that the aged binders from all RAP sources vary.

Other stiffness values and m-values of RAP mortars (A–F) and virgin binder at −12 °C and −18 °C are not shown in this paper due to the limitation of paper length. Generally similar trends can be found regardless of RAP source and test temperature in this study.

3.1.2. Low Temperature Determinations of RAP Mortars

Figure 2 shows the minimum low temperatures at a stiffness value of 300 MPa in terms of various RAP mortars blended with virgin binder PG 58-22. It can be found that an increased test temperature reduces the stiffness value of RAP mortars. In addition, a higher involved aged binder significantly results in a greater stiffness value regardless of test temperature and RAP source. Meanwhile, it can be seen that, irrespectively of RAP source, the RAP mortars containing 5% aged binder have stiffness values less than 300 MPa at the lowest temperature of −18 °C in this study. Therefore, these minimum low temperatures are definitely less than −18 °C.

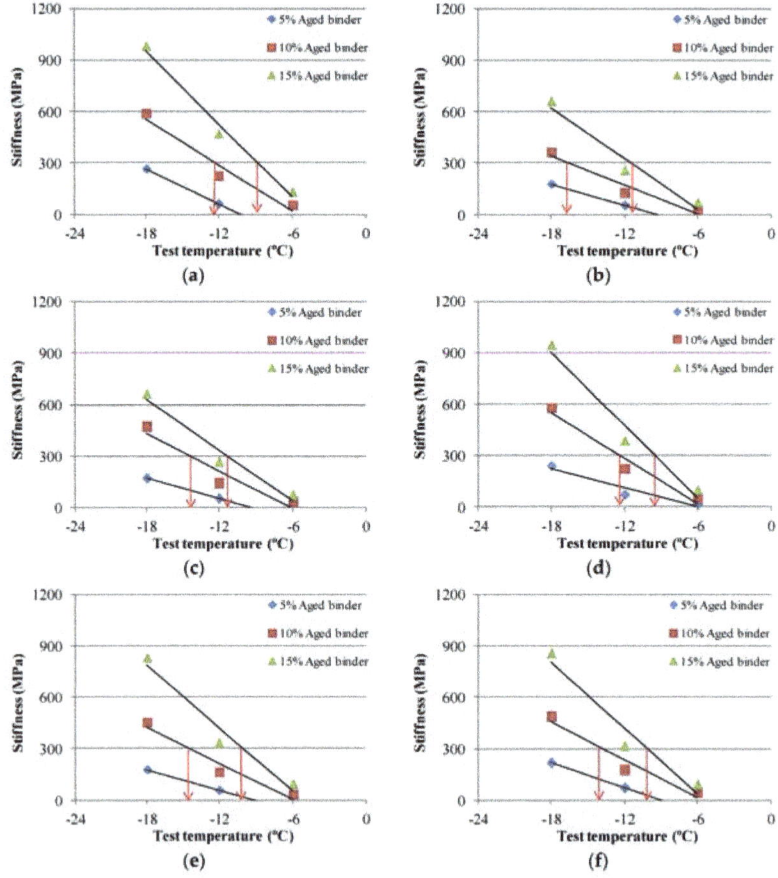

Figure 2. Low temperature determinations of RAP sources A–F with virgin binder PG 58-22 in terms of stiffness, **(a–f)** RAP sources A–F.

Additionally, Figure 2 indicates that, when the RAP mortars with 10% aged binder generally have a stiffness value of 300 MPa, their corresponding low temperatures are typically less than −12 °C. However, when the used aged binder is greater than 15%, the minimum low temperature is usually greater than −12 °C.

The minimum low temperature determinations of the RAP mortars with PG 58-28 with respect to m-values are shown in Figure 3. It can be observed that the m-values are greater than 0.300 at a temperature greater than −18 °C regardless of RAP source and aged binder content because the virgin binder PG 58-28 is generally quite soft. Therefore, the low temperatures of these RAP mortars were mainly determined by the stiffness values of these binders.

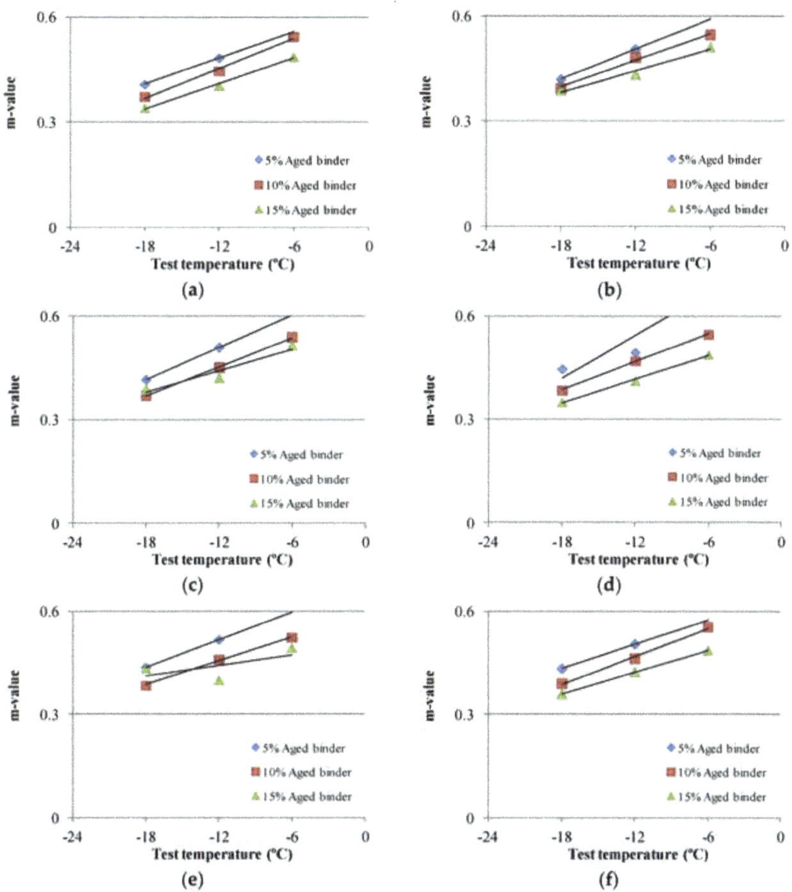

Figure 3. Low temperature determinations of RAP sources A–F with virgin binder PG 58-28 in terms of m-value, (**a**–**f**) RAP sources A–F.

In order to obtain the minimum low temperatures of various RAP mortars from various RAP sources and aged binder contents, Table 2 presents the minimum low temperatures based on those determined values from the stiffness values and m-values, derived from the conducted regression analysis. A higher temperature was selected as a minimum low temperature in this study because this would be able to satisfy the demand of the asphalt binder to resist the pavement cracking at a low performance temperature.

Table 2. Minimum low temperatures of RAP sources A–F with virgin binder PG 58-28.

Min. Temp (°C)	Stiffness			m-Value			Low Temperature Determination		
	Aged Binder Percentage			Aged Binder Percentage			Aged Binder Percentage		
	5%	10%	15%	5%	10%	15%	5%	10%	15%
A	−21	−12.2	−8.8	<−24	−22.7	−20.8	**−21**	**−12.2**	**−8.8**
B	<−24	−16.5	−11.7	<−24	<−24	<−24	**<−24**	**−16.5**	**−11.7**
C	<−24	−14.2	−11.3	<−24	−22.5	<−24	**<−24**	**−14.2**	**−11.3**
D	−21.3	−12.2	−9.7	−23.4	<−24	−21.9	**−21.3**	**−12.2**	**−9.7**
E	<−24	−14.1	−10	<−24	<−24	<−24	**<−24**	**−14.1**	**−10**
F	−21.8	−13.7	−10.1	<−24	<−24	−23.8	**−21.8**	**−13.7**	**−10.1**

As shown in Table 2, it can be seen that the minimum low temperatures are generally close to −12 °C for all RAP mortars when using 5% aged binder. However, these minimum low temperatures rise to approximately −6 °C when 15% aged binder was utilized to produce the BBR samples. Obviously, the increase of aged binder results in the remarkable increase of minimum low temperatures of these RAP mortars. Additionally, the RAP source only has a slight impact on the minimum low temperatures when using a higher aged binder content, but had a medium influence as lower aged binders were employed.

3.2. *RAP Mortars Mixed with RTFO Binder*

3.2.1. Stiffness Values and m-Values of RAP Mortars

This section presents the test results of the RAP mortars mixed with various RAP sources and a short-term aged (RTFO) binder of PG 58-28. The fabricated BBR samples were tested at three temperatures (−6 °C, −12 °C, and −18 °C). The main test results are shown in Figure 4.

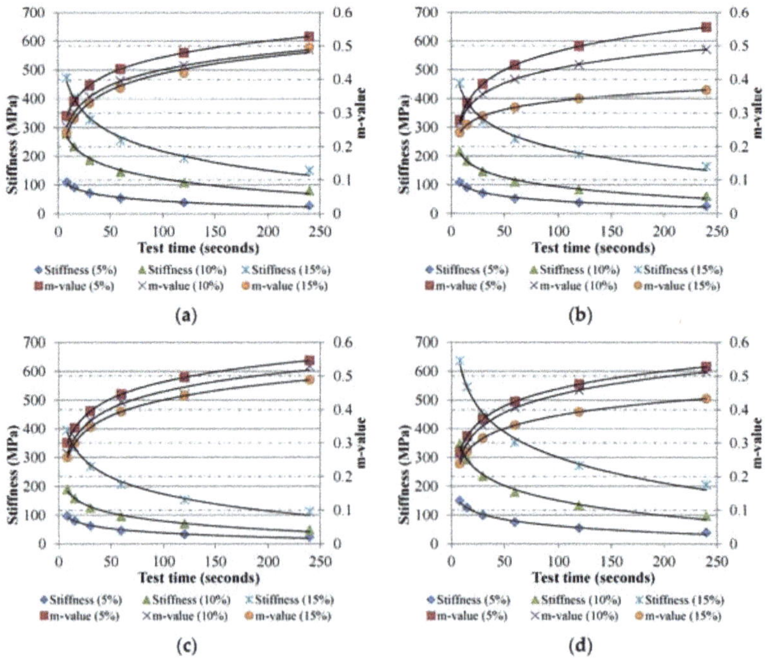

Figure 4. *Cont.*

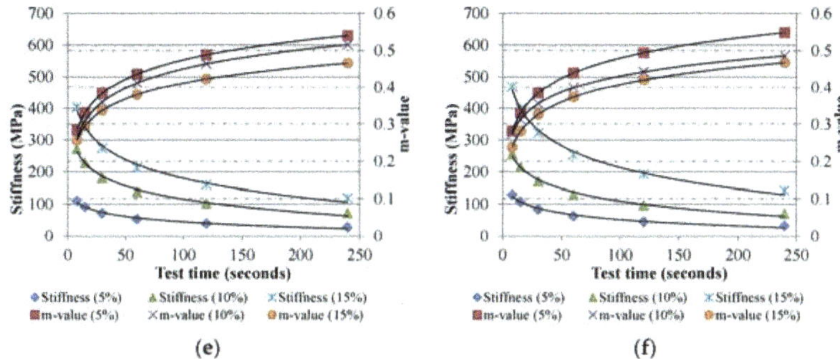

Figure 4. Stiffness values and m-values of RAP sources A–F modified with reclaimed asphalt pavement (RTFO) binder PG 58-28 at −6 °C, (**a–f**) RAP sources A–F.

As shown in Figure 4, it can be found that, at −6 °C, all RAP mortars mixed with sieved RAP and RTFO aged PG 58-28 binder showed increased m-values and the decreased stiffness values during a loading process. It is also noted that, as expected, RAP mortars with a higher aged binder content containing the aged soft binder have a higher stiffness and a lower m-value, following logarithmic trends regardless of RAP source and test time. In addition, different from the RAP mortars mixed with virgin binder PG 58-28, the RAP mortar with a 5% aged binder mixed with RTFO binder can show stiffness values and m-values during a loading procedure. Moreover, these stiffness values are significantly higher compared to those values of the modified binders mixed with virgin binder due to the RTFO aged binders.

3.2.2. Low Temperature Determinations of RAP Mortar

The previous data indicated that the cracking resistance at a low temperature of a modified binder is based on the stiffness and m-values at various test temperatures. In this section, the minimum low temperature determinations of the RAP mortars mixed with RAPs (A–F) and RTFO binders are summarized.

In Figure 5, it can be found that the RAP mortar with a higher aged binder content has a higher low temperature when its stiffness value is 300 MPa. In other words, the aged binder results in a higher stiffness regardless of RAP type. However, various RAP mortars generally have different low temperature values, dependent on RAP type.

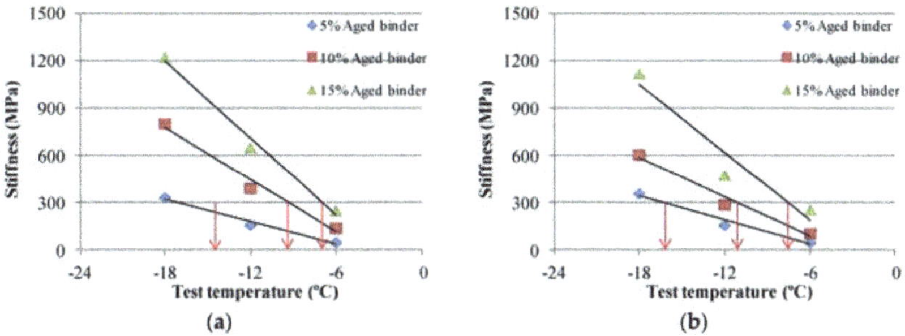

Figure 5. *Cont.*

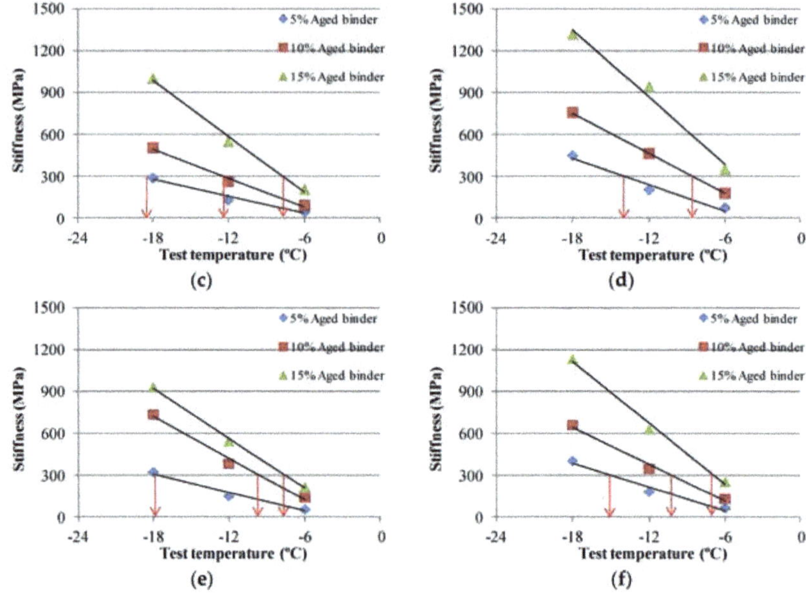

Figure 5. Low temperature determinations of RAP sources A–F with RTFO binder PG 58-22 in terms of stiffness, (**a**–**f**) RAP sources A–F.

In accordance with m-values of the RAP mortars, it can be noted that, in Figure 6, in some cases, m-values are greater than 0.300 when the test temperature is lower than −18 °C. Therefore, these RAP mortars can resist a low temperature of −18 °C or even lower. In addition, it can be noted that a higher aged binder content results in a higher low temperature regardless of RAP type.

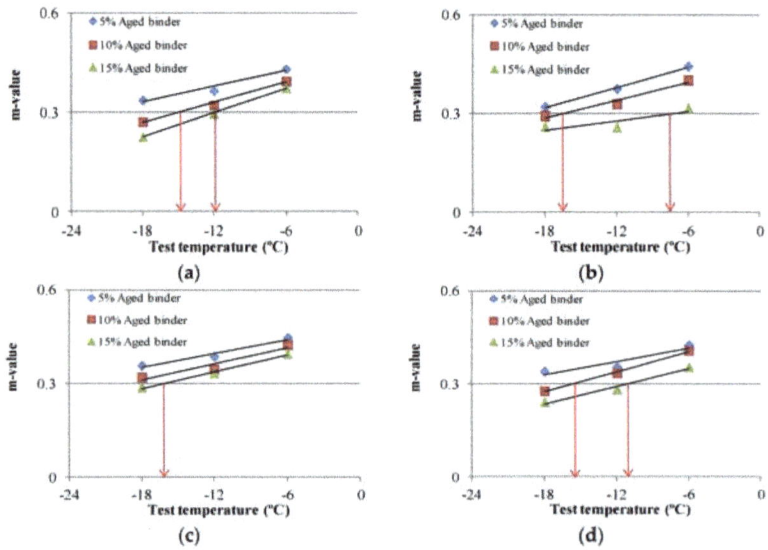

Figure 6. *Cont.*

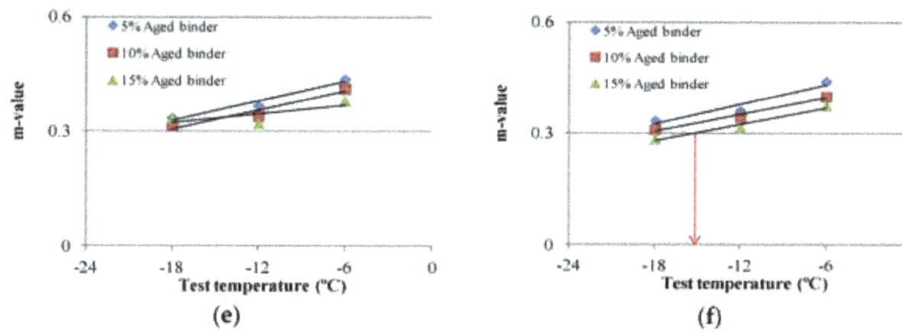

Figure 6. Low temperature determinations of RAP sources A–F with RTFO binder PG 58-22 in terms of m-value, **(a–f)** RAP sources A–F.

The minimum low temperatures of various RAP mortars mixed with various RAP sources and aged binder contents are shown in Table 3. It can be observed that the low temperatures derived from the conducted regression analysis were summarized from stiffness values and m-values. As before, a higher temperature was selected as a minimum low temperature in this study because this could satisfy the demand of the asphalt binder to resist the pavement cracking.

Table 3. Minimum low temperatures of RAP sources A–F with RTFO binder PG 58-28.

Min. Temp. (°C)	**Stiffness**			**m-Value**			**Low Temperature Determination**		
	Aged Binder Percentage			**Aged Binder Percentage**			**Aged Binder Percentage**		
	5%	10%	15%	5%	10%	15%	5%	10%	15%
A	−17	−9.4	−6.9	−21.4	−14.8	−11.9	**−17**	**−9.4**	**−6.9**
B	−16.2	−11.1	−7.8	−19.8	−16.4	−7.3	**−16.2**	**−11.1**	**−7.3**
C	−18.9	−12.2	−7.9	<−24	−19.2	−7.3	**−18.9**	**−12.2**	**−7.3**
D	−13.8	−8.7	−4.6	−21.8	−15.6	−12.1	**−13.8**	**−8.7**	**−4.6**
E	−17.7	−9.6	−7.8	−20.8	−18.5	<−24	**−17.7**	**−9.6**	**−7.8**
F	−15.2	−10.1	−6.9	−20.4	−19.1	−15.4	**−15.2**	**−10.1**	**−6.9**

3.3. *RAP Mortar Mixed with PAV binder*

3.3.1. Stiffness Values and m-Values of RAP Mortar

As shown before, the summarized figures present the stiffness values and m-values of the RAP mortars mixed with RAPs A–F and PAV aged binders. These values are shown in Figure 7, which presents the stiffness values and m-values of the RAP mortars with PAV PG 58-28. As described before, the aged binder concentration and RAP source affect the stiffness values and m-values of RAP mortars.

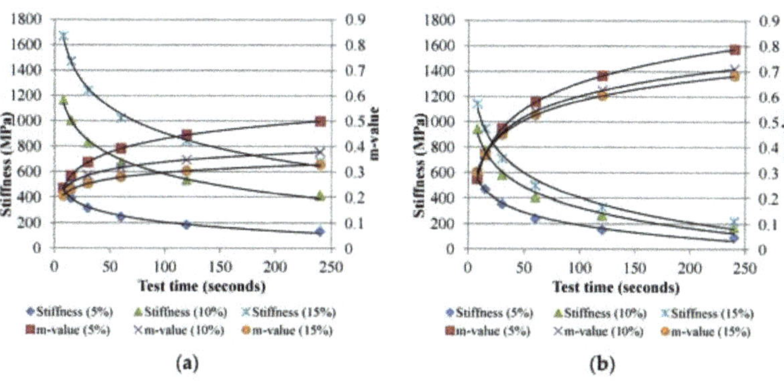

Figure 7. *Cont.*

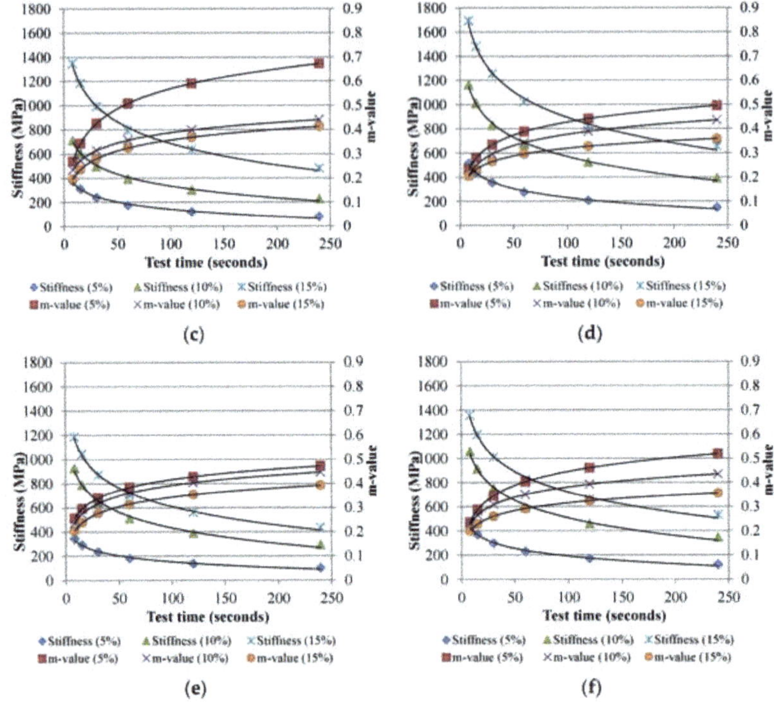

Figure 7. Stiffness and m-values of RAP sources A–F modified with PAV binder PG 58-22, (**a**–**f**) RAP sources A–F.

3.3.2. Low Temperature Determinations of Mortar

Similar to virgin and RTFO binders, the stiffness values and m-values of the RAP mortars mixed with PAV binder can determine the minimum low temperatures of various binders with a specified value of stiffness equaling to 300 MPa and a m-value of 0.300. These determined values can be found in Figure 8.

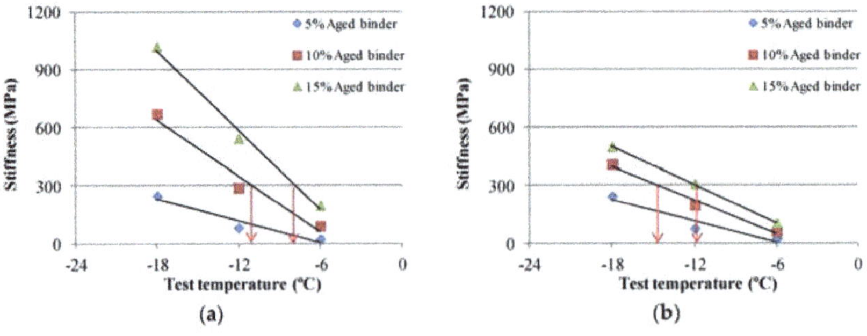

Figure 8. *Cont.*

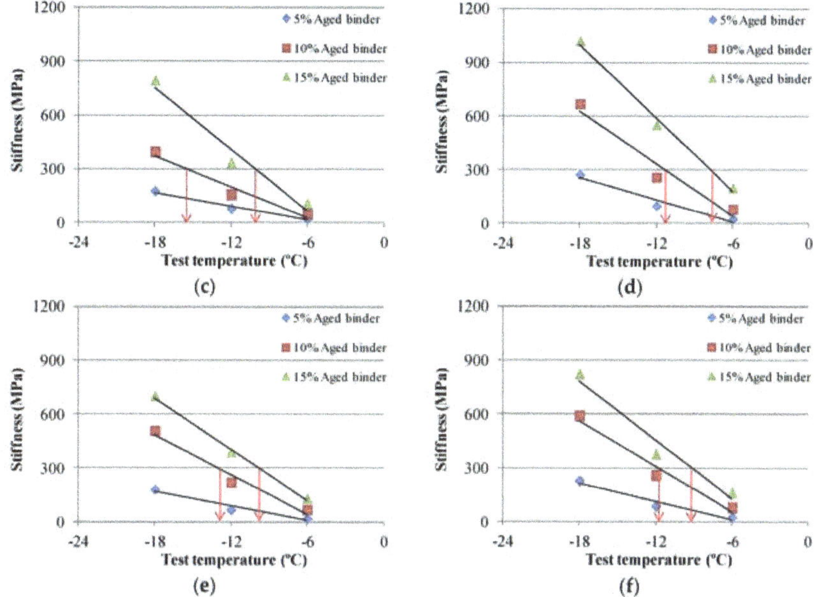

Figure 8. Minimum low temperature determinations of RAP sources A–F with PAV binder PG 58-28 in terms of stiffness, (**a–f**) RAP sources A–F.

In addition, these minimum low temperatures also can be determined by the m-values of these RAP mortars, based on the m-values greater than 0.300. As expected, Figure 9 indicates that the RAP mortars blended with PAV aged PG 58-28 binder and a lower aged binder have higher m-values. In addition, a higher test temperature results in a greater m-value.

Table 4 summarized the minimum low temperatures of various RAP mortars mixed with various RAP sources and aged binder contents, derived from the conducted regression analysis. As described before, a higher temperature was selected as a minimum low temperature in this study because this could satisfy the demand of the asphalt binder to resist the pavement cracking. Obviously, these minimum low temperatures from PAV binders are higher than those minimum low temperatures from RTFO binders, followed by those values from virgin binders.

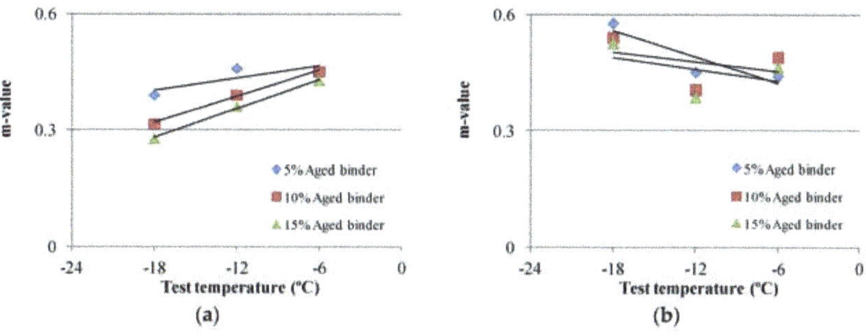

Figure 9. *Cont.*

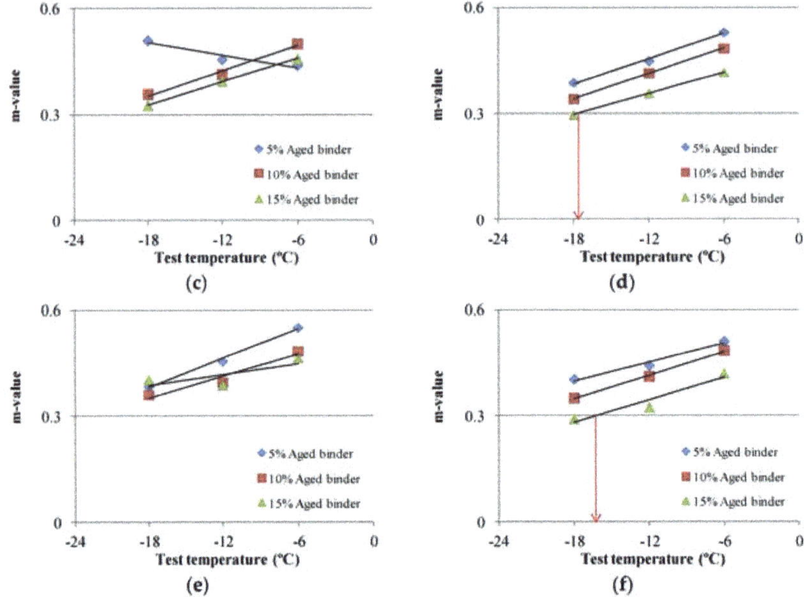

Figure 9. Minimum low temperature determinations of RAP sources A–F with PAV binder PG 58-28 in terms of m-value, (**a–f**) RAP sources A–F.

Table 4. Minimum low temperatures of RAP sources A–F containing PAV binder PG 58-28.

Min. Temp. (°C)	Stiffness			m-Value			Low Temperature Determination		
	Aged Binder Percentage			Aged Binder Percentage			Aged Binder Percentage		
	5%	10%	15%	5%	10%	15%	5%	10%	15%
A	−21.2	−11	−7.9	<−24	−20	−16.3	**−21.2**	**−11**	**−7.9**
B	−22.2	−14.8	−11.9	-	-	-	**−22.2**	**−14.8**	**−11.9**
C	<−24	−15.8	−10.1	-	−22.3	−19.8	-	**−15.8**	**−10.1**
D	−19.7	−11.3	−7.8	<−24	−21.7	−17.8	**−19.7**	**−11.3**	**−7.8**
E	<−24	−12.9	−9.8	−23.8	−22.7	<−24	**−23.8**	**−12.9**	**−9.8**
F	−22.5	−11.9	−9.3	<−24	−22.4	−16.2	**−22.5**	**−11.9**	**−9.3**

3.4. Minimum Low Temperature Comparisons

The minimum low temperature results of RAP mortars containing virgin binder, RTFO binder and PAV binder simulate the low temperature resistances of RAP mixture during and after construction and after long-term performance, respectively. As shown in Figure 10a, minimum low temperatures of RAP mortars with 5% aged binder and virgin soft binder are generally less than −18 °C, but these low temperatures are only less than −12 °C when blended with RTFO and PAV binders. Thus, as expected, the short- and long-term aging procedures can result in the increase of minimum low temperatures regardless of RAP type.

Similarly, as shown in Figure 10b,c, it can be noted that when using 10% and 15% aged binders, the RAP mortars with virgin soft binder have the best low temperature resistance, followed by the RAP mortars with RTFO and PAV binders. Therefore, it is necessary to use soft binder to modify the RAP mixture to achieve a better low temperature resistance.

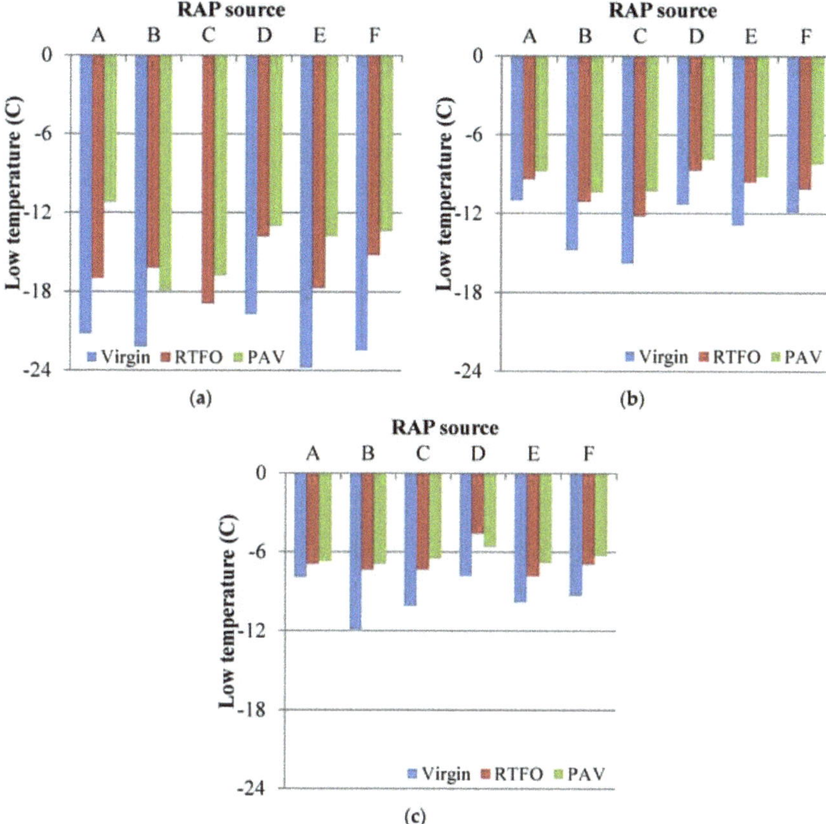

Figure 10. Minimum low temperatures of RAP mortars in terms of aged binder percentage and binder aged states, (**a**) 5% aged binder; (**b**) 10% aged binder; (**c**) 15% aged binder.

4. Conclusions

The RAP mortars containing six RAP sources blending one soft binder at three aging states were investigated with respect to their stiffness and m-values at three minimum test temperatures of −6 °C, −12 °C, and −18 °C. The following conclusions can be drawn:

1. For the sieved RAP containing a high percentage of aged binder over 15% it was generally not easy to conduct the BBR test, and thus it was recommended to use a low-aged binder of less than 15%.
2. The conducted BBR tests for RAP mortars were effective and no modifications were needed to test RAP mortars, which were combined with sieved fine RAP and asphalt binder.
3. The stiffness values and m-values at 60 s from BBR tests could be utilized to explore the minimum low temperatures based on the stiffness value of 300 MPa and m-value of 0.300.
4. RAP mortar with a higher aged binder content had a higher minimum low temperature regardless of RAP source. RAP mortars with virgin soft binder had the best low temperature resistance followed by the RAP mortars with RTFO and PAV binders.
5. The source of RAP did not play a crucial role in determining the low temperature performance characteristics of RAP mortars.

Author Contributions: Feipeng Xiao and Serji Amirkhanian set up the experimental designs and wrote the paper; Ruoyu Li summarized tested data and contributed the data analysis; Henglong Zhang performed the tests.

Conflicts of Interest: The authors declare no conflict of interest.

References

1. Newcomb, D.E.; Brown, E.R.; Epps, J.A. *Designing HMA Mixtures with High RAP Content: A Practical Guide*; Quality Improvement Series 124; National Asphalt Pavement Association: Lanham, MD, USA, 2007.
2. Swiertz, D.; Mahmoud, E.; Bahia, H.U. Estimating the Effect of Recycled Asphalt Pavements and Asphalt Shingles on Fresh Binder, Low-Temperature Properties without Extraction and Recovery. *J. Transp. Res. Board* **2011**, *2208*, 48–55. [CrossRef]
3. Mogawer, W.S.; Austerman, A.J.; Bonaquist, R. Determining the Influence of Plant Type and Production Parameters on Performance of Plant-Produced Reclaimed Asphalt Pavement Mixtures. *J. Transp. Res. Board* **2012**, *2268*, 71–81. [CrossRef]
4. Xiao, F.; Hou, X.; Amirkhanian, S.; Kim, K. Superpave Evaluation of Higher RAP Contents Using WMA Technologies. *Constr. Build. Mater.* **2016**, *112*, 1080–1087. [CrossRef]
5. Hansen, K.; Copeland, A. *Annual Asphalt Pavement Industry Survey on Recycled Materials and Warm-Mix Asphalt Usage: 2009–2013*; Information Series 138; National Asphalt Pavement Association: Lanham, MD, USA, 2013; Available online: https://www.asphaltpavement.org/PDFs/IS138/IS138-2013_RAP-RAS-WMA_Survey_Final.png (accessed on 15 January 2017).
6. West, R.; Kvasnak, A.; Tran, N.; Powell, B.; Turner, P. Testing of Moderate and High Reclaimed Asphalt Pavement Content Mixes Laboratory and Accelerated Field Performance Testing at the National Center for Asphalt Technology Test Track. *J. Transp. Res. Board* **2009**, *2126*, 100–108. [CrossRef]
7. Hong, F.; Chen, D.; Mikhail, M.M. Long-Term Performance Evaluation of Recycled Asphalt Pavement Results from Texas Pavement Studies Category 5 Sections from the Long-Term Pavement Performance Program. *Transp. Res. Rec. J. Transp. Res. Board* **2010**, *2180*, 58–66. [CrossRef]
8. Daniel, J.S.; Pochily, J.L.; Boisvert, D.M. Can More Reclaimed Asphalt Pavement Be Added? Study of Extracted Binder Properties from Plant-Produced Mixtures with up to 25% Reclaimed Asphalt Pavement. *Transp. Res. Rec. J. Transp. Res. Board* **2010**, *2180*, 19–29. [CrossRef]
9. Attia, M.; Abdelrahman, M. Modeling the Effect of Moisture on Resilient Modulus of Untreated Reclaimed Asphalt Pavement. *Transp. Res. Rec. J. Transp. Res. Board* **2010**, *2167*, 30–40. [CrossRef]
10. Roque, R.; Yan, Y.; Cocconcelli, C.; Lopp, G. *Perform an Investigation of the Effects of Increased Reclaimed Asphalt Pavement (RAP) Levels in Dense Graded Friction Courses*; University of Florida: Gainesville, FL, USA, 2015; Available online: http://www.dot.state.fl.us/research-center/completed_Proj/Summary_SMO/FDOT-BDU77-rpt.png (accessed on 16 January 2017).
11. National Cooperative Highway Research Program (NCHRP); Stroup-Gardiner, M. *Use of Reclaimed Asphalt Pavement and Recycled Asphalt Shingles in Asphalt Mixtures (No. Project 20-05, Topic 46-05)*; National Cooperative Highway Research Program: Washington, DC, USA, 2016.
12. Li, X.; Clyne, T.R.; Marasteanu, M.O. *Recycled Asphalt Pavement (RAP) Effects on Binder and Mixture Quality*; Report Mn/DOT 2005-02; Minnesota Department of Transportation: Maplewood, MN, USA, 2004.
13. National Cooperative Highway Research Program (NCHRP). *Synthesis of Highway Practice 54: Recycling Materials for Highways*; TRB, National Research Council: Washington, DC, USA, 1978; p. 1.
14. Xiao, F.; Amirkhanian, S.N.; Juang, H.C. Rutting Resistance of Rubberized Asphalt Concrete Pavement Containing Reclaimed Asphalt Pavement Mixtures. *J. Mater. Civ. Eng.* **2007**, *19*, 475–483. [CrossRef]
15. West, R.; Willis, J.R.; Marasteanu, M. *NCHRP Report 752: Improved Mix Design, Evaluation, and Materials Management Practices for Hot Mix Asphalt with High Reclaimed Asphalt Pavement Content*; Transportation Research Board of the National Academies: Washington, DC, USA, 2013.
16. Li, X.; Marasteanu, M.O.; Williams, R.C.; Clyne, T.R. Effect of Reclaimed Asphalt Pavement (Proportion and Type) and Binder Grade on Asphalt Mixtures. *Transp. Res. Rec. J. Transp. Res. Board* **2008**, *2051*, 90–97. [CrossRef]
17. Amirkhanian, S.; Xiao, F.; Corley, M. *Estimation of Low Temperature Properties of RAP Binder without Extraction*; Report No. FHWA-SC-16-01; Tri County Technical College: Pendleton, SC, USA, 2016.

18. Maupin, G.W.; Diefenderfer, S.D.; Gillespie, J.S. *Evaluation of Using Higher Percentages of Recycled Asphalt Pavement in Asphalt Mixes in Virginia*; Report No. VTRC 08-R22; Virginia Transportation Research Council: Charlottesville, VA, USA, 2008.
19. Scholz, T.V. *Preliminary Investigation of RAP and RAS in HMAC, Final Report SR 500-291, Kiewit Center for Infrastructure and Transportation*; School of Civil and Construction Engineering, Oregon State University: Corvallis, OR, USA, 2010.
20. Ma, T.; Mahmoud, E.; Bahia, H.U. Estimation of Reclaimed Asphalt Pavement Binder Low-Temperature Properties without Extraction. Development of Testing Procedure. *J. Transp. Res. Board* **2010**, *2179*, 58–65. [CrossRef]
21. Xiao, F.; Amirkhanian, S.N.; Wu, B. Fatigue and Stiffness Evaluations of Reclaimed Asphalt Pavement in Hot Mix Asphalt Mixtures. *J. Test. Eval.* **2011**, *39*, 50–58.

Article

Effect of Fibers on Mixture Design of Stone Matrix Asphalt

Yanping Sheng [1,*], Haibin Li [2], Ping Guo [3], Guijuan Zhao [2], Huaxin Chen [1] and Rui Xiong [1]

1. School of Materials Science and Engineering, Chang'an University, Xi'an 710064, China; chx92070@163.com (H.C.); xiongr61@126.com (R.X.)
2. School of Architecture and Civil Engineering, Xi'an University of Science and Technology, Xi'an 710054, China; lihaibin1212@126.com (H.L.); guijuanzhao@126.com (G.Z.)
3. Xi'an Highway Research Institute, Xi'an 710054, Shaanxi, China; guoping8088@163.com

* Correspondence: syp@chd.edu.cn; Tel.:+86-1899-1848-117

Academic Editors: Zhanping You, Qingli (Barbara) Dai and Feipeng Xiao
Received: 28 December 2016; Accepted: 10 March 2017; Published: 18 March 2017

Abstract: Lignin fibers typically influence the mixture performance of stone matrix asphalt (SMA), such as strength, stability, durability, noise level, rutting resistance, fatigue life, and water sensitivity. However, limited studies were conducted to analyze the influence of fibers on the percent voids in mineral aggregate in bituminous mixture (VMA) during the mixture design. This study analyzed the effect of different fibers and fiber contents on the VMA in SMA mixture design. A surface-dry condition method test and Marshall Stability test were applied on the SMA mixture with four different fibers (i.e., flocculent lignin fiber, mineral fiber, polyester fiber, blended fiber). The test results indicated that the bulk specific gravity of SMA mixtures and asphalt saturation decreased with the increasing fiber content, whilst the percent air voids in bituminous mixtures (VV), Marshall Stability and VMA increased. Mineral fiber had the most obvious impact on the bulk specific gravity of bituminous mixtures, while flocculent lignin fiber had a minimal impact. The mixture with mineral fiber and polyester fiber had significant effects on the volumetric properties, and, consequently, exhibited better VMA over the conventional SMA mixture with lignin fiber. Modified fiber content range was also provided, which will widen the utilization of mineral fiber and polyester fiber in the applications of SMA mixtures. The mixture evaluation suggested no statistically significant difference between lignin fiber and polyester fiber on the stability. The mineral fiber required a much larger fiber content to improve the mixture performance than other fibers. Overall, the results can be a reference to guide SMA mixture design.

Keywords: stone matrix asphalt; volume parameters; Marshall Stability; flocculent lignin fiber; polyester fiber; mineral fiber; fiber content

---

1. Introduction

Stone matrix asphalt (SMA) is a hot asphalt mixture in which coarse aggregate interlocks to form a stone skeleton that resists permanent deformation. SMA was first used in Europe as a mixture that would resist the wear of studded tires. Then it was used successfully in the United States in 1990, and is now widely used in China. The advantages of SMA include high resistance to rutting, excellent low-temperature performance, improved macrotexture, long service life, low tire noise, less water spray from tires, and weak light reflection on rainy nights [1–3]. However, the coarse texture of an SMA mixture may result in more internal air voids that are related to performance degradation, even when the volume of air voids is the same as that of common asphalt mixtures [4]. The coarse surface texture makes it more difficult to differentiate between mixture air voids and surface texture.

Fiber additive is important for SMA due to its oil absorptive characteristics. A certain quantity of fiber should be added into the SMA mixture in order to prevent asphalt from flowing out due to the high asphalt content. The outflow of the asphalt can result in fat spots on the pavement surface [5]. The mineral skeleton of coarse aggregate supplies the mixture with a strong particle interlock increasing the mixture resistance, and the mastic supplies the mixture with better durability. Since the fiber occupies some space, the gap between aggregates will be increased if it blocks the contact of the aggregates, and then the mixture performance will be reduced through the influence of volumetric parameters. The volumetric parameters are the direct controlling indicators in the design and preparation of the SMA mixtures.

In the early stage of hot-mix-asphalt (HMA) mix design, percent voids in mineral aggregate in bituminous mixtures (VMA) were determined and maintained throughout the mix-design procedure. VMA includes the air voids and the volume occupied by the effective asphalt content. This volumetric property is correlated to mechanical properties [6–8], e.g., small percent air voids in bituminous mixtures (VV) will cause bleeding and high VV may lead to water damage or instability in asphalt pavement. In addition to the size gradation, VMA is one of the most important HMA design criteria to obtain durable pavement, and it significantly affects the permanent deformation and fatigue performance of a compacted mix [9,10]. The use of VMA criteria for mix design is a time-honored and fairly successful tool. The VMA requirements for HMA mixtures were initially developed in the 1950s and were considered one of the most important volumetric parameters for HMA and SMA mixtures [11,12]. Then other influence factors of VMA, such as aggregate factors and volumetric basis, were pointed out, and VMA specifications were strongly emphasized during the process of asphalt mixture design and analysis [12–15].

In order to determine VMA, the bulk density, percent air voids in bituminous mixtures (VV), and percent voids in mineral aggregate that are filled with asphalt in bituminous mixtures (VFA) have to be obtained first because they are critical parameters to obtain proper VMA in design and practice. Studies have reported the difficulty of meeting the minimum VMA requirement in an efficient manner [10,16,17]. It indicated that the minimum VMA should be based on the minimum asphalt film thickness rather than the minimum asphalt content [18]. Although both Bailey's method and the NCHRP 9-33 manual have provided suggestions for adjusting the mix design to achieve the target VMA, the determination of VMA still requires a large amount of experimental testing [15,19]. As another point of view, VMA was to incorporate at least the minimum permissible asphalt content into the mixture to ensure its durability. VMA and the shape of aggregate particles influence workability, shear resistance, fatigue, and durability of the mixture [20–24].

The most commonly adopted fibers in SMA mixtures are lignin fibers. The success in SMA mixtures spurred the adoption of the fiber for many major highway projects. Then, lignin fiber, glass fiber, and mineral fiber have been studied in asphalt mixtures [25,26]. From then on, other types of fibers, such as carpet fiber, polyester fiber, waste tires, cellulose oil palm fiber, waste glass fiber, and coconut fiber, were used to study the service properties of the HMA mixture and SMA mixture [26–30]. The studies focused on the mixture to obtain a better performance, such as strength, stability, durability, reduction of noise, rutting resistance, fatigue life, and water sensitivity. However, fiber types may influence oil absorption and fiber content will affect the VV. Then it can affect the VMA directly. Limited studies were conducted to analyze the influence of fibers on the VMA during the mixture design.

Therefore, this study investigated the effects of four different fibers on the mixture volume parameter during the SMA mixture design, with the goal of identifying the adaptability of polyester fiber and mineral fiber for satisfactory binder performance. The mineral fiber has a similar density with aggregates and smaller oil absorption and specific surface area, which means it cannot absorb much asphalt binder to fill the mineral outside space, and is less sensitive to the content change. Polyester fiber has better asphalt absorption and higher ductility; therefore, it can form much more space in the SMA mixture [31]. High content of polyester fiber means low asphalt content, which potentially reduces the adhesion between the aggregate and asphalt binder.

Empirical binder tests were conducted to identify volume parameters and appropriate contents of different fibers. The bulk specific gravity of bituminous mixtures, VV, VFA, and VMA, were studied with a surface-dry condition method test. Then, the Marshall Stability of the SMA mixtures with optimized mixing procedures was evaluated to check the effect of the fibers on the mechanical performance. The flocculent lignin fiber was used as a control fiber. The suggested fiber content of this study for different fibers could provide better performance of the SMA mixture. The results provide effective references for the SMA mixture design.

2. Materials and Methods

2.1. Materials

2.1.1. Asphalt Binder

A modified asphalt binder, i.e., styrene-butadiene-styrene (SBS) (I-C), which has been regularly used in pavement engineering, was selected in this study. Table 1 shows the measured technical indicators of the SBS asphalt binder. Basic binder tests, such as the penetration, softening point, and ductility were conducted to evaluate the fundamental characteristics of SBS asphalt binder which may influence the SMA mixture.

Table 1. Technical indicators of the asphalt binder.

Test Properties		Unit	Test Results	Specification Requirements
Penetration (25 °C, 100 g, 5 s)		0.1 mm	71.7	60~80
Softening point		°C	97	≥55
Ductility (5 °C, 5 cm/min)		cm	33.4	≥30
Penetration index		-	0.21	≥−0.4
Density (15 °C, g/cm^3)		-	1.032	-
Viscosity (135 °C)		Pa·s	1.83	≤3
Flash point		°C	328	≥230
Solubility (Trichloroethylene)		%	99.37	≥99
Segregation, 48 h D-value of Softening point		°C	2.1	≤2.5
Elastic recovery (25 °C)		%	98	≥65
Short-term oven aging, 163 °C, 75 min	Mass change	%	0.006	≤±1.0
	Penetration ratio, 25 °C	%	76.4	≥60
	Ductility, 5 °C	cm	29.3	≥20

The asphalt penetration test is used to evaluate the asphalt's soft and hard levels and its shear resistance. The test reflects the asphalt's relative viscosity. The softening point test is used to determine the temperature at which the asphalt becomes soft and achieves a certain viscosity. Ductility is mainly about deformability of asphalt and indirectly reflects low-temperature anti-cracking property. It is an important index that can be used to evaluate asphalt plasticity such that the larger the ductility value, the better plasticity of the asphalt. All of these are part of the basic performance index to evaluate the asphalt binder.

2.1.2. Aggregate

The diabase gravel and the limestone sand were chosen as coarse aggregate and fine aggregate in this study. Some important technical indicators are listed in Tables 2 and 3, respectively. To create a better adhesion between the aggregate and the asphalt binder during the mixing procedure, the aggregates were first cleaned and then dried well. The required amounts of aggregates and fillers were placed into an oven at 105 °C for 5 h, and then the temperature rose to 180 °C for mixing.

Table 2. Technical indicators of coarse aggregate.

Test Properties		Unit	Test Results	Specification Requirements
Apparent relative density		-	2.927	≥2.60
Crushing value		%	8.3	≤26
Sturdiness		%	9.8	≤12
LA abrasion value		%	10.2	≤28
Water absorption		%	0.48	≤2.0
Adhesion with asphalt		Grade	5	5
<0.075 mm Grain content		%	0.3	≤1
Soft stone content		%	1.1	≤3
Needle and plate particle content	Mixture	%	6.2	≤15
	>9.5 mm	%	5.3	≤12
	<9.5 mm	%	9.3	≤18

Table 3. Technical indicators of fine aggregate.

Test Properties	Unit	Test Results	Specification Requirements
Sturdiness (>0.3 mm)	%	14	≥12
Apparent relative density	-	2.745	≥2.50
Methylene blue value (g/Kg)	%	10.6	≤25
Angularity (flow time)	s	43.1	≥30

2.1.3. Mineral Filler

The mineral filler was produced by limestone. Some important technical indicators of mineral filler were shown in Table 4.

Table 4. Technical indicators of mineral filler.

Test Properties	Unit	Test Results	Specification Requirements
Apparent density	t/m^3	2.726	≥2.50
Hydrophilic coefficient	-	0.7	<1
Plasticity index	%	2.3	<4
Water content	%	-	≤1.0

2.1.4. Fiber

Four different fibers, i.e., flocculent lignin fiber, mineral fiber, polyester fiber, and blended fiber, were selected in order to analyze the effect on VMA at different fiber contents. The blended fiber was made up of flocculent lignin fiber and polyester fiber with mass ratio of 2:1. Some important technical indicators of these fibers are listed in Table 5.

Table 5. Technical indicators of different fibers.

Test Properties	Unit	Flocculent Lignin Fiber	Polyester Fiber	Mineral Fiber
Relative density	g/cm^3	1.813	1.390	2.720
Length	mm	5	6	4
Thickness	mm	0.047	-	-
Diameter	μm	-	20	5
Ash content (by weight)	%	16	-	8.8
PH value	-	6.9	-	-
Water content rate (by weight)	%	3	2.43	-
Oil absorption rate	times	6.5	4.1	-
Melting points	°C	-	260	>1000
Tensile strength	MPa	-	570	935

2.2. Test Methods

2.2.1. Mixing Proportion Determination

A typical SMA mixture, i.e., SMA-13 with a nominal maximum aggregate size of 13.2 mm, which has been regularly used in asphalt pavement construction, was selected to study volumetric parameters and mixture stability. Table 6 and Figure 1 show the gradation of the SMA-13. As an additive, the fiber was added to make mixture specimens.

Table 6. Gradation of SMA-13 asphalt mixture.

Composite	Percentage	Mesh Size (mm/%)									
		16.0	13.2	9.5	4.75	2.36	1.18	0.6	0.3	0.15	0.075
9.5~16 mm	44%	100	85.1	12.3	0.40	0.40	0.40	0.40	0.40	0.40	0.40
4.75~9.5 mm	33%	100	100	97.9	8.6	0.5	0.50	0.50	0.50	0.50	0.50
0~2.36 mm	13%	100	100	100	100	92.6	62.0	36.1	19.9	13.6	9.9
Mineral filler	10%	100	100	100	100	100	100	100	99.3	95.4	85.3

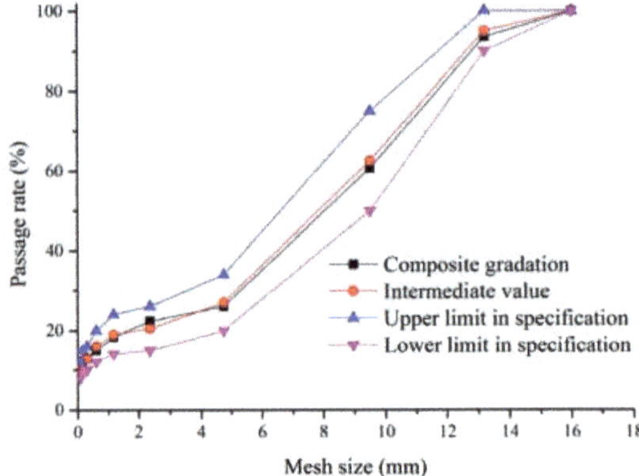

Figure 1. Gradation of the SMA-13 mixture.

Using the materials and the aggregate gradation described above, the SMA mixtures with different fibers and 0.3% content (by the weight of total mix) was prepared for laboratory testing. The optimal asphalt aggregate ratio of SMA mixtures with different fibers shown in Table 7 were determined according to the Marshall mixture design method of the Chinese technical specification for construction of highway asphalt pavements (JTG F40-2004).

Table 7. Asphalt aggregate ratio of SMA mixtures with different fibers.

Different Fibers	Flocculent Lignin Fiber	Mineral Fiber	Polyester Fiber	Blended Fiber
Asphalt aggregate ratio (%)	5.9	5.5	5.7	5.9

2.2.2. SMA Sample Preparation

In this study, SMA specimens were prepared using the compaction method. The dimensions were 101.6 mm × 63.5 mm. According to the standard requirement and field construction experience, the fiber content (by the weight of the total mix) used in the SMA mixture was selected, which was 0.1%, 0.2%, 0.3%, 0.4%, 0.5%, and 0.6%, respectively. The fiber types and contents are shown in Table 8.

Table 8. Different fiber types and fiber content in SMA mixtures.

Type	Flocculent Lignin Fiber					Mineral Fiber					
Content (%)	0.1	0.2	0.3	0.4	0.5	0.1	0.2	0.3	0.4	0.5	0.6
Type	Blended fiber					Polyester fiber					
Content (%)	0.1	0.2	0.3	0.4	0.5	0.1	0.2	0.3	0.4	0.5	-

The density was determined with the surface-dry condition method (T0705-2011/JTG E20-2011), which was very similar to ASTM D2726-14. There were mainly two differences. First, ASTM D2726-14 explained if the temperature of the specimen differs from the temperature of the water bath by more than 2 °C (3.6 °F), the specimen should be immersed in the water bath for 10 to 15 min; instead of 3 to 5 min. The immersed time in the water bath was only 3 to 5 min in the T0705-2011. Secondly, ASTM D2726-14 required, after determining the mass in water and in a saturated-surface dry condition, thoroughly drying the specimen to a constant mass at 110 ± 5 °C (230 ± 9°F). While it only required making the specimens thoroughly dry, it did not illustrate the temperature in the T0705-2011.

The loose fibers (by the weight of total mix) were first blended with the hot aggregates to prevent the asphalt binder from draining during the mixing procedure. The heated asphalt binder was added after mixing 1 to 1.5 min. Then the heated filler was added until all of the aggregate was completely covered. The total mixing time was 3 min. During the mixing process, the temperature was kept between 170 °C and 180 °C. Finally, the Marshall specimens were made with dimensions of 101.6 mm × 63.5 mm using the compaction method. For note, the final temperature was greater than 145 °C.

2.2.3. Marshall Stability and Flow Tests

The Marshall stability and flow tests were conducted to evaluate the resistance of asphalt mixtures to distortion, displacement, rutting, and shearing stresses. The stability test measures the maximum load sustained by the specimen at a loading rate of 50.8 mm/min. Basically, the applied testing load increases until the specimen splits into two pieces, then the loading is finished and the maximum load is recorded as the Marshall Stability.

3. Results and Discussion

3.1. *Effect of Fiber on Bulk Specific Gravity, VV, and VFA*

Figure 2 displayed the bulk specific gravity of SMA mixtures with different fibers and fiber contents. The data show an inverse correlation between bulk specific gravity of SMA mixtures and fiber content for fiber types. The SMA mixture with mineral fiber had the largest bulk specific gravity values, followed by the mixture with polyester fiber. The mixture with flocculent lignin fiber had the smallest values. All of the bulk specific gravity values were between 2.465 and 2.523. The bulk specific gravity of SMA mixtures with polyester fiber and blended fiber decreased with fiber content increasing from 0.1% to 0.4%, and then it maintained a slight decrease with fiber content from 0.4% to 0.5%. The mixture with mineral fiber had a similar variety, with the only difference being the relatively smaller reduction in fiber content from 0.5% to 0.6%.

These results indicated that it was not necessarily true that the larger bulk specific gravity of the SMA mixture results from a higher fiber content. The higher fiber content results in lower bulk specific gravity. In terms of the four different fibers, the mineral fiber has a density very close to that of the aggregate, and it was much easier to combine with the asphalt binder than other fibers. Under the same compaction effort, the SMA mixture with mineral fiber can reach a larger dry mass per unit volume. Therefore, it appeared to have a larger bulk specific gravity value.

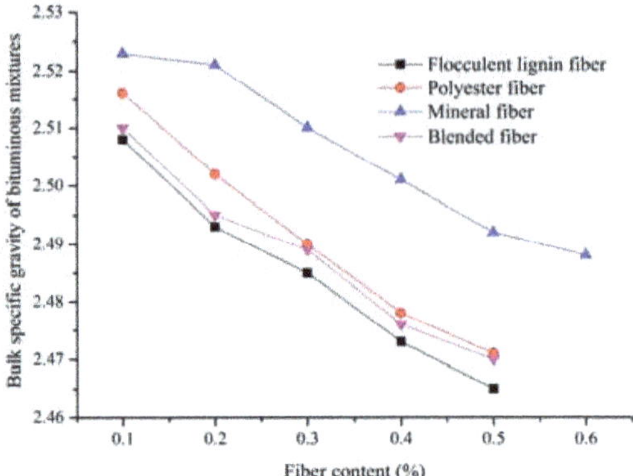

Figure 2. Bulk specific gravity of SMA mixtures with four different fibers.

Figure 3 displayed the percent air voids in SMA mixtures with different fibers and fiber contents. The data showed a positive correlation between VV and fiber content. Higher fiber content resulted in better asphalt absorption and adsorption in the SMA mixture. The measured VV values of different fiber types and contents were between 2.9% to 4.4%.

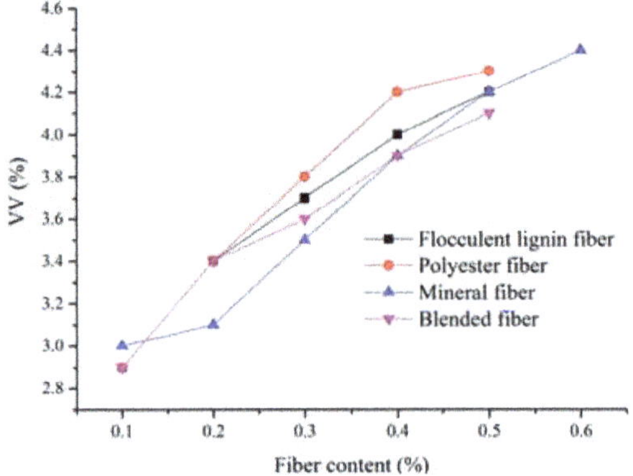

Figure 3. Percent air voids in SMA mixtures with four different fiber types.

It was found from the results that the polyester fiber had an important impact on the VV values when the content was up to 0.2%. It indicated that polyester fiber prevented the aggregates contacting each other and formed much more space in the mixture due to its higher ductility. The VV with polyester fiber, flocculent lignin fiber, and blended fiber increased with the fiber content increasing from 0.1% to 0.4%, and then it maintained a slight increase with the fiber content from 0.4% to 0.5%, and the increments of flocculent lignin fiber, polyester fiber, and blended fiber were 4.5%, 1.9%, and 4.9%, respectively. However, mineral fiber had little impact on VV values when the content was less than 0.2% due to the negligible impact on the asphalt absorption. However, when the content was more than 0.2%, mineral fiber began to show its effect on absorptive action and the mixture's adsorption, for which the VV showed a significant increase. At the same content, the mixture with polyester fiber had the largest VV value.

In practice, the optimum fiber type and content should be selected to achieve satisfactory performance of the SMA mixture in the production process. This study suggests that the minimum content is above 0.3% for mineral fiber and above 0.2% for the other fibers. However, the specific value may depend more on other factors, such as cost considerations, availability, and ease of field construction application because of the difference between lab tests and field construction.

Figure 4 showed the percentage of voids in mineral aggregate (VFA) that are filled with asphalt in SMA mixtures with different fibers and fiber contents. The data from Figure 4 indicated that fiber types and contents had an obvious impact on VFA, which decreased as fiber content increased. All of the VFA values were between 73% and 82%. The percentage of voids in mineral aggregate that are filled with asphalt in SMA mixtures with blended fiber, flocculent lignin fiber, and polyester fiber decreased with the fiber content increasing from 0.1% to 0.4%, and then it maintained a slight decrease with fiber content from 0.4% to 0.5%. When the additive was mineral fiber, it had similar variety, with the only difference being the relatively smaller reduction in the content from 0.1% to 0.2% and 0.5% to 0.6%.

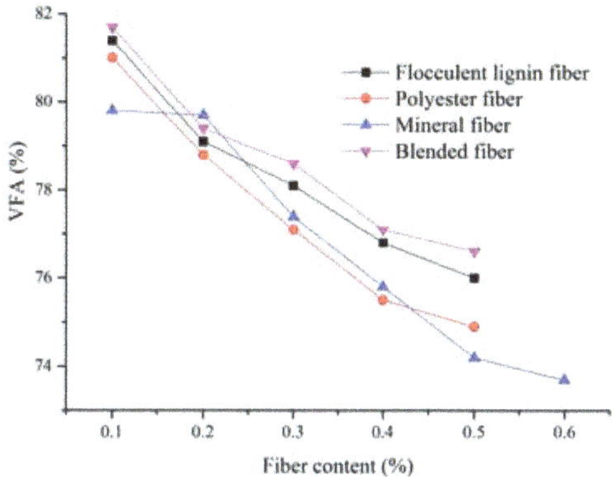

Figure 4. VFA of SMA specimens with four different fiber types.

The content change of mineral fiber had a negligible impact on the VFA values due to its large density and low oil absorption rate when the content was less than 0.2%. Then VFA showed a significant decrease after the content was more than 0.2%. However, when the mineral fiber content was up to 0.4%, it began to show its absorptive action and the mixture's adsorption, which made the VFA much smaller than that with polyester fiber. The three other fibers had obvious impacts on the VFA value, and the SMA mixture with polyester fiber had a much smaller VFA than that with blended fiber and flocculent lignin fiber.

The results indicated that the mineral fiber and polyester fiber had better asphalt absorption and adsorption in the SMA mixture. To obtain the same VFA value, lower content is needed for mineral fiber and polyester fiber. Therefore, the fiber selection should comprehensively consider both absorptive action and the mixture's adsorption. Higher contents of mineral fiber and polyester fiber were not the best choice. Fiber with higher adsorption, but lower absorption, will improve the SMA mixture volume index.

3.2. *Effect of Fiber on Percent Voids in Mineral Aggregate in Bituminous Mixtures (VMA)*

Figure 5 displayed the percent voids in mineral aggregate in SMA mixtures. The VMA increased with fiber content increasing. The measured VMA values were between 15% and 18%. At the same fiber content, the VMA value had the minimum gap between lignin fiber and blended fiber. It was only 16.4% even when the mineral fiber content was up to 0.5%. The values were more than 17.3% for the other fibers. This indicated that the mineral fiber had the minimum impact on VMA. The same trend was found for Marshall Stability. Compared with the previous literature, the mix designer was able to judge the proper VMA requirement for each kind of fiber.

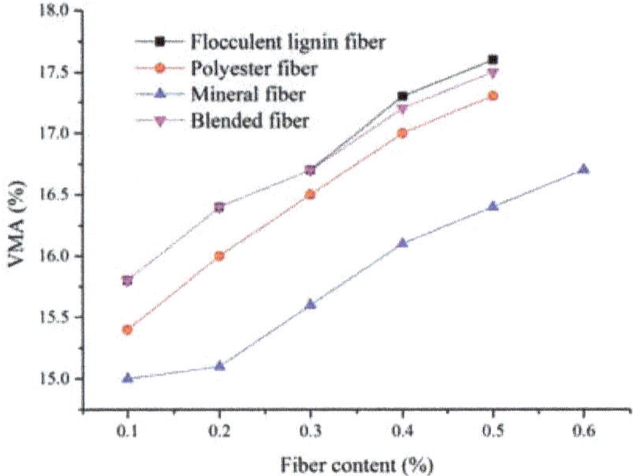

Figure 5. VMA of SMA specimens with four different fiber types.

The VMA with polyester fiber, flocculent lignin fiber and blended fiber increased with fiber content increasing from 0.1% to 0.4%, and then the increment gradually reduced with fiber content from 0.4% to 0.5%. However, the content change of mineral fiber had a negligible impact on VMA when the content was less than 0.2%. The mixture with mineral fiber had a similar variety with the only difference being the relatively lower VMA values with fiber contents from 0.2% to 0.6%. When the content of flocculent lignin fiber, blend fiber, and polyester fiber was 0.3%, and mineral fiber content was 0.4%, the VMA values of SMA mixtures were 16.7%, 16.7%, 16.5%, and 16.1%, respectively, which explained much more mineral fiber content was needed to achieve a target VMA value in the SMA mixture design.

Furthermore, mineral fiber had a similar density with aggregate and smaller oil absorption and smaller specific surface area; it cannot absorb much asphalt binder to fill the outside space of the mineral fiber. Therefore, effective asphalt content was not sensitive to the content change. Therefore, the SMA mixture with mineral fiber can obtain a lower VMA under the same compaction effort. These test findings can be a reference and used to promote the use of fibers in SMA mixtures.

3.3. *Effect of Fiber on Marshall Stability*

Figure 6 displayed the Marshall Stability of SMA mixtures. It had the similar trend with VV. All of the Marshall Stability values were between 7 and 14 kN. Compared with the previous literature, this was larger than with cellulose oil palm fibers. The data from Figure 6 indicated that fiber types and contents had evident correlations between fiber content and Marshall Stability, which increased with fiber content increasing. When the fiber content was less than 0.3%, the SMA mixture with polyester fiber showed better stability. When the fiber content was larger than 0.3%, the SMA mixture with blended fiber showed better stability than that with the other three fibers.

The SMA mixture with the mineral fiber had the lowest Marshall Stability values within the content ranges. The amount of polyester fiber was the smallest at the same Marshall Stability when the fiber content was larger than 0.2%, followed by blended fiber, flocculent lignin fiber, and mineral fiber. Meanwhile, the Marshall Stability of SMA mixtures with polyester fiber and flocculent lignin fiber increased with the fiber content increasing from 0.1% to 0.3%. The Marshall Stability then maintained a slight increase with the fiber content from 0.3% to 0.5%, and 0.4% to 0.6% for the blended fiber and the mineral fiber.

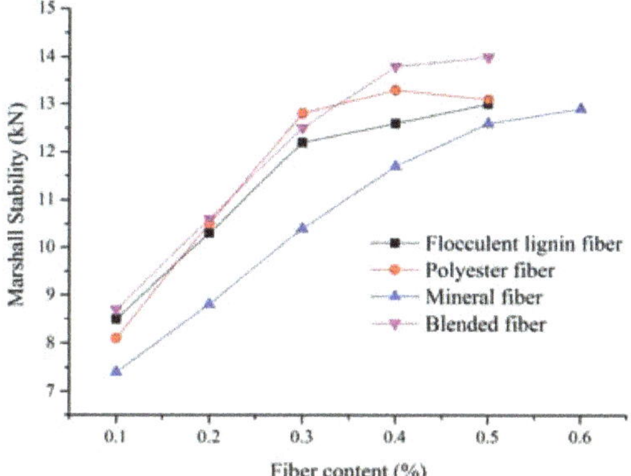

Figure 6. Marshall Stability of the SMA mixtures with four different fibers.

4. Conclusions

HMA and SMA are commonly used pavement materials which are composed of aggregates, fillers, binders, and fibers. These loose materials are mixed together by controlling the volumetric properties to obtain an optimum mechanical performance. Thus, the volumetric properties are the direct indicators in design and construction practice. This paper analyzed the effects of four fibers on the volumetric parameters and Marshall Stability (mechanical performance) of SMA mixtures. As a control fiber, the flocculent lignin fiber was compared with the other three fibers.

This paper found that both the polyester fiber and mineral fiber had an obvious impact on VV, VFA, VMA, and Marshall Stability of the SMA mixture compared to the regular flocculent lignin fiber. The experiment data indicated that increasing the content of polyester fiber and mineral fiber benefitted the mechanical performance of SMA mixtures, which potentially enlarged the fibers' application range. This study suggests that the optimum fiber content is to be larger than 0.3% for flocculent lignin fiber, blended fiber, and polyester fiber, and larger than 0.4% for mineral fiber.

When fiber content was up to 0.5%, the VMA of the sample with the mineral fiber and the polyester fiber was both larger than 16.4%. However, it was more than 17.5% for lignin fiber and blended fiber. Therefore, SMA mixtures with mineral fiber and polyester fiber will be better in heavy traffic sections or hot areas with larger high-temperature stability. They can provide more options in SMA mixture design, which indicates the significant potential for their application. Fiber with higher adsorption, but lower absorption, is a better choice as a potential alternative for lignin fiber in SMA mixture design and construction.

The study proved the adaptability of polyester fiber and mineral fiber in SMA mixture design and provided proper content ranges of the fibers. However, other factors (e.g., cost, availability, and ease of field construction application) should be considered in practice. The fiber type and fiber content should be selected according to field construction experience. The results obtained from the lab tests are a reference to bring convenience to field projects.

In the future, microstructure analysis will be the research emphasis. The dispersion and dissolution of mineral fiber or polyester fiber in the asphalt binder will be characterized with micro-analytical instruments. Then, the chemical reactions between polyester fiber and the asphalt binder can be evaluated. Finally, the high-temperature stability and cracking performance (load-induced, fatigue, top-down, etc.,) of SMA mixtures with mineral fiber and polyester fiber will be evaluated in different areas. Through synthetic consideration of microscopic analysis and field construction, the application range of mineral fiber and polyester fiber will be further expanded in pavement industries and create economic benefits.

Acknowledgments: The project was supported by the National Natural Science Foundation of China (No. 51608046), Qinghai Transportation Science and Technology Project (No. 2014-GX-A2A), the Special Fund for Basic Scientific Research of Central Colleges Chang'an University (No. 310831171009 and 310831171001) and China Postdoctoral Science Foundation (No. 2015T81000 and 2015M582590). Y. Sheng also appreciates the funding from the Chinese Scholarship Council (CSC) of the Ministry of Education, P.R. China.

Author Contributions: Yanping Sheng and Haibin Li conceived and designed the experiments; Ping Guo and Guijuan Zhao performed the experiments; Huaxin Chen and Rui Xiong analyzed the data; Yanping Sheng and Haibin Li wrote the paper.

Conflicts of Interest: The authors declare no conflict of interest.

References

1. Cooley, L.A., Jr.; Hurley, G.C. Potential of using stone matrix asphalt (SMA) in Mississippi. Available online: http://mdot.ms.gov/documents/research/Reports/Interim%20and%20Final%20Reports/Using%20Stone%20Matrix%20Asphalt%20(SMA)%20In%20Mississippi.pdf (accessed on 11 March 2017).
2. Dong, Y.; Tan, Y. Mix design and performance of crumb rubber modified asphalt SMA. *J. Mater. Civil Eng.* **2011**, *10*, 78–86.
3. Xue, Y.J.; Hou, H.B.; Zhu, S.J.; Zha, J. Utilization of municipal solid waste incineration ash in stone mastic asphalt mixture: Pavement performance and environmental impact. *Constr. Build. Mater.* **2009**, *23*, 989–996. [CrossRef]
4. Xie, H.B.; Watson, D. Determining air voids content of compacted stone matrix asphalt mixtures. *Transp. Res. Record: J. Transp. Res. Board* **2014**. [CrossRef]
5. Micheal, L.; Burke, G.; Schwartz, C.W. Performance of stone matrix asphalt pavements in Maryland. *Asph. Paving Technol.: Assoc. Asph. Paving Technol.* **2003**, *72*, 287–314.
6. Christensen, D.W.; Pellinen, T.; Bonaquist, R.F. Hirsch model for estimating the modulus of asphalt concrete. *J. Assoc. Asph. Paving Technol.* **2003**, *72*, 97–121.
7. You, Z.P.; Dai, Q.L. Dynamic complex modulus predictions of hot-mix asphalt using a micromechanical-based finite element model. *Can. J. Civil Eng.* **2007**, *34*, 1519–1528. [CrossRef]
8. Goh, S.W.; You, Z.P.; Williams, R.; Li, X.J. Preliminary dynamic modulus criteria of HMA for field rutting of asphalt pavements: Michigan's experience. *J. Transp. Eng.* **2011**, *137*, 37–45. [CrossRef]
9. Harvey, J.; Eriksen, K.; Sousa, J.; Monismith, C. Effects of laboratory specimen preparation on aggregate–asphalt structure, air-voids contents measurement, and repetitive simple shear test results. *Transp. Res. Board* **1994**, 113–122.
10. Hinrichsen, J.; Heggen, J. Minimum VMA in HMA based on gradation and volumetric properties. In Proceedings of the 75th Transportation Research Board Annual Meeting, Washington, DC, USA, 7–11 January 1996; Volume 1545, pp. 75–79.
11. McLeod, N.W. *Void Requirements for Dense-Graded Bituminous Paving Mixtures*; American Society for Testing and Materials (ASTM): Philadelphia, PA, USA, 1959. [CrossRef]
12. Roberts, F.L.; Kandhal, P.S.; Brown, E.R.; Lee, D.-Y.; Kennedy, T.W. *Hot Mix-Asphalt Materials, Mixture Design, and Construction*, 2nd ed.; NAPA Research and Education Foundation: Lanham, MD, USA, 1991.
13. Hislop, W.; Corree, B. A Laboratory Investigation Into the Effects of Aggregate-Related Factors of Critical VMA in Asphalt Paving Mixtures. Available online: http://publications.iowa.gov/11672/1/tr415.pdf (accessed on 11 March 2017).
14. Sengoz, B.; Topal, A. Minimum voids in mineral aggregate in hot-mix asphalt based on asphalt film thickness. *Build. Environ.* **2007**, *42*, 3629–3632. [CrossRef]
15. Christensen, D.W. Draft final report to the National Cooperative Highway Research Program (NCHRP) on Project NCHRP 9-33: A mix design manual for hot mix asphalt. Advanced Asphalt Technologies: Sterling, VA, USA, 2009; Available online: http://www.trb.org/Main/Blurbs/161248.aspx (accessed on 17 March 2017).
16. Anderson, R.M.; Bahia, H.U. Evaluation and selection of aggregate gradations for asphalt mixtures using Superpave. *Transp. Res. Rec.: Transp. Res. Board* **1997**, *1583*, 91–97. [CrossRef]
17. Mallick, R.B.; Buchanan, M.S.; Kandhal, P.S.; Bradbury, R.L.; McClay, W. Rational approach of specifying the voids in the mineral aggregate for dense-graded hot-mix asphalt. In Proceedings of the 79th Annual Meeting of the Transportation Research Board, Washington, DC, USA, 9 January 2000.

18. Kandhal, P.S.; Foo, K.; Mallick, R. Critical review of voids in mineral aggregate requirements in Superpave. *Transp. Res. Rec.: Transp. Res. Board* **1998**, *1609*, 28–35. [CrossRef]
19. Vavrik, W.R.; Huber, G.; Pine, W.J.; Carpenter, S.H.; Bailey, R. *Bailey Method for Gradation Selection in Hot-Mix-Asphalt Mixture Design*; Transportation Research E-Circular, E-C044; Transportation Research Board: Washington, DC, USA, 2002; Available online: www.trb.org/publications/circulars/ec044.pdf (accessed on 17 March 2017).
20. Asphalt Institute Superpave Mix Design Series No.2 (SP-2). Lexington, Kentucky, 1996. Available online: http://www.asphaltinstitute.org/superpave-documents (accessed on 17 March 2017).
21. Coree, B.J.; Hislop, W.P. A laboratory investigation into the effect of aggregate related factors of critical VMA in asphalt paving mixture. *J. Assoc. Asph. Paving Technologists* **2001**, *70*, 70–131.
22. Hargett, E.R. Effect of size, surface texture, and shape of aggregate particles on the properties of bituminous mixtures. *Highw. Res. Board Special Rep.* **1970**, 25–26.
23. Barksdale, R.D.; Kemp, M.A.; Sheffield, W.J.; Hubbard, J.L. Measurement of aggregate shape, surface area, and roughness. *Transp. Res. Rec.* **1991**, 107–116.
24. Adhikari, S.; You, Z.P. Investigating the sensitivity of aggregate size within sand mastic by modeling the microstructure of an asphalt mixture. *J. Mater. Civil Eng.* **2011**, *23*, 580–586. [CrossRef]
25. Brown, E.R.; Manglorkar, H. *Evaluation of Laboratory Properties of SMA Mixtures*; NCAT Rep. No. 93–5; Auburn Univ.: Auburn, AL, USA, 1993.
26. Hassan, J.H.; Israa, J.Y. The effect of using glass power filler on hot asphalt concrete mixture properties. *J. Eng. Technol.* **2010**, *29*, 44–57.
27. Fu, Z.; Dang, Y.N.; Guo, B.; Huang, Y. Laboratory investigation on the properties of asphalt mixtures modified with double-adding admixtures and sensitivity analysis. *J. Traffic Transp. Eng. (Engl. Ed.)* **2016**, *3*, 412–426. [CrossRef]
28. Putman, B.J.; Amirkhanian, S.N. Utilization of waste fiber in stone matrix asphalt mixtures. *Resour. Conserv. Recycl.* **2004**, *42*, 265–274. [CrossRef]
29. Muniandy, R.; Huat, B.B.K. Laboratory diameteral fatigue performance of stone matrix asphalt with cellulose oil palm fiber. *Am. J. Appl. Sci.* **2006**, *3*, 2005–2010. [CrossRef]
30. Al-Hadidy, A.I.; Tan, Y.Q. Mechanistic analysis of ST and SBS-modified flexible pavements. *Constr. Build. Mater.* **2009**, *23*, 2941–2950. [CrossRef]
31. Chen, H.X.; Xu, Q.W. Experimental study of fibers in stabilizing and reinforcing asphalt binder. *Fuel* **2010**, *89*, 1616–1622. [CrossRef]

Article

Ultrasonic Techniques for Air Void Size Distribution and Property Evaluation in Both Early-Age and Hardened Concrete Samples

Shuaicheng Guo [1], Qingli Dai [1,*], Xiao Sun [1], Ye Sun [2] and Zhen Liu [1]

[1] Department of Civil and Environmental Engineering, Michigan Technological University, Houghton, MI 49931-1295, USA; sguo3@mtu.edu (S.G.); xiaos@mtu.edu (X.S.); zhenl@mtu.edu (Z.L.)

[2] Department of Mechanical Engineering and Engineering Mechanics, Michigan Technological University, 1400 Townsend Drive, Houghton, MI 49931-1295, USA; yes@mtu.edu

* Correspondence: qingdai@mtu.edu; Tel.: +1-906-487-2620

Academic Editor: Faris Ali
Received: 19 January 2017; Accepted: 8 March 2017; Published: 16 March 2017

Abstract: Entrained air voids can improve the freeze-thaw durability of concrete, and also affect its mechanical and transport properties. Therefore, it is important to measure the air void structure and understand its influence on concrete performance for quality control. This paper aims to measure air void structure evolution at both early-age and hardened stages with the ultrasonic technique, and evaluates its influence on concrete properties. Three samples with different air entrainment agent content were specially prepared. The air void structure was determined with optimized inverse analysis by achieving the minimum error between experimental and theoretical attenuation. The early-age sample measurement showed that the air void content with the whole size range slightly decreases with curing time. The air void size distribution of hardened samples (at Day 28) was compared with American Society for Testing and Materials (ASTM) C457 test results. The air void size distribution with different amount of air entrainment agent was also favorably compared. In addition, the transport property, compressive strength, and dynamic modulus of concrete samples were also evaluated. The concrete transport decreased with the curing age, which is in accordance with the air void shrinkage. The correlation between the early-age strength development and hardened dynamic modulus with the ultrasonic parameters was also evaluated. The existence of clustered air voids in the Interfacial Transition Zone (ITZ) area was found to cause severe compressive strength loss. The results indicated that this developed ultrasonic technique has potential in air void size distribution measurement, and demonstrated the influence of air void structure evolution on concrete properties during both early-age and hardened stages.

Keywords: air void structure; ultrasonic scattering measurement; inverse analysis; transport ability; compressive strength; dynamic modulus

1. Introduction

As an important phase existing in the concrete, the air void content can significantly affect the physical properties of concrete, including the compressive strength [1], elastic modulus [2,3], permeability [4], and long-term freeze-thaw durability [5]. During the concrete construction, the air void content needs to be properly designed and controlled for appropriate property. Also, the influence of air void content on the concrete properties development during both early-age and hardened stages needs to be investigated.

The ultrasonic scattering method has been used to analyze hardened concrete air void characteristics by different researchers. For instance, Zhang [6] studied the microstructure evolution of cementitious material with the ultrasonic methods. Based on the ultrasonic monitoring results, the influences of mineral admixture type, steel fiber content, and curing temperature on the microstructure development were discussed. Zhu [7] analyzed the relationships between the ultrasonic wave speed and the existing air voids in cement paste. Particularly, the air voids can obviously decrease the compressional wave speed, while the influence of air void content on the shear wave speed was insignificant. Liu [8] analyzed the early-age concrete microstructure evolution with the measured ultrasonic pulse speed. The influences of the water/cement ratio, mixtures, and aggregates gradation on the microstructure were analyzed. Lai [9] applied the ultrasonic surface wave to detect the honeycomb defects existed in concrete during early stage. The results demonstrated that the honeycomb can cause obvious spectra dispersion of the surface wave. Punurai [10] analyzed the volume content of capillary pores and air voids in hardened cement pastes with ultrasonic measurement. The results were comparable to the traditional petrographic methods based on American Society for Testing and Materials (ASTM) C457 [11]. Sun [12] employed the log-normal distribution for the description of air void size distribution, and obtained the inversed air void size distribution based on the scattering attenuation calculation. Yim [13] quantified the air void size distribution by ultrasonic attenuation calculation with the Roney equation [14]. The three-dimensional characteristics of air voids were determined through the inverse analysis, with the result verified by the ASTM C457 test [11]. Guo [15] compared different distribution functions for the description of hardened concrete air void size distribution. Based on the verification with the ASTM C457 test, the combination of normal distribution for small air voids and log-normal distribution for large air voids performs better. Currently, the application of ultrasonic measurements on the air void content evolution during early stages has not been conducted to the best of the authors' knowledge.

Besides the application on concrete air void structure analysis, the ultrasonic scattering technique has also been employed to investigate concrete properties evolution during both early-age and hardened stages. Boumiz [16] applied ultrasonic measurements to study the properties of early-age cement pastes and mortars. Particularly, the elastic modulus and Poisson's ratio were determined acoustically. Jerome [17] monitored the setting process of mortar samples with the ultrasonic technique. The shear wave velocity was found more appropriate to determine the final setting time than the compression wave velocity. The method was further applied on the concrete samples [18], and the effect of aggregate on the setting process was also analyzed. Subramaniam [19] studied the steel reinforced concrete with the ultrasonic measurement. The wave reflection factor at the steel–concrete interface was calculated and its relationship to strength gain was also analyzed. Qixian [20] predicted the dynamic modulus of concrete with the measured surface wave velocity. The calculated dynamic modulus was in good accordance with the measured results through the resonance method. Trtnik [21] proposed the TG (Trtnik Gregor) parameter for the strength gain prediction, which is the peak value ratio between the high frequency and low frequency range of the transmission signal in the frequency domain. The TG parameter was found to be strongly linearly correlated to the strength gain during the first eight hours. The TG parameter also became stable after the first several curing hours; thus, it cannot be used for the strength gain prediction during the early stages (1–28 days). The ultrasonic pulse velocity (UPV) has also been used to monitor the hardening behavior. The UPV-based models were found to be especially suitable for the early age property prediction [22].

The influence of the air void structure on the concrete properties has already been widely studied [23], but the study on the influence of the air void structure evolution of concrete performance is relatively limited. Wong [24] studied the influence of entrained air voids on the concrete transport ability. It is reported that the entrained air void can decrease or increase the concrete transport ability, depending on saturation level. Gutmann [25] studied the strength loss due to the entrained air bubbles in concrete. The compressive strength was found to decrease, with [26] analyzing the effect of air void clustering on concrete strength loss. It is found that the strength loss was mainly caused by the total air content, rather than the air void clustering; this is contradictory to the findings of Hover [27]. Currently, research on how the transport ability is evaluated during curing age is limited, and the effect of air void clustering on compressive strength is still not clear.

This paper aims to monitor the air void structure evolution and the change of concrete properties during the early-age and hardened stages. Three concrete samples with the same mix design but different air entrainment agent dosages were prepared. The ultrasonic scattering measurements were conducted during early-age and hardened stages to detect the air void structure evolution. The experimental attenuation was obtained based on the wave spectral ratio of the first and the second transmission waves in the frequency domain. The method is following the former study [15] on hardened concrete samples by employing the combination of normal and log-normal distribution. The theoretical attenuation was then calculated with the obstacle scattering theory, by integrating the pre-determined cement paste, aggregate attenuation, and air void distribution effects. The inverse analyses were conducted to determine air void distribution parameters by achieving the minimum error between theoretical and experimental attenuations. The ultrasonic measured air void distribution was then verified with the ASTM C 457 [11] test results at Day 28. The influence of air void structure evolution on concrete properties were also evaluated in this study, including transport ability, compressive strength, and dynamic modulus. The transport ability was evaluated with the electrical resistivity measurement. The experimental measurements on compressive strength and ultrasonic parameters were fitted with linear relations at early stages. The dynamic modulus predicted with ultrasonic parameters was also compared with ASTM C 215 measurement of hardened concrete.

2. Attenuation-Based Ultrasonic Scattering Measurement Method

2.1. Theoretical Attenuation

The obstacle scattering theory built by Ying [28] was applied for the theoretical ultrasonic scattering calculation. This theory is based on a motion equation, which describes the wave function in homogeneous solid materials. Particularly, the wave function can be obtained by solving the motion equation, and represented with the Bessel function. The total energy loss can be obtained by the integration of the scattered wave and represented as the item scattering cross section γ^{Sca}. In this research, both the aggregates and the air voids are considered as spherical obstacles. Then, assuming the obstacles inside concrete do not interact with each other [12], the total theoretical attenuation can be depicted as Equation (1).

$$\alpha_f = (1-\phi)\alpha_{a,f} + \frac{1}{2}\sum_{i=1}^{m1} n_{i1}\gamma_{i1,f}^{Sca} + \frac{1}{2}\sum_{i=1}^{m2} n_{i2}\gamma_{i2,f}^{Sca} \quad (1)$$

where α_f is the total concrete sample attenuation coefficient (Nepper/m); ϕ is the combined volume fraction of the air voids and aggregate (%); $\alpha_{a,f}$ is the attenuation coefficient of the viscoelastic cement paste (Nepper/m); n_{i1} and n_{i2} are the numbers of a certain size aggregate and air void per volume respectively; $\gamma_{i1,f}^{Sca}$ and $\gamma_{i2,f}^{Sca}$ represent the corresponding scattering cross sections of the aggregate and air voids respectively.

2.1.1. Attenuation Generated by Cement Paste and Aggregates

The attenuation generated by the cement paste matrix $\alpha_{a,f}$ can be directly measured with the ultrasonic scattering measurement on the separately prepared cement paste with the same water/cement ratio. The aggregate size distribution $G(a_{i1})$ can be determined with the designed aggregate aggregation. Then, the corresponding amount of a certain size aggregate per unit volume n_i can be deduced as shown in Equation (2). Introducing Equation (2) into Equation (1), the theoretical attenuation generated by the aggregate can be determined as depicted in Equation (3).

$$n_{i1} = \left[\left[G(a_{i1}) - G(a_{i1-1})\right]\phi_1\right] \Big/ \frac{4}{3}\pi a_{i1}^3 \tag{2}$$

$$\alpha_{f,aggregate} = \frac{1}{2}\sum_{i=1}^{m1}\left[\left[G(a_{i1}) - G(a_{i1-1})\right]\phi_{i1}\gamma_{i1,f}^{Sca}\right] \Big/ \frac{4}{3}\pi a_{i1}^3 \tag{3}$$

where ϕ_1 is the volume fraction of the total aggregate (%); a_{i1}, ϕ_{i1} and n_{i1} represent the size, volume fraction and count per volume of a certain type aggregate; $\alpha_{f,aggregate}$ is the theoretical attenuation of the aggregate (Nepper/m).

2.1.2. Attenuation Generated by Air Voids

The normal and log-normal distribution are desirable to characterize the size distribution of small air voids and large air voids respectively [15]. The combination of the normal and log-normal distribution function is depicted in Equation (4). The detailed normal and log-normal distributions are shown as Equations (5) and (6) respectively

$$f = \phi_{21} \times f_{norm}(r_{norm}; \mu_{norm}, \sigma_{norm}) + \phi_{22} \times f_{log-norm}\left(r_{log-norm}; \mu_{log-norm}, \sigma_{log-norm}\right) \tag{4}$$

where f represents the distribution function for the whole air void structure; r_{norm}, μ_{norm}, σ_{norm} are the radius of the air void, mean value, and standard deviation of the norm size distribution, respectively; $r_{log-norm}$, $\mu_{log-norm}$, $\sigma_{log-norm}$ represent the corresponding parameters for the log-normal distribution. The volume fractions of small air voids and large air voids are ϕ_{21} and ϕ_{22}, respectively ($\phi_2 = \phi_{12} + \phi_{22}$).

$$f_{norm}(r_{norm}; \mu_{norm}, \sigma_{norm}) = \left(1/\sqrt{2\pi\sigma_{norm}^2}\right)\exp\left[-(r_{norm} - \mu_{norm})^2/2\sigma_{norm}^2\right] \tag{5}$$

$$f_{log-norm}\left(r_{log-norm}; \mu_{log-norm}, \sigma_{log-norm}\right) = \left(1/\sqrt{2\pi\sigma_{log-norm}^2}\right)\exp[-\left(\ln r_{log-norm} - \mu_{log-norm}\right)^2 \Big/ 2\sigma_{log-norm}^2 \tag{6}$$

where $f_{norm}(r_{norm}; \mu_{norm}, \sigma_{norm})$ and $f_{log-norm}\left(r_{log-norm}; \mu_{log-norm}, \sigma_{log-norm}\right)$ represent the normal distribution and log-normal distribution separately, and r_{norm}, μ_{norm}, σ_{norm} are the radius of the air void, mean value, and standard deviation of the norm size distribution, respectively, while $r_{log-norm}$, $\mu_{log-norm}$, $\sigma_{log-norm}$ represent those parameters for the log-normal distribution.

Then, the corresponding cumulative function for a certain size air void can be represented as $F_2(a_{i2})$ through the integration of distribution function $f_2(a_{i2})$. Similar to the analysis of aggregate attenuation above, the amount of a certain size air void per unit volume n_{i2} can be deduced as shown in Equations (7) and (8). Particularly, the air void size distribution function is divided by $F_2(a_{N2}) - F_2(a_{12})$ for normalization, where a_{N2} and a_{12} represent the largest and smallest air void size respectively.

$$\phi_{i2} = \left[F_2(a_{i2}) - F_2(a_{i2-1})\right]\phi_2 \Big/ \left[F_2(a_{N2}) - F_2(a_{12})\right] \tag{7}$$

$$n_{i2} = \left[F_2(a_{i2}) - F_2(a_{i2-1})\right]\phi_2 \Big/ \left[F_2(a_{N2}) - F_2(a_{12})\right]\frac{4}{3}\pi a_{i2}^2 \tag{8}$$

where ϕ_2 is the volume fractions of the air void ($\phi_1 + \phi_2 = 1 - \phi$).

Introducing Equation (9) into Equation (1), the total theory attenuation can be linked to the air voids' size distribution and aggregate gradation as illustrated in Equation (9).

$$\alpha_{f,airvoid} = \frac{1}{2}\sum_{i=1}^{m2}\left[F_2(a_{i2}) - F_2(a_{i2-1})\right]\phi_{i2}\gamma_{i2,f}^{Sca} \Big/ \left[F_2(a_{N2}) - F_2(a_{12})\right]\frac{4}{3}\pi a_{i2}^3 \tag{9}$$

2.2. Experimental Setup and Experimental Attenuation Measurement

The basic setup for ultrasonic scattering measurement is demonstrated in Figure 1. The two Olympus 5077 transducers (Olympus Co., Center Valley, PA, USA) are located on the opposite sides of the concrete specimen (Prepared by the authors in Benedict Lab, Houghton, MI, USA) as the pulser and receiver separately. The couplant B2 from Olympus (Center Valley, PA, USA) was also used to mitigate the coupling effect between the specimen surface and ultrasonic transducers. The ultrasonic wave centered at 500 kHz frequency is applied in this study, which is appropriate for the detection of the air voids and has been used in the concrete study by different researchers [29,30]. Ultrasonic waves generated by the excitation voltage can transmit through the specimen and keep being reflected at the transducer–specimen interface. The wave signal captured by the receiver transducer is recorded by an oscilloscope (Olympus Co., Center Valley, PA, USA) and displayed on a computer (Dell Co., Round Rock, TX, USA). Particularly, the first two transmission waves $S_1(t)$ and $S_2(t)$, as shown in Figure 1, are used for the experimental attenuation calculation.

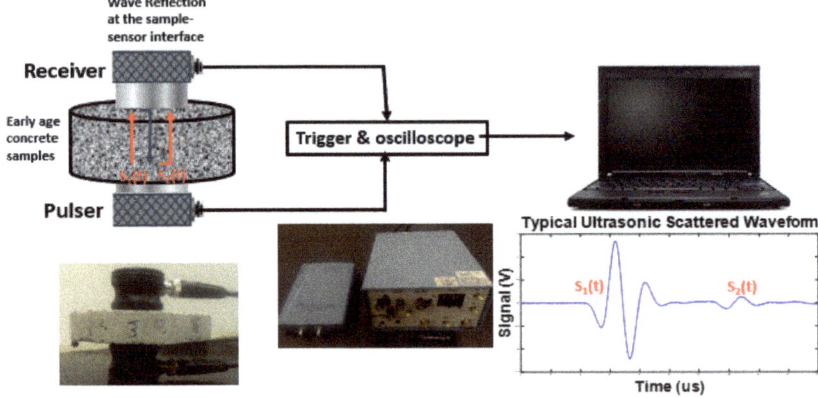

Figure 1. Demonstration of the ultrasonic scattering measurement of air void size distribution in concrete samples.

The first and second transmission waves referred to as S_1 and S_2 were first converted from the time domain to the frequency domain by Fast Fourier Transform on MATLAB (R2015a, MathWorks Co., Natick, MA, USA), as shown in Figure 2. It is clear that the amplitude of S_2 is significantly reduced along the whole frequency range compared to the first transmission wave S_1. The experimental attenuation is estimated from Equation (10) proposed by Sears [31]. The irrelevant attenuation caused by the wave diffraction was eliminated by using the diffraction correlations as shown in Equation (10).

$$\alpha(f) = \frac{1}{2d}\left[\ln\left(\frac{s_1(f)}{s_2(f)}\right) - \ln\left(\frac{D(s;d)}{D(s;3d)}\right)\right] \tag{10}$$

where $\alpha(f)$ is the experimental attenuation of the specimen, which is expressed with wave frequency f; d is the specimen thickness; and $s_1(f)$ and $s_2(f)$ are the first and second transmission waves respectively. $D(s;d)$ and $D(s;3d)$ are proposed by Rogers and Buren [32], which are the simplified closed-form expressions for the diffraction correction.

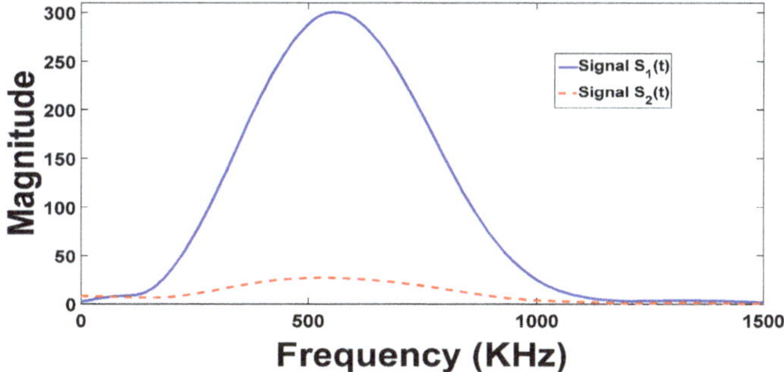

Figure 2. Magnitude spectrum of the cement paste sample ultrasonic results.

Before the investigation of ultrasonic attenuation on the concrete specimen, the attenuation measurement from the ultrasonic test setup was validated with a thin polymethylmethacrylate sample. The sample was placed between the "pulser" and "receiver" transducers, and the couplant was also used to reduce the coupling effect. The first and second transmission waves were recorded and converted to the frequency spectrum with Fourier transformation for attenuation calculation. The attenuation was calculated with Equation (10) and the measurement results are shown in Figure 3. The relationship between ultrasonic attenuation and wave frequency is demonstrated by the fitting curve of $y = 1.725 \times 10^{-5} \times x + 10.958$, where x and y represent the wave frequency (Hz) and wave attenuation (Np/m), respectively. The transmission speed can also be measured based on the arrival time of the first transmission signal and the sample thickness. The measured transmission speed is 2717 m/s; the corresponding attenuation per wavelength can then be calculated as 0.0469 NP. The results are close to the measured values (2782 m/s and 0.045 NP) in reference [10], and the published values (2750 m/s and 0.041 NP) in reference [33], which demonstrate the accuracy of attenuation measurement from this system.

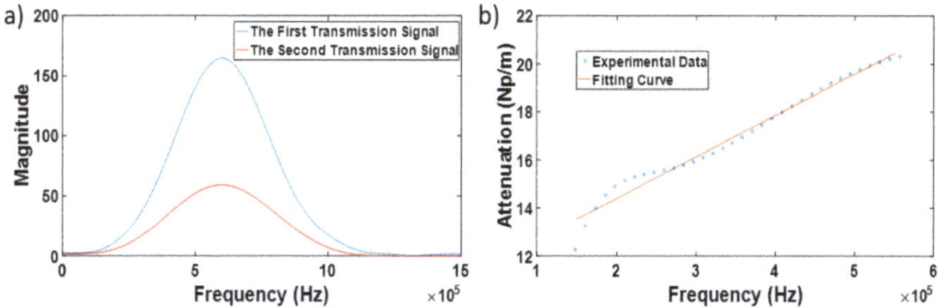

Figure 3. The attenuation measurement results with a polymethylmethacrylate material sample. (**a**) The magnitude spectrum of the first and second transmission wave after Fourier transformation; (**b**) The curve fitting based on the experimentally measured ultrasonic attenuation.

2.3. Inverse Analysis to Determine Air Void Distribution Functions

According to Equation (1), the total theoretical attenuation includes the viscoelastic property of cement paste and the obstacle effect of the aggregate and air voids. The cement paste prepared with the same water/cement ratio was measured with the same method as shown in Figure 1. The corresponding cement paste attenuation was used as the viscoelastic matrix attenuation $\alpha_{a,f}$ as shown in Equation (1). Based on Equation (6), the attenuation generated by the aggregate can be calculated with the designed aggregate gradation. Then, the total theoretical attenuation can be determined and compared with the experimental attenuation estimated from Equation (10). The air voids' size distribution parameters can be obtained by achieving the minimum error between experimental and theoretical attenuations.

3. Ultrasonic Measurement of Air Void Distribution in Early-Age Concrete

3.1. Sample Preparation and Early-Age Property Measurement

Three different types of concrete samples were prepared based on the ASTM C192/C192M-16a [34] during this study. To focus on the influence of the air entrainment agent on the air void structure, the samples were prepared with the same mix design but different air entrainment agent dosages. MasterAir AE 200 type air entrainment agent (BASF Co., Florham Park, NJ, USA) was chosen in this study. The detailed mix design for three types of concrete are shown in Table 1. The volume percentage of size three-quarters (19 mm), one-half (12.5 mm), three-eighths (9.5 mm), No. 4 (4.75 mm), No. 8 (2.36 mm) and No. 16 (1.18 mm) coarse aggregate are 5%, 25%, 25%, 20%, 16% and 9% respectively. Particularly, the fine aggregates belong to the No. 8 (2.36 mm) sieve size. After mixing, the specific gravity of each sample was measured and used for the air void volume fraction calculation based on the ASTM C138 [35]. Then, the fresh concrete was placed into cylinder molds (Humboldt Mfg. Co., Elgin, IL, USA) with 10.16 cm (4 in) diameter and 20.32 cm (8 in) height. After demolding at 24 h, the concrete samples were submerged into water and cured at room temperature for 28 days. Before the tests at each curing age, the samples were first air dried for 12 h to remove the extra water. Besides the ultrasonic scattering measurement, the compressive strength and electrical resistivity tests were also conducted at Day 3, 7 and 28. The compressive strength test was conducted based on the ASTM C39/C39M-16 [36] standard. The bulk electrical resistivity of the concrete sample was measured with the Rcon2 [37] equipment from Giatec Scientific Co. (Ottawa, ON, Canada). Considering the transmission ability of the ultrasonic wave, the concrete cylinder was cut with a diamond saw to obtain slices of around 2.54 cm (1 in) in height for measurement. The selection of thickness partly depended on the penetration ability of the ultrasonic source; the first two signals will not overlap with each other at the thickness of 2.54 cm.

Table 1. Lab concrete mixture proportional design for early-age concrete studies.

Sample	Cement Content (kg/m^3)	Coarse Aggregates (kg/m^3)	Fine Aggregates (kg/m^3)	Water (kg/m^3)	Glass Particle (kg/m^3)	Design w/c	Air entrainment Agent (mL/m^3)
Type 1	334.6 (*sg* * = 3.15)	1110.7 (*sg* = 2.75)	666.4 (*sg* = 2.65)	148.4	74.0 (*sg* = 2.5)	0.45	0
Type 2	334.6 (*sg* = 3.15)	1110.7 (*sg* = 2.75)	666.4 (*sg* = 2.65)	148.4	74.0 (*sg* = 2.5)	0.45	167.5
Type 3	334.6 (*sg* = 3.15)	1110.7 (*sg* = 2.75)	666.4 (*sg* = 2.65)	148.4	74.0 (*sg* = 2.5)	0.45	335.0

* *sg* is short for specific gravity.

3.2. Attenuation Measurement of Cement Paste for Different Ages

As mentioned before, the ultrasonic attenuation of the cement paste first needs to be determined as attenuation for the corresponding viscoelastic matrix in concrete, especially during the early ages. The cement paste samples with the same water/cement ratio (0.45) as concrete samples were separately prepared and no air entrainment agent was added during preparation. Then, the air voids in the separately prepared cement paste are mainly capillary pores due to cement hydration. The attenuation of cement paste is mainly generated from its viscoelastic property, and the contribution from the capillary air void structure is limited. The air void structure of the separately prepared cement paste and that of the cement paste in concrete is different, as the air void structure in concrete can be further affected by the added air entrainment agent, air voids entrapped by the aggregate, and the existence of the Interfacial Transition Zone (ITZ) area. Also, due to the relatively small volume fraction (around

25%) of cement paste inside concrete, the influence of the intrinsic capillary air void content of the cement paste on the concrete air void structure is further limited. The influence of the entrained air voids, entrapped air voids, and the air voids in the ITZ area on the ultrasonic attenuation, are all considered during the measurement in this study.

The cement paste properties are highly impacted by the hydration process. In this study, the ultrasonic scattering measurement was conducted at Day 1, 3, 7, 14, and 28 for both the cement paste and concrete specimen. The calculated cement paste attenuation on the frequency domain based on Equation (10) is shown in Figure 4. A parabolic equation was then used for the curve fitting between the attenuation and corresponding frequency. The fitted equations are depicted as Equations (11)–(14). These fitted equations served as the background attenuation for further inverse analysis of the characteristics of concrete air voids. As the center attenuation used in this study is 500 kHz, the signal is more stable during the range of [150, 550] kHz and this range was applied for the further analysis. It can be observed from Figure 3 that the attenuation of the cement paste matrix decreased with the curing age on the whole frequency domain.

$$\alpha_{a,f} = -222.8110e - 6 \times f^2 + 259.3297e - 3 \times f + 23.8052 \quad (11)$$

$$\alpha_{a,f} = -371.3589e - 6 \times f^2 + 362.8550e - 3 \times f - 10.4897 \quad (12)$$

$$\alpha_{a,f} = -275.0380e - 6 \times f^2 + 280.7362e - 3 \times f - 5.5200 \quad (13)$$

$$\alpha_{a,f} = -129.7629e - 6 \times f^2 + 146.8296e - 3 \times f + 20.7148 \quad (14)$$

where f is the frequency range (KHz) and $\alpha_{a,f}$ is the attenuation (Np/m).

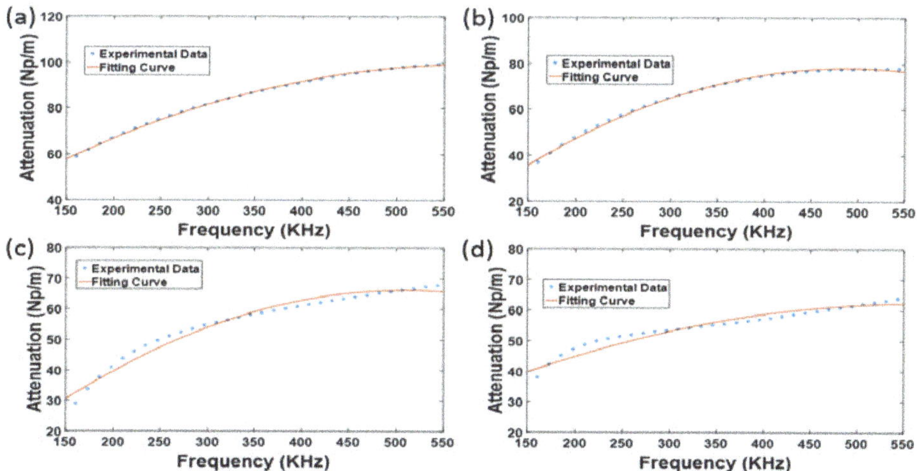

Figure 4. The curve fitting of the early-age cement paste attenuation with frequency. (**a**) Curve fitting at Day 1; (**b**) Curve fitting at Day 3; (**c**) Curve fitting at Day 7; (**d**) Curve fitting at Day 14.

3.3. Determination of Air Void Distribution at Early Stages

The inverse analysis was conducted on MATLAB code by comparing the theoretical attenuation and experimental attenuation estimated based on Equations (1) and (10) respectively. The fitted attenuation equations for cement paste shown in Section 3.2 were applied as the background attenuation $\alpha_{a,f}$. The attenuation generated by the aggregate were calculated with Equation (3), depending on the designed aggregate gradation. Based on the description equation for the air void size distribution shown as Equation (4), the attenuation generated by air voids can also be determined with

Equation (9). Combining all these three parts, the total theoretical attenuation can be determined and compared with the experimental attenuation determined with the methods mentioned in Section 2.2. Three parameters need to be determined for both the small and large air void size distribution: the volume fraction, average size, and the corresponding standard deviation. The air void size distribution parameters can be obtained by achieving the minimum error between experimental and theoretical attenuation.

The comparison between the experimental and theoretical attenuation in the frequency domain for Sample Type 1 was demonstrated in Figure 5, where (a), (b), (c) and (d) depict the comparison at Day 1, 3, 7, and 14 respectively. The corresponding inversed air void size distribution and volume fraction were shown in Figure 6, where (a) and (b) represent the change of the air void volume fraction and evolution of air void size distribution with curing ages. Particularly, the air void volume fraction at Day 0 (fresh stage) was measured by the ASTM C138 test, which is in good agreement with the ultrasonic scattering measurement results.

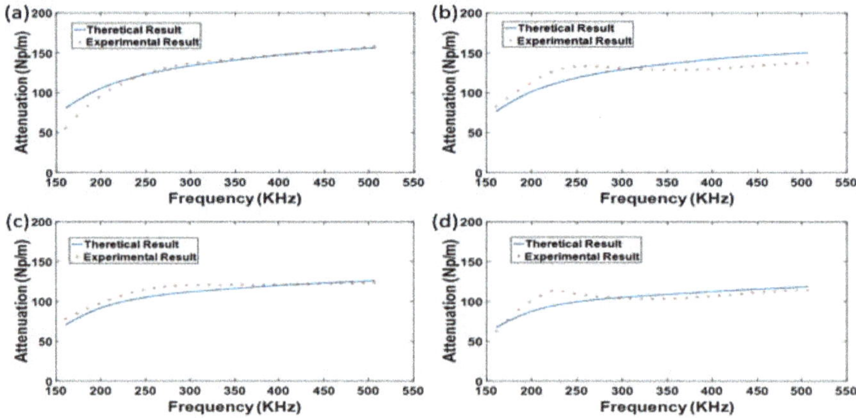

Figure 5. The comparison between theoretical and experimental attenuation during early stages (**a**) The comparison at Day 1; (**b**) The comparison at Day 3; (**c**) The comparison at Day 7; (**d**) The comparison at Day 14.

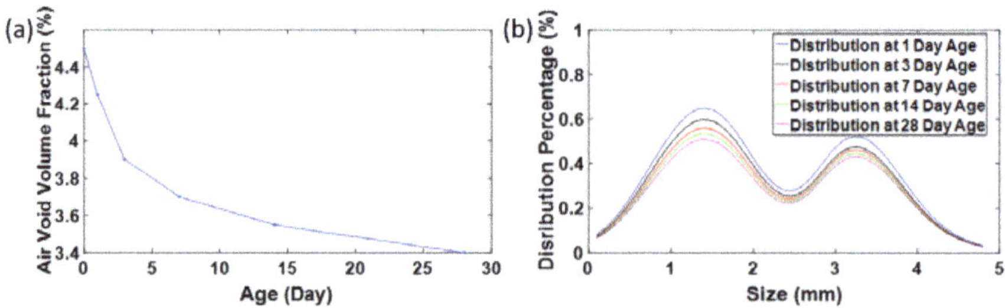

Figure 6. Air voids evolution for Sample Type 1. (**a**) The change of air void content with curing age measured with ultrasonic scattering (at early stages Days 1–28) and the ASTM C138 (Day 0); (**b**) Air void size distribution evolution with curing time.

Similar to the inverse analysis methods used for Sample Type 1, the inverse analysis results were conducted for Sample Types 2 and 3 separately. The changes of air void volume fractions with curing ages are represented in Figures 7a and 8a. Similar to the results shown in Figure 6a, the decreasing trend of the air void volume fraction was observed for all these three samples. Particularly, the air void volume fraction change after Day 14 is much slower compared to the volume change during the first 14 days. The corresponding air void size distribution evolution for Sample Types 2 and 3 are shown in Figures 7b and 8b respectively. The results demonstrate that the volume fraction of both small and large air voids decrease with the curing ages. As mentioned above, the air void volume fraction was observed to decrease with curing age. This phenomenon is in accordance with the results that the air void volume fraction can decrease during hydration age [38], especially on the ITZ area. It is also clear that the decreasing rate of the air void volume fraction declines with the curing age, which is in accordance with the cement hydration speed [39].

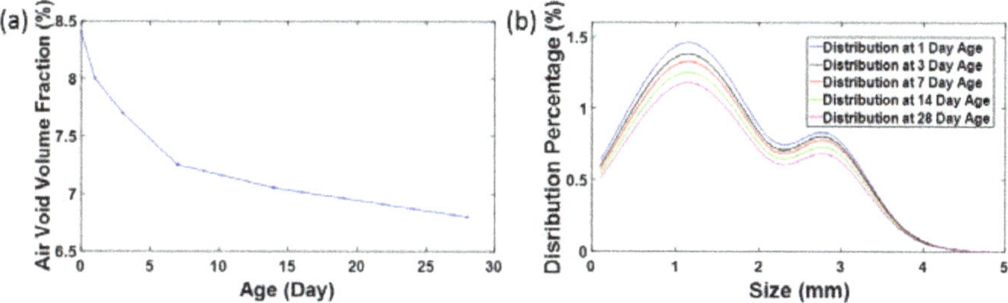

Figure 7. Air voids evolution for Sample Type 2. (**a**) The change of air void content with curing age measured with ultrasonic scattering (at early stages Day 1–28) and the ASTM C138 (Day 0); (**b**) Air void size distribution evolution with curing time.

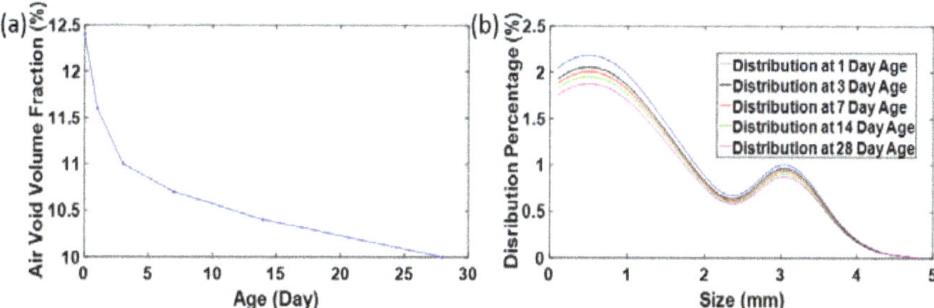

Figure 8. Air voids evolution for Sample Type 3. (**a**) Comparison of the air void content measurement with ultrasonic scattering (at early stages Day 1–28) and the ASTM C138 (Day 0); (**b**) Air void size distribution evolution with curing time.

3.4. The Correlation between Compressive Strength and Utlrasonic Parameter

Besides the analysis of air void volume shrinkage, the compressive strength gain during early age can also be possibly investigated with ultrasonic parameters. The cement hydration process can lead to the increase in concrete stiffness and the decline of viscosity [39]. The rising concrete stiffness means higher transmission speed. The decreasing viscosity can also generate a drop in cement attenuation, as demonstrated in Figure 3. In addition, the shrinkage of the air void volume fraction, as depicted in Figures 6–8, can also lead to higher transmission speed and lower attenuation. Hence, the ultrasonic transmission speed and attenuation are in positive and negative correlations with the compressive strength during the concrete early-age. The ultrasonic transmission speed is determined with the arrival time of Signal $S_1(t)$ and the sample thickness. The calculated ultrasonic transmission speeds are shown in Table 2. Similar to the study [21], this study also applied the peak value ratios for the strength gain prediction. As shown in Figure 2, the peak value of both the first and second transmission waves are located around 500 kHz. Hence, the ultrasonic attenuation value at 500 kHz was chosen to represent the change of attenuation with time, as depicted in Table 2.

Table 2. Properties of concrete samples during the early stages.

Sample	Compressive Strength (MPa)			Ultrasonic Transmission Speed (m/s)			Ultrasonic Transmission Attenuation at 500 KHz (Np/m)			Electrical Resistivity (Ω·m)		
	Day 3	Day 7	Day 28	Day 3	Day 7	Day 28	Day 3	Day 7	Day 28	Day 3	Day 7	Day 28
Type 1	35.8	42.4	48.7	4468.2	4507.8	4612.9	128.1	116.7	105.3	47.3	51.2	57.3
Type 2	29.0	37.7	45.1	4243.9	4311.7	4488.4	134.7	120.8	112.1	49.1	54.6	75.1
Type 3	20.1	25.1	30.0	3788.9	3824.1	4024.2	155.0	141.2	129.4	47.5	87.6	116.5

The compressive strength analysis, applied to the measured results of all three sample types, was applied for the correlation analysis to better demonstrate the relationship between ultrasonic parameters and the strength gain during early stages. The relationship between compressive strength with ultrasonic speed and attenuation is demonstrated in Figure 9a,b respectively. Both of these correlations were evaluated with the linear fitting, and the fitted results are demonstrated as Equations (15) and (16), respectively. The corresponding correlation coefficient between the fitted and measured results are 93.23% and 98.20%, respectively, which demonstrate that the change of ultrasonic attenuation is in stronger linear correlation with the strength gain during early stages.

$$f_c = 0.0291 \times v_p - 88.96 \tag{15}$$

where f_c is the compressive strength (MPa) and v_p is the ultrasonic transmission speed (m/s).

$$f_c = -0.61 \times \alpha_{a,f} + 112.59 \tag{16}$$

where $\alpha_{a,f}$ is the ultrasonic attenuation (Np/m).

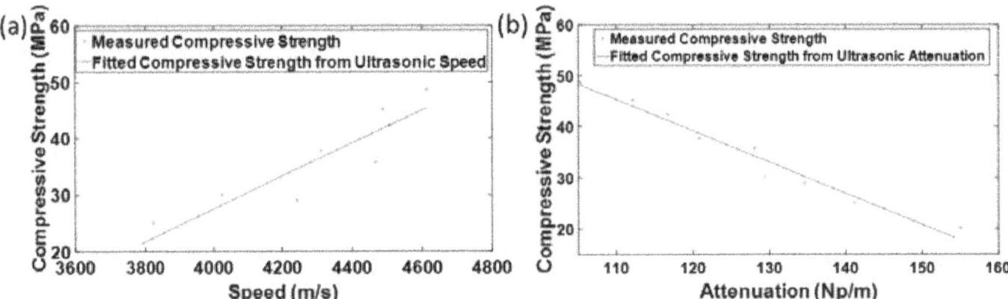

Figure 9. The comparison between the fitted and measured compressive strength. (**a**) The fit between the compressive strength and ultrasonic speed; (**b**) The fit between compressive strength and ultrasonic attenuation.

3.5. The Correlation between Transport Property and Air Void Structure

The back-calculation results demonstrate that the air void content can decrease with the curing age. The shrinkage of air voids can affect the transport ability of concrete. In this study, the transport property was evaluated with the electrical resistivity measurement; the measured results are shown in Table 2. It is clear that the electrical resistivity increases with the curing age for all three sample types. The raising of electrical resistivity can be generated from the decreasing of ion concentration in pore solution and the decline of concrete connectivity. Based on reference [40], the ion concentrations in concrete pore solution become relatively stable after Day 7. Hence the obvious increase in electrical resistivity from Day 7–28 can be mainly generated from the declining connectivity. This result is in accordance with the air void shrinkage during the curing age, as shown in Figures 6–8. In addition, the electrical resistivity can also increase with the air entrainment agent dosage as shown in Table 2. This result demonstrates that the small air voids generated by the added air entrainment agent are mainly isolated in the cement matrix. The small air voids can serve as an insulator [24], and reduce the conductivity of concrete.

4. The Air Void Size Distribution and Dynamic Modulus Analysis of the Hardened Stage

In this study, the estimated air void size distribution based on ultrasonic scattering measurement was compared with the ASTM C457 results for verification at the hardened stage. Further studies on the prediction of the dynamic modulus of hardened concrete were also conducted.

4.1. Compare Measured Air Void Size Distribution with ASTM C457 at Hardened Stages

The obtained air void size distribution and volume fraction from the ultrasonic scattering measurement were further compared with the ASTM C457 test on Day 28. The ASTM C457 tests were conducted with the flatbed scanning methods [41]. The cylinder samples were cut into slabs (7.62 cm (3 in) × 10.16 cm (4 in)) and the surface was carefully polished with grinder. The surfaces were then painted with black marker pen and powdered with calcium silicate. The prepared samples are shown in Figure 10. After processing, the white spots represent air voids existing in concrete specimens. Comparing these three samples, it is clear that the whitened areas increase with the added air entrainment agent dosage. To obtain the size distribution and volume fraction of air voids, the prepared samples were scanned and analyzed with the BubbleCounter [41] embedded in ImageJ software [42].

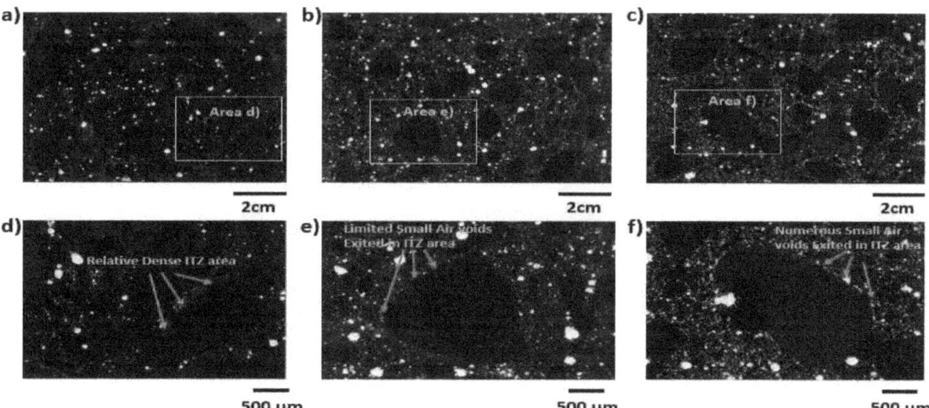

Figure 10. Scanned pattern of concrete slabs from the ASTM (American Society for Testing and Materials) C457 Test. (**a**) The scanned results of Sample Type 1; (**b**) The scanned results of Sample Type 2; (**c**) The scanned results of Sample Type 3d; (**d**) The demonstration of the Interfacial Transition Zone (ITZ) area (Area **d**)) in Sample Type 1; (**e**) The demonstration of the ITZ area (Area **e**)) in Sample Type 2; (**f**) The demonstration of the ITZ area (Area **f**)) in Sample Type 3.

The comparison between the ultrasonic scattering measurement results and the ASTM C457 results is shown in Figure 11a–c, respectively. The corresponding correlation coefficients on the whole size range are 88.86%, 96.88%, and 90.75% for Samples 1–3 separately. The distribution parameters are shown in Table 3. It is clear that the average size and standard deviation for the small air voids decrease and increase respectively with the air entrainment agent content. The air void fraction obtained with ultrasonic scattering techniques is around 3.4%, 6.8%, and 10.0% for Samples 2–4, which are very close to the ASTM C 457 measured values of 3.4%, 6.9%, and 10.1% for tested samples. The size distribution of the small air voids (0.1–1.2 mm) are shown in Figure 12, where a–c represent the non-entrained concrete (Sample Type 1), entrained concrete with regular entrainment dosage (Sample Type 2), and entrained concrete with double entrainment dosage (Sample Type 3), respectively. On the range of small air voids, the correlation coefficient between the ultrasonic scattering measurement results and ASTM C457 measured results are 93.98%, 93.75%, and 71.33%, for Sample Type 1, 2, and 3, respectively. The closeness of both the size distributions curve and volume fraction between these two results demonstrates the feasibility of the ultrasonic scattering measurement method.

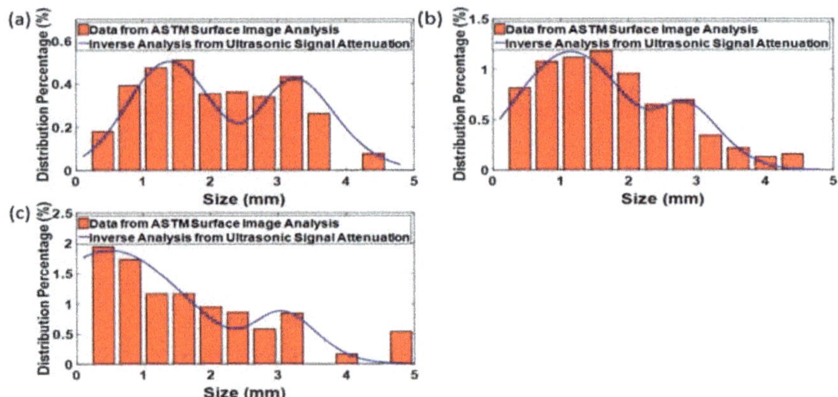

Figure 11. Comparison between the air void size distribution measurement with ultrasonic scattering and ASTM C 457 methods. (**a**) The comparison for Sample Type 1; (**b**) The comparison for Sample Type 2; (**c**) The comparison for Sample Type 3.

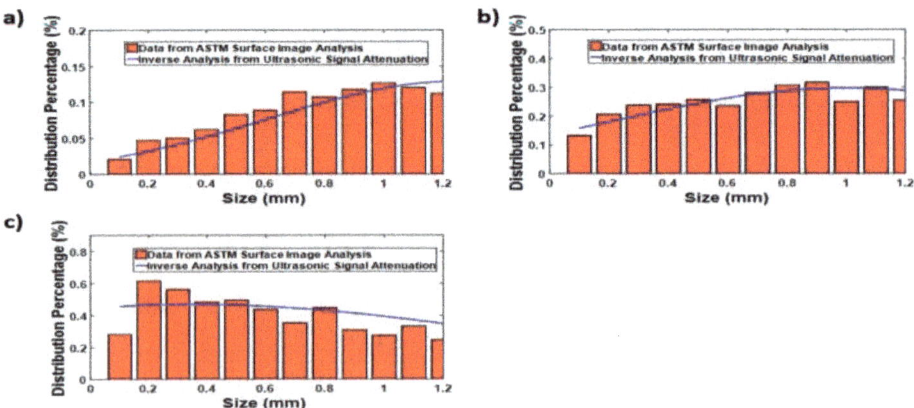

Figure 12. Comparison of the air void size distribution in the small size range. (**a**) The comparison of the experimental result and the theoretical result on the small-size air voids range for Sample Type 1; (**b**) The comparison of the experimental result and the theoretical result on the small-size air voids range for Sample Type 2; (**c**) The comparison of the experimental result and the theoretical result on the small-size air voids range for Sample Type 3.

Table 3. Inverse analysis of parameters for combined distributions.

Sample	Small Air Voids			Large Air Voids			Total air Void Content (%) ϕ
	Volume Fraction (%) ϕ_1	Average Size(mm) μ_{norm}	Standard Deviation σ_{norm}	Volume Fraction (%) ϕ_2	Average Size(mm) $\mu_{log-norm}$	Standard Deviation $\sigma_{log-norm}$	
Sample Type 1	2	1.2	6.3×10^{-4}	1.4	3.2	3×10^{-7}	3.4
Sample Type 2	5.4	0.8	8×10^{-4}	1.4	2.8	2×10^{-7}	6.8
Sample Type 3	8.6	0.3	1.1×10^{-3}	1.4	3.0	2×10^{-7}	10.0

As shown in these figures, the volume fractions of both small air voids and large air voids decrease with the curing age. Particularly, a peak shift from about 1.2 mm to about 0.3 mm can be observed for small air void distribution from Sample Type 1 to Type 3. In contrast, the peaks of the large air void distribution are relatively stable among these three samples. Also, the volume fraction of small air voids increases significantly with the added air entrainment agent, while the volume fraction of the large air voids is almost the same among these three samples. These differences indicate that almost all of the air voids introduced by AE 200 air entrainment agent are smaller than 1.2 mm.

4.2. Analysis of Compressive Strength Loss and Air Void Content

As shown in Table 2, the added air entrainment agent can lead to obvious compressive strength loss. Compared to the Type 1 sample, the compressive strength of Types 2 and 3 samples decrease by 7.39% and 38.40%, respectively. In addition, the air void content of Types 2 and 3 Samples increase by 3.4% and 6.6% separately, as compared to the Type 1 sample. Hence, for every 1% increase in air void content, the compressive strength decreases by 2.17% and 5.82% for Types 2 and 3 samples, respectively. It is reported that the compressive strength loss rate was in the range of [2%, 6%] for a 1% increase in air void content [26]. Though the strength loss rate of Types 2 and 3 samples are both in this range, the loss rate of the Type 3 sample is still much higher than the value of the Type 2 sample. Hence, the correlation between compressive strength and air void content cannot be simply treated as a linear relationship. It is considered that the air void distribution can affect the compressive strength, and not only the air void content.

In particular, the distributions of small air voids along the ITZ area are demonstrated in Figure 10d–f for Type 1, 2, and 3 samples, respectively. It is clear that the ITZ area of the Type 1 sample is relatively dense and small air voids can seldom be detected. Similarly, only limited small air voids can be found in the ITZ area of the Type 2 sample, as shown in Figure 10e. For the Type 3 sample, numerous small air voids can be found along the ITZ area as depicted in Figure 10f. The existence of small air voids in the ITZ area can lead to a weaker bonding strength between the cement paste and aggregate, which can be the main reason for the severe strength loss in sample Type 3. This finding demonstrates that clustered air voids can play an important role for the compressive strength loss.

4.3. Dynamic Modulus Analysis with Ultrasonic Parameters on Hardened Samples

The dynamic Young's modulus of elasticity was usually applied to evaluate the freeze-thaw durability of concrete [43,44], which can be determined with the resonance frequency based on ASTM C215-14 [45]. During the resonant frequency tests, the concrete sample supported by the soft plastic materials was struck with an impactor at the center of one end of the surface, while the resultant signal was captured by an accelerator on the opposite site. The resonant frequency was then determined with the captured signal, after which the dynamic modulus can be calculated. In this study, the dynamic modulus was determined with the longitudinal resonance frequency as shown in Equation (17).

$$E_{Dynamic} = D \times M \times n^2,\ D = 5.093\left(\frac{L}{d^2}\right) \text{ for a cylinder and } 4\left(\frac{L}{bt}\right) \text{ for a prism.} \quad (17)$$

where $E_{Dynamic}$ is the Young's dynamic modulus of elasticity; M is the concrete mass; n is the measured resonant frequency; L is the sample length; d is the cylinder diameter; b and t are the cross-section dimensions of the concrete beam.

The determination of the dynamic modulus with resonance frequency is based on empirical equations as shown in Equation (17). The dynamic modulus can also be calculated with the sound speed through a theoretical equation [20,46] as shown in Equation (18).

$$E_p = \frac{(1+\vartheta)(1-2\vartheta)}{(1-\vartheta)} v_p^2 \rho_{concrete} \tag{18}$$

where E_p is the concrete dynamic Young's modulus; ϑ is the dynamic Poisson's ratio of concrete; v_p is the longitudinal transverse speed of concrete; $\rho_{concrete}$ is the mass density of concrete samples.

The dynamic Poisson's ratio of concrete is in the range of [0.21, 0.28] based on reference [20]. Similarly, the average dynamic Poisson's ratio was determined as 0.24 in reference [47]. Hence, the Poisson's ratio chosen as 0.24 is used in this study. The longitudinal ultrasonic wave speed is determined based on the arrival time of signal $S_1(t)$ and the sample thickness. The mass density is obtained as the sample mass/volume ratio. The longitudinal dynamic modulus can then be determined with Equation (18). The analyzed dynamic modulus based on Equation (18) was then compared with the measured results from ASTM C215 in Table 4. The relative error between the predicted and measured results are all within 5%, which demonstrate the feasibility of this method.

Table 4. Properties of concrete samples during the early stages .

Sample	Mass Density at Hardened Stage (kg/m^3)	Ultrasonic Transmission Speed (m/s)	Predicted Dynamic Modulus (GPa)	Measured Dynamic Modulus (GPa)	Relative Errors (%)
Type 1	2.30	4612.9	41.5	43.3	4.2%
Type 2	2.23	4488.4	39.2	37.5	1.6%
Type 3	2.16	4024.2	31.6	31.7	0.3%

5. Conclusions

This study analyzed the change of the air void structure and concrete property during the early stage. The ultrasonic scattering method was applied to measure the air void size distribution and to analyze the properties of both early-age and hardened concrete. The results were favorably compared with the ASTM C457 test results for validation. Based on the measured results, the change of the air void structure during the early age, and its relationship to concrete transport ability and the compressive strength were also analyzed.

During the early stages, the decreasing trend of the cement paste attenuation can be observed with the curing ages. The air void volume fraction obtained from the ultrasonic scattering measurement was found to decrease in both the small and large air void range with the curing ages. These phenomena are caused by the change in cement paste volume during the continuous hydration process. The results obtained from the ultrasonic scattering measurement were further verified with the ASTM C 457 test results on Day 28. The correlation coefficient between the ultrasonic scattering back-calculated values and the ASTM C457 measured results on the whole size range are 88.86%, 96.88%, and 90.75%, for Sample Types 1–3, respectively. The relatively high correlation coefficients further indicate the feasibility of the ultrasonic scattering measurement methods.

The electrical resistivity of concrete samples was found to increase with the curing age, which is in accordance with the air void shrinkage during the hydration process. In addition, the correlation between ultrasonic parameters and compressive strength was evaluated. The dynamic modulus was analyzed with the ultrasonic transmission speed, and the results are in good agreement (within 5% error) with the resonance-measured results. The relationship between the air void content and strength loss was also analyzed. The clustered air voids along the ITZ area can cause more severe compressive strength loss.

This paper provides an ultrasonic scattering method to measure the air void size distribution and volume fraction during both early-age and hardened stages. The correlation between concrete properties and ultrasonic parameters was also analyzed. Thus, the ultrasonic techniques can be further developed to measure concrete air void evolution and property development during early stages for concrete construction inspection.

Acknowledgments: The authors would like to acknowledge the Michigan Tech Transportation Institute Initiative Fund program for partial financial support. The first author acknowledges the financial support from China Scholarship Council under No. 201406370141. The authors would like to thank Gerald Anzalone at Michigan Technological University for assistance in the ASTM C457 air void measurement.

Author Contributions: Qingli Dai and Shuaicheng Guo conceived and designed the experiments; Shuaicheng Guo and Xiao Sun performed the experiments and analyzed the data; Ye Sun, Zhen Liu and Shuaicheng Guo contributed the analysis tools; Shuaicheng Guo and Qingli Dai wrote the paper.

Conflicts of Interest: The authors declare no conflict of interest.

References

1. Sturrup, V.; Vecchio, F.; Caratin, H. Pulse Velocity as a Measure of Concrete Compressive Strength. *Spec. Publ.* **1984**, *82*, 201–228.
2. Du, L.; Folliard, K.J. Mechanisms of air entrainment in concrete. *Cem. Concr. Res.* **2005**, *35*, 1463–1471. [CrossRef]
3. Hansen, T.C. Influence of aggregate and voids on modulus of elasticity of concrete, cement mortar, and cement paste. *Am. Concr. Inst. J. Proc.* **1965**, *62*, 193–216.
4. Kearsley, E.; Wainwright, P. Porosity and permeability of foamed concrete. *Cem. Concr. Res.* **2001**, *31*, 805–812. [CrossRef]
5. Cordon, W.A. *Freezing and Thawing of Concrete-Mechanisms and Control*; American Concrete Institute: Farmington Hills, MI, USA, 1966.
6. Zhang, Y.; Zhang, W.; She, W.; Ma, L.; Zhu, W. Ultrasound monitoring of setting and hardening process of ultra-high performance cementitious materials. *NDT E Int.* **2012**, *47*, 177–184.
7. Zhu, J.; Kee, S.H.; Han, D.; Tsai, Y.T. Effects of air voids on ultrasonic wave propagation in early age cement pastes. *Cem. Concr. Res.* **2011**, *41*, 872–881. [CrossRef]
8. Liu, Z.; Zhang, Y.; Jiang, Q.; Sun, G.; Zhang, W. In situ continuously monitoring the early age microstructure evolution of cementitious materials using ultrasonic measurement. *Constr. Build. Mater.* **2011**, *25*, 3998–4005. [CrossRef]
9. Lai, W.; Wang, Y.H.; Kou, S.C.; Poon, C.S. Dispersion of ultrasonic guided surface wave by honeycomb in early-aged concrete. *NDT E Int.* **2013**, *57*, 7–16.
10. Punurai, W.; Jarzynski, J.; Qu, J.; Kurtis, K.E.; Jacobs, L.J. Characterization of entrained air voids in cement paste with scattered ultrasound. *NDT E Int.* **2006**, *39*, 514–524.
11. ASTM C457/C457M-12. *Standard Test Method for Microscopical Determination of Parameters of the Air-Void System in Hardened Concrete*; ASTM International: West Conshohocken, PA, USA, 2012.
12. Sun, Y.; Yu, X.B.; Liu, Z.; Liu, Y.; Tao, J. Advanced ultrasonic technology for freezing damage prevention of concrete pavement. *Int. J. Pavement Res. Technol.* **2013**, *6*, 86–92.
13. Yim, H.J.; Kim, J.H.; Bang, Y.L.; Kwak, H.G. Air voids size distribution determined by ultrasonic attenuation. *Constr. Build. Mater.* **2013**, *47*, 502–510. [CrossRef]
14. Roney, R.K. *The Influence of Metal Grain Structure on the Attenuation of an Ultrasonic Acoustic Wave*; California Institute of Technology: Pasadena, CA, USA, 1950.
15. Guo, S.; Dai, Q.; Sun, X.; Sun, Y. Ultrasonic scattering measurement of air void size distribution in hardened concrete samples. *Constr. Build. Mater* **2016**, *113*, 415–422. [CrossRef]
16. Boumiz, A.; Vernet, C.; Tenoudji, F.C. Mechanical properties of cement pastes and mortars at early ages: Evolution with time and degree of hydration. *Adv. Cem. Based Mater.* **1996**, *3*, 94–106. [CrossRef]
17. Carette, J.; Staquet, S. Monitoring the setting process of mortars by ultrasonic P and S-wave transmission velocity measurement. *Constr. Build. Mater* **2015**, *94*, 196–208. [CrossRef]
18. Carette, J.; Staquet, S. Monitoring the setting process of eco-binders by ultrasonic P-wave and S-wave transmission velocity measurement: Mortar vs. concrete. *Constr. Build. Mater* **2016**, *110*, 32–41. [CrossRef]

19. Subramaniam, K.V.; Shaw, C.K.; Subramaniam, K.V. Ultrasonic technique for monitoring concrete strength gain at early age. *Mater. J.* **2002**, *99*, 458–462.
20. Qixian, L.; Bungey, J. Using compression wave ultrasonic transducers to measure the velocity of surface waves and hence determine dynamic modulus of elasticity for concrete. *Constr. Build. Mater* **1996**, *10*, 237–242. [CrossRef]
21. Trtnik, G.; Gams, M. Ultrasonic assessment of initial compressive strength gain of cement based materials. *Cem. Concr. Res.* **2015**, *67*, 148–155. [CrossRef]
22. Carette, J.; Staquet, S. Monitoring and modelling the early age and hardening behaviour of eco-concrete through continuous non-destructive measurements: Part II. Mechanical behaviour. *Cem. Concr. Compos.* **2016**, *73*, 1–9. [CrossRef]
23. Neville, A.M. *Properties of Concrete*; Longman Group UK Limited: Harlow, UK, 1995.
24. Wong, H.; Pappas, A.M.; Zimmerman, R.W.; Buenfeld, N.R. Effect of entrained air voids on the microstructure and mass transport properties of concrete. *Cem. Concr. Res.* **2011**, *41*, 1067–1077. [CrossRef]
25. Gutmann, P.F. Bubble characteristics as they pertain to compressive strength and freeze-thaw durability. *MRS Proc.* **1987**, *114*, 128–130.
26. Vosahlik, J.; Riding, K.; Esmaeily, A.; Billinger, A.; Mcleod, H. Effects of Air Void Clustering on Concrete Compressive Strength. *ACI Mater. J.* **2016**, *113*, 759–767. [CrossRef]
27. Hover, K.C. Some recent problems with air-entrained concrete. *Cement. Concr. Aggreg.* **1989**, *11*, 67–72.
28. Ying, C.; Truell, R. Scattering of a plane longitudinal wave by a spherical obstacle in an isotropically elastic solid. *J. Appl. Phys.* **1956**, *27*, 1086–1097. [CrossRef]
29. Demčenko, A.; Visser, H.; Akkerman, R. Ultrasonic measurements of undamaged concrete layer thickness in a deteriorated concrete structure. *NDT E Int.* **2016**, *77*, 63–72.
30. Larose, E.; Hall, S. Monitoring stress related velocity variation in concrete with a 2×10^{-5} relative resolution using diffuse ultrasound. *J. Acoust. Soc. Am.* **2009**, *125*, 1853–1856. [CrossRef] [PubMed]
31. Sears, F.M.; Bonner, B.P. Ultrasonic attenuation measurement by spectral ratios utilizing signal processing techniques. *IEEE Trans. Geosci. Remote Sens.* **1981**, *2*, 95–99. [CrossRef]
32. Rogers, P.H.; van Buren, A.L. An exact expression for the Lommel-diffraction correction integral. *J. Acoust. Soc. Am.* **1974**, *55*, 724–728. [CrossRef]
33. Cheeke, J.D.N. *Fundamentals and Applications of Ultrasonic Waves*; CRC Press: New York, NY, USA, 2012.
34. ASTM C192/C192M-16a. *Standard Practice for Making and Curing Concrete Test Specimens in the Laboratory*; ASTM International: West Conshohocken, PA, USA, 2016.
35. ASTM C138/C138M-16a. *Standard Test Method for Density (Unit Weight), Yield, and Air Content (Gravimetric) of Concrete*; ASTM International: West Conshohocken, PA, USA, 2016.
36. ASTM C39/C39M-16. *Standard Test Method for Compressive Strength of Cylindrical Concrete Specimens*; ASTM International: West Conshohocken, PA, USA, 2016.
37. Giatec Scientific Inc. *RCON2 User Manual*; Giatec Scientific Inc.: Ottawa, ON, Canada, 2016.
38. Scrivener, K.L.; Crumbie, A.K.; Laugesen, P. The interfacial transition zone (ITZ) between cement paste and aggregate in concrete. *Interface Sci.* **2004**, *12*, 411–421. [CrossRef]
39. Mehta, P.K. *Concrete. Structure, Properties and Materials*; Prentice-Hall: Englewood Cliffs, NJ, USA, 1986.
40. Vollpracht, A.; Lothenbach, B.; Snellings, R.; Haufe, J. The pore solution of blended cements: A review. *Mater. Struct.* **2016**, *49*, 3341–3367. [CrossRef]
41. Peterson, K.K.W.; Anzalone, G.C.; Nezami, S.; Oh, C.Y.S.; Lu, H. Robust Test of the Flatbed Scanner for Air-Void Characterization in Hardened Concrete. *Evaluation* **2016**, *44*, 1–16. [CrossRef]
42. Schneider, C.A.; Rasband, W.S.; Eliceiri, K.W. NIH Image to ImageJ: 25 years of image analysis. *Nat. Methods* **2012**, *9*, 671–675. [CrossRef] [PubMed]
43. Martins, L.; Vasconcelos, G.; Lourenço, P.B.; Palha, C. Influence of the Freeze-Thaw Cycles on the Physical and Mechanical Properties of Granites. *J. Mater. Civ. Eng.* **2015**, *28*, 04015201. [CrossRef]
44. ASTM C666/C666M-15. *Standard Test Method for Resistance of Concrete to Rapid Freezing and Thawing*; ASTM International: West Conshohocken, PA, USA, 2015.
45. ASTM C215-14. *Standard Test Method for Fundamental Transverse, Longitudinal, and Torsional Resonant Frequencies of Concrete Specimens*; ASTM International: West Conshohocken, PA, USA, 2014.

46. Kinsler, L.E.; Frey, A.R.; Coppens, A.B.; Sanders, J.V. Fundamentals of acoustics. In *Fundamentals of Acoustics*, 4th ed.; Kinsler, L.E., Frey, A.R., Coppens, A.B., James, V., Sanders, J.V., Eds.; Wiley-VCH: Weinheim, Germany, 1999; p. 560.
47. Teller, L. Elastic properties. *Signif. Tests Prop. Concr. Concr. Aggreg.* **1956**, *169*, 94.

Article

Thermal and Fatigue Evaluation of Asphalt Mixtures Containing RAP Treated with a Bio-Agent

Karol J. Kowalski [1,*], Jan B. Król [1], Wojciech Bańkowski [2], Piotr Radziszewski [1] and Michał Sarnowski [1]

1. Faculty of Civil Engineering, Warsaw University of Technology, Warsaw 00-637, Poland; j.krol@il.pw.edu.pl (J.B.K.); p.radziszewski@il.pw.edu.pl (P.R.); m.sarnowski@il.pw.edu.pl (M.S.)
2. The Road and Bridge Research Institute, Warsaw 03-302, Poland; wbankowski@ibdim.edu.pl

* Correspondence: k.kowalski@il.pw.edu.pl; Tel.: +48-22-234-5674

Academic Editors: Zhanping You, Qingli (Barbara) Dai and Feipeng Xiao
Received: 31 December 2016; Accepted: 20 February 2017; Published: 23 February 2017

Abstract: Environment conservation and diminishing natural resources caused an increase in popularity of the application of renewable bio-origin resources for the construction of road pavement. Currently, there are known additions of bio-origin materials for bitumen modification. Such material is also used as a flux additive for bitumen or as a rejuvenator once working with reclaimed asphalt pavement (RAP). This paper presents research dealing with asphalt mixtures with RAP modified with a bio-agent of rapeseed origin. The main idea of the conducted research was to apply more RAP content directly to the batch mix plant without extra RAP heating. The RAP used in this study was milled from a base asphalt layer; the addition of RAP stiffens new asphalt mixtures. A bio-agent, due to its fluxing action, was used to support the asphalt mixing process and to decrease the over-stiffening of the mixture caused by RAP addition. This research includes bitumen and mixture tests. For the bitumen study, three different bitumens (35/50, 50/70, and 70/100) were tested in a dynamic shear rheometer (DSR) for complex modulus G* and for phase angle $|\delta|$ in the temperature range 0–100 °C. The reference mixture and mixtures with 2.5% bio-agent were tested to assess the influence of RAP and the bio-agent addition on the asphalt mixture properties. Low temperature behavior (TSRST), stiffness, and fatigue resistance (4PB) were tested. Based on the bitumen test, it was determined that even a low rate of bio-agent (2.5%) beneficially changes bitumen properties at a low temperature; moreover, polymerization processes occurring in the second stage of the process improves bitumen properties at a high operational temperature. The research with these asphalt mixtures demonstrates that the bio-origin flux acts as a rejuvenator and allows for an application of 30% cold RAP. Thermal cracking resistance of the mixture with RAP and 2.5% bio-agent improved. The bio-agent removes unfavorable stiffening of RAP and increases the fatigue resistance of the asphalt mixture.

Keywords: bioadditive; reclaim asphalt pavement (RAP); warm mix asphalt (WMA); sustainable roads

---

1. Introduction

Durable road pavement is a necessary element for economic development in most countries. Recently, the application of renewable bio-origin resources for the construction of road pavement has become more popular as a result of both environment conservation and diminishing natural resources. Currently, there are known additions of bio-binders or bio-origin polymers for bitumen modification. Bio-derived oil is also used as a flux additive for bitumen or as a rejuvenator for reclaimed asphalt pavement (RAP). A growing demand for paving materials requires the application of recycling materials. Material from milled asphalt pavement is used for the construction of new pavement, in hot, warm, or cold technology. The recycling process can be accomplished on-site, in a remote location, or in an asphalt plant. The main reason for recycling milled asphalt mixtures is the decreased need for new material (aggregate and bitumen) in the production of new asphalt mixtures. It should be highlighted that reused bitumen serves as a fully functional, albeit aged, bitumen [1–7].

The bio-derived materials are produced using various raw materials and are combined with a bituminous binder in a wide range of proportions [8–10]. As studied earlier [11], such materials may also change the binder and asphalt mixture stiffness, allowing for rejuvenating action. Advanced rheological methods can be accomplished to study rejuvenating effects [12].

Gawel et al. [13] proposed solutions to the oxidation conditions for rapeseed and linseed oils and the corresponding methyl esters in order to obtain environmentally friendly bitumen fluxes. A new generation of bituminous binders fluxed with rapeseed oil methyl esters with siccative exhibit lower consistency during the asphalt mixture production process and rebuild consistency during pavement operation, as a result of the polymerization reaction [14].

A critical issue related to the application of the bio-derived additive fluxing bitumen is the final viscosity of the bitumen and stiffness of the mixture placed in the pavement. While fluxing is appreciated during mixture production, placement, and compaction, it is no longer desired once road is open to traffic [15,16].

This paper presents research on an asphalt mixture with RAP, modified with a bio-agent of rapeseed origin. The main idea of the conducted research was to apply more RAP content directly to the batch mix plant without extra RAP heating. The RAP used in this study was milled from a base asphalt layer; the addition of RAP stiffens the new asphalt mixture. A bio-agent, due to its fluxing action, was used to support the asphalt mixing process and to decrease over-stiffening of the mixture caused by RAP addition. During the study, a reference mixture and a mixture with 2.5% addition (based on bitumen weight) of bio-agent were tested. In order to assess the influence of RAP and the bio-agent on the asphalt mixture properties, low temperature behavior (Thermal Stress Restrained Specimen Test (TSRST)), stiffness, and fatigue resistance (four-point bending (4PB)) were studied.

The main objective of this research was to apply cold RAP during asphalt production without compromising the functional properties of the mixture. The research demonstrates that the bio-origin flux acts as a rejuvenator and allows for the application of 30% cold RAP. Thermal cracking resistance of the mixture with RAP and 2.5% bio-agent improved. The bio-agent removes unfavorable stiffening of the asphalt mixture with RAP and increases its fatigue resistance.

2. Materials and Methods

Road bitumen 35/50 (according to the European Specification EN) refined from Ural crude oil was used for the research. As a modifier, a bio-agent of fatty acid methyl esters (FAMEs) based on rapeseed oil (RME—rapeseed methyl esters) with a cobalt catalyst in the amount of 0.1% m/m as calculated for metal and with a reaction initiator of cumene hydroperoxide in the amount of 1% m/m was used. Details of the production process, composition, and chemical reaction connected to bio-agent processing are described elsewhere [17]. The bio-modifier was produced by an application of an oxidation reaction promoter to the RME and then by allowing this mixture to be oxidized at a temperature of 20 ± 5 °C.

The bio-agent used in this study is a bi-functional material. During the first stage, after being added to the bitumen, it fluidized bitumen by lowering its viscosity. During the second stage, due to the presence of double unsaturated bonds and cobalt catalyzer in the RME, a slow polymerization process started in the bio-agent. Polymerization occurred in the presence of oxygen and partially recovered initial bitumen viscosity.

The research presented in this paper was conducted observing the bitumen specimens and the asphalt mixture samples.

The following samples were subjected to bitumen tests: clean unmodified bitumen 35/50 and bitumen modified with bio-modifier at the amounts of 1.25%, 2.5%, 3.75% and 5.0% of addition, in terms of bitumen weight. In addition, as reference points, unmodified bitumens 50/70 and 70/100 were also tested. Road bitumen 35/50 was mixed with a bio-modifier for one minute at a temperature of 150 °C. FAME demonstrated full blendability to the bitumen binder. In order to determine the characteristics of the polymerization process in time, bitumens modified with the bio-agent were conditions in the thin layer of 1 mm for up to 56 days at room temperature. The tests were conducted directly after modification (Day 0) and after 7, 14, 28, and 56 days of conditioning. Bitumens were tested in a dynamic shear rheometer (DSR) for complex modulus G* and for phase angle $|\delta|$. Tests were accomplished with a constant frequency of 10 rad/s in the temperature range between 0 °C and 100 °C, using two parallel plates with a 25 mm diameter and a 1 mm gap.

For asphalt mixture tests, asphalt concrete (AC) with a 16 mm maximum aggregate sieve size was used. Two types of asphalt mixtures were tested: a reference AC with 30% RAP with 35/50 bitumen and AC with 30% RAP, 35/50 bitumen, and 2.5% bio-agent (bio-agent content was determined based on the bitumen testing as explained later in this paper). Figure 1 presents an aggregate blend gradation curve, while basic properties of asphalt concrete are shown in Table 1. For air void determination, asphalt mixtures were compacted in a Marshall compactor with 2 × 75 blows for each sample, according to EN 12697-30 Bituminous mixtures. Test methods for hot mix asphalt. Specimen preparation by impact compactor. Compaction level was determined based on the air void determination according to EN 12697-8 Bituminous mixtures. Test methods for hot mix asphalt. Determination of void characteristics of bituminous specimens. During compaction, all samples demonstrated a temperature similar to 135 ± 5 °C. The reference mixture exhibited air void content of 4.6%. The bio-agent application resulted in better mixture compaction (3.2% air voids). The asphalt mixture with bio-agent was not optimized for air void content due to the Polish requirements for asphalt mix design. According to requirements WT-2 [18], the mixture was designed for the required air void range. Once air void content is within the range (3%–6%), further optimization is not required. In addition, in this research, there was an attempt to determine the influence of bio-agent on the compactibility of asphalt mixture and those on its properties.

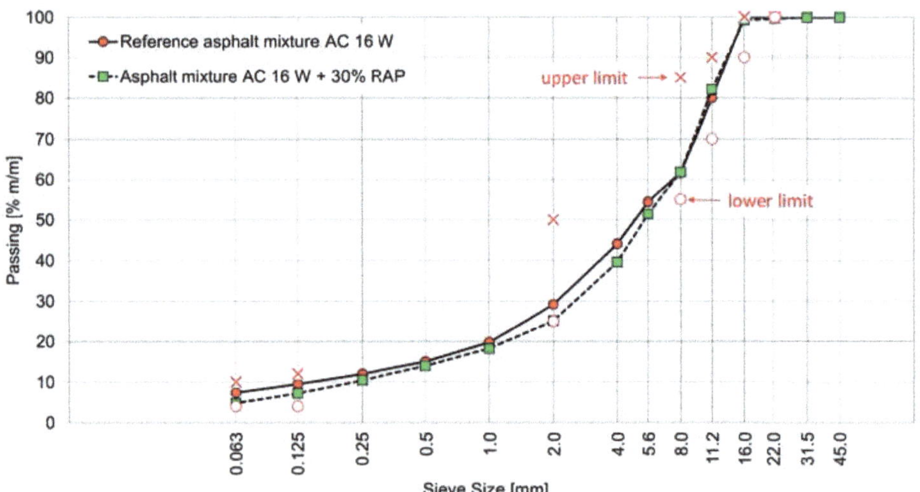

Figure 1. Gradation of asphalt mixture AC 16 containing 30% reclaimed asphalt pavement (RAP); the presented limit points are according to Polish specifications WT-2: 2014.

Table 1. Physical properties of asphalt mixtures.

Properties	Units	AC 16 + 30% RAP 35/50	AC 16 + 30% RAP 35/50 + 2.5% Bio-Agent
Lime filler content	%	2	2
Fine aggregate content	%	15	15
Coarse aggregate content	%	53	53
Aggregate from RAP	%	30	30
Bio-agent	%	0	2.5
Whole binder content	%	4.6	4.6
Virgin binder	%	3.1	3.1
Binder from RAP	%	1.5	1.5
Binder replacement factor	%	32	32
Air voids (75 blows)	%	4.6	3.2

Asphalt mixtures were tested for 4PB stiffness and fatigue durability (resistance) as well as for low-temperature behavior. During the specimen preparation process, plate samples were compacted in the roller compactor according to EN 12697-33 Bituminous mixtures. Test methods for hot mix asphalt. Specimen prepared by roller compactor, reaching 98%–100% of the compaction index.

Stiffness and fatigue tests were conducted in the machine equipped with the 4PB bending module. Stiffness modulus $|E^*|$ was determined according to App. B of specification EN 12697-26 Bituminous mixtures. Test methods for hot mix asphalt. Stiffness. The test consisted of cyclic bending of beam with constant strain amplitude. While testing, load, beam deflection, phase angle, and cycle number are registered and stiffness modulus $|E^*|$ is calculated. The following test conditions were observed: a temperature of 10 °C, an amplitude of 1, 2, 5, 8, 10, and 15 Hz, and a strain of 50 mm/mm. A fatigue test was conducted in the same apparatus according to EN 12697-24 Bituminous mixtures. Test methods for hot mix asphalt. Resistance to fatigue. Tests were conducted with controlled strains, at temperature of 10 °C, and a 10 Hz amplitude. Six samples with three levels of deflection were tested. Based on the fatigue test results obtained for various strain levels, material fatigue characteristics were determined using linear regression, for which correlation coefficient r was calculated. The correlation r between fatigue and macrostrain was tested for statistical significance. The following was assumed: hypothesis H0: $p = 0$, meaning that a correlation does not exist and hypothesis H1: $p \neq 0$, meaning that a correlation does exist. Once the correlation coefficient r was calculated, values of a Student *t*-test were determined. Then, the probability of a critical value transgression by variable t was calculated and compared with statistical significance at the level of $\alpha = 0.05$. Once $p < \alpha$, the correlation is significant; for $p > \alpha$, the hypothesis (correlation between tested values) is not rejected.

To determine the low temperature cracking susceptibility of asphalt mixes, the TSRST was accomplished according to EN 12697-46:2012 Bituminous mixtures. Test methods for hot mix asphalt. Low temperature cracking and properties by uniaxial tension tests. Tests were conducted with the MTS test setup. The tested specimens were rectangular with dimensions of 50 × 50 × 250 mm. For each mixture, four rectangular specimens were prepared. The initial test temperature was 10 °C. This temperature decreased while testing with a rate of 10 °C/h. A special frame was used in order to position specimens in such way that any strain of the specimen is prohibited. In such a condition, inside tested sample thermal tensile stresses are induced. During the testing temperature, force and strain are registered. The end of the test is determined once the sample breaks. A test result is determined as a stress prior to sample breaking (cryogenic stress) and corresponding temperature (failure temperature).

3. Results

3.1. Binder Properties

The complex modulus $|G^*|$ and phase angle $|\delta|$ are good measures of the viscoelastic properties of bituminous binders. Once the temperature increases, the complex modulus decreases and the phase angle increases towards the viscous stage. A similar effect can be seen once the fluxing agent is added to the road bitumen, as a viscous part of the bitumen complex modulus increases. The bio-agent obtained from RME, as compared to regular oils, presents lower viscosity and good fluxing action in bitumen. Figure 2 shows complex modulus curves with respect to temperature for reference binders and for binders with various amounts of bio-agent. Based on the curve location, the desired bio-agent content was determined for further tests on asphalt mixtures. It can be seen that, in all tested specimens in the temperature range, an unmodified binder shows the highest stiffness. Together with the bio-agent content increment (ranging from 1.25% to 5.0%), stiffness of the fluxed bitumen decreases (refer to Figure 2). Comparison between bitumen modified with various amounts of bio-agent with reference 50/70 and 70/100 bitumens allows us to conclude that bitumen 35/50 with 2.5% bio-agent has similar viscoelastic properties to road bitumen 50/70.

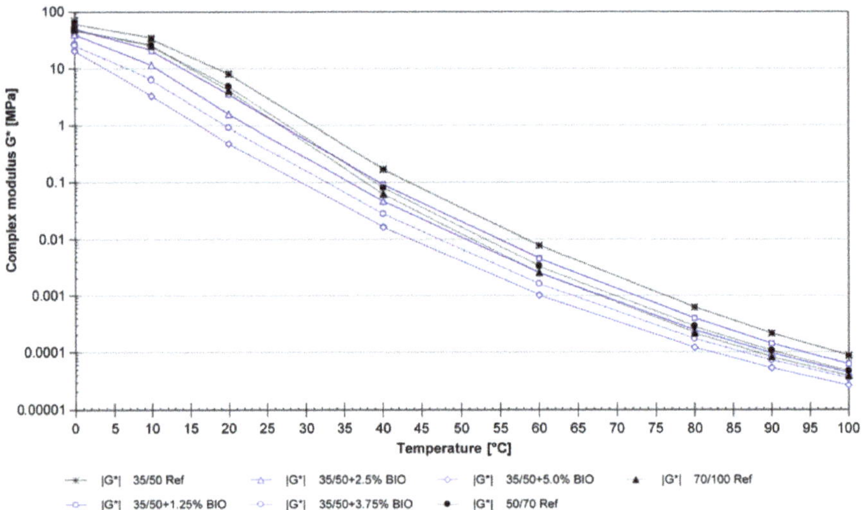

Figure 2. Complex modulus |G*| for reference bitumens and 35/50 bitumen modified with bio-agent.

Phase angle changes as a function of temperature are shown in Figure 3 for reference binders and for binders with various amounts of bio-agent. It can be seen that, in the range of medium road operational temperature between 0 °C and 30 °C, bitumen 35/50 with bio-agent in the amount of 1.25% shows similar properties to reference bitumens 50/70 and 70/100. As the temperature increases above 40 °C, the behavior of reference bitumens 50/70 and 70/100 corresponds to properties of 35/50 road bitumen with bio-agent content above 2.5%. The influence of bio-agent is very positive due to the variable fluxing effect depending on the temperature. At a low temperature, bitumen 35/50 with 2.5% bio-agent shows higher flexibility than reference bitumens 50/70 and 70/100, which makes it more resistant to cracking, but at a high temperature presents lower flexibility, which is related to a higher resistance to permanent deformation (rutting). It can be concluded that bitumen 35/50 modified with 2.5% bio-agent demonstrates better visco-elastic properties than does non-modified 50/70 bitumen.

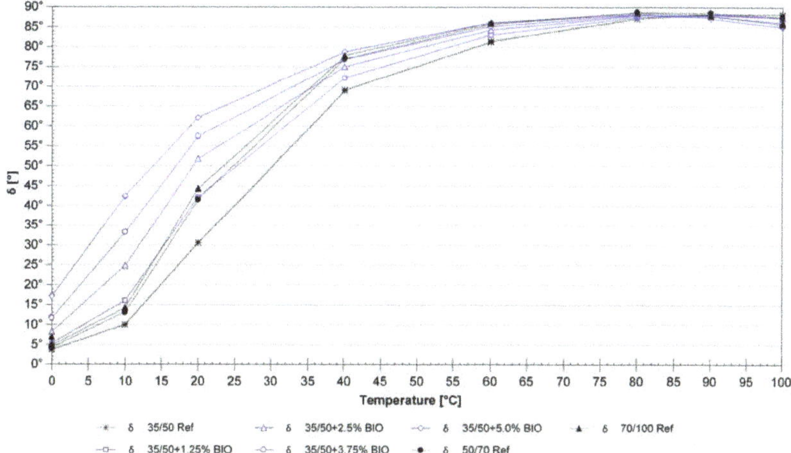

Figure 3. Phase angle |δ| for reference bitumens and 35/50 bitumen modified with bio-agent.

As was previously determined, modification of road bitumen with bio-agent may have a positive influence on the properties of hard bitumen and causes its fluxing during the technological processes of asphalt mixture production and placement. During the second stage of asphalt mixture life, once opened to traffic, as a result of bio-agent polymerization in bitumen, partial bitumen recovery to its properties before modification is expected. As shown in Figure 4, curves of a complex modulus in the temperature function for bitumen 35/50 modified with 2.5% bio-agent, in time, demonstrate expected beneficial properties. It can be seen that the polymerization reaction of the bio-agent occurs in the bitumen in a permanent way and changes the bitumen stiffness at all temperatures. It also can be observed that the complex modulus curve for bitumen 35/50 shortly after its modification is the most shifted from the reference bitumen 35/50. In time, bitumen modified with the bio-agent increases in stiffness.

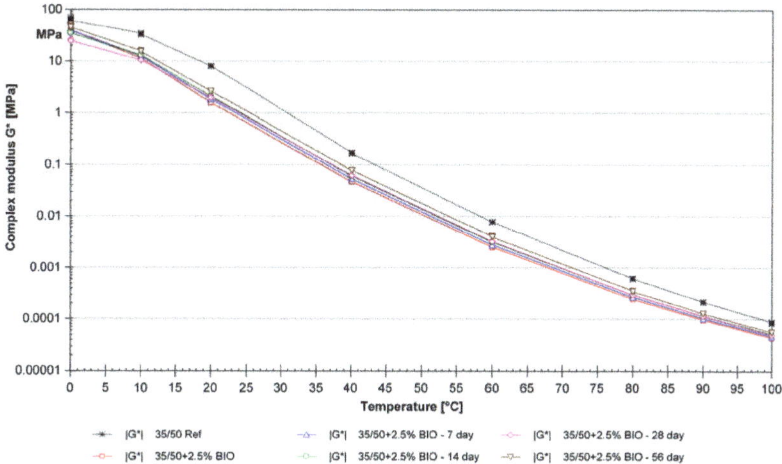

Figure 4. Changes of complex modulus |G*| in time for reference bitumen and 35/50 bitumen modified with 2.5% bio-agent.

Figure 5 shows curves of the phase angle in a function of temperature for road bitumen 35/50 modified with 2.5% bio-agent in time. Based on the presented data, it can be seen that bitumen 35/50 after fluxidation presents higher flexibility as expressed by the higher phase angle value. This flexibility effect then decreases in time once polymerization occurs.

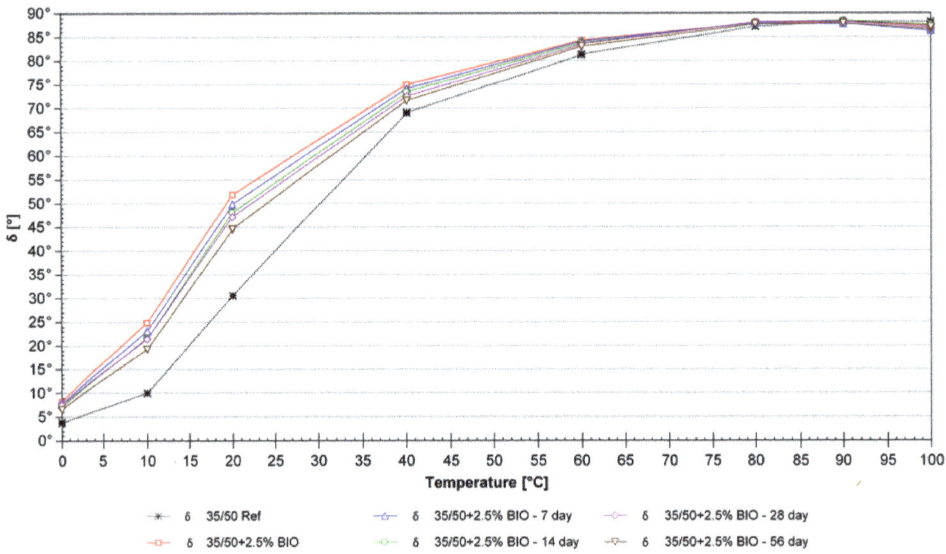

Figure 5. Changes of phase angle |δ| in time for reference bitumen and 35/50 bitumen modified with 2.5% bio-agent.

3.2. Asphalt Mixture Properties

Complex modulus and phase angle test results are presented in Figure 6 for 4PB tests in the frequency function for AC with bitumen 35/50 and bio-agent. Curves shown in Figure 6 are an average from six measurements conducted for six different specimens of the same asphalt mixture. On plots, there are shown error bars as a value of standard deviation for the average of each repetition. Based on curve analysis, it can be concluded that the mixtures with the bio-agent demonstrate lower stiffness as compared to the mixture with unmodified binder. The aforementioned can be seen in all frequency range. The bio-agent presence in the asphalt mixture causes a higher value of phase angle, which is a beneficial phenomenon in the medium range of temperatures to which the road can be subjected.

Fatigue characteristics of both asphalt mixtures are shown in Figure 7. Based on a statistical analysis of correlation, it can be assumed that correlations describing fatigue characteristics of both asphalt mixture are true ($p < 0.05$). Based on the location of both fatigue curves, it can be concluded that mixture with bio-agent demonstrates higher fatigue durability (resistance) as compared to the one without bio-agent. Curve for the mixture with bio-agent is parallel-shifted to the top as compared to the one with unmodified bitumen. Such shift is a sign of the fact that the mixture with comparable strain is more resistant to the fatigue and pavement will reach end-of-life in a longer timer period. On the other hand, based on the mixture properties at the same number of cycles, it can be found that mixture with bio-agent allows for an application of load with higher frequency strain. The asphalt mixture with RAP and bitumen without modification allows for a million cycles with a strain of 122 μm/m, while the mixture with RAP and bitumen modified with the bio-agent allows for a million cycles with a strain of about 10% higher (i.e., 134 μm/m).

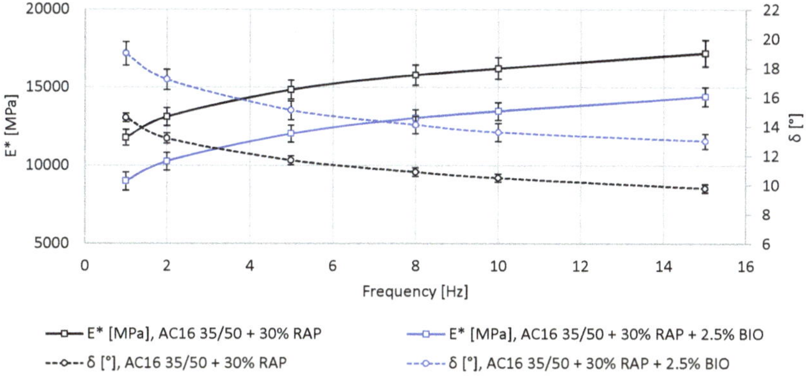

Figure 6. Complex modulus |E*| and phase angle |δ| in function of frequencies at 10 °C for asphalt mixtures.

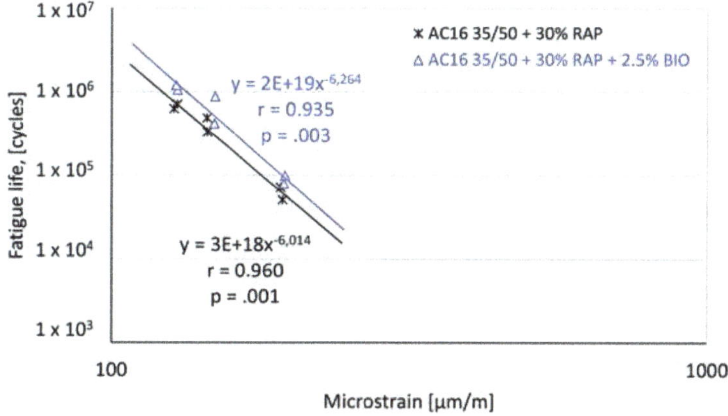

Figure 7. Fatigue characteristics of asphalt mixtures.

Figure 8 presents results of low temperature resistance according to the TSRST. On Figure 8, there are shown error bars as a value of standard deviation for average from each of four repetitions. It can be seen that the mixture with the bio-agent added exhibited a 5 °C lower breaking temperature compared with the AC without the bio-agent (refer to Figure 8a). Asphalt specimen destruction during the TSRST occurs as a result of thermal contraction once temperature decreases. As visible, the difference between test results of the mixes is higher than standard deviation, which implies the important influence of the bio-agent on low-temperature behavior. A comparison between cryogenic stresses suggests that the asphalt mixture with the bio-agent exhibits higher tensile strength occurring with low-temperature contraction (refer to Figure 8b). Once taking into account the test results, including the error bars as a value of standard deviation, the differences in cryogenic stress between mixtures are not visible. From a statistical point of view, such a correlation is still possible and should be verified with a greater population.

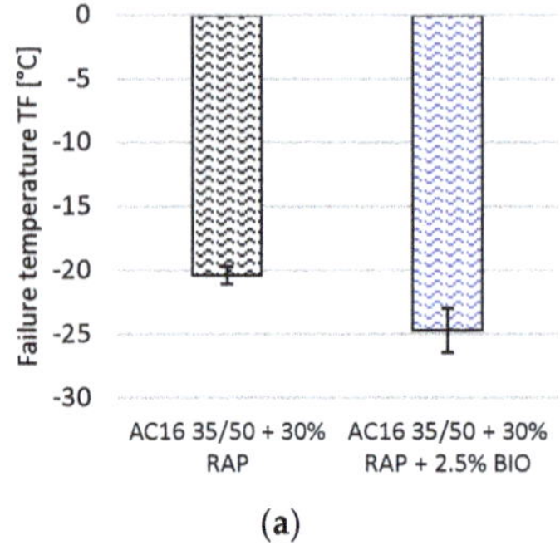

(a)

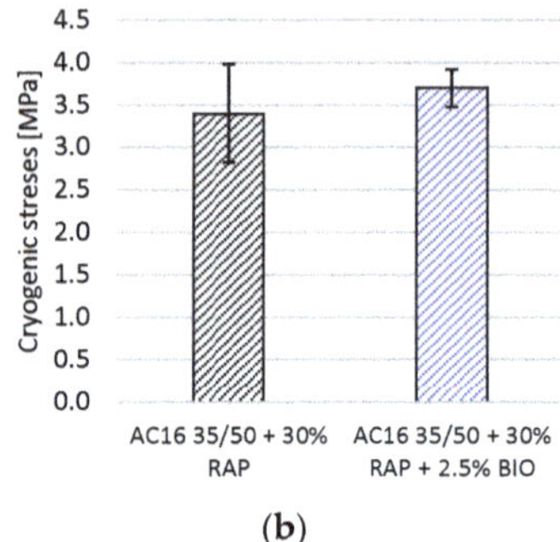

(b)

Figure 8. Low temperature characteristic of asphalt mixtures: (**a**) failure temperature; (**b**) cryogenic stress.

4. Discussion and Conclusions

Bio-materials, allowing for the use of renewable resources, are increasingly applied in industry due to sustainability policy. Once used, they cannot detrimentally change properties of the material, instead they should make it more ecologically friendly. The effect of initial fluxidation of bitumen with the bio-agent and then the gradual polymerization suggests the potential of its application as a bi-functional additive in warm mix asphalt (WMA) technology or asphalt mixes with reclaimed asphalt pavement (RAP).

Asphalt mixture production is a sensitive process requiring high quality control. During application of RAP directly to the mixer there is a potential risk related to the mixture uniformity. Application of bio-agent with fluxing function lowers bitumen viscosity and helps to reach better homogeneity in the mixtures containing RAP. It also should be highlighted that, in asphalt mixtures containing RAP there are two types of bitumen binder: binder from RAP and virgin bitumen. It is commonly assumed that during asphalt mixture production both bitumens will mix resulting with blended bitumen. The tests conducted shows that even low rate of bio-agent (2.5%) beneficially changes bitumen properties in low temperature; moreover polymerization processes occurring in second stage of the process improves bitumen properties in high operational temperature. The testing of complex modulus $|G^*|$ and phase angle $|\delta|$ demonstrated that it is possible to select optimal bio-agent content depending on the desired consistency of bitumen. As mentioned earlier, bio-agent polymerization occurring during the second stage of the process partially recovers initial properties of the bitumen. Unrecovered parts of the fluxing effect compensate, in the blended RAP + virgin bitumen, the stiffening of the binder caused by the application of aged bitumen from RAP.

Due to the fact that bio-additive modifies visco-elastic properties of bitumen in a wide temperature range and has a fluxing effect, it also improves the compaction process of the asphalt mixture. As shown in this research, the bio-agent influences the amount of air voids in the compacted mixture. Based on the observed compaction of mixtures with RAP as well as with RAP and bio-agent, it can be concluded that the design process of such mixes should include an expected effect of fluxing in order to obtain the desired air void content and compaction level.

The fluxing effect of bitumen binder and the changes in its viscoelastic properties permanently influence the behavior of asphalt mixtures. Initially, before bio-agent polymerization, improvement in compactibility can be seen. Next, after polymerization, the mixture exhibits higher fatigue durability, accommodates higher strains, and shows improved low-temperature properties. Those effects are most likely caused by an improvement in both virgin binder properties and aged (RAP origin) binder properties. Asphalt mixtures with bio-agents added in the amount of 2.5% shows 10% higher fatigue durability (4PB) and a 5 °C lower failure temperature (TSRST).

As was demonstrated in another paper [15] (more chemically oriented) produced based on this study, the polymerization process of the bio-agent occurs until the saturation of double unsaturated bonds, whose number is limited—typically, bio-agent polymerization occurs during the first month (with a complete stop after two months). For this reason, over-stiffening of the mixture during pavement in-service life is not expected.

The bio-agent used in this research is a material obtained from RME with an average polymerization potential due to the limited number of unsaturated bonds. In future studies, another vegetable-origin material such as tung oil, linseed oil, and sunflower oil can be applied in order to determine its applicability for road bitumen modification and to verify the relationship between the polymerization mechanism and the durability of asphalt mixtures.

Acknowledgments: The research leading to these results has received funding from the European Union's Seventh Program for research, technological development, and demonstration under grant agreement No. 603862. This work was also supported by funds for science for years 2014–2017 granted by the Polish Ministry of Science and Higher Education to support international projects. This article reflects only the author's views, and the European Union is not liable for any use that may be made of the information contained. The authors wish to express their gratitude to Irena Gaweł and Jerzy Piłat for their earlier inspiration.

Author Contributions: Karol J. Kowalski, Jan B. Król, and Piotr Radziszewski conceived and designed the experiments; Michał Sarnowski performed the binder experiments; Wojciech Bańkowski performed the asphalt mixtures experiments; Jan B. Król and Karol J. Kowalski analyzed the data; Jan B. Król and Karol J. Kowalski wrote the paper.

Conflicts of Interest: The authors declare no conflict of interest. The founding sponsors had no role in the design of the study; in the collection, analyses, or interpretation of data; in the writing of the manuscript; or in the decision to publish the results.

References

1. Król, J.; Kowalski, K.J.; Radziszewski, P.; Sarnowski, S. Rheological behaviour of n-alkane modified bitumen in aspect of Warm Mix Asphalt technology. *Constr. Build. Mater.* **2015**, *93*, 703–710. [CrossRef]
2. Sarnowski, M. Rheological properties of road bitumen binders modified with SBS polymer and polyphosphoric acid. *Roads Bridges Drogi Mosty* **2015**, *14*, 47–65.
3. Iwański, M.; Chomicz-Kowalska, A.; Maciejewski, K. Application of synthetic wax for improvement of foamed bitumen parameters. *Constr. Build. Mater.* **2015**, *83*, 62–69. [CrossRef]
4. Guarin, A.; Khan, A.; Butt, A.A.; Birgisson, B.; Kringos, N. An extensive laboratory investigation of the use of bio-oil modified bitumen in road construction. *Constr. Build. Mater.* **2015**, *106*, 133–139. [CrossRef]
5. Dhasmana, H.; Ozer, H.; Al-Qadi, I.L.; Zhang, Y.; Schideman, L.; Sharma, B.K.; Chen, W.; Minarick, M.J.; Zhang, P. Rheological and Chemical Characterization of Biobinders from Different Biomass Resources. *Transp. Res. Rec. J. Transp. Res. Board* **2015**, *2505*, 121–129. [CrossRef]
6. Fini, E.H.; Al-Qadi, I.L.; You, Z.; Zada, B.; Mills-Beale, J. Partial Replacement of Asphalt Binder with Bio-binder: Characterization and Modification. *Int. J. Pavement Eng.* **2012**, *13*, 515–522. [CrossRef]
7. Mohammad, L.N.; Elseifi, M.A.; Cooper, S.B.; Challa, H.; Naidoo, P. Laboratory Evaluation of Asphalt Mixtures that Contain Biobinder Technologies. *Transp. Res. Rec. J. Transp. Res. Board* **2013**, *2371*, 58–65. [CrossRef]
8. Yang, X.; You, Z.; Dai, Q.; Mills-Beale, J. Mechanical performance of asphalt mixtures modified by bio-oils derived from waste wood resources. *Constr. Build. Mater.* **2013**, *51*, 424–431. [CrossRef]
9. Zaumanis, M.; Mallick, R.B.; Poulikakos, L.; Frank, R. Influence of six rejuvenators on the performance properties of Reclaimed Asphalt Pavement (RAP) binder and 100% recycled asphalt mixtures. *Constr. Build. Mater.* **2014**, *71*, 538–550. [CrossRef]
10. Chen, C.; Podolsky, J.H.; Hernandez, N.; Hohmann, A.; Williams, R.C.; Cochran, E.W. Use of Bioadvantaged Materials for Use in Bituminous Modification. *Transp. Res. Procedia* **2016**, *14*, 3592–3600. [CrossRef]
11. Sun, Z.; Yi, J.; Huang, Y.; Feng, D.; Guo, C. Properties of asphalt binder modified by bio-oil derived from waste cooking oil. *Constr. Build. Mater.* **2015**, *102*, 496–504. [CrossRef]
12. Podolsky, J.H.; Buss, A.; Williams, R.C.; Hernández, N.; Cochran, E.W. Effects of aging on rejuvenated vacuum tower bottom rheology through use of black diagrams, and master curves. *Fuel* **2016**, *185*, 34–44. [CrossRef]
13. Gaweł, I.; Piłat, J.; Radziszewski, P.; Niczke, Ł.; Król, J.; Sarnowski, M. Bitumen fluxes of vegetable origin. *Polimery* **2010**, *55*, 55–60.

14. Wexler, H. Polymerization of drying oils. *Chem. Revient* **1964**, *64*, 591–611. [CrossRef]
15. Simonen, M.; Blomberg, T.; Pellinen, T.; Valtonen, J. Physicochemical Properties of Bitumens Modified with Bioflux. *Road Mater. Pavement Des.* **2013**, *14*, 36–48. [CrossRef]
16. Simonen, M.; Blomberg, T.; Pellinen, T.; Makowska, M.; Valtonen, J. Curing and Ageing of Biofluxed Bitumen: A Physicochemical Approach. *Road Mater. Pavement Des.* **2013**, *14*, 159–177. [CrossRef]
17. Król, J.B.; Kowalski, K.J.; Niczke, Ł.; Radziszewski, P. Effect of bitumen fluxing using a bio-origin additive. *Constr. Build. Mater.* **2016**, *114*, 194–203. [CrossRef]
18. General Directorate for National Roads and Motorways (GDDKiA). WT-2 "Asphalt Pavement for National Roads. Asphalt Mixtures. Technical Requirement". 2014. Available online: https://www.gddkia.gov.pl/userfiles/articles/z/zarzadzenia-generalnego-dyrektor_13901/zalacznik%20do%20zarz%2047.pdf (accessed on 31 December 2017).

Article

Numerical Study on the Asphalt Concrete Structure for Blast and Impact Load Using the Karagozian and Case Concrete Model

Jun Wu [1,2], Liang Li [1,*], Xiuli Du [1] and Xuemei Liu [3]

1 The Key Laboratory of Urban Security and Disaster Engineering, Ministry of Education, Beijing University of Technology, Beijing 100124, China; cvewujun@sues.edu.cn (J.W.); duxiuli@bjut.edu.cn (X.D.)
2 School of Urban Railway Transportation, Shanghai University of Engineering Science, Shanghai 201620, China
3 School of Civil Engineering and Built Environment, Queensland University of Technology, Brisbane 4001, Australia; x51.liu@qut.edu.au
* Correspondence: liliang@bjut.edu.cn; Tel.: +86-10-6739-2430

Academic Editor: Zhanping You
Received: 23 November 2016; Accepted: 14 February 2017; Published: 17 February 2017

Abstract: The behaviour of an asphalt concrete structure subjected to severe loading, such as blast and impact loadings, is becoming critical for safety and anti-terrorist reasons. With the development of high-speed computational capabilities, it is possible to carry out the numerical simulation of an asphalt concrete structure subjected to blast or impact loading. In the simulation, the constitutive model plays a key role as the model defines the essential physical mechanisms of the material under different stress and loading conditions. In this paper, the key features of the Karagozian and Case concrete model (KCC) adopted in LSDYNA are evaluated and discussed. The formulations of the strength surfaces and the damage factor in the KCC model are verified. Both static and dynamic tests are used to determine the parameters of asphalt concrete in the KCC model. The modified damage factor is proposed to represent the higher failure strain that can improve the simulation of the behaviour of AC material. Furthermore, a series test of the asphalt concrete structure subjected to blast and impact loadings is conducted and simulated by using the KCC model. The simulation results are then compared with those from both field and laboratory tests. The results show that the use of the KCC model to simulate asphalt concrete structures can reproduce similar results as the field and laboratory test.

Keywords: asphalt concrete; constitutive model; impact loading; blast loading; numerical simulation

1. Introduction

The behaviour of structures or infrastructures under extreme loadings has become a hot topic in the area of civil, mechanical and material engineering. Critical infrastructures, such as runway pavement designed for normal aircraft landing and taking off, are expected to have adequate resistance when subjected to extreme loadings, such as impact or blast loadings (e.g., heavy airplane landing or taking off, air plane crash or terrorist attack). Asphalt concrete (AC) is made of bitumen binder and coarse aggregate. It is usually used as the surface course for both highway and runway flexible pavement [1]. The dynamic load in the daily application for AC pavement normally corresponds to a strain rate less than 10^{-1} s^{-1}. Tashman et al. [2] conducted experiments of AC under triaxial compressive loading at the strain rate from 10^{-6} s^{-1} to 10^{-3} s^{-1}. The results showed that the failure stress increased with the increase of the applied strain rate. Seibi et al. [3] studied AC subjected to uniaxial compressive loading with strain rate from 0.064 s^{-1} to 0.28 s^{-1}. It was found that the yield stress was significantly dependent on the strain rate. Park et al. [4] carried out tests on AC under uniaxial and triaxial compression with the strain rate changing from 10^{-4} s^{-1} to 0.07 s^{-1}. The results showed that with the increase of the applied strain rates, the yield stress and failure stress increased, and the strain rate dependency was clearly showing up at the higher strains. It was also found that the viscous behaviour of AC decreased with the increase of the strain rate. However, when the pavement structure is subjected to the impact loading from further heavy traffic loading or aircraft loading, the corresponding strain rate exceeds 10^{-1} s^{-1}, especially for heavy aircraft loading, the related strain rate reaches around 100 s^{-1}. Based on previous research [5,6], the compressive strength of AC material can be enhanced with the increase of strain rate, and AC material exhibits high plastic behaviour at the high strain rate. However, it is very expensive to conduct field test to investigate the actual behaviour of AC under severe dynamic loading, the numerical simulation is an effective alternative solution.

There are many factors that influence the reliability of numerical simulation. Among these factors, the material model plays a key role because it should reproduce the essential physical mechanisms of the material under various loading conditions. Seibi et al. [3] and Park et al. [4] used the Drucker–Prager yield function to simulate the compressive behaviour of AC under dynamic strain loading (strain rate from 0.0001 s^{-1} to 0.0701 s^{-1}). Tashman et al. [2] developed a microstructure-based viscoplastic continuum model to take into account the effect of temperature in AC material with the strain rate ranging from 10^{-6} s^{-1} to 10^{-3} s^{-1}. Since the late 1990s, microstructural-based discrete element models (DEM) have been used for better understanding of asphalt pavement concrete (e.g., [7–10]). In DEM models, the assumption of elastic or viscoelastic behaviour was employed for the simulation of the static or creep behaviour of AC under normal traffic loading. However, it was found that when the pavement structure was under blast or high impact load, plastic deformation and severe damage would occur. Thus, a robust material model should be developed to consider the strain rate effect, strain hardening, strain softening and damage of the AC material under the severe dynamic loading. Recently, several concrete-like material models subjected to dynamic loadings have been developed, such as the Riedel-Hiermaier-Thoma (RHT) model [11], the Advanced Fundamental Concrete (AFC) model [12,13], the Karagozian and Case concrete model (KCC) [14] and the Holmquist-Johnson-Cook (HJC) concrete model [15,16]. These robust material models are capable of capturing varying concrete-like materials' behaviour under different loading conditions. When subjected to dynamic loading, such as blast loading or high impact loading, concrete-like materials show a highly non-linear response. Besides, due to the general complexity of the constitutive models, the determination of the parameters (i.e., residual strength, failure strain and failure criteria in model) also plays an important role in achieving the actual performance of the concrete-like materials. This requires sufficient understanding of the modelling formulation and the associated considerations. In the current study, the KCC model [14] is used to simulate the AC material. This model is capable of capturing the varied concrete-like material behaviours under different loading conditions. However, it should be noticed that this model cannot consider the temperature effect. In this study, the dynamic behaviour of AC under a high strain rate is investigated, and the temperature effect is not considered. Hence, the KCC can be used; otherwise, the temperature issue should come up.

In this paper, several key features of the KCC model are firstly discussed, and then, the determination of the parameters in the KCC model for AC is provided. An application example on the AC structure under blast and impact loading is also illustrated and validated based on a field blast test and a laboratory drop weight impact test.

2. Review on KCC Model

When subjected to blast loading or high impact loading, concrete or other concrete-like materials shows a highly non-linear response. They usually exhibit pressure hardening and strain hardening under static loading and strain rate hardening in tension and compression under dynamic loading. A number of material models has been developed to model concrete-like materials recently [11–16]. Among them, the KCC model is widely used to analyse concrete-like materials' response to blast and impact loading due to its simple implementation. In addition, the KCC model can capture the non-linear behaviour of the material under dynamic loading [14]. The key features of the model are discussed briefly in the following section.

2.1. Strength Surface in KCC Model

The KCC model decouples stress into the hydrostatic pressure and deviatoric stress as shown in Equation (1):

$$\sigma_{ij} = s_{ij} + \frac{1}{3}\sigma_{ii}\delta_{ij} \tag{1}$$

where σ_{ij} is the stress tensor, s_{ij} is the deviatoric stress tensor and σ_{ii} is the hydrostatic pressure tensor. It should be noted that stress is positive in tension, and pressure is positive in compression. The hydrostatic pressure is related to the volumetric change of material, while the deviatoric stress is related to shear resistance of the material and is usually expressed by the second invariant of the deviatoric stress tensor, J_2:

$$J_2 = \frac{1}{2}s_{ij}s_{ji} = \frac{s_1^2 + s_2^2 + s_3^2}{2} \tag{2}$$

The KCC model has three independent strength surfaces: maximum strength surface, yield surface and residual strength surface, which are shown in Figure 1. The general formation of strength surfaces can be written as:

$$\Delta\sigma = \sqrt{3J_2} = f(p, J_2) \tag{3}$$

in which $\Delta\sigma$ is the principal stress difference and p is the hydrostatic pressure. Usually, the above Equation (3) refers to the compressive meridian. The whole failure curve can be obtained through rotation of the compressive meridian around the hydrostatic pressure axis by multiplying $r_3(\theta_L)$, which has the formation:

$$\Delta\sigma = r_3(\theta_L) \cdot \sqrt{3J_2} = f(p, J_2, J_3) \tag{4}$$

$$r_3(\theta_L) = \frac{r}{r_c} = \frac{2(1-\psi^2)\cos\theta_L + (2\psi - 1)\sqrt{4(1-\psi^2)\cos^2\theta_L + 5\psi^2 - 4\psi}}{4(1-\psi^2)\cos^2\theta_L + (1-2\psi)^2} \tag{5}$$

where $\psi = {}^{r_t}/_{r_c}$ and r_t and r_c are the radius of tensile and compressive meridian, respectively. According to Equation (5), it can be found that the $r_3(\theta_L)$ depended on ψ and θ_L. The parameter ψ in turn relies on the hydrostatic pressure. For the concrete-like material, the value of ψ varies from $\frac{1}{2}$ at negative (tensile) pressures to unity at high compressive pressures [14]. The value of Lode angle θ_L can be obtained from:

$$\cos\theta_L = \frac{\sqrt{3}}{2}\frac{s_1}{\sqrt{J_2}} \quad \text{or} \quad \cos 3\theta_L = \frac{3\sqrt{3}}{2}\frac{J_3}{J_2^{3/2}} \tag{6}$$

During the initial increase of hydrostatic pressure p, the deviatoric stresses $\Delta\sigma$ remains in the elastic region until the yield surface is reached. Deviatoric stress can be further developed until the maximum strength surface is reached, and the material will subsequently start to fail. After failure is initiated, the material will gradually lose its load-carrying capacity and reaches its residual strength surface. The formations of these three surfaces are also given in Equations (7)–(9).

$$\text{Yield surface } \Delta\sigma_y = a_{0y} + \frac{p}{a_{1y} + a_{2y}p} \tag{7}$$

$$\text{Maximum strength surface } \Delta\sigma_m = a_0 + \frac{p}{a_1 + a_2 p} \tag{8}$$

$$\text{Residual strength surface } \Delta\sigma_r = \frac{p}{a_{1f} + a_{2f}p} \tag{9}$$

where the eight parameters, namely, $a_0, a_1, a_2, a_{1f}, a_{2f}, a_{0y}, a_{1y}$ and a_{2y}, for three surfaces can be obtained from the experimental data (e.g., triaxial compression test, biaxial compression test or uniaxial tension/compression test) [14].

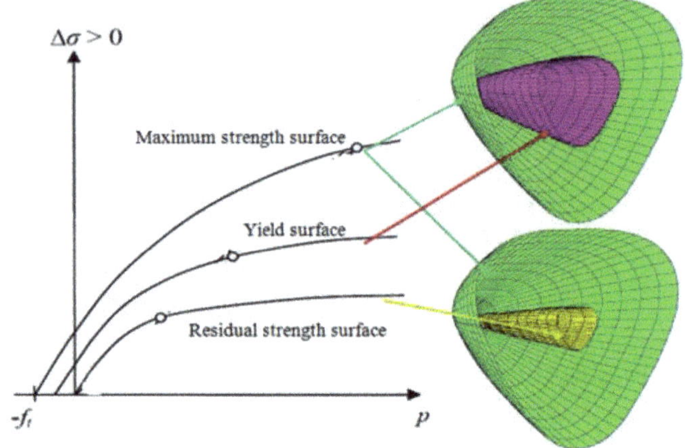

Figure 1. Strength surfaces for the Karagozian and Case concrete model (KCC) material model [14].

2.2. Damage Factor in the KCC Model

After reaching the initial yield surface, but before reaching the maximum strength surface, the current surface can be obtained as a linear interpolation between yield surface $\Delta\sigma_y$ and maximum strength surface $\Delta\sigma_m$:

$$\Delta\sigma = \eta(\Delta\sigma_m - \Delta\sigma_y) + \Delta\sigma_y \tag{10}$$

After reaching the maximum strength surface, the current failure is interpolated between the maximum strength surface $\Delta\sigma_m$ and the residual strength surface $\Delta\sigma_r$, which is similar to the above computation:

$$\Delta\sigma = \eta(\Delta\sigma_m - \Delta\sigma_r) + \Delta\sigma_r \tag{11}$$

where η varies from zero to one depending on the accumulated effective plastic strain parameter λ. The value of η normally starts at zero and increases to unity at $\lambda = \lambda_m$, then decreases back to zero at some larger value of λ. λ_m is the plastic strain at maximum strength surface. The accumulated effective plastic strain λ can be expressed as follows:

$$\lambda = \int_0^{\bar{\varepsilon}^p} \frac{d\bar{\varepsilon}^p}{r_f[1 + p/(r_f f_t)]^{b_1}} \quad \text{for} \quad p \geq 0 \tag{12}$$

$$\lambda = \int_0^{\bar{\varepsilon}^p} \frac{d\bar{\varepsilon}^p}{r_f[1 + p/(r_f f_t)]^{b_2}} \quad \text{for} \quad p < 0 \tag{13}$$

where f_t is the quasi-static tensile strength, $d\bar{\varepsilon}^p$ is effective plastic strain increment and r_f is the dynamic increase factor (DIF) of the material under dynamic loading. The damage factors b_1 and b_2 define the softening behaviour due to compression ($P \geq 0$) and tension ($P < 0$), respectively. Parameter b_1 can be determined by considering compressive energy G_c (area under the compressive stress-strain curve) obtained from the uniaxial compression test in single element simulation. It is obtained iteratively until the area under the stress-stain curve from single element simulation coincides with G_c/h, where h is the element size. The damage factor b_2 in Equation (13) is related to tensile softening of the material and determined from experimental data. The fracture energy G_f can be obtained from the uniaxial tensile test or three-point notched beam test. Then, the single element simulation of uniaxial tensile test is employed to obtain the stress-strain curve from the numerical analysis. Hence, the value of b_2 can be obtained until the area under the tensile stress-stain curve from a single element coincides with G_f/w_c, where w_c is the localization width, and typically, w_c is normally taken as one- to six-times the maximum aggregate size [14].

Based on Equations (10) and (11), the stress softening factors η and λ are governed by the accumulation of effective plastic strain. However, when the stress path is very close to the negative hydrostatic pressure axis, i.e., isotropic tension, wherein the hydrostatic pressure would decrease from zero to $-f_t$, where no deviatoric stress occurred, no damage accumulation would occur based on these equations. However, in such concrete-like materials, damage cannot be avoided even at this state. Therefore, the above condition has to be modified by including pressure-softening effects near or after tensile failures. In this case, a volumetric damage increment is calculated and added to the total damage factor λ whenever the stress path is close to the triaxial tensile path.

2.3. Strain Rate Effect

The material model KCC also includes a radial rate enhancement on the material failure surface. This is because experimental data for concrete-like materials are typically obtained along radial paths from the origin in deviatoric stresses versus hydrostatic pressure via unconfined compressive and tensile tests. It is well known that the strain rate effect is important for concrete-like materials under severe dynamic loading. A typical DIF-strain rate curve for concrete-like materials can be obtained from the servo hydraulic fast loading tests and the split Hopkinson pressure bar (SHPB) test.

2.4. Equation of State

In addition to the strength surface model, an equation of state (EOS) is needed to describe the relationship between hydrostatic pressure and volume change of the material subject to dynamic load. EOS is usually determined using a fly impact (i.e., for steel) test or triaxial compressive test (i.e., for concrete or geomaterials). The isotropic compression portion of the KCC material model consists of pairs of hydrostatic pressure P and corresponding volume strain μ. It is implemented as a piece-wise curve in this model.

3. Determination of Parameters for Asphalt Concrete Material

The KCC model is employed to simulate AC material to capture dynamic response under impact and blast loading. This section describes the determination of the model parameter for the asphalt concrete with the compressive strength of 4.6 MPa, which is used in the following application example. For the asphalt concrete with a different compressive strength, this method can be used as a reference.

3.1. Strength Surface

As mentioned in Section 2.1, the KCC material model has three strength surfaces: strength, residual strength and yield surfaces. These three surfaces can be obtained through curve fitting of suitable experimental data. In this study, due to the few data for triaxial compressive tests of asphalt concrete, available data are extracted from Park et al. [4] with the compressive strength f_c = 0.311 MPa for AC. Figure 2 presents the determination of the three surfaces by curve fitting for AC with f_c = 0.311 MPa. The intersection point of maximum strength surface and residual strength surface is the so-called brittle-to-ductile point. This point can be determined by experimental data under high confining pressure. However, it is difficult to determine this point in the strength surface as no experimental data are available for AC materials. Based on the experimental data for concrete [17], this point was usually taken as p/f_c = 3.878. Considering that the size and strength of aggregates used in AC and concrete material are almost the same, the brittle-to-ductile point for AC is taken to be the same as that for concrete. This value might be conservative for AC due to the higher content of coarse aggregate in AC. However, for simulation purpose, this value is acceptable.

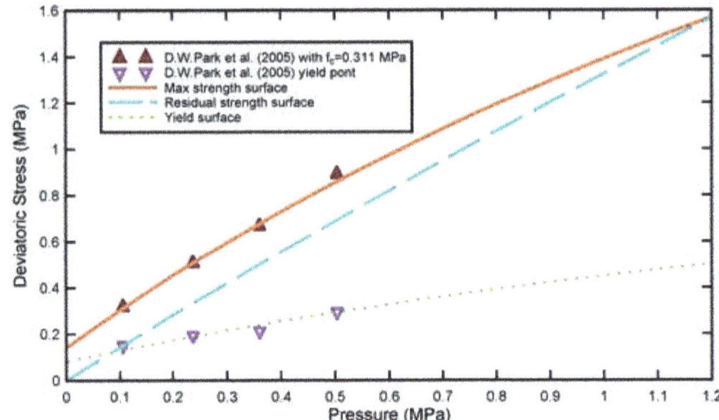

Figure 2. Determination of parameters in KCC from experimental data.

If new asphalt concrete with known unconfined compression strength $f'_{c,new}$ is to be modelled, but its strength surfaces are otherwise unknown, then one way of scaling data from a known material is proposed as follows [14]:

$$r = \frac{f'_{c,new}}{f'_{c,old}} \tag{14}$$

where $f'_{c,old}$ is the unconfined compressive strength for a previously modelled AC. Then, the new material strength surface can be taken as:

$$\Delta\sigma_n = a_{0n} + \frac{p}{a_{1n} + a_{2n}p} \tag{15}$$

in which $a_{0n} = a_0 r, a_{1n} = a_1, a_{2n} = a_2/r$.

The new asphalt concrete with unconfined compressive strength f_c = 0.8 MPa [2] is used to validate the parameters obtained from the scaling method. Figure 3 shows the maximum strength surface determined by the scaling method. It can be seen that the maximum strength surface fit very well with the experimental data, and thus, it can be concluded that the parameters for AC with different compressive strengths can be obtained by the scaling method.

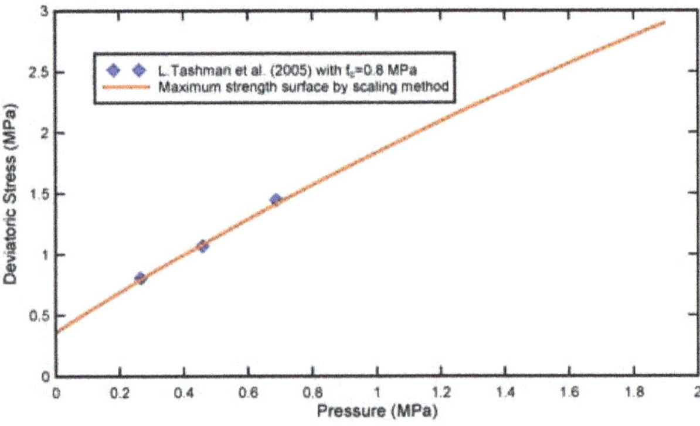

Figure 3. Validation of the failure surface from the scaling law.

In this study, the unconfined compressive strength for AC material is 4.6 MPa, and the tensile strength is 0.7 MPa at 35 °C. By scaling the data from the established curves given in Figure 2 [11], the appropriate strength surface of the current materials can be determined; the strength parameters are given in Table 1.

Table 1. Parameters for asphalt concrete (AC) material with f_c = 4.6 MPa.

Parameter	Value
a_0	2.071
a_1	0.6
a_2	0.0135
a_{0y}	1.183
a_{1y}	2.00
a_{2y}	0.0473
a_{1f}	0.70
a_{2f}	0.0037

3.2. Damage Factor

The strain hardening and softening pairs (η, λ) in Equations (10) and (11) describe the material behaviour transmitted from the yield surface to the maximum strength surface and from the maximum strength surface to the residual strength surface, respectively. During the transmission, parameter η varies from zero to one, depending on the accumulated effective plastic strain parameter λ. However, it is found that the original damage factor pairs (η, λ) in the KCC model are only suitable for concrete and not for the AC material due to AC having higher plastic failure strain. Thus, the input for accumulated effective plastic strain λ should be modified. Based on the uniaxial compressive test for AC, it was found that at peak stress, the corresponding strain was approximately 0.018, and the final failure strain was about 0.1; while for normal concrete, the corresponding strain at peak stress was around 0.0022. Hence, the λ is modified to give the high failure strain for AC in the current study. Additionally, it is found that when λ is adjusted to 10-times the original λ value, the numerical results seemed to show good agreement with the experimental results from the unconfined compressive test for AC. Figure 4 shows the modified and original series of (η, λ) pairs. It can be seen that the modified damage factor provided smoother descending than the original damage factor and had a higher failure strain that matched the behaviour of AC very well.

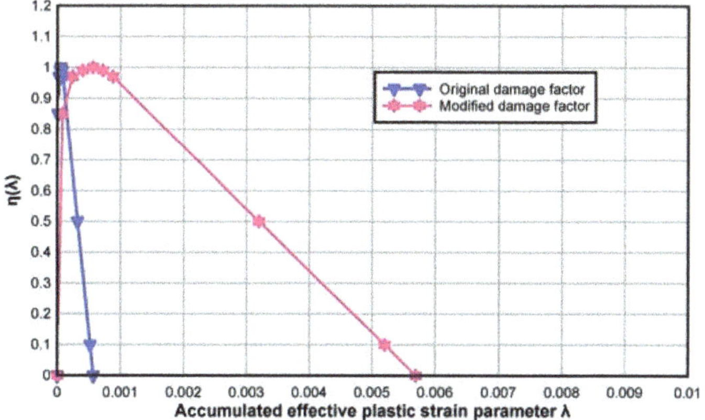

Figure 4. Damage factor used for AC material.

3.3. Equation of State

There are limited EOS data for AC material. The available EOS data are for AC with compressive strength of f_c = 3.8 MPa [5]. In this study, the compressive strength for AC is f_c = 4.6 MPa. Thus, the pressure-volume pairs can be calculated using the volumetric scaling method [18]. In this method, assuming that new data are obtained at the same volumetric strains, thus, the new corresponding pressure (pc_{new}) can be:

$$pc_{new} = pc_{old}\sqrt{r} \tag{16}$$

and the new corresponding unloading bulk modulus ($k\mu_{new}$) is:

$$ku_{new} = ku_{old}\sqrt{r} \tag{17}$$

where r is the scaling factor, which is the ratio of compression strength for new material to the compression strength of the previous material modelled. Hence, the EOS data for f_c = 4.6 MPa are calculated based on Equations (16) and (17), and the EOS inputted in the numerical model is shown in Figure 5.

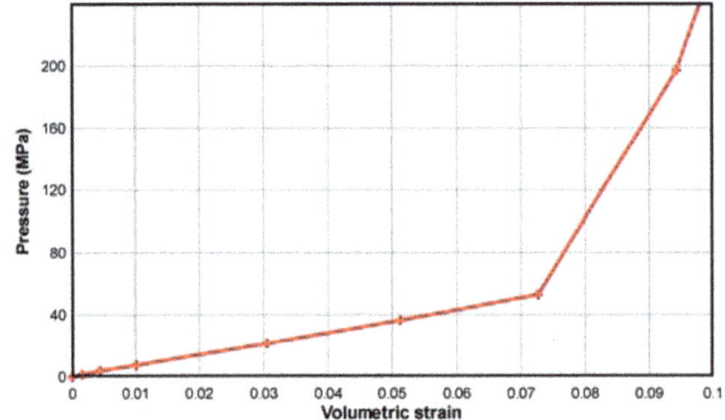

Figure 5. Equation of state (EOS) for AC with f_c = 4.6 MPa.

3.4. Softening Parameters b_1, b_2

The softening parameters (b_1, b_2) shown in Equations (12) and (13) control the material softening behaviour after peak stress. These parameters are obtained from experiments, as detailed below.

- Value of b_1 from the uniaxial compressive test:

  The uniaxial compressive test was conducted for AC according to ASTM 1074. The detailed test results and setup can be further referred to [11]. Based on the test results, it is found that the corresponding strain at peak stress (f_c = 4.6 MPa) was about 0.018, and the final failure strain was about 0.1, which was higher than that of concrete. The Young's modulus obtained from stain gauges attached at the middle height of the sample was 598 MPa. Based on experimental results, the compressive energy G_c was calculated at 15.1 MPa·mm. Hence, for example, the b_1 value for a 10-mm mesh size was calculated as 3.45 using the method stated in Section 2.1.
- Value of b_2 from the fractural test:

  The value of b_2 is determined by fracture energy G_f, which can be obtained from the uniaxial tensile test or the three-point single-edge notched beam test (SNB). In the current study, the SNB test was carried out to evaluate fracture energy G_f for the AC material. The detailed theory about the SNB test can be found in the established literature [19]. Therefore, only the test result is presented here. In the SNB test, the compacted AC beam was fabricated with a dimension of

$400 \times 100 \times 100$ mm^3 depth. A mechanical notch was sawn with a depth of 20 mm, which gave a ratio of notch to beam depth of 0.2. The simply supported sample with a span length of 340 mm was tested under a 35 °C temperature. From the test, fracture toughness K_{IC} can be obtained according to the formula suggested by Karihaloo and Nallathambi [19]. Then, the fracture energy G_f is calculated using:

$$G_f = \frac{(1 - v^2)K_{IC}^2}{E} \tag{18}$$

in which E is the elastic modulus and ν is Poisson's ratio.

The parameter b_2 is further determined by assigning fracture energy G_f in the use of single element simulation of the uniaxial tensile test. The b_2 is then obtained via an iterative procedure until the area under the stress-stain curve from the single element simulation coincides with the value of G_f/w_c. The parameters obtained from SNB and single element simulation for AC (f_c = 4.6 MPa) are summarized in Table 2.

Table 2. Parameters from the single-edge notched beam test (SNB) and single element simulation.

Parameters	Unit	Value
K_{IC}	MPa·mm $^{1/2}$	12.2
ν	-	0.35
E	MPa	598
G_f	MPa·mm	0.221
w_c	mm	40
G_f/w_c	-	0.00554
f_t	MPa	0.7
b_2	-	0.2

3.5. Strain Rate Effect

The DIF curve for AC under different strain rates was obtained using servo hydraulic fast loading tests and the split Hopkinson pressure bar (SHPB) test in the current study. The strain rate produced by the servo hydraulic machine was approximately 10^{-5} to 1 s^{-1}, and the higher strain rate loading was obtained through SHPB testing. The detailed setup and procedure for SHPB and the hydraulic test for AC can be referred to Wu [20]. The DIF value for compressive (DIF_c) and tensile (DIF_t) strength obtained from the test are given as:

$$DIF_c = \frac{f_{dc}}{f_{sc}} = 3.18 + 1.098\log_{10}(\dot{\varepsilon}) + 0.1397{\log_{10}}^2(\dot{\varepsilon}) \quad \text{for} \quad \dot{\varepsilon} \leq 100s^{-1}$$

$$DIF_c = \frac{f_{dc}}{f_{sc}} = 21.39\log_{10}(\dot{\varepsilon}) - 36.76 \quad \text{for} \quad 100s^{-1} < \dot{\varepsilon} \leq 200s^{-1} \tag{19}$$

$$DIF_t = \frac{f_{dt}}{f_{st}} = 1.86 + 0.1432\log_{10}(\dot{\varepsilon}) \quad \text{for} \quad \dot{\varepsilon} \leq 15s^{-1}$$

$$DIF_t = \frac{f_{dt}}{f_{st}} = 6.06\log_{10}(\dot{\varepsilon}) - 5.024 \quad \text{for} \quad 15s^{-1} \leq \dot{\varepsilon} \leq 100s^{-1} \tag{20}$$

However, for the compressive DIF curve, a numerical modelling of the SHPB test adopting this DIF curve found that the initial segment of this curve matched the experimental results very well, while the numerical model results for strain rate larger than 100 s^{-1} seemed to overestimate the stress. This can be due to the "double counting" of the inertia effect in the numerical modelling when the strain rate exceeded 100 s^{-1}. Hence, in the current model, the second segment in compressive DIF (Equation (19)) is ignored when the strain rate exceeds 100 s^{-1}. Beyond this, the DIF is assumed to remain a constant value. For the tensile DIF curves, in the macro-level numerical model, the

KCC material model cannot capture the aggregate interlocking that propagates the micro-cracking and energy dissipation beyond the localization zone [20–22]. Therefore, the above tensile DIF curve (Equation (20)) with two branches is used in the numerical model. The tensile and compressive DIF curves of asphalt concrete used in numerical model are summarized in Figure 6.

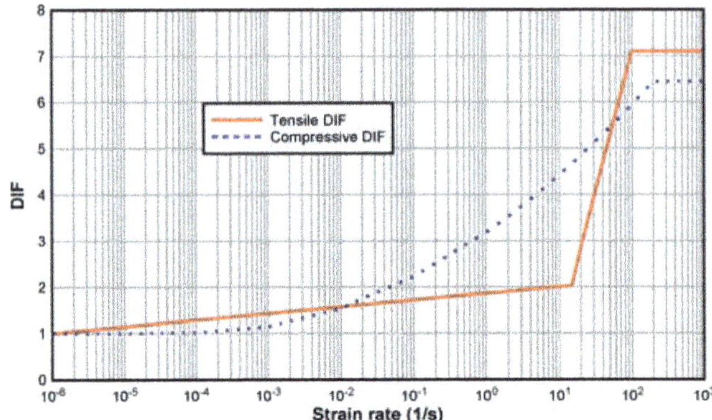

Figure 6. Tensile and compressive dynamic increase factor (DIF) curve used in the numerical model for asphalt concrete.

4. Application Example

4.1. Case for Blast Loading

4.1.1. Field Blast Test

In this section, the AC material is used in the multi-layer pavement. Figure 7 shows the cross-sectional view of the multi-layer pavement slab including 100 mm-thick engineering cementitious composite (ECC) at the bottom, 100 mm-thick high strength concrete (HSC) in the middle and 75 mm-thick AC at the top. The geogrid (GST) was placed at the middle of the AC layer to reinforce the AC material. The numerical model for the multi-layer pavement system under blast load is developed based on the configuration of the full-scale field blast test [23]. Selected key features of the field blast test and numerical model are presented below.

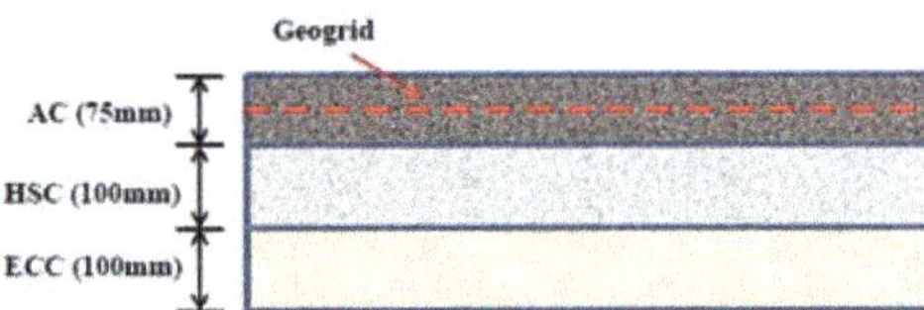

Figure 7. Cross-section of the multi-layer pavement slab. HSC, high strength concrete; ECC, engineering cementitious composite.

In the field blast test, a bomb with the equivalent of a 7.3 kg Trinitrotoluene (TNT) charge weight was selected for testing and placed at about 170 mm above the surface of the pavement. The charge weight was evaluated based on the typical terrorist weapon attack. This multi-layer pavement slab was cast at site with a dimension of 2800 mm × 2800 mm × 275 mm (width × depth × thickness). The pavement slab was anchored to the ground to simulate the practical boundary condition, and a vertical anchor was installed at each corner. Figure 8 presents the multi-layer pavement slab before the blast load.

Figure 8. Plan view of the multi-layer pavement slab before the blast event.

Various instruments were installed onto the slab to measure its responses during blast loading. Figure 9 shows the instrumentation installed on the pavement slab. Four accelerometers were installed at the middle of the side of the slab to measure both vertical (V1 and V2 in Figure 9) and horizontal accelerations (H1 and H2 in Figure 9). The accelerometers were mounted onto steel frames that were cast together with the slab. Three total pressure cells (TPC) (TPC1, TPC 2 and TPC3 in Figure 9) were buried in the soil just below the slab to measure the pressure transferred from the pavement slab.

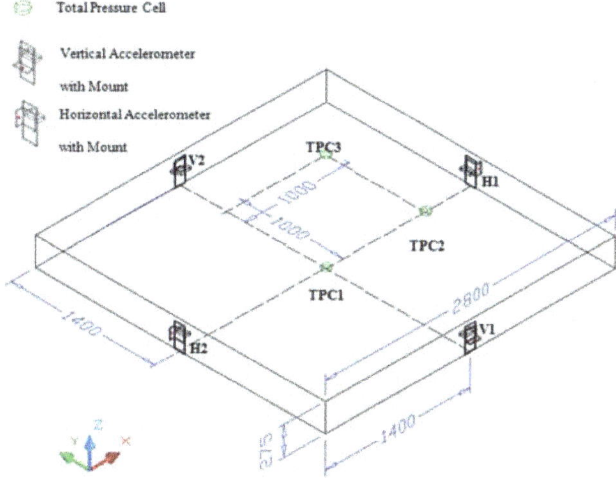

Figure 9. Layout of instrumentation for field blast test. TPC, total pressure cell.

A numerical model is established using LSDYNA package [24]. In this model, the slab and foundation soil are discretised in space with one point Gauss integration eight-node hexahedron Lagrange element. Only a quarter of the slab is modelled with symmetric boundary conditions (as shown in Figure 10). The geogrid is spatially discretised with the shell element, and it is assumed that the geogrid is fully bonded within the AC layer. The anchors on the pavement slab are simulated as the fixed points (fixed in vertical direction) in the corresponding position in the numerical model. The non-reflection boundary is applied on the side and bottom of the foundation soil to model semi-infinite space. Based on the mesh study on the numerical model, the 10-mm element size is adopted for the pavement slab, geogrid and soil mass. The blast pressure is extracted from AUTODYN and used for numerical model analysis. The detailed process of applying pressure to the pavement surface can be referred to elsewhere [20,25].

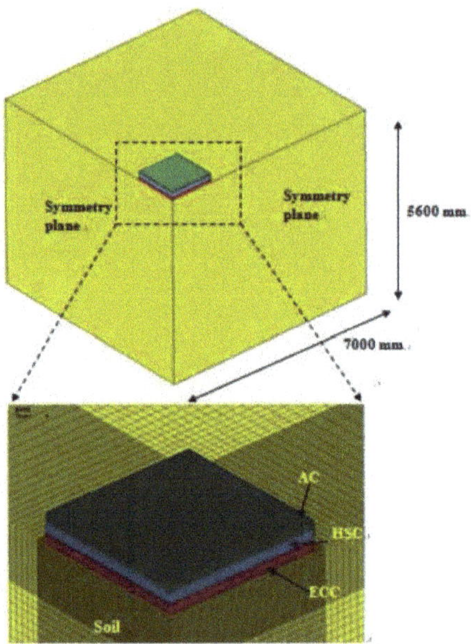

Figure 10. Finite element model of the multi-layer pavement slab.

In the numerical model, AC, ECC and HSC are all grouped as concrete-like materials and modelled by the KCC model [20,23]. The basic parameters for these three materials are listed in Table 3. The parameters for AC material in the KCC model can be found in Section 3. While the process of the determination of HSC and ECC parameters for KCC model is the same as that mentioned in Section 3, the detailed parameters can be referred to Wu [20]. The Drucker–Prager model and plastic-kinematic model [17] are employed to model the foundation soil and geogrid, respectively. The AUTOMATIC_SURFACE_TO_SURFACE contact algorithm is employed to model the interaction between pavement slab and soil. The TIEBREAK contact algorithm in LSDYNA is used to simulate the interface behaviour between HSC and AC layers. The parameters for the foundation soil, geogrid and interface property are in Tables 4–6, respectively.

Table 3. Basic properties of materials in the multi-layer pavement for the blast load.

Parameters	AC	HSC	ECC
Young's modulus E (MPa)	598	33,000	18,000
Compressive strength f_c (MPa)	4.6	55	64
Tensile strength f_t (MPa)	0.7	4.35	5
Poisson ratio v	0.35	0.2	0.22

Table 4. Parameters for Geogrid MG-100 using the plastic-kinematic model.

Parameters	Symbol	Units	Value
Density	ρ	kg/m^3	1030
Young's modulus	E	MPa	500
Poisson's ratio	v	-	0.3
Yield stress	σ_y	MPa	7.5
Tangent modulus	E_t	MPa	333
Thickness	t	mm	2.4
Erosion strain	ε_s	-	0.038

Table 5. Material properties of the soil mass.

Parameters	Symbol	Units	Value
Density	ρ	kg/m^3	2100
Shear modulus	G	MPa	13.8
Poisson's ratio	ν	-	0.3
Cohesion	c	kPa	62
Friction angle	ϕ	°	26

Table 6. Parameters for the interface simulation.

Parameters	Value
Contact type	TIEBREAK
Friction for static	0.71
Friction for dynamic	0.56

4.1.2. Numerical Result

The results of the numerical modelling of the multi-layer pavement under blast loading, with the incorporation of the above-mentioned material models, are summarized and compared with the blast test results. In the numerical results, the fringe level in the damage contour is the value for the scaled damage indicator δ, which is defined to describe the damage level of the material [14,20,23]. A scaled damage indicator δ is related to the effective plastic strain λ in the material: (i) at the yield surface, $\lambda = 0$, leading to $\delta = 0$; (ii) at the maximum strength surface, $\lambda = \lambda_m$, leading to $\delta = 1$; and (iii) at the residual strength surface, $\lambda = \lambda_r >> \lambda_m$, leading to $\delta = 1.99 \approx 2$. Thus, the δ value moving from 0 to 1 to 2 indicates that the failure surface migrates from the yield surface to the maximum strength surface and to the residual strength surface, respectively, as the material being stressed. In this study, when the residual strength of material reduces to 20% of its peak strength, the material seems to suffer severe failure. The plastic strain corresponded to that residual strength used to calculate the delta value. Both the laboratory and field test for AC material indicated that when δ was greater than 1.8, the material would be severely damaged.

The damage situation for the multi-layer pavement slab in field blast test is shown in Figure 11. Figure 11a shows that the blast pressure destroyed the upper half of the AC layer above the GST reinforcement. It is also noted that only the centre of the GST piece was burned off during the blast event. Figure 11b shows the resulting damage on the HSC layer after removing the top layer of asphalt. From this figure, it could be seen that the crater was very shallow and did not punch through the whole layer, and a crater of around 700 mm in diameter and a depth of 10 mm was formed on the HSC layer.

(a)

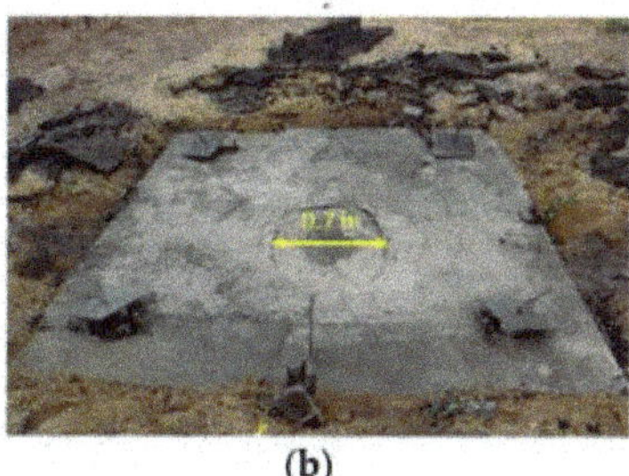
(b)

Figure 11. Damage of the multi-layer pavement after blast loading: (**a**) damage pattern of the AC layer in field blast test; (**b**) damage pattern of the HSC layer in field blast test.

The damage contour for the AC layer in numerical model is shown in Figure 12a. Comparing Figures 11a and 12a, it is observed that the damage pattern in the numerical model is symmetrical, while that in the field measurement was skewed. This is because the bomb in the field was not placed at the centre of the slab, and one side of the AC layer was more severely damaged than the other. Shear cracking near the anchor point was observed in the numerical model, which was similar to the experimental observations in the field test. It could be concluded that the basic failure pattern given by the numerical model agrees well with the results obtained from the field-testing. Figure 12b shows the damage pattern for the HSC layer. Comparing Figure 11b with Figure 12b, the damage pattern for HSC is very consistent between field measurement and numerical results. The diameter of the crater was about 750 mm in the numerical model, which is very close to that of the blast test result. As shown in Figure 12b, shear cracks are also observed near the anchor points. Based on the damage pattern in the field blast test, the crater on the top face of the HSC is shown to be shallow and with a thickness of less than 10 mm. However, after cracking occurred at the bottom of the HSC layer, the numerical model shows that the bottom of the HSC layer has experienced severe cracking. This might be due to the combination of the bending of the HSC layer under the blast load and the reflection of the stress wave at the bottom interface. In the numerical model, the interface between HSC and ECC is assumed to be fully bonded. However, ECC is more flexible than HSC, and thus, it would cause tensile stress at the bottom of the HSC layer when deformed together. The compression stress wave from the top face would also travel within the HSC layer and reflect as a tension stress at the interface, which could cause spalling. Based on the damage pattern in the numerical model, the HSC layer might be considered having failed, while the field observation suggests that HSC may have partially failed.

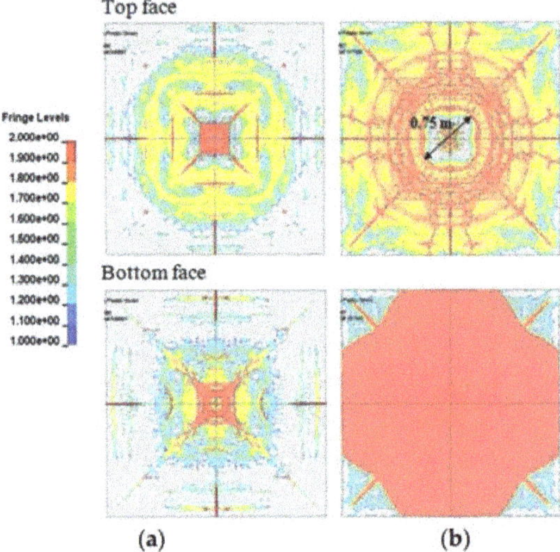

Figure 12. Damage contour for the AC and HSC layer in the multi-layers pavement: (**a**) damage contour of the AC layer in the numerical simulation; (**b**) damage contour of the HSC layer in the numerical simulation.

In the field blast test, the four accelerometers were installed at the mid-side of pavement slab (as shown in Figure 9). These accelerometers were used to measure the vertical and horizontal acceleration of the pavement slab subjected to blast loading. For the horizontal acceleration, due to the centre of the charge being closer to one side of the pavement slab; there were two different horizontal acceleration readings; while in the numerical model, it was assumed that the explosive occurred in the centre of the pavement slab. Thus, in this section, only the vertical acceleration from the field blast test was compared with that of the numerical model. In the numerical model, the raw nodal acceleration contained considerable numerical noise. The ELEMENT_SEATBELT_ACCELEROMETER could be used to eliminate numerical noise and obtain more accurate node acceleration. The vertical acceleration from the blast testing is compared with that of the numerical model as shown in Table 7. The results from both the blast testing and the numerical simulation are comparable. The maximum difference of vertical acceleration between the blast testing and the numerical model is about 10%, and the numerical model predicted slightly higher in the vertical acceleration than that of the blast test.

Table 7. Vertical acceleration of the multi-layer pavement slab.

Item	Field Trial Test	Numerical Result	Deviation from Field Trial Test
Max. vertical acceleration (m/s^2)	35,400	38,870	10%

The pressure values in the corresponding points in the numerical model are compared with pressures obtained from the blast test, as summarized in Table 8. The layout of the total pressure cell in the blast could be referred to Figure 9. The pressure values from the numerical simulation are shown to be close to that from the blast test for TPC2; while for TPC3, it has a 20% discrepancy with the numerical simulation considering the inherent variation in the blast test. TPC1 was damaged during the blast test, and hence, no pressure reading was recorded from it. The numerical model predicts that the pressure might be as high as 13 MPa at that point, which is far beyond the maximum measurement capacity of the pressure cell installed. That can explain why TPC1 was destroyed due to the overwhelming blast loading.

Table 8. Peak reading for the total pressure cell.

Item	Field Blast Test (kPa)	Numerical Result (kPa)	Deviation from Field Trial Test
TPC1	Destroyed	13,393	Sensor destroyed as pressure >> range
TPC2	273	267	2%
TPC3	200	241	20%

4.2. *Case for Impact Loading*

4.2.1. Laboratory Drop Weight Impact

The multi-layer pavement slab was subjected to an 1181-kg drop weight impact. The cross-section of the multi-layer pavement is the same as that given in Figure 7. The basic mechanical properties of the, HSC, ECC and AC are determined according to ASTM standards, and the results are summarized in Table 9. The drop weight was a cylindrical projectile with a hemispheric head (100 mm in diameter and made with high strength steel), and the pavement slab was subjected to two times of impact from the same drop height of 1.5 m. During the test, the multi-layer pavement slab was placed on the top of compacted soil/sand in a steel strong box. Directly below the slab was the geocell (MiraCell MC-100) which was filled with compacted soil/sand. This was to enhance the strength of the soil/sand layer and to provide a solid sub-base as to simulate the practical condition. The setup for the multi-layer pavement slab is given in Figure 13. Various instruments were also installed to monitor the response of the pavement during the drop weight test. Figure 14 shows the positioning of the potentiometers. A photodiode system was used to trigger the data acquisition system during the test. It consists of two photodiodes and two laser sources placed 100 mm vertically apart. The data acquisition system would be triggered when the falling projectile crosses the top laser emitter. Impact velocity could be determined using the time interval that the projectile took to cross the second laser emitter.

Table 9. Basic properties of the materials in the multi-layer pavement in the drop weight test.

Parameters	AC	HSC	ECC
Young's modulus E (MPa)	598	40,000	18,000
Compressive strength f_c (MPa)	4.6	90	80
Tensile strength f_t (MPa)	0.7	4.35	5
Poisson ratio ν	0.35	0.2	0.22

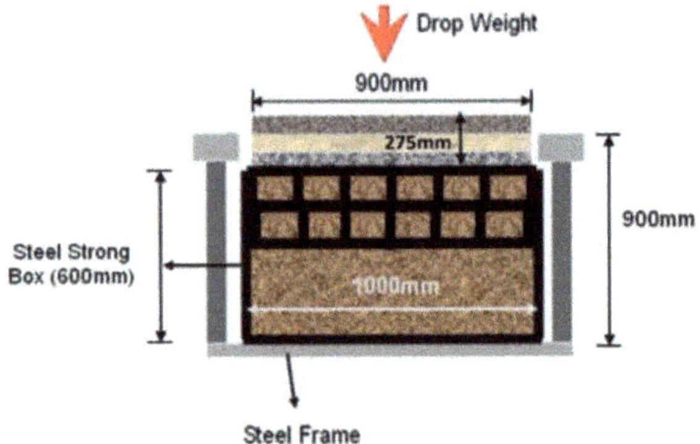

Figure 13. Setup for the impact test.

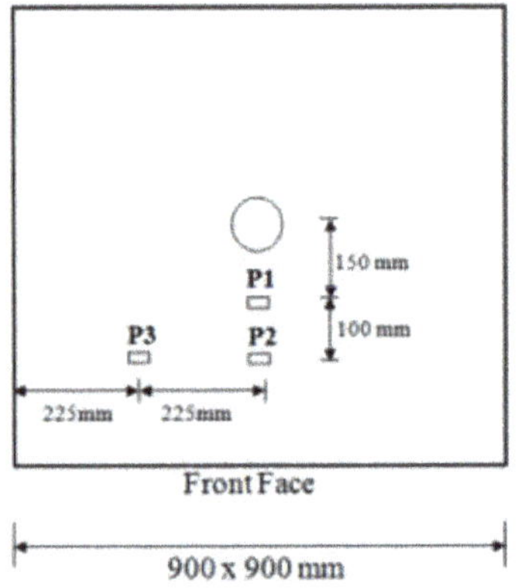

Figure 14. Positioning of the potentiometers.

In the numerical model, the AUTOMATIC_SURFACE_TO_SURFACE contact algorithm is employed to model the interaction between pavement slab and soil. The contact algorithm TIEBREAK is also used to simulate the interface behaviour between the AC and HSC layer, which is the same as that in the numerical model for blast loading as mentioned in Section 4.1.2. As the impactor might penetrate the AC layer, the erosion method is employed in the simulation, in which the maximum plastic failure strain of 0.2 is used to delete the distorted element once the actual strain exceeds this predefined failure strain. The multi-layer pavement slab, the drop weight head and soil mass are discretised in space with one point Gauss integration eight-node hexahedron elements. The geogrid is simulated with the four-node Belytschko–Tsay shell element that allows no bending resistance. In addition, the PLASTIC_KINEMATIC material model is employed to describe the bi-linear behaviour of the geogrid under tensile loading [23]. The parameters used for the geogrid are the same as given in Table 4.

The Drucker–Prager model is used to simulate soil mass. In the laboratory drop weight test, the upper soil layers were compacted and reinforced with geocell material, which would enhance the strength of the soil, and the lower layer has no reinforcement. Hence, in the numerical model, it is necessary to consider the function of the geocell material. From the laboratory test on the geocell-encased sand [26], it is observed that the geocell confinement did not change the friction angle of soil while significant cohesion occurred in the granular soil, which indicated that for the geocell-reinforced sand layer, the strength and stiffness behaviour of the soil would be enhanced. However, in the numerical model, it is difficult to model and mesh the geocell material due to its complex geometry. Hence, it would be preferable to use the composite model to consider the enhancement of the shear strength and stiffness of the geocell-reinforced sand layer. Madhavi et al. [27] purposed an empirical equation to calculate Young's modulus of the geocell-reinforced sand using the secant tensile modulus of the geocell material and Young's modulus parameter of the unreinforced sand, which could be expressed as:

$$E_r = 4(\sigma_3)^{0.7}\left(K_u + 200M^{0.16}\right) \tag{21}$$

in which E_r is the Young's modulus of the geocell-reinforced sand, M is the secant modulus of the geocell material at axial strain 2.5% in kN/m and σ_3 is the confining pressure from the geocell in kPa. k_u is the dimensionless modulus parameter of the unreinforced sand, which is a modulus number in the hyperbolic model developed by Duncan and Chang [28]. The confining pressure σ_3 could be calculated as:

$$\sigma_3 = \frac{2M}{D_0}\left(\frac{1-\sqrt{1-\varepsilon_a}}{1-\varepsilon_a}\right) \tag{22}$$

where D_0 is the initial diameter of the geocell and ε_a is the axial strain of the geocell at failure; the induced cohesion in the geocell-reinforced sand is then related to the increase in the confining pressure σ_3:

$$c_r = \frac{\sigma_3}{2}\sqrt{K_P} \tag{23}$$

in which c_r is the enhanced cohesion and k_p is the coefficient of passive earth pressure.

In the current study, the geocell MC-100 from Polyfelt is used. The geocell dimension is a rhombus with two diagonal lengths of 203 mm and 244 mm. Thus, the equivalent diameter is calculated as about 177.5 mm. The secant modulus M for the geocell is obtained as 278 kN/m from the tensile test [20]. Additionally, the value of ε_a is taken as 4.8%. The modulus parameter k_u for unreinforced sand in the current study is taken as 727 MPa according to the curve fitting from the triaxial test, and hence, the confining pressure, enhanced cohesion and Young's modulus for geocell-reinforced sand could be calculated based on Equations (21)–(23). The parameters for the unreinforced sand and geocell reinforced sand are summarized in Table 10.

Table 10. Material properties of the foundation soil.

Parameters	Reinforced Sand Layer	Unreinforced Sand Layer
Density, ρ (kg/m^3)	1600	1600
Young's modulus E (MPa)	103.5	40
Shear modulus, G (MPa)	39.8	15.4
Poisson's ratio, ν	0.3	0.3
Cohesion, c (MPa)	0.089	0.001
Friction angle, ϕ (°)	40	40

During the impact test, the deformation of the drop weight head was negligible compared to the deformation of the pavement slab. Hence, the drop weight head is modelled with a rigid body in the current study. For the configuration of the drop weight head, the simple cylindrical shape is modelled instead of modelling the head with weight mass (as shown in Figure 15). The simple cylindrical head has a diameter of 100 mm with a length of 1292 mm. The total mass for the simple cylindrical head is about 1181 kg, from which the density of the drop head would be obtained. The properties for the drop head are listed in Table 11. The convergence study is conducted, and it was found that a 5-mm element size gave a stable response, which is therefore applied for the simulation. The numerical model of the multi-layer pavement slab under drop weight impact load is given in Figure 16.

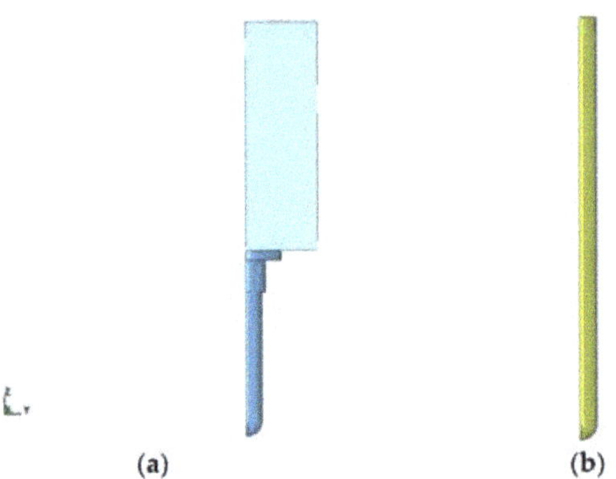

Figure 15. Configuration of the drop weight head (quarter model): (**a**) standard configuration; (**b**) simple configuration.

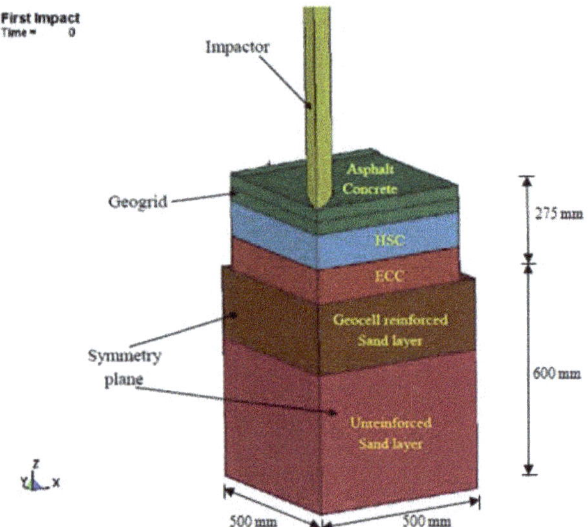

Figure 16. Numerical model for a multi-layer pavement slab under drop weight impact (quarter model).

Table 11. Properties of the drop weight head.

Parameters	Drop-Weight Head
Young's modulus, E (GPa)	207
Yield stress, f_y (MPa)	500
Poisson ratio, ν	0.3
Density, ρ (kg/m^3)	118,000

For the 1.5-m drop weight impact, the drop weight head is assigned with 5.02 m/s for the first impact. For the second impact, due to the penetration of the AC layer in the first impact, the distance between the laser diode system and the face of the pavement slab is increased, while in the numerical model, the impact head is just placed at the position right before reaching the surface. Hence, the initial velocity in numerical model for the second impact is determined as the sum of the experimental recorded velocity and velocity caused by gravity acceleration. Thus, the velocity is calculated as 5.06 m/s for the second impact with the gravity acceleration of 9.8 m/s^2.

After the first impact, the fully restarted method in LSDYNA is used to conduct the second impact simulation. At the beginning of the second impact, the stress and residual velocity within the pavement slab and the velocity from the first impact are set to be zero. The damage factor and plastic strain is retained in order to check the accumulated damaged behaviour after the second impact. The simulation of the second impact is carried out when the downward velocity of impactor reached zero, as it is very time consuming to continue to simulate the vibration of pavement slab after impact. It should be noticed that the pavement slab stopped rebounding, which might bring numerical errors (i.e., energy unbalance).

4.2.2. Numerical Result

The damaged situations for the multi-layer slab under the first drop weight impact for the experiment and numerical model are presented in Figure 17. It is found from the numerical simulation that the AC layer is penetrated through, and the drop weight head is impeded by the HSC layer due to the high compressive strength. Such findings from the numerical model are consistent with the observation in the physical test. It is observed that no severe and moderate damage happened in the ECC layer. Additionally, the integrity of both HSC and ECC layers is kept after the first impact.

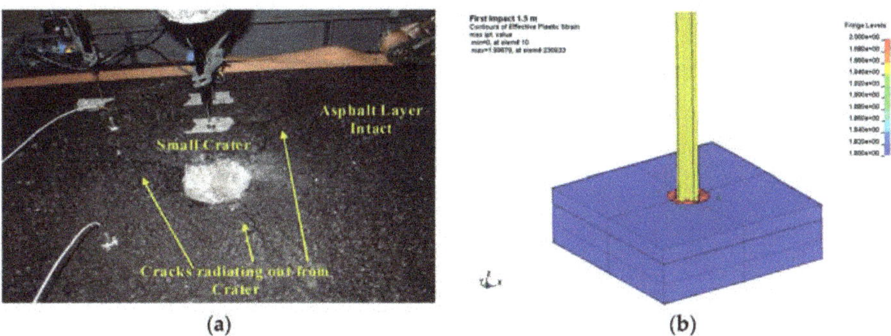

Figure 17. Damage of the multi-layer slab after the first impact: (**a**) damage pattern of slab in the laboratory test; (**b**) damage pattern of slab in the numerical simulation.

The damage situations for the multi-layer pavement slab under the second drop weight impact and the numerical model are given in Figure 18. The simulation results indicate that the HSC layer stopped the impactor; however, severe damage happens at the top surface and in the middle at the side of the HSC layer. Severe damage occurs at the rear surface of the HSC layer, as well. For the ECC layer, it is shown that the cracking occurs and propagates from the centre at the top surface. However, the damage area is smaller than that of the HSC layer. The cracks are also found in the middle at the side of the ECC layer, and similar to the experiment, the crack does not propagate through the thickness of the layer. Figure 19 compares the numerical results with the experimental results about the rear surface of ECC layer after the second impact for the multi-layer slab. It is found that the damaged contour and cracking pattern were similar. Such results further indicate the reasonable close agreement between the numerical analysis and the experimental results using the KCC model for AC material.

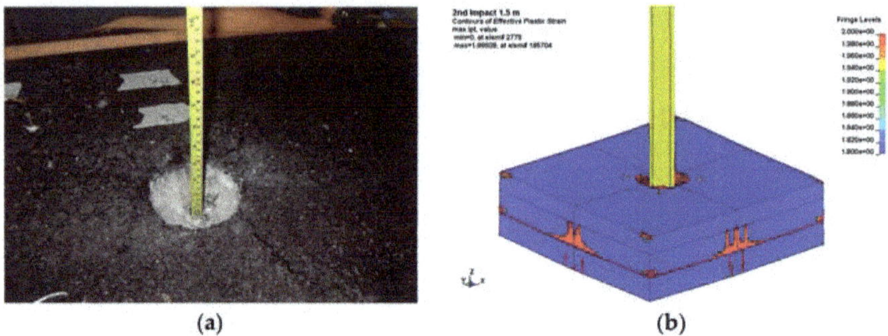

Figure 18. Damage of multi-layer slab after the second impact: (**a**) damage pattern of slab in the laboratory test; (**b**) damage pattern of slab in the numerical simulation.

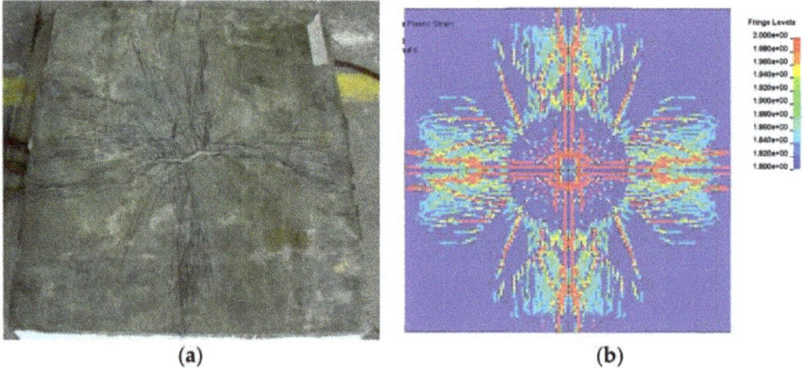

Figure 19. ECC bottom after the second impact: (**a**) damage pattern of the bottom of ECC layer in the laboratory test; (**b**) damage pattern of the bottom of ECC layer in the numerical simulation.

Table 12 compares the numerical results with the experimental data for the multi-layer slab under two times of the drop weight impact. After the first impact, the settlement from the numerical results is close to that obtained from the experiment at P2. A big deviation is found in between the numerical results and experiment results on the settlement at P3. This might be attributed to the dislocation of the potentiometer at P3 upon impact, which could lead to a misreporting of the settlement at P3. In addition, the settlement at P1 is lower in the numerical results compared with the experimental results. This could be attributed to the erosion process in the numerical model or, more specifically, the energy released phenomena. At the second impact, since no erosion technology is employed for the HSC layer, the numerical results are more reasonable. In the impact simulation, the HSC layer suffered deformation instead of being penetrated, and hence, no erosion technique was adopted to trigger the energy released phenomena. Thus, the settlement profile of the pavement slab decreased with the increase of the radial distance from the centre, which indicated that the whole pavement slab bent during the impact. In general, the numerical models provide reasonable estimations on the settlements of the multi-layer slab under impacts, in particular under the second impact load, though discrepancies exist at certain regions.

Table 12. Summary of the settlements of the multi-layer slab from both numerical and experiment results.

Potentiometer	1st Impact			2nd Impact		
	Test Result	FE Results	FE/Test	Test Result	FE Results	FE/Test
P1 (mm)	—	−1.54	—	−3.62	−7.62	2.10
P2 (mm)	−1.81	−1.70	0.94	−7.60	−6.63	0.87
P3 (mm)	−5.27	−1.86	0.35	−5.58	−4.83	0.87
			−ve → downwards			

5. Conclusions

In this paper, the formulations of strength surfaces and damage factor in the KCC model are discussed and evaluated. It is found that the KCC model would reproduce the real behaviour of concrete-like material under severe loadings, such as blast and impact loadings. Then, the key parameters controlling the behaviour of AC material under dynamic loadings are discussed, and the parameters of AC materials using the KCC model are determined by laboratory static and dynamic tests. The proposed damage factor in this paper results in smoother descending than the original damage factor and also a higher failure strain that could improve the simulation of the behaviour of AC material. Furthermore, the numerical modelling of multi-layer pavement, in which the AC material served as the surface layer, under blast and impact loading is conducted using the KCC model. These results are compared with that of the field blast test and the laboratory drop weight impact tests, and it is found that the KCC model for the AC material can reproduce the real material behaviour under blast and impact loading.

Acknowledgments: Part of this research was sponsored by the National Basic Research Program of China (No. 2015CB058003), the Beijing Municipal Natural Science Foundation (No. 8172010) and the Youth Teacher Training Scheme from the Shanghai Education Committee (No. ZZGCD15053). Thanks are given here.

Author Contributions: Jun Wu conceived of and designed the experiments. Jun Wu performed the experiments and analysed the data. Liang Li developed the numerical model and prepared the paper. Xiuli Du discussed the numerical results and provided a suggestion on the overall organization of the paper. Xuemei Liu conducted the numerical model of the SHPB test for concrete-like materials and revised the manuscript.

Conflicts of Interest: The authors declare no conflict of interest.

References

1. Huang, Y.H. *Pavement Analysis and Design*; Pearson Education, Inc.: Upper Saddle River, NJ, USA, 2004.
2. Tashman, L.; Masad, E.; Little, D.; Zbib, H. A Microstructure-Based Viscoplastic Model for Asphalt Concrete. *Int. J. Plast.* **2005**, *21*, 1659–1685. [CrossRef]
3. Seibi, A.C.; Sharma, M.; Ali, G.; Kenis, W.J. Constitutive Relations for Asphalt Concrete under High Rates of Loading. *Transp. Res. Rec.* **2001**, *1767*, 111–119. [CrossRef]
4. Park, D.W.; Martin, A.E.; Lee, H.S.; Masad, E. Characterization of Permanent Deformation of an Asphalt Mixture Using a Mechanistic Approach. *KSCE J. Civ. Eng.* **2005**, *9*, 213–218. [CrossRef]
5. Tang, W.; Ding, Y.; Yuan, X. The HJC Model Parameters of an Asphalt Mixture. In Proceedings of the 9th International Conference on the Mechanical and Physical Behaviour of Materials under Dynamic Loading (DYMAT 2009), Brussels, Belgium, 7–11 September 2009; pp. 1419–1423.
6. Tekalur, S.A.; Shukla, A.; Sadd, M.; Lee, K.W. Mechanical Characterization of a Bituminous Mix under Quasi-Static and High-Strain Rate Loading. *Constr. Build. Mater.* **2009**, *23*, 1795–1802. [CrossRef]
7. Chang, G.K.; Meegoda, J.N. Micromechanical Simulation of Hot Mixture Asphalt. *J. Eng. Mech.* **1997**, *123*, 495–503. [CrossRef]
8. Collop, A.C.; McDowell, G.R.; Lee, Y. Use of the Distinct Element Method to Model the Deformation Behavior of an Idealized Asphalt Mixture. *Int. J. Pavement Eng.* **2004**, *5*, 1–7. [CrossRef]
9. Liu, Y.; Dai, Q.; You, Z. Viscoelastic Model for Discrete Element Simulation of Asphalt Mixtures. *J. Eng. Mech.* **2009**, *135*, 324–333. [CrossRef]

10. You, Z.; Liu, Y.; Dai, Q. Three-Dimensional Microstructural-Based Discrete Element Visoelastic Modeling of Creep Compliance Tests for Asphalt Mixtures. *J. Mater. Civ. Eng.* **2010**, *23*, 79–87. [CrossRef]
11. Riedel, W.; Hiermaier, K.T.S. Penetration of reinforced concrete by BETA-B-500-numerical analysis using a new macroscopic concrete model for hydrocodes. In Proceedings of the 9th International Symposium on Interaction of the Effect of Munitions with Structures, Berlin, Germany, 3–7 May 1999; pp. 315–322.
12. Adley, M.D.; Frank, A.Q.; Danielson, K.T.; Akers, S.A.; O'Daniel, J.L.; United States Army Corps of Engineers; Engineer Research and Development Center (U.S.); Geotechnical and Structures Laboratory (U.S.). *The Advanced Fundamental Concrete (AFC) Model*; Technical Report ERDC/GSL TR-10-51; Army Engineer Research and Development Center: Vicksburg, MS, USA, 2010.
13. Sherburn, J.A.; Hammons, M.I.; Roth, M.J. Modeling Finite Thickness Slab Perforation Using a Coupled Eulerian–Lagrangian Approach. *Int. J. Solids Struct.* **2014**, *51*, 4406–4413. [CrossRef]
14. Malvar, L.J.; Crawford, J.E.; Wesevich, J.W.; Simons, D. A plasticity concrete material model for DYNA3D. *Int. J. Impact Eng.* **1997**, *19*, 847–873. [CrossRef]
15. Polanco-Loria, M.; Hopperstad, O.S.; Borvik, T.; Berstad, T. Numercial Predictions of Ballistic Limit for Concrete Slabs Using a Modified Version of the HJC Concrete Model. *Int. J. Impact Eng.* **2008**, *35*, 290–303. [CrossRef]
16. Holmquist, T.J.; Johnson, G.R.; Cook, W.H. A computational constitutive model for concrete subjected to large strains, high strain rates, and high pressures. In Proceedings of the 14th International Symposium on Ballistics, Quebec, QC, Canada, 26–29 Sepetember 1993; pp. 591–600.
17. Chen, W.F. *Constitutive Equations for Engineering Materials*; John Wiley & Sons: Hoboken, NJ, USA, 1982.
18. Malvar, L.J.; Crawford, J.E.; Wesevich, J.W. *A New Concrete Material Model for DYNA3D Release II: Shear Dilation and Directional Rate Enhancements*; Defense Nuclear Agency: Alexandria, VA, USA, 1996.
19. Karihaloo, B.L.; Nallathambi, P. Effective Crack Model for the Determination of Fracture Toughness (Kice) of Concrete. *Eng. Fract. Mech.* **1990**, *35*, 637–645. [CrossRef]
20. Wu, J. Development of Advanced Pavement Materials System for Blast Load. Ph.D. Thesis, National University of Singapore, Singapore, 2012.
21. Li, Q.M.; Meng, H. About the dynamic strength enhancement of concrete-like materials in a split Hopkinson pressure bar test. *Int. J. Solids Struct.* **2003**, *40*, 343–360. [CrossRef]
22. Lu, Y.B.; Li, Q.M. About the dynamic uniaxial tensile strength of concrete-like materials. *Int. J. Impact Eng.* **2011**, *38*, 171–180. [CrossRef]
23. Wu, J.; Chew, S.H. Field Performance and Numerical Modelling of Multi-Layer Pavement System Subject to Blast Load. *Constr. Build. Mater.* **2014**, *52*, 177–188. [CrossRef]
24. Livermore Software Technology Corporation (LSTC). *LS-DYNA Keyword User's Manual*; Livermore Software Technology Corporation (LSTC): Livermore, CA, USA, 2007.
25. ANSYS Inc. *AUTODYN Theory Manual*; Century Dynamics; ANSYS Inc.: Canonsburg, PA, USA, 2005.
26. Rajagopal, K.; Krishnaswamy, N.R. Behavior of Sand Confined with Single and Multiple Geocells. *Geotext. Geomembr.* **1999**, *17*, 171–181. [CrossRef]
27. Madhavi, L.G.; Rajagopal, K. Parametric Finite Element Analyses of Geocell-Support Embankments. *Can. Geotech. J.* **2007**, *44*, 917–927. [CrossRef]
28. Duncan, J.M.; Chang, C. Nonlinear Analysis of Stress and Strain in Soils. *J. Soil Mech. Found. Div.* **1970**, *96*, 1629–1653.

Article

Steady-State Creep of Asphalt Concrete

Alibai Iskakbayev [1], Bagdat Teltayev [2,*] and Cesare Oliviero Rossi [3]

1. Department of Mechanics, Al-Farabi Kazakh National University, Almaty 050040, Kazakhstan; iskakbayeva@inbox.ru
2. Kazakhstan Highway Research Institute, Almaty 050061, Kazakhstan
3. Department of Chemistry and Chemical Technologies, University of Calabria, Rende 87036, Italy; cesare.oliviero@unical.it

* Correspondence: bagdatbt@yahoo.com; Tel.: +7-701-760-6701

Academic Editors: Zhanping You, Qingli (Barbara) Dai and Feipeng Xiao
Received: 11 November 2016; Accepted: 25 January 2017; Published: 4 February 2017

Abstract: This paper reports the experimental investigation of the steady-state creep process for fine-grained asphalt concrete at a temperature of 20 ± 2 °C and under stress from 0.055 to 0.311 MPa under direct tension and was found to occur at a constant rate. The experimental results also determined the start, the end point, and the duration of the steady-state creep process. The dependence of these factors, in addition to the steady-state creep rate and viscosity of the asphalt concrete on stress is satisfactorily described by a power function. Furthermore, it showed that stress has a great impact on the specific characteristics of asphalt concrete: stress variation by one order causes their variation by 3–4.5 orders. The described relations are formulated for the steady-state of asphalt concrete in a complex stressed condition. The dependence is determined between stress intensity and strain rate intensity.

Keywords: asphalt concrete; creep curve; steady-state creep; strain rate; viscosity; stress intensity; strain intensity

---

1. Introduction

Asphalt concrete is one of the main materials used for highway pavements. Mechanical properties of an asphalt concrete are highly dependent on temperature and time of loading [1–3]. Therefore, the determination of the mechanical behavior of an asphalt concrete, taking into account the variation of the above-mentioned factors, has important practical value.

It is known that the basic methods for evaluating the mechanical behavior of viscoelastic materials are tests on creep and relaxation [4–6]. Technically, creep test investigations are easy to conduct, and their results make it possible to construct creep and long-term strength curves. Relaxation curves can be obtained from the creep curves by using known methods [6,7]. The long-term strength curves enable us to determine the service life of an asphalt concrete pavement.

The work in Reference [8] analyzes experimentally the process of uniaxial and triaxial creep for two asphalt concrete types (dense bitumen macadam and hot rolled asphalt) at a temperature range of 10 to 40 °C. It was determined that the creep curve of asphalt concretes has three characteristic stages, the second of which (steady-state creep stage) has a constant strain rate. The viscosity of asphalt concretes for this stage of creep curve depends nonlinearly on the stress.

As the result of experimental tests for four asphalt concrete types, Hassan [9] approximated the first and the second stages of the creep curve by power function. However, an attempt to establish a linear correlation relationship between an exponent of power function and an exponent of Paris law was not successful.

Soleimanbeigi et al. [10] obtained results experimentally that demonstrated the creep strain rate of recycled asphalt shingles with bottom ash (RAS-BA) increased the most with the increase of applied stress value. The relationship of creep strain rate with stress is described by the power function and showed that the strain and the strain rate of creep increased with a temperature increase.

The asphalt concrete test for uniaxial creep at four different temperatures and three levels of stress, with the purpose of permanent strain evaluation was performed by Mahan [11]. Test results showed that both stress and temperature impact greatly on creep strain and permanent strain. Having tested the cylindrical specimens of two asphalt concrete types for uniaxial creep during compression at the temperatures of 25, 40, and 60 °C and three levels of stress, Zhigang et al. [12] also determined that the creep curve of asphalt concrete had three stages, where the second stage had a constant strain rate. The authors defined creep strain as a viscous flow deformation, and also called the gradual reduction of deformation on the first stage and maintenance of strain rate as a constant on the second stage as the "consolidation effect". Either temperature or stress greatly impacts creep strain. The greater the temperature, the less the asphalt concrete resists strain. The greater the stress at a similar temperature, the faster the damage occurs. The time period for the test was 5000 s. A simple model of creep with five parameters was determined from a generalized model of Kelvin, and was proposed for the same maximum duration of deformation.

The work in Reference [13] used the test results of asphalt concretes based on the three point bending test at different temperatures and levels of stress to develop master curves of stiffness modulus.

In this paper, test results of hot fine-grained asphalt concrete samples on creep are presented. Creep tests were carried out by a direct tensile scheme until complete fracture of the asphalt concrete samples was obtained. The test temperature was 20 ± 2 °C. The applied stress was changed from 0.055 to 0.311 MPa. Creep curves under different loads of the asphalt concrete were constructed. Three characteristic stages of creep curves—the unsteady-state, the steady-state, and the accelerating creep stages—are shown. The first creep curve stage of asphalt concrete was satisfactorily approximated by Rabotnov's fractional exponential function [14]. A description of the third creep curve stage should be carried out on the basis of continuum damage mechanics approach [15,16], which is currently work in progress. Therefore, the second stage of the asphalt concrete creep curve is described.

2. Materials and Methods

2.1. Bitumen

Bitumen of grade 100–130, which met the requirements of the Kazakhstan standard [17], was used in this study. The bitumen grade on Superpave is PG (Performance Grade) 64–40 [18]. Basic standard indicators of the bitumen are shown in Table 1. Bitumen is produced by the Pavlodar processing plant from the crude oil of Western Siberia (Oil processing plant, Omsk, Russia) by the direct oxidation method.

Table 1. Basic standard indicators of the bitumen, ST RK: Standard of the Republic of Kazakhstan.

Indicator	Measurement Unit	Requirements of ST RK 1373	Value
Penetration, 25 °C, 100 gr, 5 s	0.1 mm	101–130	104
Penetration Index PI	-	−1.0, ... ,+1.0	−0.34
Tensility at temperature:	cm	-	-
25 °C	-	≥90	140
0 °C	-	≥4.0	5.7
Softening point	°C	≥43	46.0
Fraas point	°C	≤−22	−25.9
Dynamic viscosity, 60 °C	Pa·s	≥120	175.0
Kinematic viscosity	mm^2/s	≥180	398.0

2.2. Asphalt Concrete

Hot dense asphalt concrete of type B that met the requirements of the Kazakhstan standard [19] was prepared with the use of aggregate fractions of 5–10 mm (20%); 10–15 mm (13%); and 15–20 mm

(10%) from the Novo-Alekseevsk rock pit (Almaty region, Kazakhstan); sand of fraction 0–5 mm (50%) from the plant "Asphaltconcrete-1" (Almaty city, Kazakhstan); and activated mineral powder (7%) from the Kordai rock pit (Zhambyl region, Kazakhstan).

The bitumen content of grade 100–130 in the asphalt concrete was 4.8% by weight of dry mineral material. Basic standard indicators of the aggregate and the asphalt concrete are shown in Tables 2 and 3, respectively. A granulometric composition curve for the mineral part of asphalt concrete is shown in Figure 1.

Table 2. Basic standard indicators of the crushed stone.

Indicator	Measurement Unit	Requirements of ST RK 1284 [20]	Value	
			Fraction 5–10 mm	Fraction 10–20 mm
Average density	g/cm^3	-	2.55	2.62
Elongated particle content	%	≤25	13	9
Clay particle content	%	≤1.0	0.3	0.2
Bitumen adhesion	-	-	satisf.	satisf.
Water absorption	%	-	1.93	0.90

Table 3. Basic standard indicators of the asphalt concrete.

Indicator	Measurement Unit	Requirements of ST RK 1225	Value
Average density	g/cm^3	-	2.39
Water saturation	%	1.5–4.0	2.3
Voids in mineral aggregate	%	≤19	14
Air void content in asphalt concrete	%	2.5–5.0	3.8
Compression strength at temperature	MPa	-	-
0 °C	-	≤13.0	7.0
20 °C	-	-	3.4
50 °C	-	≥1.3	1.4

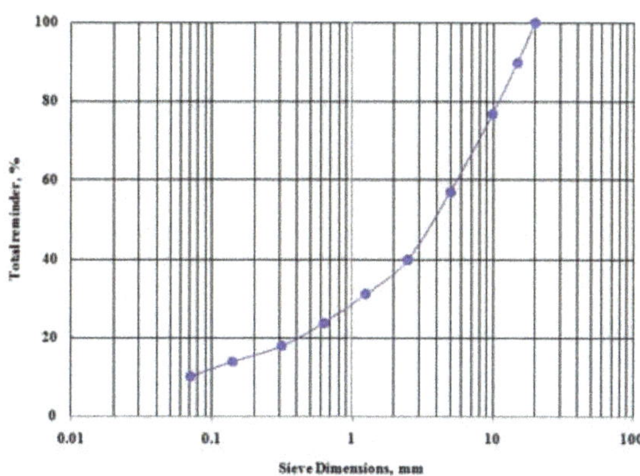

Figure 1. Granulometric curve of mineral part of the asphalt concrete.

2.3. Sample Preparation

Samples of the hot asphalt concrete in the form of a rectangular prism with dimensions 15,050 × 50 mm (Figure 2) were manufactured as follows. First, the asphalt concrete samples were prepared in the form of a square slab (Figure 3) using a Cooper compactor (model CRT-RC2S, Cooper, Nottingham, UK) (Figure 4) according to the standard in [21]. The samples were then cut from the asphalt concrete slabs in the form of a prism. Deviations in sizes of the beams did not exceed two millimeters.

Figure 2. Samples of the asphalt concrete with dimension 150 × 50 × 50 mm.

Figure 3. A square slab with dimension 305 × 305 × 50 mm.

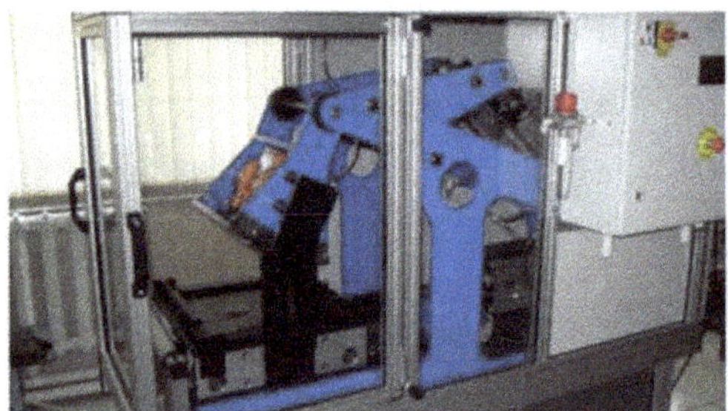

Figure 4. The Cooper compactor CRT-RC2S.

2.4. Test

Tests on creep were carried out on hot asphalt concrete samples in the form of a rectangular prism, according to the direct tensile scheme until complete failure was reached. The test temperature was equal to 20 ± 2 °C, stress was variable from 0.055 to 0.311 MPa. The tests were carried out in a specially assembled installation (Figure 5). The sample strain was measured by means of two clock type indicators while data were recorded on a video camera.

Figure 5. A specially assembled installation for creep test.

3. Results and Discussion

3.1. Creep Curve

Previous works [14,22,23] showed that an asphalt concrete creep curve—as with most viscoelastic materials—had three characteristic stages: stage I of unsteady-state creep with decreasing rate; stage II of steady-state creep with a constant (minimum) rate; and stage III of accelerating creep with increasing rate which precedes failure. The above studies presented test results of asphalt concrete for creep with relatively narrow ranges of stress variation. This work includes test results for seven values of stress from 0.055 MPa to 0.311 MPa. Practically, five samples of asphalt concrete were tested for each value of stress.

Figure 6 shows the creep curve for asphalt concrete at stress 0.117 MPa. It is clearly seen that the creep curve contains all three stages. It is important to underline that since the beginning of loading to the failure moment, the asphalt concrete passes three stages of deformation.

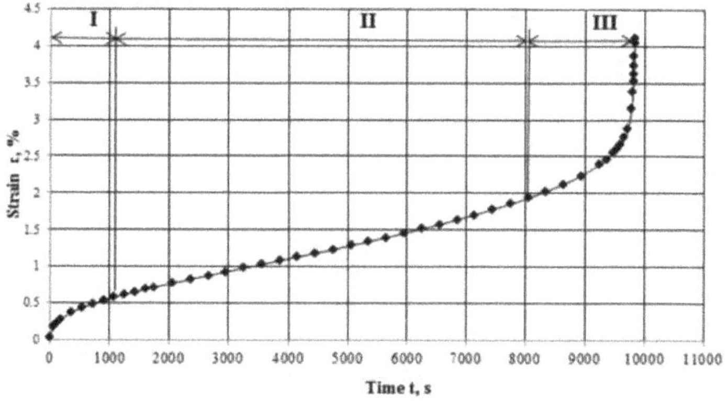

Figure 6. The asphalt concrete creep curve at stress 0.117 MPa.

3.2. Steady-State Creep

To solve practical problems, it is important to describe a creep curve obtained experimentally. Usually the third stage of the creep curve for the material is not considered in engineering calculations, as within its range the intensive accumulation of damage occurs, which results in a short life [16]. Therefore, the first and second stages of the creep curve are of main interest for mathematical description. The first stage of the creep curve for asphalt concrete was described with the use of Rabotnov's fractional exponential function [16] and was reported previously by the authors of Reference [14].

Analyzing the results for the creep experiment, we can represent the second stage of the creep curve as a straight line under all applied stresses with high accuracy, i.e., the deformation of asphalt concrete under constant stress occurs at a constant rate $\dot{\varepsilon}_2$ (Figure 7). Figure 8 shows that the strain rate depends on the stress, and its dependence is satisfactorily described by a power function. It should be emphasized that stress impacts greatly on strain rate: the increase of the stress by one order causes the increase of strain rate approximately by four orders.

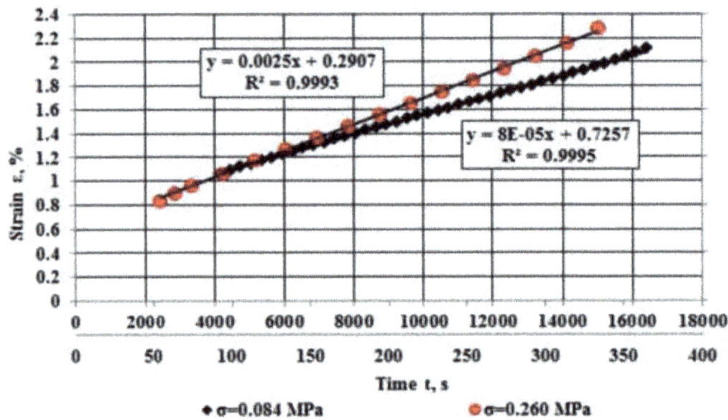

Figure 7. Stage II of asphalt concrete creep curves under stress between 0.084 and 0.260 MPa.

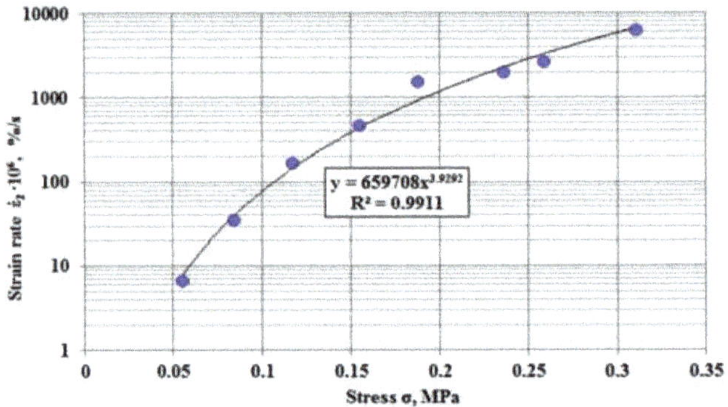

Figure 8. Dependence of steady-state creep rate on stress.

It is very important to know when steady-state creep starts and finishes, and the length of duration. As Figures 9–11 show, the specific time characteristics of asphalt concrete depend on stress and are satisfactorily approximated by a power function. It was found that the stress also impacts greatly on start point, end point, and steady-state creep duration, where the increase of stress for one order increases these time characteristics for 4.3–4.5 orders.

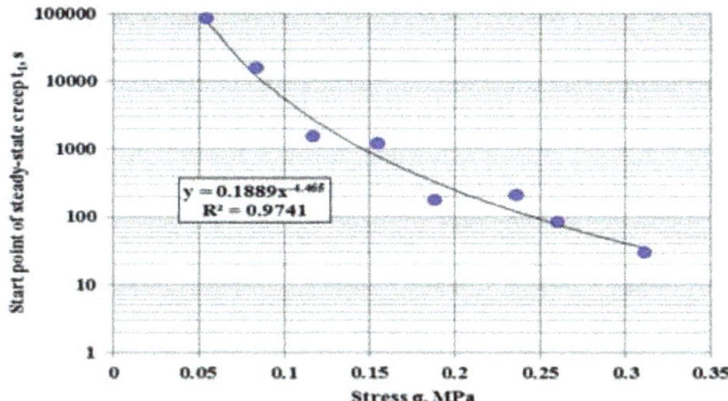

Figure 9. Dependence of start point for the stage of steady-state creep on stress.

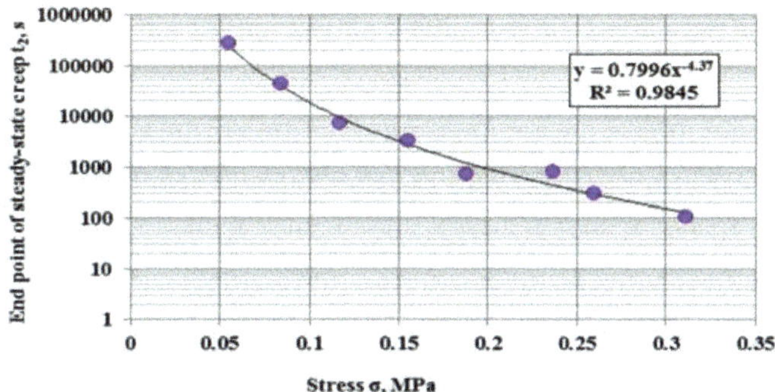

Figure 10. Dependence of end point for the stage of steady-state creep on stress.

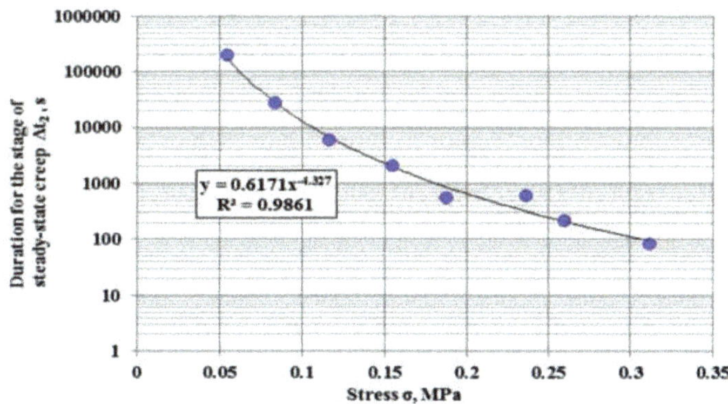

Figure 11. Dependence of duration for the stage of steady-state creep.

To show the full strain reached by the start of stage II of the creep curve through ε_1, then the strain of creep for this stage at any time t ,is calculated as:

$$\varepsilon_2(t) = \varepsilon_1 + \dot{\varepsilon}_2 \cdot t, \quad (t_1 < t \leq t_2), \tag{1}$$

where $\varepsilon_2(t)$ is strain of creep at time moment t, %; ε_1 is the strain at the start of stage II for creep curve, %; $\dot{\varepsilon}_2$ is the steady-state creep rate, %/s; t is time, s.

As a result of processing by the least square method, the following dependence was obtained between the steady-state creep rate of asphalt concrete and stress (Figure 8):

$$\dot{\varepsilon}_2 = 0,6597 \cdot \sigma^{3,9292}, \tag{2}$$

where σ is stress, MPa.

Having substituted the dependence in Equation (2) into Equation (1), we have:

$$\varepsilon_2(t) = \varepsilon_1 + 0,6597 \cdot \sigma^{3,9292} \cdot t, \quad (t_1 < t \leq t_2). \tag{3}$$

As mentioned above, the dependences of start point t_1 and end point t_2 of steady-state creep on stress are approximated by power functions, and in particular by the following ones (Figures 9 and 10):

$$t_1 = 0,1889 \cdot \sigma^{-4,465}, \tag{4}$$

$$t_2 = 0,7996 \cdot \sigma^{-4,370}. \tag{5}$$

The strain ε_1 can be calculated by the methods described in [14]. Equations (3)–(5) allow the determination of the creep strain of asphalt concrete for the stage of steady-state creep (stage II) at the time moment t ($t_1 < t \leq t_2$).

Similar to Newton's law for ideal viscous liquid, in our case for each creep curve, we can write [24]:

$$\sigma = \eta \times \dot{\varepsilon}_2, \tag{6}$$

where σ is stress; $\dot{\varepsilon}_2$ is steady-state creep (flow) rate; η is viscosity of steady-state flow.

Additionally, for all considered stress variation limit, we can write the Equation (6) in the following form:

$$\sigma = \eta(\sigma) \cdot \dot{\varepsilon}_2(\sigma). \tag{7}$$

From Equation (7), we obtain viscosity:

$$\eta(\sigma) = \frac{\sigma}{\dot{\varepsilon}_2(\sigma)}. \tag{8}$$

The dependence of viscosity of asphalt concrete on stress, constructed under the Equation (8) with the use of experimental results, is shown in Figure 12, which is satisfactorily approximated by a power function. The stress impacts greatly on the viscosity of asphalt concrete, the increase of stress by one order reduces the viscosity by three orders.

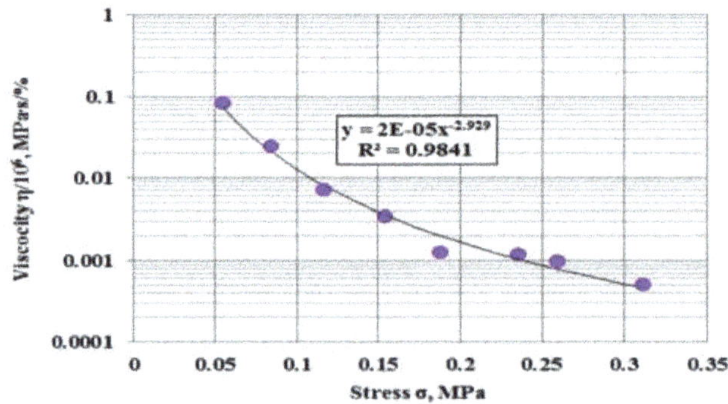

Figure 12. Dependence of viscosity for asphalt concrete on stress.

3.3. Defining Relations for Steady-State Creep of Asphalt Concrete at Complex Stressed Condition

In the reality of pavement structure, the points of asphalt concrete pavement during the loading impact of vehicle wheels are in a complex stressed and strained condition [3,25]. Experimental tests for determining the mechanical characteristics of asphalt concrete, as a rule, are carried out under simple loading and strain schemes (uniaxial tension, uniaxial compression, three-point bending, four-point bending) [2]. Therefore, the integration of these experimental results, which are carried out under simple loading and strain schemes for cases of complex stress and strained condition, is of significant importance for use in practice [24].

By assuming that (x, y, z) is a Cartesian coordinate system, the stressed condition in point of asphalt concrete pavement is described by stress tensor [26]:

$$\begin{pmatrix} \sigma_{xx} & \sigma_{xy} & \sigma_{xz} \\ \sigma_{yx} & \sigma_{yy} & \sigma_{yz} \\ \sigma_{zx} & \sigma_{zy} & \sigma_{zz} \end{pmatrix} = \begin{pmatrix} \sigma_0 & 0 & 0 \\ 0 & \sigma_0 & 0 \\ 0 & 0 & \sigma_0 \end{pmatrix} + \begin{pmatrix} \sigma_{xx}-\sigma_0 & \sigma_{xy} & \sigma_{xz} \\ \sigma_{yx} & \sigma_{yy}-\sigma_0 & \sigma_{yz} \\ \sigma_{zx} & \sigma_{zy} & \sigma_{zz}-\sigma_0 \end{pmatrix}, \tag{9}$$

where $\begin{pmatrix} \sigma_{xx}-\sigma_0 & \sigma_{xy} & \sigma_{xz} \\ \sigma_{yx} & \sigma_{yy}-\sigma_0 & \sigma_{yz} \\ \sigma_{zx} & \sigma_{zy} & \sigma_{zz}-\sigma_0 \end{pmatrix}$ is the deviator of stress; and $\sigma_0 = \frac{1}{3}(\sigma_{xx}+\sigma_{yy}+\sigma_{zz})$ is the mean stress.

Strained condition in the point is determined by strain rate tensor:

$$\begin{pmatrix} \dot{\varepsilon}_{xx} & \dot{\varepsilon}_{xy} & \dot{\varepsilon}_{xz} \\ \dot{\varepsilon}_{yx} & \dot{\varepsilon}_{yy} & \dot{\varepsilon}_{yz} \\ \dot{\varepsilon}_{zx} & \dot{\varepsilon}_{zy} & \dot{\varepsilon}_{zz} \end{pmatrix} = \begin{pmatrix} \dot{\varepsilon}_0 & 0 & 0 \\ 0 & \dot{\varepsilon}_0 & 0 \\ 0 & 0 & \dot{\varepsilon}_0 \end{pmatrix} + \begin{pmatrix} \dot{\varepsilon}_{xx}-\dot{\varepsilon}_0 & \dot{\varepsilon}_{xy} & \dot{\varepsilon}_{xz} \\ \dot{\varepsilon}_{yx} & \dot{\varepsilon}_{yy}-\dot{\varepsilon}_0 & \dot{\varepsilon}_{yz} \\ \dot{\varepsilon}_{zx} & \dot{\varepsilon}_{zy} & \dot{\varepsilon}_{zz}-\dot{\varepsilon}_0 \end{pmatrix}, \tag{10}$$

where $\begin{pmatrix} \dot{\varepsilon}_{xx}-\dot{\varepsilon}_0 & \dot{\varepsilon}_{xy} & \dot{\varepsilon}_{xz} \\ \dot{\varepsilon}_{yx} & \dot{\varepsilon}_{yy}-\dot{\varepsilon}_0 & \dot{\varepsilon}_{yz} \\ \dot{\varepsilon}_{zx} & \dot{\varepsilon}_{zy} & \dot{\varepsilon}_{zz}-\dot{\varepsilon}_0 \end{pmatrix}$ is the deviator of strain rates; and $\dot{\varepsilon}_0 = \frac{1}{3}(\dot{\varepsilon}_{xx}+\dot{\varepsilon}_{yy}+\dot{\varepsilon}_{zz})$ is the mean strain.

As seen in Figure 13, the strain of asphalt concrete at the start ε_1 and final ε_2 of the steady-state creep depends on the stress and at a minimum stress of 0.055 MPa, does not exceed 1.2% and 1.9% respectively, i.e., they are very small.

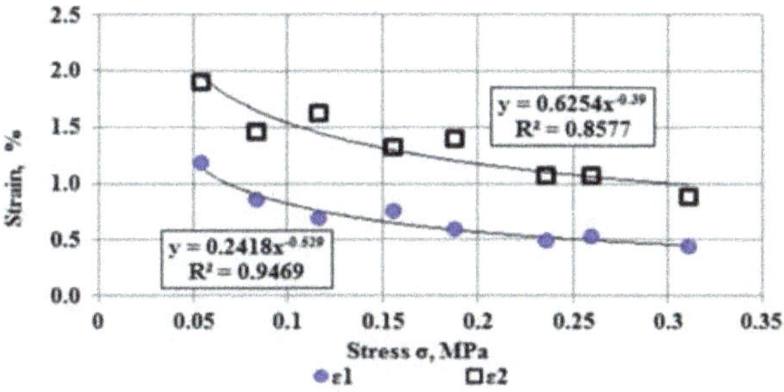

Figure 13. Dependence of strains ε_1 and ε_2 on stress.

To describe the steady-state creep of asphalt concrete, we assumed the following three hypotheses: (1) Material is uncompressible, i.e., the incompressibility condition is true [24]:

$$\dot{\varepsilon}_{xx}+\dot{\varepsilon}_{yy}+\dot{\varepsilon}_{zz}=0. \tag{11}$$

(2) The deviator of stresses is proportional to deviator of strain rates [24]:

$$\begin{aligned} \sigma_{xx} - \sigma_0 &= \psi\dot{\varepsilon}_{xx}, \quad \sigma_{xy} = \psi\dot{\varepsilon}_{xy}, \\ \sigma_{yy} - \sigma_0 &= \psi\dot{\varepsilon}_{yy}, \quad \sigma_{yz} = \psi\dot{\varepsilon}_{yz}, \\ \sigma_{zz} - \sigma_0 &= \psi\dot{\varepsilon}_{zz}, \quad \sigma_{zx} = \psi\dot{\varepsilon}_{zx}, \end{aligned} \tag{12}$$

where ψ is the parameter of proportionality depending on stress and strain rates.

Stress intensity:

$$\sigma_i = \frac{1}{\sqrt{2}}\sqrt{(\sigma_{xx}-\sigma_{yy})^2 + (\sigma_{yy}-\sigma_{zz})^2 + (\sigma_{zz}-\sigma_{xx})^2 + 6({\sigma_{xy}}^2 + {\sigma_{yz}}^2 + {\sigma_{zx}}^2)}. \tag{13}$$

Strain rate intensity:

$$\dot{\varepsilon}_i = \frac{\sqrt{2}}{3}\sqrt{(\dot{\varepsilon}_{xx}-\dot{\varepsilon}_{yy})^2 + (\dot{\varepsilon}_{yy}-\dot{\varepsilon}_{zz})^2 + (\dot{\varepsilon}_{zz}-\dot{\varepsilon}_{xx})^2 + 6(\dot{\varepsilon}_{xy}^2 + \dot{\varepsilon}_{yz}^2 + \dot{\varepsilon}_{zx}^2)}. \tag{14}$$

Having substituted the Equation (12) into Equation (13), we have

$$\sigma_i = \frac{1}{\sqrt{2}}\sqrt{\psi^2\left[(\dot{\varepsilon}_{xx}-\dot{\varepsilon}_{yy})^2 + (\dot{\varepsilon}_{yy}-\dot{\varepsilon}_{zz})^2 + (\dot{\varepsilon}_{zz}-\dot{\varepsilon}_{xx})^2 + 6(\dot{\varepsilon}_{xy}^2 + \dot{\varepsilon}_{yz}^2 + \dot{\varepsilon}_{zx}^2)\right]}. \tag{15}$$

Comparison of Equation (15) with Equation (14) shows that

$$\sigma_i = \frac{3}{2}\psi\dot{\varepsilon}_i. \tag{16}$$

From Equation (16), we obtained

$$\psi = \frac{2}{3}\frac{\sigma_i}{\dot{\varepsilon}_i}. \tag{17}$$

Then, we write Equation (12) again considering Equation (17):

$$\begin{aligned} \sigma_{xx} - \sigma_0 &= \tfrac{2}{3}\tfrac{\sigma_i}{\dot{\varepsilon}_i}\dot{\varepsilon}_{xx}, \quad \sigma_{xy} = \tfrac{2}{3}\tfrac{\sigma_i}{\dot{\varepsilon}_i}\dot{\varepsilon}_{xy}, \\ \sigma_{yy} - \sigma_0 &= \tfrac{2}{3}\tfrac{\sigma_i}{\dot{\varepsilon}_i}\dot{\varepsilon}_{yy}, \quad \sigma_{yz} = \tfrac{2}{3}\tfrac{\sigma_i}{\dot{\varepsilon}_i}\dot{\varepsilon}_{yz}, \\ \sigma_{zz} - \sigma_0 &= \tfrac{2}{3}\tfrac{\sigma_i}{\dot{\varepsilon}_i}\dot{\varepsilon}_{zz}, \quad \sigma_{zx} = \tfrac{2}{3}\tfrac{\sigma_i}{\dot{\varepsilon}_i}\dot{\varepsilon}_{zx}. \end{aligned} \tag{18}$$

(3) There is a functional dependence between stress intensity and strain rate intensity, which does not depend on the type of stressed condition:

$$\dot{\varepsilon}_i = f(\sigma_i). \tag{19}$$

For uniaxial tension $\sigma_{xx} > 0$, $\sigma_{yy} = \sigma_{zz} = \sigma_{xy} = \sigma_{yz} = \sigma_{zx} = 0$. Therefore, from Equation (13), we have

$$\sigma_i = \sigma_{xx} \tag{20}$$

Considering the condition of incompressibility for the material from Equation (11) for uniaxial tension is also true, $\dot{\varepsilon}_{xx} > 0$, $\dot{\varepsilon}_{xy} = \dot{\varepsilon}_{yz} = \dot{\varepsilon}_{zx} = 0$, $\dot{\varepsilon}_{yy} = \dot{\varepsilon}_{zz} = -\frac{1}{2}\dot{\varepsilon}_{xx}$.

Then, from Equation (14), we have

$$\dot{\varepsilon}_i = \dot{\varepsilon}_{xx}. \tag{21}$$

Taking into account Equations (19)–(21) for the considered asphalt concrete, the relation between strain rate intensity and stress intensity will have the following form:

$$\dot{\varepsilon}_i = 0.6597 \cdot \sigma_i^{3.9292}, \quad (22)$$

where $\dot{\varepsilon}_i$ is strain rate intensity, %/s; σ_i is stress intensity, MPa.

Comparing Equations (19) and (22), we have

$$f(\sigma_i) = 0.6597 \cdot \sigma^{3.9292}. \quad (23)$$

The previous paper in Reference [22] showed that asphalt concrete is deformed as a plastic material at 20 °C during cyclic loading under scheme "loading—deforming under constant load—rest". Thus, keeping the applied stress equal to 0.138 MPa constant for 120 s and keeping the asphalt concrete sample without load for the following 300 s in each cycle showed that even after four to five cycles the elastic (recovered) strain was only 5%–6% (Figure 14), i.e., asphalt concrete was deformed plastically.

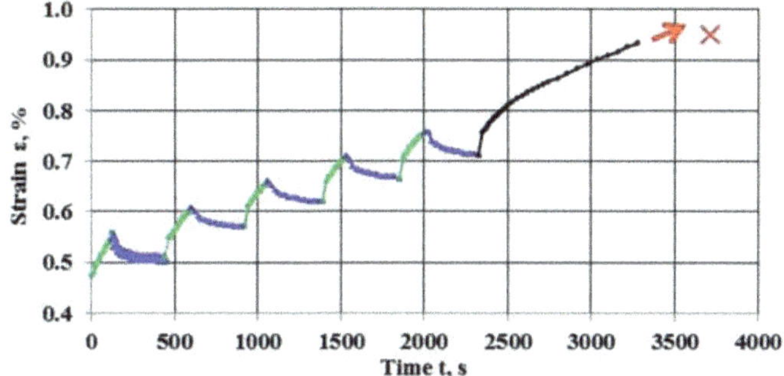

Figure 14. Curve of asphalt concrete strain during cyclic loading, green line: creep under load; blue line: recovery of strain after unloading; red cross: failure.

The above data emphasized the importance of defining relations for the steady-state creep of asphalt concrete as seen in Equations (18) and (22); and should also be considered as defining relations for steady-state plastic flow. By using them we can set the problems and solve them for the modeling of rutting in asphalt concrete layers of highways (Figure 15).

Figure 15. Plastic strains (the depth of the rut is more than 60 mm) in asphalt concrete layers of the highway "Ekaterinburg-Almaty" (km 1903.1—Karaganda Region, Kazakhstan).

4. Conclusions

The results of the experimental investigation into fine-grained asphalt concrete at static loading at a temperature of 20 ± 2 °C and at stress between 0.055 and 0.311 MPa showed the following:

- For creep curve stage II, the asphalt concrete deformation occurred at a constant rate. The strain rate for this stage is dependent on the stress, and this dependence is satisfactorily described by a power function. The stress has a great influence on the strain rate where the increase of stress by one order increases the strain rate approximately by four orders;
- The dependences were constructed for the start point, end point, and the duration of the stage of steady-state creep on the stress. The stress also impacts greatly on the specified time characteristics, where the increase of stress by one order increases these characteristics for 4.3–4.5 orders;
- The values of viscosity for asphalt concrete were determined at various stresses. The dependence was defined for viscosity on the stress and can also be satisfactorily described by a power function. In particular, the increase of stress by one order reduces the viscosity by three orders;
- Assuming that asphalt concrete is an incompressible material, then the stress deviator is proportional to the strain rate deviator. Hence, there is a functional relationship between the stress intensity and the strain rate intensity, which does not depend on the type of stress condition. The defining relations were formulated for the steady-state creep of asphalt concrete under complex stressed conditions.

Acknowledgments: This research is supported by the Road Committee of the Ministry for Investments and Development of the Republic of Kazakhstan (Agreement No. 36 dated 21 July 2016).

Author Contributions: Alibai Iskakbayev and Bagdat Teltayev conceived and designed the experiments; Alibai Iskakbayev and Bagdat Teltayev performed the experiments; Alibai Iskakbayev, Bagdat Teltayev and Cesare Oliviero Rossi analysed the data; Alibai Iskakbayev, Bagdat Teltayev and Cesare Oliviero Rossi contributed reagents/materials/analysis tools; Alibai Iskakbayev, Bagdat Teltayev and Cesare Oliviero Rossi wrote the paper.

Conflicts of Interest: The authors declare no conflict of interest.

References

1. *The Asphalt Handbook (MS-4)*, 7th ed.; Asphalt Institute: Lexington, MA, USA, 2008.
2. Papagiannakis, A.T.; Masad, E.A. *Pavement Design and Materials*; John Wiley & Sons, Inc.: New Jersey, NJ, USA, 2008.
3. Yoder, E.J.; Witczak, M.W. *Principles of Pavement Design*; John Wiley & Sons, Inc.: New Jersey, NJ, USA, 1975.
4. Cristensen, R.M. *Theory of Viscoelasticity: An Introduction*; Academic Press: New York, NY, USA, 1971.
5. Ferry, J.D. *Viscoelastic Properties of Polymers*; John Wiley & Sons, Inc.: New York, NY, USA, 1980.
6. Tschoegl, N.W. *The Phenomenological Theory of Linear Viscoelastic Behavior. An Introduction*; Springer: Berlin, Germany, 1989.
7. Hopkins, I.L.; Hamming, R.W. On creep and relaxation. *J. Appl. Phys.* **1957**, *28*, 906–909. [CrossRef]
8. Taherkhani, H. Compressive creep behavior of asphalt mixtures. *Eng. Procedia* **2011**, *10*, 583–588. [CrossRef]
9. Hassan, M.M. Relationship between creep time dependent index and Paris low parameters for bituminous mixtures. *J. S. Afr. Inst. Civ. Eng.* **2013**, *2*, 8–11.
10. Soleimanbeigi, A.; Edil, T.B.; Benson, C.H. Creep response of recycled asphalt shingles. *Can. Geotech. J.* **2013**, *51*, 103–114. [CrossRef]
11. Mahan, H.M. Behavior of permanent deformation in asphalt concrete pavements under temperature variation. *Al-Qadisiya J. Eng. Sci.* **2013**, *1*, 62–73.
12. Zhou, Z.G.; Feng, L.; Yuan, X.X.; Xiong, H. Study of the creep damage properties of asphalt mixture under static load. In Proceedings of the 13th International Conference on Fracture, Beijing, China, 16–21 June 2013.
13. Jaczewski, M.; Judycki, J. Effects of deviations from thermo-rheologically simple behavior of asphalt mixtures in creep on developing of master curves of their stiffness modulus. In Proceedings of the 9th International Conference on Environmental Engineering, Vilnius, Lithuania, 22–23 May 2014.

14. Iskakbayev, A.; Teltayev, B.; Alexandrov, S. Determination of the creep parameters of linear viscoelastic materials. *J. Appl. Math.* **2016**, *2016*, 6568347. [CrossRef]
15. Kachanov, L. *Introduction to Continuum Damage Mechanics*; Martinus Nijhoff Publishers: Dordrecht, The Netherlands, 1986.
16. Rabotnov, Y.N. *Introduction to Fracture Mechanics*; Nauka: Moscow, Russia, 1987.
17. *Bitumens and Bitumen Binders. Oil Road Viscous Bitumens (ST RK 1373)*; Technical Specifications: Astana, Kazakhstan, 2013.
18. *Performance Graded Asphalt Binder Specification and Testing (Superpave Series No. 1)*; Asphalt Institute: Lexington, MA, USA, 2003.
19. *Hot Mix Asphalt for Roads and Airfields (ST RK 1225)*; Technical Specifications: Astana, Kazakhstan, 2013.
20. *Crushed Stone and Gravel of Dense Rock for Construction Works (ST RK 1284)*; Technical Specifications: Astana, Kazakhstan, 2004.
21. *Bituminous Mixtures. Test Methods for Hot Mix Asphalt. Part 33: Specimen Prepared by Roller Compactor (EN 12697-33)*; European Committee for Standardization: Brussels, Belgium, 2003.
22. Iskakbayev, A.; Teltayev, B.; Andriadi, F.; Estayev, K.; Suppes, E.; Iskakbayeva, A. Experimental research of creep, recovery and fracture processes of asphalt concrete under tension. In Proceedings of the 24th International Congress on Theoretical and Applied Mechanics, Montreal, QC, Canada, 21–26 August 2016.
23. Teltayev, B.B.; Iskakbayev, A.I.; Rossi, C.O. Regularities of Creep and Long-Term Strength of Hot Asphalt Concrete under Tensile. In Proceedings of the 4th Chinese-European Workshop "Functional Pavement Design", Delft, The Netherlands, 29 June–1 July 2016.
24. Nadai, A. *Theory of Flow and Fracture of Solids*; MeGraw-Hill: New York, NY, USA, 1963.
25. Huang, Y.H. *Pavement Analysis and Design*, 2nd ed.; Pearson Education, Inc.: Upper Saddle River, NJ, USA, 2004.
26. Talreja, R.; Sing, C.V. *Damage and Failure of Composite Materials*; Cambridge University Press: Cambridge, UK, 2012.

MDPI
St. Alban-Anlage 66
4052 Basel, Switzerland
Tel. +41 61 683 77 34
Fax +41 61 302 89 18
http://www.mdpi.com

Applied Sciences Editorial Office
E-mail: applsci@mdpi.com
http://www.mdpi.com/journal/applsci

www.ingramcontent.com/pod-product-compliance
Lightning Source LLC
Chambersburg PA
CBHW051705210326
41597CB00032B/5374

* 9 7 8 3 0 3 8 4 2 8 8 9 3 *